SECOND EDITION

Reliability Engineering and Risk Analysis

A PRACTICAL GUIDE

SECOND EDITION

Reliability Engineering and Risk Analysis

A PRACTICAL GUIDE

Mohammad Modarres
Mark Kaminskiy
Vasiliy Krivtsov

CRC Press is an imprint of the
Taylor & Francis Group, an **informa** business

CRC Press
Taylor & Francis Group
6000 Broken Sound Parkway NW, Suite 300
Boca Raton, FL 33487-2742

Printed in the United States of America on acid-free paper
10 9 8 7 6 5 4 3 2 1

International Standard Book Number: 978-0-8493-9247-4 (Hardback)

Library of Congress Cataloging-in-Publication Data

Modarres, M. (Mohammad)
Reliability engineering and risk analysis : a practical guide / Mohammad Modarres, Mark Kaminskiy, Vasiliy Krivtsov. -- 2nd ed.
p. cm.
Includes bibliographical references and index.
ISBN 978-0-8493-9247-4 (hardcover : alk. paper)
1. Reliability (Engineering) 2. Risk assessment. I. Kaminskiy, Mark, 1946- II. Krivtsov, Vasiliy, 1963- III. Title.

TA169.M627 2010
620'.00452--dc22 2009029623

Visit the Taylor & Francis Web site at
http://www.taylorandfrancis.com

and the CRC Press Web site at
http://www.crcpress.com

Contents

Preface ... xiii
Authors ... xv

Chapter 1 Reliability Engineering in Perspective ... 1
1.1 Why Study Reliability? ... 1
1.2 Failure Models ... 1
1.2.1 Stress–Strength Model ... 2
1.2.2 Damage–Endurance Model ... 2
1.2.3 Challenge–Response Model ... 2
1.2.4 Performance–Requirement Model (or Tolerance–Requirement Model) ... 3
1.3 Failure Mechanisms ... 3
1.4 Performance Measures ... 7
1.5 Formal Definition of Reliability ... 11
1.6 Definition of Availability ... 12
1.7 Definition of Risk ... 12
References ... 13

Chapter 2 Basic Reliability Mathematics: Review of Probability and Statistics ... 15
2.1 Introduction ... 15
2.2 Elements of Probability ... 15
2.2.1 Sets and Boolean Algebra ... 15
2.2.2 Basic Laws of Probability ... 18
2.2.2.1 Classical Interpretation of Probability (Equally Likely Concept) ... 19
2.2.2.2 Frequency Interpretation of Probability ... 19
2.2.2.3 Subjective Interpretation of Probability ... 19
2.2.2.4 Calculus of Probability ... 20
2.2.3 Bayes's Theorem ... 23
2.3 Probability Distributions ... 27
2.3.1 Random Variable ... 27
2.3.2 Some Basic Discrete Distributions ... 27
2.3.2.1 Discrete Uniform Distribution ... 28
2.3.2.2 Binomial Distribution ... 28
2.3.2.3 Hypergeometric Distribution ... 30
2.3.2.4 Poisson Distribution ... 31
2.3.2.5 Geometric Distribution ... 32
2.3.3 Some Basic Continuous Distributions ... 33
2.3.3.1 Normal Distribution ... 35
2.3.3.2 Lognormal Distribution ... 36
2.3.3.3 Exponential Distribution ... 38
2.3.3.4 Weibull Distribution ... 39

2.3.3.5 Gamma Distribution ... 39
2.3.3.6 Beta Distribution ... 41
2.3.4 Joint and Marginal Distributions ... 42
2.4 Basic Characteristics of Random Variables ... 44
2.5 Estimation and Hypothesis Testing ... 49
2.5.1 Point Estimation ... 50
2.5.1.1 Method of Moments ... 50
2.5.1.2 Maximum Likelihood Method ... 51
2.5.2 Interval Estimation and Hypothesis Testing ... 52
2.5.2.1 Hypothesis Testing ... 53
2.6 Frequency Tables and Histograms ... 54
2.7 Goodness-of-Fit Tests ... 56
2.7.1 Chi-Square Test ... 56
2.7.2 Kolmogorov–Smirnov Test ... 59
2.8 Regression Analysis ... 62
2.8.1 Simple Linear Regression ... 63
Bibliography ... 70

Chapter 3 Elements of Component Reliability ... 71
3.1 Concept of Reliability ... 71
3.1.1 Reliability Function ... 71
3.1.2 Failure Rate ... 73
3.1.3 Some Useful Bounds and Inequalities for IFR (DFR) ... 76
3.1.3.1 Bounds Based on a Known Quantile ... 76
3.1.3.2 Bounds Based on a Known Mean ... 76
3.1.3.3 Inequality for Coefficient of Variation ... 76
3.2 Common Distributions in Component Reliability ... 78
3.2.1 Exponential Distribution ... 78
3.2.2 Weibull Distribution ... 78
3.2.3 Gamma Distribution ... 79
3.2.4 Normal Distribution ... 81
3.2.5 Lognormal Distribution ... 81
3.2.6 Extreme Value Distributions ... 81
3.2.6.1 Some Basic Concepts and Definitions ... 82
3.2.6.2 Order Statistics from Samples of Random Size ... 82
3.2.6.3 Asymptotic Distributions of Maxima and Minima ... 83
3.2.6.4 Three Types of Limit Distributions ... 84
3.3 Component Reliability Model Selection ... 85
3.3.1 Graphical Nonparametric Procedures ... 86
3.3.1.1 Small Samples ... 86
3.3.1.2 Large Samples ... 87
3.3.2 Probability Plotting ... 89
3.3.2.1 Exponential Distribution Probability Plotting ... 89
3.3.2.2 Weibull Distribution Probability Plotting ... 90
3.3.2.3 Normal and Lognormal Distribution Probability Plotting ... 92
3.3.3 Total-Time-on-Test Plots ... 94
3.4 Maximum Likelihood Estimation of Reliability Distribution Parameters ... 96
3.4.1 Censored Data ... 97
3.4.1.1 Left and Right Censoring ... 97
3.4.1.2 Type I Censoring ... 97

3.4.1.3 Type II Censoring 97
3.4.1.4 Types of Reliability Tests 98
3.4.1.5 Random Censoring 98
3.4.2 Exponential Distribution Point Estimation 98
3.4.2.1 Type I Life Test with Replacement 98
3.4.2.2 Type I Life Test without Replacement 99
3.4.2.3 Type II Life Test with Replacement 99
3.4.2.4 Type II Life Test without Replacement 99
3.4.3 Exponential Distribution Interval Estimation 100
3.4.4 Lognormal Distribution 103
3.4.5 Weibull Distribution 104
3.4.6 Binomial Distribution 106
3.5 Classical Nonparametric Distribution Estimation 108
3.5.1 Confidence Intervals for cdf and Reliability Function for Complete and Singly Censored Data 108
3.5.2 Confidence Intervals for cdf and Reliability Function for Multiply Censored Data 110
3.6 Bayesian Estimation Procedures 113
3.6.1 Estimation of the Parameter of Exponential Distribution 115
3.6.1.1 Selecting Parameters of Prior Distribution 116
3.6.1.2 Uniform Prior Distribution 118
3.6.2 Bayesian Estimation of the Parameter of Binomial Distribution 120
3.6.2.1 Standard Uniform Prior Distribution 121
3.6.2.2 Truncated Standard Uniform Prior Distribution 122
3.6.2.3 Beta Prior Distribution 123
3.6.2.4 Lognormal Prior Distribution 125
3.6.3 Bayesian Estimation of the Weibull Distribution 126
3.6.3.1 Prior Distribution 129
3.6.3.2 Posterior Distributions (Posterior Distribution of Weibull cdf at a Fixed Exposure) 130
3.6.3.3 Posterior Distribution of Weibull Distribution Parameters 130
3.6.3.4 Particular Case 132
3.6.4 Bayesian Probability Papers 134
3.6.4.1 Classical Simple Linear Regression 135
3.6.4.2 Classical Probability Papers 136
3.6.4.3 Bayesian Simple Linear Regression and Bayesian Probability Papers 137
3.6.4.4 Including Prior Information about Model Parameters 138
3.6.4.5 Including Prior Information about the Reliability or cdf .. 141
3.7 Methods of Generic Failure Rate Determination 142
Bibliography 151

Chapter 4 System Reliability Analysis 153
4.1 Reliability Block Diagram Method 153
4.1.1 Series System 153
4.1.2 Parallel Systems 155
4.1.3 Standby Redundant Systems 157
4.1.4 Load-Sharing Systems 159
4.1.5 Complex Systems 161
4.2 Fault Tree and Success Tree Methods 164

4.2.1 Fault Tree Method 164
4.2.2 Evaluation of Logic Trees 168
4.2.2.1 Boolean Algebra Analysis of Logic Trees 169
4.2.2.2 Combinatorial (Truth Table) Technique for Evaluation of Logic Trees 172
4.2.2.3 Binary Decision Diagrams 175
4.2.3 Success Tree Method 179
4.3 Event Tree Method 180
4.3.1 Construction of Event Trees 181
4.3.2 Evaluation of Event Trees 182
4.4 Master Logic Diagram 183
4.5 Failure Mode and Effect Analysis 189
4.5.1 Types of FMEA 190
4.5.2 FMEA/FMECA Procedure 190
4.5.3 FMEA Implementation 191
4.5.3.1 FMEA for Aerospace Applications 191
4.5.3.2 FMEA for Transportation Applications 194
4.5.4 FMECA Procedure: Criticality Analysis 197
References 217

Chapter 5 Reliability and Availability of Repairable Components and Systems 219
5.1 Repairable System Reliability 219
5.1.1 Basics of Point Processes 219
5.1.2 Homogeneous Poisson Process 222
5.1.3 Renewal Process 222
5.1.4 Nonhomogeneous Poisson Process 225
5.1.5 General Renewal Process 228
5.1.6 Probabilistic Bounds 231
5.1.7 Nonparametric Data Analysis 236
5.1.7.1 Estimation of the Cumulative Intensity Function 236
5.1.7.2 Estimation Based on One Realization (Ungrouped Data) 236
5.1.7.3 Estimation Based on One Realization (Grouped Data) 237
5.1.7.4 Estimation Based on Several Realizations 237
5.1.7.5 Confidence Limits for Cumulative Intensity Function 240
5.1.8 Data Analysis for the HPP 243
5.1.8.1 Procedures Based on the Poisson Distribution 243
5.1.8.2 Procedures Based on the Exponential Distribution of Time Intervals 245
5.1.9 Data Analysis for the NHPP 246
5.1.9.1 Regression Analysis of Time Intervals 246
5.1.9.2 Maximum Likelihood Procedures 248
5.1.9.3 Laplace's Test 250
5.1.10 Data Analysis for GRP 254
5.1.10.1 Example Based on Simulated Data 254
5.1.10.2 Example Based on Real Data 256
5.2 Availability of Repairable Systems 256
5.2.1 Instantaneous (Point) Availability 257
5.2.2 Limiting Point Availability 259
5.2.3 Average Availability 260

5.3 Use of Markov Processes for Determining System Availability 260
5.4 Use of System Analysis Techniques in the Availability Calculations of Complex Systems 265
References 277

Chapter 6 Selected Topics in Reliability Modeling 279
6.1 Probabilistic Physics-of-Failure Reliability Modeling 279
6.1.1 Stress–Strength Model 280
6.1.2 Damage–Endurance Model 283
6.1.3 Performance–Requirement Model 288
6.2 Software Reliability Analysis 290
6.2.1 Introduction 290
6.2.2 Software Reliability Models 290
6.2.2.1 Classification 291
6.2.3 Software Life-Cycle Models 295
6.2.3.1 Waterfall Model 295
6.3 Human Reliability 295
6.3.1 HRA Process 296
6.3.2 HRA Models 299
6.3.2.1 Simulation Methods 300
6.3.2.2 Expert Judgment Methods 300
6.3.2.3 Analytical Methods 301
6.3.3 Human Reliability Data 305
6.4 Measures of Importance 306
6.4.1 Birnbaum Measure of Importance 306
6.4.2 Criticality Importance 307
6.4.3 Fussell–Vesely Importance 308
6.4.4 RRW Importance 309
6.4.5 RAW Importance 309
6.4.6 Practical Aspects of Importance Measures 312
6.5 Reliability-Centered Maintenance 314
6.5.1 History and Current Procedures 314
6.5.2 Optimal Preventive Maintenance Scheduling 315
6.5.2.1 Economic Benefit of Optimization 317
6.6 Reliability Growth 317
6.6.1 Graphical Method 318
6.6.2 Duane Method 318
6.6.3 Army Material Systems Analysis Activity (AMSAA) Method 320
References 324

Chapter 7 Selected Topics in Reliability Data Analysis 327
7.1 Accelerated Life Testing 327
7.1.1 Basic AL Notions 327
7.1.1.1 Time Transformation Function for the Case of Constant Stress 327
7.1.1.2 AL Model 327
7.1.1.3 Cumulative Damage Models and AL Model 329
7.1.1.4 PH Model 329
7.1.2 Some Popular AL (Reliability) Models 329
7.1.3 AL Data Analysis 330

7.1.3.1 Exploratory Data Analysis (Criteria of Linearity of Time Transformation Function for Constant Stress) 330
7.1.3.2 Two-Sample Criterion 330
7.1.3.3 Checking the Coefficient of Variation 333
7.1.3.4 Logarithm of Time-to-Failure Variance 333
7.1.3.5 Quantile–Quantile Plots 333
7.1.3.6 Reliability Models Fitting: Constant Stress Case 334
7.1.4 AL Model for Time-Dependent Stress 335
7.1.4.1 AL Reliability Model for Time-Dependent Stress and Palmgren–Miner's Rule 336
7.1.5 Exploratory Data Analysis for Time-Dependent Stress 338
7.1.5.1 Statistical Estimation of AL Reliability Models on the Basis of AL Tests with Time-Dependent Stress 338
7.1.6 PH Model Data Analysis 339
7.1.6.1 Automotive Tire Reliability 339
7.1.6.2 Reliability Test Procedure 340
7.1.6.3 PH Model and Data Analysis 340
7.2 Analysis of Dependent Failures 342
7.2.1 Single-Parameter Models 345
7.2.2 Multiple-Parameter Models 347
7.2.2.1 Multiple Greek Letter Model 347
7.2.2.2 α-Factor Model 349
7.2.2.3 BFR Model 349
7.2.3 Data Analysis for CCFs 350
7.3 Uncertainty Analysis 352
7.3.1 Types of Uncertainty 352
7.3.1.1 Parameter Uncertainties 353
7.3.1.2 Model Uncertainties 353
7.3.1.3 Completeness Uncertainty 354
7.3.2 Uncertainty Propagation Methods 354
7.3.2.1 Method of Moments 354
7.3.3 System Reliability Confidence Limits Based on Component Failure Data 358
7.3.3.1 Lloyd–Lipow Method 358
7.3.4 Maximus Method 360
7.3.4.1 Classical Monte Carlo Simulation 360
7.3.4.2 Bayes's Monte Carlo Simulation 361
7.3.4.3 Bootstrap Method 361
7.3.5 Graphic Representation of Uncertainty 365
7.4 Use of Expert Opinion for Estimating Reliability Parameters 366
7.4.1 Geometric Averaging Technique 368
7.4.2 Bayesian Approach 369
7.4.3 Statistical Evidence on the Accuracy of Expert Estimates 369
7.5 Probabilistic Failure Analysis 370
7.5.1 Detecting Trends in Observed Failure Events 372
7.5.2 Failure Rate and Failure Probability Estimation for Data with No Trend 372
7.5.2.1 Parameter Estimation when Failures Occur by Time 372
7.5.2.2 Parameter Estimation when Failures Occur on Demand (Binomial Model) 372

7.5.3 Failure Rate and Failure Probability Estimation for Data with Trend 372
7.5.4 Evaluation of Statistical Data 372
7.5.4.1 Evaluation of Data with No Trend 372
7.5.4.2 Evaluation of Data with Trend 373
7.5.5 Root-Cause Analysis 374
References 381

Chapter 8 Risk Analysis 385
8.1 Determination of Risk Values 385
8.2 Formalization of Quantitative Risk Assessment 388
8.2.1 Identification of Hazards 388
8.2.2 Identification of Barriers 388
8.2.3 Identification of Challenges to Barriers 388
8.2.4 Estimation of Hazard Exposure 389
8.2.5 Consequences of Evaluation 389
8.3 Probabilistic Risk Assessment 389
8.3.1 Strengths of PRA 390
8.3.2 Steps in Conducting a PRA 390
8.3.2.1 Objectives and Methodology 390
8.3.2.2 Familiarization and Information Assembly 391
8.3.2.3 Identification of Initiating Events 392
8.3.2.4 Sequence or Scenario Development 393
8.3.2.5 Logic Modeling 394
8.3.2.6 Failure Data Collection, Analysis, and Performance Assessment 395
8.3.2.7 Quantification and Integration 396
8.3.2.8 Uncertainty Analysis 397
8.3.2.9 Sensitivity Analysis 398
8.3.2.10 Risk Ranking and Importance Analysis 398
8.3.2.11 Interpretation of Results 400
8.4 Compressed Natural Gas Powered Buses: A PRA Case Study 400
8.4.1 Primary CNG Fire Hazards 400
8.4.2 PRA Approach 401
8.4.3 System Description 401
8.4.4 Natural Gas Supply 401
8.4.5 Compression and Storage Station 401
8.4.6 Storage Cascade 401
8.4.7 Dispensing Facility 402
8.4.8 CNG Bus 402
8.4.9 Gas Release Scenarios 402
8.4.10 Fire Scenario Description 404
8.4.11 Gas Release 404
8.4.12 Gas Dispersion 405
8.4.13 Ignition Likelihood 405
8.4.14 Consequence Determination 405
8.4.15 Fire Location 406
8.4.16 Risk Value Determination 406
8.4.17 Summary of PRA Results 406
8.4.18 Overall Risk Results 407

8.4.19 Uncertainty Analysis ... 407
8.4.20 Sensitivity and Importance Analysis ... 408
8.5 A Simple Fire Protection Risk Analysis ... 409
8.6 Precursor Analysis ... 416
8.6.1 Introduction ... 416
8.6.2 Basic Methodology ... 418
8.6.3 Differences between APA and PRAs ... 419
References ... 422

Appendix A: Statistical Tables ... 425

Appendix B: Generic Failure Probability Data ... 439

Index ... 445

Preface

This book provides a practical and comprehensive overview of reliability and risk analysis techniques. It is written for both engineering students at the undergraduate and graduate levels, and for practicing engineers. The book mainly concentrates on reliability analysis. In addition, elementary performance and risk analysis techniques are also presented. Since reliability analysis is a multidisciplinary subject, the scope is not limited to any one engineering discipline; rather, the material is applicable to most engineering disciplines. This second edition has additional topics in probability plotting; maximum likelihood estimation of censored data; generalized renewal with applications; more detailed Bayesian estimation methods; estimation of probability bounds of availability of repairable units; models for the physics-of-failure approach to life estimation; an expanded discussion of uncertainty analysis; and further modifications, updates, clarifications, and discussions on more than half of the book's topics. The contents of the book have been primarily based on the materials used in three courses at the undergraduate and graduate levels at the University of Maryland, College Park. These courses have been offered over the past 25 years. This book has greatly benefited from the contribution of many talented students who actively participated in gathering and updating practical and useful materials. Therefore, the book presents a large number of examples to clarify the technical subjects. Additionally, there are many end-of-chapter exercises that are mainly based on the prior exams and homework problem sets of the reliability and risk analysis courses at the University of Maryland.

The emphasis of the book is on the introduction and explanation of the practical methods and techniques used in reliability and risk studies, and discussions of their uses and limitations rather than detailed derivations. These methods and techniques cover a wide range of topics that are used in routine reliability engineering and risk analysis activities. The book assumes that the readers have little or no background in probability and statistics. Thus, following an introductory chapter (Chapter 1) that defines reliability, availability, and risk analysis, Chapter 2 provides a detailed review of probability and statistics essential to understanding the reliability methods discussed in the book.

We have structured the book so that reliability methods applied to component reliability are described first, in Chapter 3. This is because components are the most basic building blocks of engineering systems. The techniques discussed in Chapter 3 provide a comprehensive overview of the state of the art in transforming basic field data into estimates of component reliability. Next, in Chapter 4, these analytical methods are described in the context of a more complex engineering unit, that is, a system containing many interacting components. Chapter 4 introduces new analysis methods to use the results of the component reliability analysis described in Chapter 3 to calculate estimates of the reliability of the whole system composed of these components. The materials in Chapters 1 through 4 are appropriate for an advanced undergraduate course in reliability engineering or an introductory graduate-level course.

Chapters 3 and 4 assume that the components (or systems) are "replaceable." That is, upon a failure, the component is replaced with a new one. However, many components are "repairable." That is, upon a failure, they are repaired and placed back in to service. In this case, availability as a metric becomes the key measure of performance. The techniques for availability and reliability analyses of repairable components and systems are discussed in Chapter 5. This chapter also explains the corresponding use of the analytical methods discussed in Chapters 3 and 4 when performing availability analyses of components and engineering systems.

Chapter 6, discusses a number of important methods frequently used in modeling reliability, availability, and risk problems. For example, in Section 6.2, we discuss the concept of uncertainty, sources of uncertainty, parameter and model uncertainty, and probabilistic methods for quantifying and propagating parameter uncertainties in engineering systems (or models). Similar to Chapter 6, Chapter 7 describes special topics related to reliability and availability data analyses. For example, Section 7.1 describes accelerated life testing methods. Examples clarifying the uses of the modeling and data analysis methods and their shortcomings are also presented in Chapters 6 and 7.

In Chapter 8, we discuss the method of risk analysis. A number of analytical methods explained in the preceding chapters have been integrated to perform risk assessment or risk management. Recently, probabilistic risk assessment (PRA) has been a major topic of interest in light of the hazards imposed by many engineering designs and processes. Steps involving the performance of a PRA are discussed in this chapter.

A complete solution set booklet has been developed by the late W. M. Webb and M. Modarres in a bounded volume. This booklet may be provided to educators and industrial users by sending a written request to Prof. M. Modarres.

The book could have not been completed without the help and corrections of our students and colleagues at the University of Maryland. It would be difficult to name all, but some names to mention include A. Mosleh, N. Kaceci, C. Smidts, H. Dezfuli, R. Azarkhail, and Z. Mohaghegh. We would also like to acknowledge J. Case of Ford Motor Company for his review and valuable suggestions on the FMEA and "Reliability Growth" sections of the book. Special thanks go to Y.-S. Hu for his unfailing technical and organizational support, without which this work would not have been possible. Also, the editorial help of E. Waterford and the typing and graphical support of W. M. Webb are highly appreciated.

Mohammad Modarres
Mark Kaminskiy
Vasiliy Krivtsov

Authors

Mohammad Modarres is a professor of nuclear engineering and reliability engineering. His research areas are system reliability modeling, probabilistic risk analysis, probabilistic physics of failure, and uncertainty modeling and analysis. He is a consultant to several government and private organizations as well as national laboratories. Prof. Modarres has published more than 200 papers in professional journals and proceedings of conferences; three books; and a number of book chapters, edited books, and handbooks. He is a University of Maryland Distinguished Scholar-Teacher, a fellow of the American Nuclear Society, and has received a number of other awards in reliability engineering and risk assessment. Prof. Modarres received his PhD in nuclear engineering from the Massachusetts Institute of Technology (MIT) in 1979 and his MS in mechanical engineering from MIT in 1977.

Mark Kaminskiy is the chief statistician at the Center of Technology and Systems Management, University of Maryland (College Park). Dr. Kaminskiy is a researcher and consultant in reliability engineering, life data analysis, and risk analysis of engineering systems. He has conducted numerous research and consulting projects funded by the government and industrial companies such as the Department of Transportation, the Coast Guard, the Army Corps of Engineers, the Navy, the Nuclear Regulatory Commission, the American Society of Mechanical Engineers, Ford Motor Company, Qualcomm Inc., and several other engineering companies. He has taught several graduate courses on reliability engineering at the University of Maryland. Dr. Kaminskiy is the author or coauthor of over 50 publications in journals, conference proceedings, and reports.

Vasiliy Krivtsov is a senior staff technical specialist in reliability and statistical analysis with Ford Motor Company. He holds MS and PhD degrees in electrical engineering from Kharkov Polytechnic Institute, Ukraine, and a PhD in reliability engineering from the University of Maryland. Dr. Krivtsov is the author or coauthor of over 40 professional publications, including a book on reliability engineering and risk analysis, nine patented inventions, and three Ford trade secret inventions. He is an editor of *Reliability Engineering and System Safety* journal and is a member of the IEEE Reliability Society. Prior to Ford, Dr. Krivtsov held the position of associate professor of electrical engineering in Ukraine and that of a research affiliate at the Center for Reliability Engineering, University of Maryland. Further information on Dr. Krivtsov's professional activity is available at http://www.krivtsov.net.

1 Reliability Engineering in Perspective

1.1 WHY STUDY RELIABILITY?

Engineering systems, components, and structures are not perfect. A perfect design is one that remains operational and attains its objective without failure during a preselected life. This is the deterministic view of an engineering system. This view is idealistic, impractical, and economically infeasible. Even if technical knowledge is not a limiting factor in designing, manufacturing, constructing, and operating a perfect design, the cost of development, testing, materials, and engineering analysis may far exceed economic prospects for such a system. Therefore, practical and economic limitations dictate the use of imperfect designs. Designers, manufacturers, and end users, however, strive to minimize the occurrence and recurrence of failures. In order to minimize failures in engineering designs, the designer must understand why and how failures occur. In order to maximize system performance and efficiently use resources, it is also important to know how often such failures may occur. This involves *predicting* the occurrence of failures.

The prevention of failures and the process of understanding why and how failures occur involve appreciation of the physics of failure. Failure mechanisms are the means by which failures occur. To effectively minimize the occurrence of failures, the designer should have an excellent knowledge of applicable failure mechanisms that may be associated with the design, or that can be introduced from outside of the system (e.g., by users or maintainers). When failure mechanisms are known and appropriately considered in design, manufacturing, construction, production, and operation, their impact or occurrence rate can be minimized, or the system can be protected against them through careful engineering and economic analysis. This is generally a deterministic view of reliability analysis process.

Most failure mechanisms, their interactions, and process of degradation in a particular design are generally not well understood. Accordingly, the prediction of failures involves uncertainty and thus is inherently a probabilistic problem. Therefore, reliability analysis, whether using physics of failure or historical occurrences of failures, is a probabilistic process. It is the strength of the evidence and knowledge about the failure events and processes that determine the confidence about reliability of the design. This book deals with the reliability analyses involving prediction of failures and deals with it probabilistically. First, a brief review of leading failure models and mechanisms is presented in Sections 1.2 and 1.3.

1.2 FAILURE MODELS

Failures are the result of the existence of source *challenges* and conditions occurring in a particular failure process or scenario. The system has an inherent *capacity* to withstand such challenges. Both the challenges and the capacity may be reduced by specific internal or external conditions. When challenges surpass the capacity of the system, a failure occurs.

Specific models use different definitions and metrics for capacity and challenge. "Adverse Conditions" generated artificially or naturally, internally or externally, may increase or induce challenges to the system, or/and reduce the capacity of the item to withstand challenges.

Figure 1.1 depicts elements of a framework to construct failure models. Several simple failure models, discussed by Dasgupta and Pecht [1], are consistent with the framework presented in

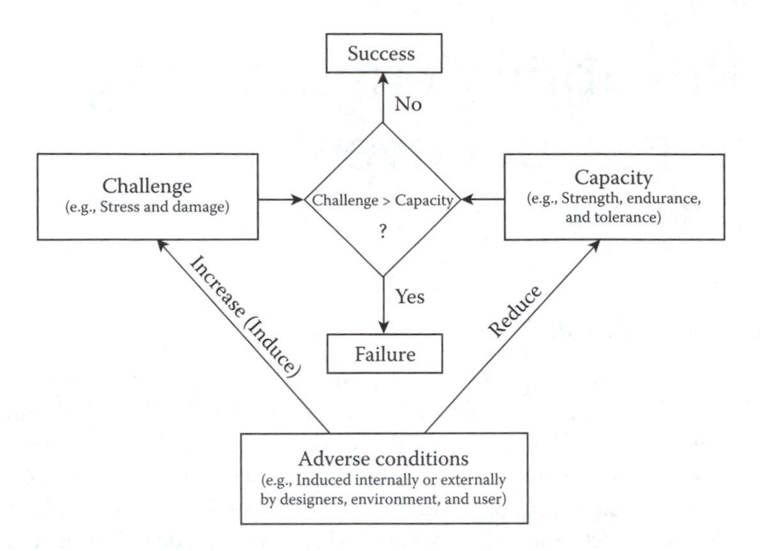

FIGURE 1.1 Framework for modeling failure.

Figure 1.1. A summary of these models has been provided below and will be discussed in more detail in Section 6.1.

1.2.1 Stress–Strength Model

The item (e.g., a system, structure, or component) fails if the challenge (i.e., stress-mechanical, thermal, etc.) exceeds the capacity (i.e., strength, yielding point, and melting point). The *stress* represents an aggregate of the challenges and external conditions. This failure model may depend on environmental conditions, applied loads, and the occurrence of critical events, rather than the mere passage of time or cycles. Strength is often treated as a random variable (r.v.) representing the effect of all conditions affecting the strength, or lack of knowledge about the item's strength (e.g., the item's capability, mechanical strength, and dexterity). Similarly stress may be considered as a r.v. Two examples of this model are (a) a steel bar in tension, and (b) a transistor with a voltage applied across the emitter–collector.

1.2.2 Damage–Endurance Model

This model is similar to the stress–strength model, but the scenario of interest is that *stress* (load) causes damage that accumulates irreversibly, as in corrosion, wear, embrittlement, creep, and fatigue. The aggregate of challenges and external conditions leads to the metric represented as cumulative damage (e.g., crack size in case of fatigue-based fracture). The cumulative damage may not degrade performance; the item fails only when the cumulative damage exceeds the endurance (i.e., the damage accumulates until the endurance of the item is reached). As such, an item's capacity is measured by its tolerance of damage endurance (such as fracture toughness). Accumulated damage does not disappear when the *stresses* are removed, although sometimes treatments such as annealing are possible. Damage and endurance may be treated as a r.v.

Similar to the stress–strength model, endurance is an aggregate measure for effects of challenges and external conditions on the item's capability to withstand cumulative stresses.

1.2.3 Challenge–Response Model

This model closely resembles the framework shown in Figure 1.1. An element of the system may have failed, but only when that element is challenged (needed) does it cause the system to fail.

A common consumer example is the emergency brake of a car. Most computer program (software) failures are of this type. Telephone switching systems also resemble this failure model. This failure model depends on when critical events happen in the environment, rather than the mere passage of time or cycles.

1.2.4 Performance–Requirement Model (or Tolerance–Requirement Model)

A system performance characteristic is satisfactory if it falls within acceptable tolerance limit (or accepted margin, such as a safety margin). Examples of this are a copier machine and a measuring instrument where gradual degradation eventually results in a user deciding that performance quality is unacceptable. Another example is the capability of an emergency safety injection system at a nuclear power plant. The system's performance may degrade over time, but should remain below the minimum safety margin requirements to safely cool a reactor, if needed.

In the models discussed above, challenges are caused by failure-inducing agents (sometimes referred to as failure agents). Examples of two of the most important failure-inducing agents are stress and time. Stress can be created due to mechanical (e.g., cyclic load), thermal (e.g., thermal cycling), electrical (e.g., voltage), chemical (e.g., salt), and radiation (e.g., neutron) agents. For example, turning on and off a standby component may cause a thermal or mechanical load cycling to occur. Passage of time, on the other hand, gives more opportunity for the failure agents to act, for damage to accumulate, and for a failure to occur without regard to the underlying phenomena of failure described by the mechanisms activated or promoted by such agents. As such, time may be viewed as an "aggregate" agent of failure.

A comprehensive consideration of reliability requires analysis of the two failure-inducing agents of stress and time. Both time and stress may be analyzed deterministically (e.g., by identifying the sources of stress), or probabilistically (e.g., by treating stress and time as r.v.s). In either case, it is necessary to understand why and how such stresses lead to a failure. This requires studying the physics of failure in general and failure mechanisms in particular.

The main body of this book addresses the probabilistic treatment of time and stress as agents of failure. Equally important, however, is understanding the failure mechanisms through the physics of failure. It is also important to note that models representing failure mechanisms (ones that are related stress to damage or life) are also uncertain. This means that the physics of failure do not necessarily eliminate the need to treat reliability probabilistically. Rather, they only reduce the burden for the amount of data needed to predict or describe the reliability of an item. In the next section, we will discuss a brief summary of important failure mechanisms. For further readings on the physics of failure, see Dasgupta and Pecht [1], Pecht [2], Collins [3], and Amerasekera and Campbell [4].

It is important to differentiate between stress-inducing and stress-increasing mechanisms. Stress-induced mechanisms elevate the stresses applied to an item indirectly, by motivating or persuading creation of stress. On the other hand, stress-increasing mechanisms directly cause added stress. In the former case the agent of failure acts as a helper for degradation, but in the latter it is the subject of the failure. For example, the failure mechanism impact may deform an item leading to elevated stress due to added forces applied from adjacent items. Therefore, the added stress is not a direct cause of impact, but the impact has caused a condition (deformation) that has led to additional stress. Similarly, the failure mechanism impact may cause direct stresses due to the forces applied to the item itself. Note that some failure mechanisms may be considered both stress-induced and stress-increased, depending on the way the added stress has been established. The strength-reduced mechanism lowers the capacity of an item to withstand normal stresses, resulting in a failure.

1.3 FAILURE MECHANISMS

Failure mechanisms are physical processes whose occurrence either lead to or is caused by stress, and may deteriorate the capacity (e.g., strength or endurance) of an item. Since failure mechanisms for mechanical and electronic/electrical equipment are somewhat different, these mechanisms are discussed separately.

Mechanical Failure Mechanisms can be divided into three classes: stress-induced, strength-reduced, and stress-increased. Stress-induced mechanisms refer to mechanisms that cause or are the result of localized stress (permanent or temporary). For example, elastic deformation may be the result of a force applied on the item that causes deformation (elastic) that disappears when the applied force is removed. Strength-reducing mechanisms are those that lead (indirectly) to a reduction of the item's strength or endurance to withstand stress or damage. For example, radiation may cause material embrittlement, thus reducing the material's capacity to withstand cracks or other damage. Stress-increasing mechanisms are those whose direct effect is an increase in the applied stress. For example, fatigue-induced crack growth could cause increased stress intensity in a structure. Table 1.1 shows a breakdown of each class of mechanism.

Table 1.2 summarizes the cause, effect, and physical processes involving common mechanical failure mechanisms.

Electrical Failure Mechanisms tend to be more complicated than those of purely mechanical failure mechanisms. This is caused by the complexity of the electrical items (e.g., devices) themselves. In integrated circuits, a typical electrical device, such as a resistor, capacitor, or transistor, is manufactured on a single crystalline chip of silicon, with multiple layers of various metals, oxides, nitrides, and organics on the surface, deposited in a controlled manner. Often a single electrical device is composed of several million elements, compounding any reliability problem present at the single element level. Once the electrical device is manufactured, it must be packaged, with electrical connections to the outside world. These connections, and the packaging, are as vital to the proper operation of the device as the electrical elements themselves.

Failure mechanisms for electrical devices are usually divided into three types: *electrical stress*, *intrinsic*, and *extrinsic*. These are discussed below.

- *Electrical stress failure* occurs when an electrical device is subjected to voltage levels higher than design constraints, damaging the device and degrading electrical characteristics. This failure mechanism is often a result of human error. Also known as electrical overstress (EOS), uncontrolled currents in the electrical device can cause resistive heating or localized melting at critical circuit points, which usually results in catastrophic failure but has also been known to cause latent damage. Electrostatic discharge (ESD) is one common way of imparting large, undesirable currents into an electrical device.
- *Intrinsic failure mechanisms* are related to the electrical element itself. Most failure mechanisms related to the semiconductor chip and electrically active layers grown on its surface are in this

TABLE 1.1
Categorization of Failure Mechanisms

Stress-Induced Failure Mechanisms	Strength-Reduced Failure Mechanisms	Stress-Increased Failure Mechanisms
Brittle fracture	Wear	Fatigue
Buckling	Corrosion	Radiation
Yield	Cracking	Thermal shock
Impact	Diffusion	Impact
Ductile fracture	Creep	Fretting
Elastic deformation	Radiation damage	
	Fretting	

TABLE 1.2
Mechanical Failure Mechanisms

Mechanism	Causes	Effect	Description
Buckling	Compressive load application Dimensions of the items	Item deflects greatly Possible complete loss of load-carrying ability	When load applied to items such as struts, columns, plates, or thin-walled cylinders reaches a critical value, a sudden major change in geometry, such as bowing, winking, or bending, occurs
Corrosion	Chemical action on the surface of the item Contact between two dissimilar metals in electrical contacts (galvanic corrosion) Improper welding of certain copper, chromium, nickel, aluminum, magnesium, and zinc alloys Abrasive or viscid flow of chemicals over the surface of an item Collapsing of buckles and cavities adjacent to pressure walls Living organisms in contact with the item High stress in a chemically active environment (stress-corrosion) causing cracking	Reduction in strength Cracking Fracture Geometry changes	Undesired deterioration of the item as a result of chemical or electrochemical interaction with the environment. Corrosion closely interacts with other mechanisms such as cracking, wear, and fatigue
Impact	Sudden load from dropping an item or having been struck	Localized stresses Deformation Fracture	Failure occurs by the interaction of generated dynamic or abrupt loads that result in large local stresses and strains
Fatigue	Fluctuating force (loads)	Cracking leading to deformation and fracture	Application of fluctuating normal loads (far below the yield point) causing pitting, cracking. Fatigue is a progressive failure phenomenon that initiates and propagates cracks
Wear	Solid surfaces in rubbing contact Particles (sometimes removed from the surface) entrapped between rubbing surfaces Corrosive environment near rubbing contacts and loose particles entrapped between rubbing surfaces	Cumulative change in dimensions Deformation and strength reduction	Wear is not a single process. It can be a complex combination of local shearing, plowing, welding, and tearing, causing gradual removal of discrete particles from contacting surfaces in motion. Particles entrapped between mating surfaces. Corrosion often interacts with wear processes and changes the character of the surfaces

continued

TABLE 1.2 (continued)

Mechanism	Causes	Effect	Description
Creep	Loading, usually at high temperature, leading to gradual plastic deformation	Deformation of item rupture	Plastic deformation in an item accrues over a period of time under stress until the accumulated dimensional changes interfere with the item's ability to properly function
Thermal shock	Rapid cooling, heating Large differential temperature	Yield fracture Embrittlement	Thermal gradients in an item causing major differential thermal strains that exceed the ability of the material to withstand without yielding or fracture
Yield	Large static force Operational load or motion	Geometry changes Deformation Break	Plastic deformation in an item occurs by operational loads or motion
Radiation Damage	Nuclear radiation	Changes in material property Loss of ductility	Radiation causes rigidity and loss of ductility. Polymers are more susceptible than metals. In metals, radiation reduces ductility, resulting in other failure mechanisms

category. Intrinsic failures are related to the basic electrical activity of the device and usually result from poor manufacturing or design procedures. Intrinsic failures cause both reliability and manufacturing yield problems. Common intrinsic failure mechanisms are gate oxide breakdown, ionic contamination, surface charge spreading, and hot electrons.

- *Extrinsic failure mechanisms* are external failure mechanisms for electrical devices that stem from problems with the device packaging and interconnections. Most extrinsic failure mechanisms are mechanical in nature. Often deficiencies in the electronic device and packaging manufacturing process cause these mechanisms to occur, though operating environment has a strong effect on the failure rate also. In recent years semiconductor technology has reached a high level of maturity, with a corresponding high level of control over intrinsic failure mechanisms. As a result, extrinsic failures have become more critical to the reliability of the latest generation of electronic devices.

Many electrical failure mechanisms are interrelated. Often, a partial failure due to one mechanism will ultimately evolve to another. For example, oxide breakdown may be caused by poor oxide processing during manufacturing, but it may also be exasperated by ESD, damaging an otherwise intact oxide layer. Corrosion and ionic contamination may be initiated when a packaging failure allows unwanted chemical species to contact the electronic devices, and then failure can occur through trapping, piping, or surface charge spreading. Many intrinsic failure mechanisms may be initiated by an extrinsic problem: once the package of an electrical device is damaged, there are a variety of intrinsic failure mechanisms that may manifest themselves in the chip itself.

Tables 1.3 through 1.5 summarize the cause, effect, and physical processes involving common electrical stress and intrinsic and extrinsic failure mechanisms.

TABLE 1.3
Electrical Stress Failure Mechanisms

Mechanism	Causes	Effect	Description
EOS	Improper application of handling	Localized melting Gate oxide breakdown	Device is subjected to voltages higher than design constraints
ESD	Common static charge buildup	Localized melting Gate oxide breakdown	Contact with static charge buildup during device fabrication or later handling results in high voltage discharge into device

1.4 PERFORMANCE MEASURES

The overall performance of an item (component, device, product, subsystem, or system) is dependent on the implementation of various programs that ultimately improve the performance of the item. Historically, these programs have been installed through a trial-and-error approach. For example, they are sometimes established based on empirical evidence gathered during an investigation of failures. An example of such programs is a root-cause failure analysis program. It is worthwhile, at this point, to understand why a well-established reliability analysis and engineering program can influence the performance of today's items. For this reason, let us first define the important constituents of performance.

TABLE 1.4
Intrinsic Failure Mechanisms

Mechanism	Causes	Effect	Description
Gate oxide breakdown	EOS ESD Poor gate oxide processing	Degradation in current–voltage (I–V) characteristics	Oxide layer that separates gate metal from semiconductor is damaged or degrades with time
Ionic contamination	Undesired ionic species are introduced into semiconductor	Degradation in I–V characteristics Increase in threshold voltage	Undesired chemical species can be introduced to device through human contact, processing materials, improper packaging, and so on
Surface charge spreading	Ionic contamination or excess surface moisture	Short-circuiting between devices Threshold voltage shifts or parasitic formation	Undesired formation of conductive pathways on surfaces alters electrical characteristic of device
Slow trapping	Poor interface quality	Threshold voltage shifts	Defects at gate oxide interface trap electrons, producing undesired electric fields
Hot electrons	High electric fields in conduction channel	Threshold voltage shifts	High electric fields create electrons with sufficient energy to enter oxide
Piping	Crystal defects Phosphorus or gold diffusion	Electrical shorts in emitter or collector	Diffusion along crystal defects in the silicon during device fabrication causes electrical shorts

TABLE 1.5
Extrinsic Failure Mechanisms

Mechanism	Causes	Effect	Description
Packaging failures	Most mechanical failure mechanisms can cause electrical device packaging failures	Usually increased resistance or open circuits	See section on mechanical failure mechanisms
Corrosion	Moisture, DC operating voltages, and Na or Cl ionic species	Open circuits	The combination of moisture, DC operating voltages, and ionic catalysts causes electrochemical movement of material, usually the metallization
Electromigration	High current densities Poor device processing	Open circuits	High electron velocities become sufficient to impact and move atoms, resulting in altered metallization geometry and, eventually, open circuits
Contact migration	Uncontrolled material diffusion	Open or short circuits	Poor interface control causes metallization to diffuse into the semiconductor. Often this occurs in the form of metallic "spikes"
Microcracks	Poorly processed oxide steps	Open circuits	Formation of a metallization path on top of a sharp oxide step results in a break in the metal or a weakened area prone to further damage
Stress migration	High mechanical stress in electrical device	Short circuits	Metal migration occurs to relieve high mechanical stress in device
Bonding failures	Poor bond control	Open circuits	Electrical contact to device package (bonds) are areas of high mechanical instability, and can separate if processing is not strictly controlled
Die attachment failures	Poor die attach integrity or corrosion	Hot spots Parametric shifts in circuit Corrosion mechanical failures	Corrosion or poor fabrication causes voids in die attach, or partial or complete deadhesion
Particulate contamination	Poor manufacturing and chip breakage	Short circuits	Conductive particles may be sealed in a hermetic package or may be generated through chip breakage
Radiation	Trace radioactive elements in device or external radiation source	Can cause various degrading effects	High energy radiation can create hot electron–hole pairs that can interfere with and degrade device performance

The performance of an item can be described by four elements:

- Capability: the item's ability to satisfy functional needs.
- Efficiency: the item's ability to effectively and easily realize functional needs.
- Reliability: the item's ability to start and continue to operate.
- Availability: the item's ability to quickly become operational following a failure.

The first two measures are influenced by the design, construction, production, or manufacturing of the item. Capability and efficiency reflect the levels to which the item is designed and built. For example, the designer ensures that design levels are adequate to meet the functional requirements. On the other hand, reliability is an operation-related issue and is influenced by the item's potential to remain operational. For a repairable item, the ease with which the item is maintained, repaired, and returned to operation is measured by its maintainability. Capability, efficiency, reliability, and availability may be measured deterministically or probabilistically. For more discussions on this subject see Modarres [5]. Based on the above definitions, it would be possible to have an item that is highly reliable but does not achieve a high performance. Examples include items that do not fully meet their stated design objectives. Humans play a major role in the design, construction, production, operation, and maintenance of the item. These roles can significantly influence the values of the four performance measures. The role of the human is often determined by various programs and activities that support the four elements of performance, proper implementation of which leads to a *quality* item.

To put all of these factors in perspective, consider the development of a high-performance product in an integrated framework. For this purpose, let's consider the so-called diamond tree conceptually shown in Figure 1.2. In this tree, the top goal is acceptable performance during the life cycle of the item and is hierarchically decomposed into various goals, functions, activities, programs, and organizations. By looking down from the top of this structure, one can describe *how* various goals and subgoals are achieved, and by looking up, one can identify *why* a goal or function is necessary. Figure 1.2 shows only typical goals but also reflects the general goals involved in designing and operating a high-performance item. For a more detailed description of the diamond tree the readers are referred to Hunt and Modarres [6].

Reliability and availability play a key role in the overall framework shown in Figure 1.2. Of the four elements of performance, in this book we are mainly interested in reliability and availability. Only the repairable aspects of a maintainable system are of interest to us. Therefore, we will only discuss reliability and availability as two important measures of performance.

A more detailed look at the goals of improving reliability, in an integrated manner, would yield a better perspective on the role of reliability and availability analysis as shown by the hierarchy depicted in Figure 1.3.

From this figure, one can put into a proper context the role of reliability and availability analysis. Clearly, reliability is a key element of achieving high performance since it directly and significantly influences the item's performance and ultimately its life-cycle cost and economics. Poor reliability directly causes increased warranty costs, liabilities, recalls, and repair costs.

In this book we are also interested in risk analysis. However, risk associated with an item is not a direct indicator of performance. Risk is the item's potential to cause a loss (e.g., loss of other systems, loss of human safety, environmental damage, or economic loss). However, a quantitative measure of risk can be an important metric for identifying and highlighting items that are risk-significant (i.e., those that may be associated with a potentially significant loss). This metric, however, is useful to set adequate performance levels for risk-significant items. Conversely, performance may highly influence an item's risk. For example, a highly reliable item is expected to fail less frequently, resulting in smaller risk. On the other hand, the risk of an item may be used to identify items that should attain a high performance. Accordingly, risk and performance of an item synergistically influence each other. This concept is depicted in Figure 1.4.

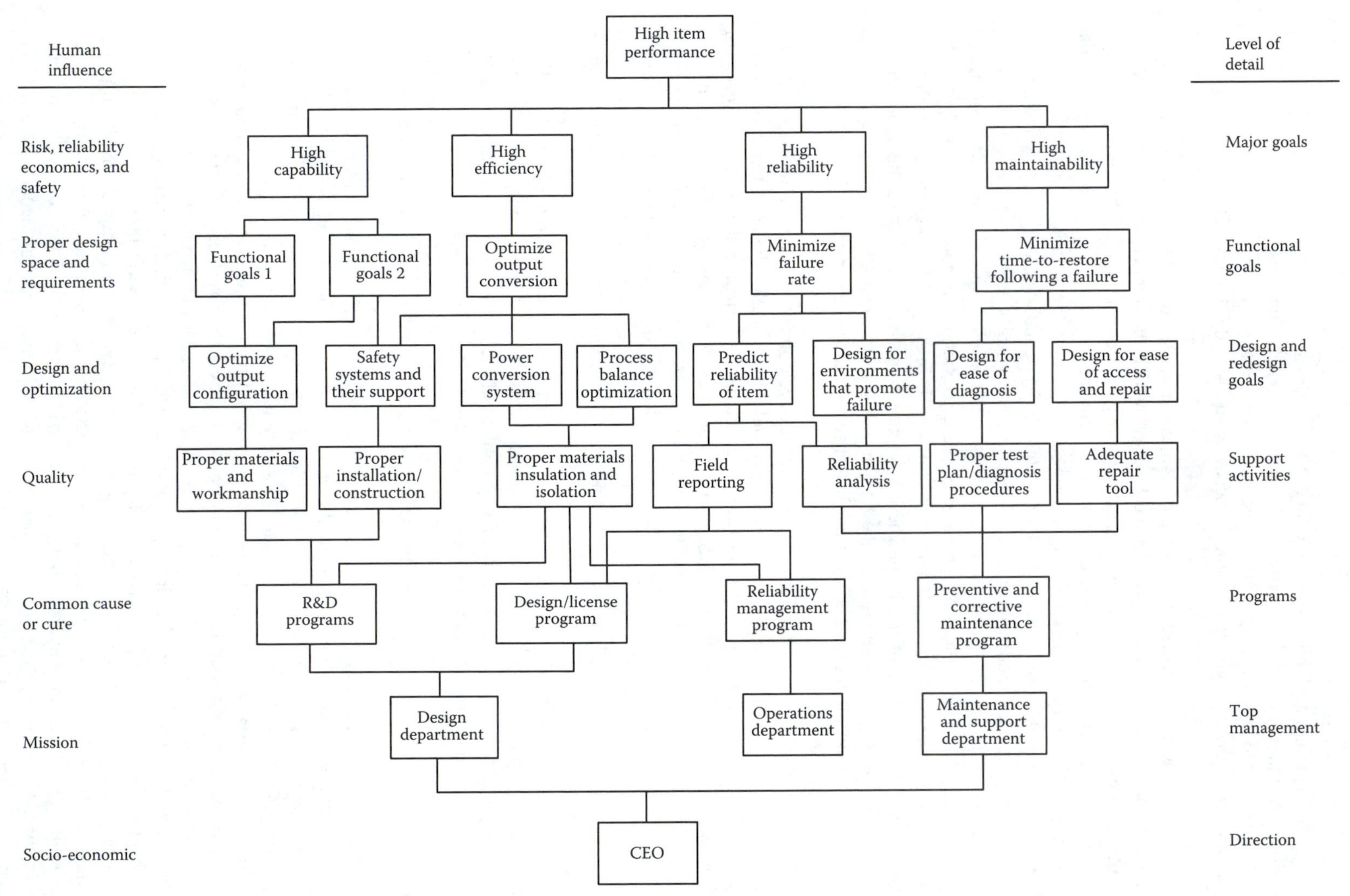

FIGURE 1.2 Conceptual diamond tree representation for achieving high performance.

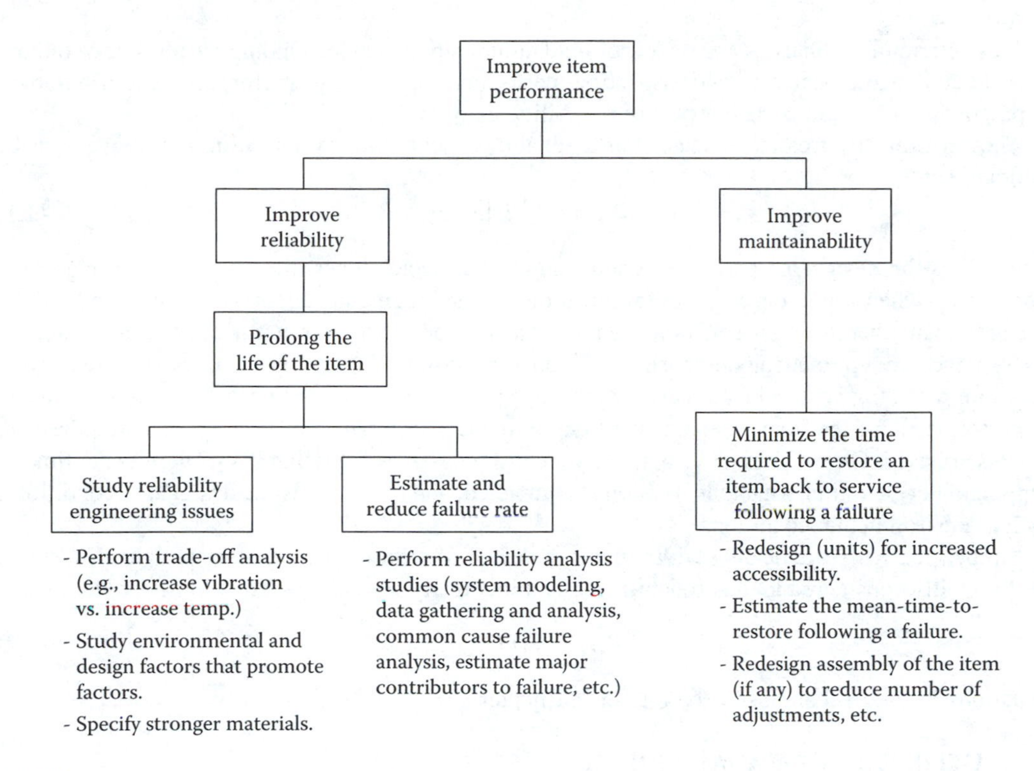

FIGURE 1.3 Conceptual hierarchy for improving performance.

For more reading on formal risk and analysis the readers are referred to Modarres [7] and Bedford and Cooke [8].

1.5 FORMAL DEFINITION OF RELIABILITY

As we discussed earlier, reliability has two connotations. One is probabilistic in nature; the other is deterministic. In this book, we generally deal with the probabilistic characterization, and we consider the two aggregate agents of failure: time or cycle of use. Based on these assumptions, let us first define what we mean by reliability. The most widely accepted definition of *reliability* is the ability of an item (a product or a system) to operate under designated operating conditions for a designated period of time or number of cycles. The *ability* of an item can be designated through a probability (the probabilistic connotation) or can be designated deterministically. The deterministic approach, as indicated in Section 1.1, deals with understanding how and why an item fails, and how it can be designed and tested to prevent such failures from occurrence or recurrence. This includes analyses

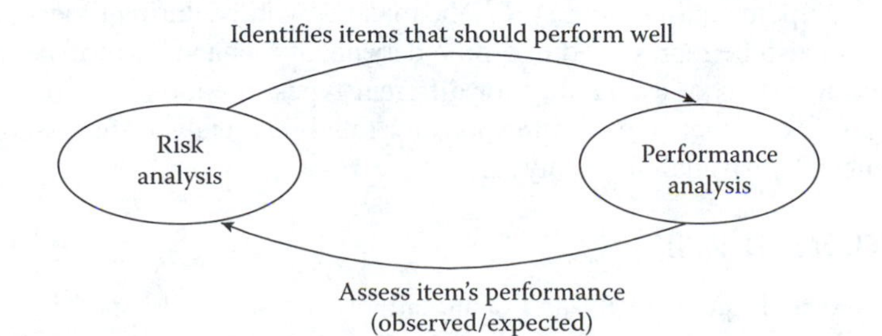

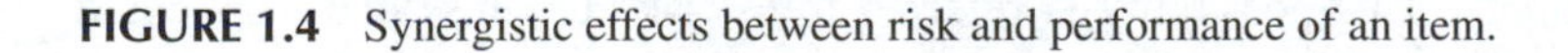
FIGURE 1.4 Synergistic effects between risk and performance of an item.

such as deterministic analysis and review of field failure reports, understanding the physics of failure, the role and degree of testing and inspection, performing redesign, or performing reconfiguration. In practice, this is an important aspect of reliability analysis.

The probabilistic treatment of an item's reliability, according to the definition above, can be summarized by

$$R(t) = \Pr(T \geq T'|c_1, c_2, \ldots), \tag{1.1}$$

where T' is the designated period of time or number of cycles for the item's operation (e.g., mission time) when time or cycle of application is the aggregate agent of failure and is the strength, endurance limit, or performance requirement when stress–strength, damage–tolerance, or performance–requirement models are used. T' can be a constant or a r.v. T is the time to failure or cycle to failure when time or application cycle is the agent of failure and is the stress, amount of damage, or performance of the item when stress–strength, damage–tolerance, or performance–requirement models are used. Clearly T is a r.v., but rarely a constant. $R(t)$ is the reliability of the item at time or application cycle t after which the mission is completed, and $c_1, c_2, \ldots$ is the designated conditions, such as environmental conditions.

Simply put, if we are dealing with time as the agent of failure, then often, in practice, $c_1, c_2, \ldots$ are implicitly considered in the probabilistic reliability analysis, and thus Equation 1.1 reduces to

$$R(t) = \Pr(T \geq t). \tag{1.2}$$

Equations 1.1 and 1.2 are discussed further in Chapter 3.

1.6 DEFINITION OF AVAILABILITY

Availability analysis is performed to verify that an item has a satisfactory probability of being operational, so it can achieve its intended objectives. In Figure 1.3, an item's availability can be thought of as a combination of its reliability and maintainability. Accordingly, when no maintenance or repair is performed (e.g., in nonrepairable items), reliability can be considered as instantaneous availability.

Mathematically, the availability of an item is a measure of the fraction of time that the item is in operating condition in relation to total or calendar time. There are several measures of availability, namely, inherent availability, achieved availability, and operational availability. For further definition of these availability measures, see Ireson and Coombs [9]. Here, we describe inherent availability, which is the most common definition used in the literature.

A more formal definition of availability is the probability that an item, when used under stated conditions in an ideal support environment (i.e., ideal spare parts, personnel, diagnosis equipment, procedures, etc.), will be operational at a given time. Based on this definition, the mean availability of an item during an interval of time T can be expressed by

$$A = \frac{u}{u+d}, \tag{1.3}$$

where u is the mean uptime during time T, d is the mean downtime during time T, and $T = u + d$.

Equation 1.3 can also be represented in a time-dependent manner. Time-dependent expressions of availability and measures of availability for different types of equipment are discussed in more detail in Chapter 5. The mathematics and methods for reliability analysis discussed in this book are also equally applicable to availability analysis.

1.7 DEFINITION OF RISK

Risk can be viewed both qualitatively and quantitatively. Qualitatively speaking, when there is a source of danger (hazard), and when there are no safeguards against exposure of the hazard, there is a possibility of loss or injury. This possibility is referred to as risk. The loss or injury could result from business, social, or military activities; operation of equipment; or investment. *Risk* can be formally defined as the potential of loss (e.g., material, human, or environmental losses) resulting from exposure to a hazard.

In complex engineering systems, there are often safeguards against exposure of hazards. The higher the level of safeguards, the lower the risk is. This also underlines the importance of highly reliable safeguard systems and shows the roles of and the relationship between reliability analysis and risk analysis.

In this book, we are concerned with quantitative risk analysis. Since quantitative risk analysis involves an estimation of the degree or probability of loss, risk analysis is fundamentally intertwined with the concept of probability of occurrence of hazards. Formal risk analysis consists of answers to the following questions [10]:

1. What can go wrong that could lead to an outcome of hazard exposure?
2. How likely is this to happen?
3. If it happens, what consequences are expected?

To answer question one, a list of outcomes (or scenarios of events leading to the outcome) should be defined. The likelihood of these scenarios should be estimated (answer to question two), and the consequence of each scenario should be described (answer to question three). Therefore, risk can be defined, quantitatively, as the following set of triplets:

$$R = \langle S_i, P_i, C_i \rangle, \quad i = 1, 2, \ldots, n, \tag{1.4}$$

where S_i is a scenario of events that lead to hazard exposure, P_i is the likelihood or frequency of scenario i, and C_i is the consequence (or evaluation measure) of scenario i, for example, a measure of the degree of damage or loss.

Since Equation 1.4 involves an estimation of the likelihood of occurrence of events (e.g., failure of safeguard systems), most of the methods described in Chapters 2 through 7 become relevant and useful. However, we have specifically devoted Chapter 8 to a detailed, quantitative description of these methods as applied to risk analysis. For more discussions on risk analysis the reader is referred to Modarres [7].

REFERENCES

1. Dasgupta, A. and M. Pecht, Materials failure mechanisms and damage models, *IEEE Trans. Reliab.*, 40 (5), 531–536, 1991.
2. Pecht, M., eds., *Handbook of Electronic Package Design*, CALCE Center for Electronic Packaging, University of Maryland, College Park, MD, Marcel Dekker, New York, 1991.
3. Collins, J. A., *Failure of Materials in Mechanical Design, Analysis, Prediction, and Prevention*, Wiley, New York, 1993.
4. Amerasekera, E. A. and D. S. Campbell, *Failure Mechanisms in Semiconductor Devices*, Wiley, New York, 1987.
5. Modarres, M., Technology-neutral nuclear power plant regulation: Implications of a safety goals-driven performance-based regulation, *Nucl. Eng. Technol.*, 37 (3), 221–230, 2005.
6. Hunt, R. N. and M. Modarres, *A Use of Goal Tree Methodology to Evaluate Institutional Practices and Their Effect on Power Plant Hardware Performance*, American Nuclear Society Topical Meeting on Probabilistic Safety Methods and Applications, San Francisco, CA, 1985.
7. Modarres, M., *Risk Analysis in Engineering*, CRC Press, Boca Raton, FL, 2006.
8. Bedford, T. and R. Cooke, *Probabilistic Risk Analysis: Foundations and Methods,* Cambridge University Press, London, 2001.
9. Ireson, W. G. and C. F. Coombs, eds., *Handbook of Reliability Engineering and Management*, McGraw-Hill, New York, 1988.
10. Kaplan, S. and J. Garrick, On the quantitative definition of risk, *Risk Anal.*, 1 (1), 11–27, 1981.

2 Basic Reliability Mathematics
Review of Probability and Statistics

2.1 INTRODUCTION

In this chapter, we discuss the elements of mathematical theory that are relevant to the study of reliability of physical objects. We begin with a presentation of basic concepts of probability. Then we briefly consider some fundamental concepts of statistics that are used in reliability analysis.

2.2 ELEMENTS OF PROBABILITY

Probability is a concept that people use formally and casually every day. Weather forecasts are probabilistic in nature. People use probability in their casual conversations to show their perception of the likely occurrence or nonoccurrence of particular events. Odds are given for the outcomes of sport events and are used in gambling. The formal use of probability concepts is widespread in science, in astronomy, biology, and engineering, for example. In this chapter, we discuss the formal application of probability theory in the field of reliability engineering.

2.2.1 SETS AND BOOLEAN ALGEBRA

To perform operations associated with probability, it is often necessary to use sets. A set is a collection of items or elements, each with some specific characteristics. A set that includes all items of interest is referred to as a *universal set*, denoted by Ω. A *subset* refers to a collection of items that belong to a universal set. For example, if set Ω represents the collection of all pumps in a power plant, then the collection of electrically driven pumps is a subset E of Ω. Graphically, the relationship between subsets and sets can be illustrated through Venn diagrams. The Venn diagram in Figure 2.1 shows the universal set Ω by a rectangle and subsets E_1 and E_2 by circles. It can also be seen that E_2 is a subset of E_1. The relationship between subsets E_1 and E_2 and the universal set can be symbolized by $E_2 \subset E_1 \subset \Omega$.

The *complement* of a set E, denoted by $\bar{E}$ and called E *not*, is the set of all items (or, more specifically, events) in the universal set that do not belong to set E. In Figure 2.1, the nonshaded area outside of the set E_2 bounded by the rectangle represents $\bar{E}_2$. It is clear that the sets E_2 and $\bar{E}_2$ together comprise Ω.

The *union* of two sets, E_1 and E_2, is a set that contains all items that belong to E_1 or E_2. The union is symbolized either by $E_1 \cup E_2$ or by $E_1 + E_2$ and is read E_1 or E_2. That is, the set $E_1 \cup E_2$ represents all elements that are in E_1, E_2, or both E_1 and E_2. The shaded area in Figure 2.2 shows the union of sets E_1 and E_2.

Suppose E_1 and E_2 represent positive odd and even numbers between 1 and 10, respectively. Then

$$E_1 = \{1, 3, 5, 7, 9\} \tag{2.1}$$

and

$$E_2 = \{2, 4, 6, 8, 10\}.$$

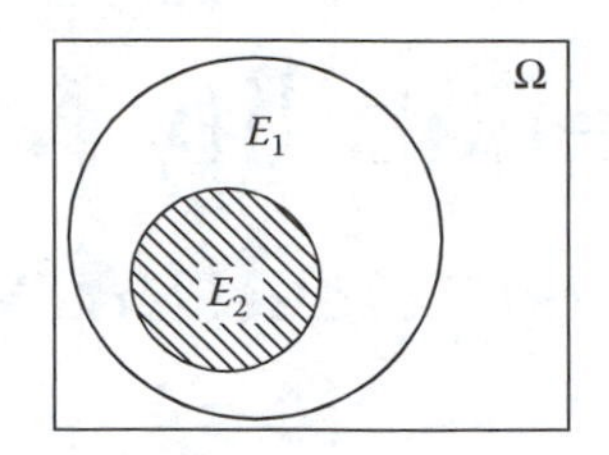

FIGURE 2.1 Venn diagram.

The union of these two sets is

$$E_1 \cup E_2 = \{1, 2, 3, 4, 5, 6, 7, 8, 9, 10\}, \tag{2.2}$$

or, if $E_1 = \{x, y, z\}$ and $E_2 = \{x, t, z\}$, then

$$E_1 \cup E_2 = \{x, y, z, t\}. \tag{2.3}$$

Note that element x is in both sets E_1 and E_2.

The *intersection* of two sets, E_1 and E_2, is the set of items that are common or shared by both E_1 and E_2. This set is symbolized by $E_1 \cap E_2$ or $E_1 \cdot E_2$ and is read E_1 and E_2. In Figure 2.3, the shaded area represents the intersection of E_1 and E_2.

Suppose E_1 is a set of manufactured devices that operate for $t > 0$ but fail prior to 1000 h of operation. If set E_2 represents a set of devices that operate between 500 and 2000 h, then $E_1 \cap E_2$ represents devices that work between 500 and 1000 and belong to both sets and can be expressed as follows:

$$\begin{aligned} E_1 &= \{t|0 < t < 1000\}, \\ E_2 &= \{t|500 < t < 2000\}, \\ E_1 \cap E_2 &= \{t|500 < t < 1000\}. \end{aligned} \tag{2.4}$$

Also, if sets $E_1 = \{x, y, z\}$ and $E_2 = \{x, t, z\}$, then

$$E_1 \cap E_2 = \{x, z\}. \tag{2.5}$$

Note that the first two sets representing time in this example represent continuous elements, and the second two sets represent discrete elements. This concept will be discussed in more detail later in this chapter.

A *null* or empty set, Ø, refers to a set that contains no items. One can easily see that the complement of a universal set is a null set, and vice versa. That is,

$$\begin{aligned} \bar{\Omega} &= \emptyset \\ \Omega &= \bar{\emptyset}. \end{aligned} \tag{2.6}$$

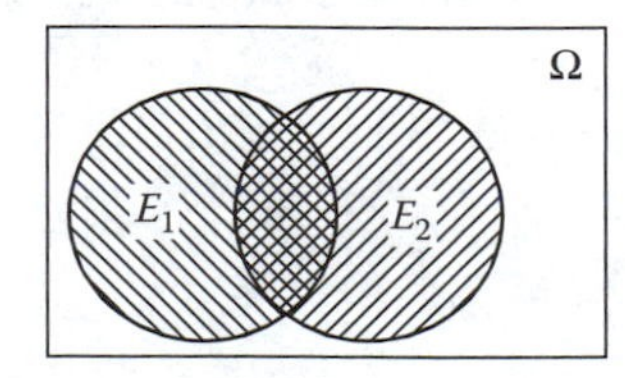

FIGURE 2.2 Union of two sets, E_1 and E_2.

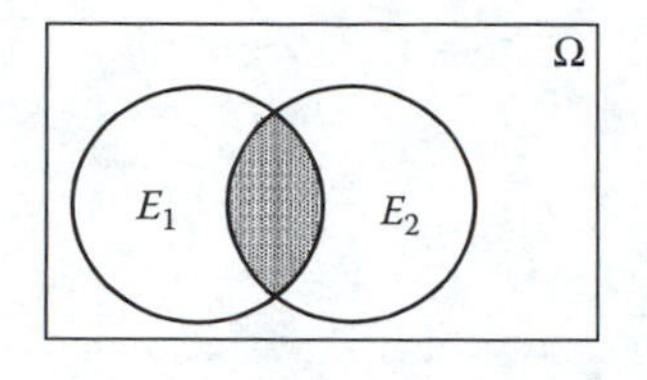

FIGURE 2.3 Intersection of two sets, E_1 and E_2.

Two sets, E_1 and E_2, are termed *mutually exclusive* or *disjoint* when $E_1 \cap E_2 = \emptyset$. In this case, there are no elements common to E_1 and E_2. Two mutually exclusive sets are illustrated in Figure 2.4.

From the discussions thus far, as well as from the examination of the Venn diagram, the following conclusions can be drawn:

The intersection of set E and a null set is a null set:

$$E \cap \emptyset = \emptyset. \tag{2.7}$$

The union of set E and a null set is E:

$$E \cup \emptyset = E. \tag{2.8}$$

The intersection of set E and the complement of E is a null set:

$$E \cap \bar{E} = \emptyset. \tag{2.9}$$

The intersection of set E and a universal set is E:

$$E \cap \Omega = E. \tag{2.10}$$

The union of set E and a universal set is the universal set:

$$E \cup \Omega = \Omega. \tag{2.11}$$

The complement of the complement of set E is E:

$$\bar{\bar{E}} = E. \tag{2.12}$$

The union of two identical sets E is E:

$$E \cup E = E. \tag{2.13}$$

The intersection of two identical sets E is E:

$$E \cap E = E. \tag{2.14}$$

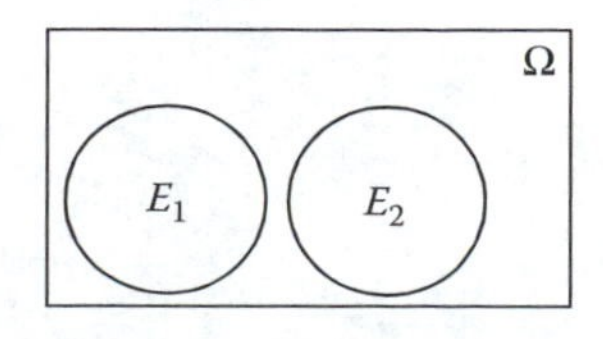

FIGURE 2.4 Mutually exclusive sets, E_1 and E_2.

Example 2.1

Simplify the following expressions:

$$E \cap (E \cap \Omega).$$

Solution: Since $E \cap \Omega = E$ and $E \cap E = E$, then the expression reduces to E.

Boolean algebra provides a means of evaluating sets. The rules are fairly simple. The sets of axioms in Table 2.1 provide all the major relations of interest in Boolean algebra, including some of the expressions discussed in Equations 2.6 through 2.14.

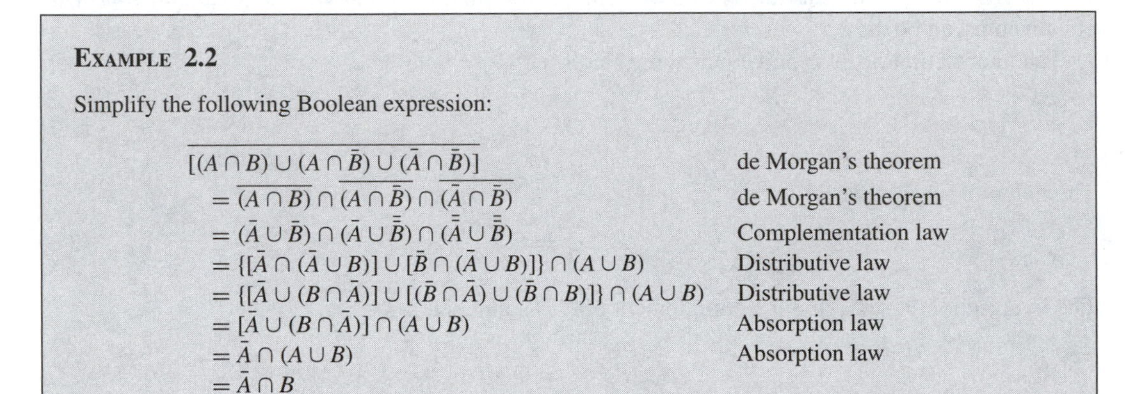

Example 2.2

Simplify the following Boolean expression:

$\overline{[(A \cap B) \cup (A \cap \bar{B}) \cup (\bar{A} \cap \bar{B})]}$	de Morgan's theorem
$= \overline{(A \cap B)} \cap \overline{(A \cap \bar{B})} \cap \overline{(\bar{A} \cap \bar{B})}$	de Morgan's theorem
$= (\bar{A} \cup \bar{B}) \cap (\bar{A} \cup \bar{\bar{B}}) \cap (\bar{\bar{A}} \cup \bar{\bar{B}})$	Complementation law
$= \{[\bar{A} \cap (\bar{A} \cup B)] \cup [\bar{B} \cap (\bar{A} \cup B)]\} \cap (A \cup B)$	Distributive law
$= \{[\bar{A} \cup (B \cap \bar{A})] \cup [(\bar{B} \cap \bar{A}) \cup (\bar{B} \cap B)]\} \cap (A \cup B)$	Distributive law
$= [\bar{A} \cup (B \cap \bar{A})] \cap (A \cup B)$	Absorption law
$= \bar{A} \cap (A \cup B)$	Absorption law
$= \bar{A} \cap B$	

2.2.2 Basic Laws of Probability

In probability theory, the elements that comprise a set are outcomes of an experiment. Thus, the universal set Ω represents the mutually exclusive listing of all possible outcomes of the experiment and is referred to as the *sample space* of the experiment. In examining the outcomes of rolling a die, the sample space is $S = \{1, 2, 3, 4, 5, 6\}$. This sample space consists of six items (elements) or *sample points*. In probability concepts, a combination of several sample points is called an event. An event

TABLE 2.1
Laws of Boolean Algebra

$X \cap Y = Y \cap X$	Commutative law
$X \cup Y = Y \cup X$	
$X \cap (Y \cap Z) = (Y \cap X) \cap Z$	Associative law
$X \cup (Y \cup Z) = (Y \cup X) \cup Z$	
$X \cap (Y \cup Z) = (X \cap Y) \cup (X \cap Z)$	Distributive law
$X \cap X = X$	Idempotent law
$X \cup X = X$	
$X \cap (X \cup Y) = X$	Absorption law
$X \cup (X \cap Y) = X$	
$X \cap \bar{X} = \emptyset$	Complementation law
$X \cup \bar{X} = \Omega$	
$\bar{\bar{X}} = X$	
$\overline{(X \cap Y)} = \bar{X} \cup \bar{Y}$	de Morgan's theorem
$\overline{(X \cup Y)} = \bar{X} \cap \bar{Y}$	

is, therefore, a subset of the sample space. For example, the event of "an odd outcome when rolling a die" represents a subset containing sample points 1, 3, and 5.

Associated with any event E of a sample space S is a probability shown by $\Pr(E)$ and obtained from the following equation:

$$\Pr(E) = \frac{m(E)}{m(S)}, \tag{2.15}$$

where $m(\cdot)$ denotes the number of elements in the set $(\cdot)$ and refers to the size of the set.

The probability of getting an odd number when tossing a die is determined by using $m(\mathit{odd\ outcomes}) = 3$ and $m(\mathit{sample\ space}) = 6$. In this case, $\Pr(\mathit{odd\ outcomes}) = 3/6 = 0.5$.

Note that Equation 2.15 describes a comparison of the relative size of the subset represented by the event E to the sample space S. This is true when all sample points are equally likely to be the outcome. When all sample points are not equally likely to be the outcome, the sample points may be weighted according to their relative frequency of occurrence over many trials or according to subjective judgment. These lead to three fundamentally different interpretations of probability.

At this point, it is important that the reader appreciates some intuitive differences between the three major conceptual interpretations of probability described below.

2.2.2.1 Classical Interpretation of Probability (Equally Likely Concept)

In this interpretation, the probability of an event E can be obtained from Equation 2.15, provided that the sample space contains N equally likely and different outcomes, that is, $m(S) = N$, n of which have an outcome (event) E, that is, $m(E) = n$. Thus $\Pr(E) = n/N$. This definition is often inadequate for engineering applications. For example, if failures of a pump to start in a process plant are observed, it is unknown whether all failures are equally likely to occur. Nor is it clear if the whole spectrum of possible events is observed. That case is not similar to rolling a perfect die, with each side having an equal probability of 1/6 at any time in the future.

2.2.2.2 Frequency Interpretation of Probability

In this interpretation, the limitation of the lack of knowledge about the overall sample space is remedied by defining the probability as the limit of n/N as N becomes large. Therefore, $\Pr(E) = \lim_{N\to\infty}(n/N)$. Thus, if we have observed 2000 starts of a pump in which 20 failed, and if we assume that 2000 is a large number, then the probability of the pump's failure to start is $20/2000 = 0.01$.

The frequency interpretation is the most widely used classical probability definition today. However, some argue that because it does not cover cases in which little or no experience (or evidence) is available, or cases where estimates concerning the observations are intuitive, a broader definition is required. This has led to the third interpretation of probability.

2.2.2.3 Subjective Interpretation of Probability

In this interpretation, $\Pr(E)$ is a measure of the degree of belief one holds in a specified event E. To better understand this interpretation, consider the probability of improving a system by making a design change. The designer believes that such a change will result in a performance improvement in one out of three missions in which the system is used. It would be difficult to describe this problem through the first two interpretations. That is, the classical interpretation is inadequate since there is no reason to believe that performance is as likely to improve as to not improve. The frequency interpretation is not applicable because no historical data exist to show how often a design change resulted in improving the system. Thus, the subjective interpretation provides a broad definition of the probability concept.

2.2.2.4 Calculus of Probability

The basic rules used to combine and treat the probability of an event are not affected by the interpretations discussed above; we can proceed without adopting any of them. (There is much dispute among probability scholars regarding these interpretations. See [1].)

In general, the axioms of probability can be defined for a sample space S as follows:

1. $\Pr(E) \geq 0$, for every event E such that $E \subset S$.
2. $\Pr(E_1 \cup E_2 \cup \cdots \cup E_n) = \Pr(E_1) + \Pr(E_2) + \cdots + \Pr(E_n)$, where the events $E_1, E_2, \ldots, E_n$ are such that no two have a sample point in common (i.e., they are disjoint or mutually exclusive).
3. $\Pr(S) = 1$.

It is important to understand the concept of independent events before attempting to multiply and add probabilities. Two events are independent if the occurrence or nonoccurrence of one does not depend on or change the probability of the occurrence of the other. Mathematically, this can be expressed by

$$\Pr(E_1|E_2) = \Pr(E_1), \tag{2.16}$$

where $\Pr(E_1|E_2)$ reads "the probability of E_1, given that E_2 has occurred." To better illustrate, let's consider the result of a test on 200 manufactured identical parts. It is observed that 23 parts fail to meet the length limitation imposed by the designer and 18 fail to meet the height limitation. Additionally, 7 parts fail to meet both length and height limitations. Therefore, 152 parts meet both of the specified requirements. Let E_1 represent the event that a part does not meet the specified length and E_2 represent the event that the part does not meet the specified height. According to Equation 2.15, $\Pr(E_1) = (7 + 23)/200 = 0.15$ and $\Pr(E_2) = (18 + 7)/200 = 0.125$. Furthermore, among 25 parts $(7 + 18)$ that have at least event E_2, 7 parts also have event E_1. Thus, $\Pr(E_1|E_2) = 7/25 = 0.28$. Since $\Pr(E_1|E_2) \neq \Pr(E_1)$, events E_1 and E_2 are dependent.

We shall now discuss the rules for evaluating the probability of the simultaneous occurrence of two or more events, that is, $\Pr(E_1 \cap E_2)$. For this purpose, we recognize two facts. First, when E_1 and E_2 are independent, the probability that both E_1 and E_2 occur or exist simultaneously is simply the multiplication of the probabilities that E_1 and E_2 occur individually. That is,

$$\Pr(E_1 \cap E_2) = \Pr(E_1) \cdot \Pr(E_2). \tag{2.17}$$

Second, when E_1 and E_2 are dependent, the probability that both E_1 and E_2 occur or exist simultaneously is obtained from the following expression:

$$\Pr(E_1 \cap E_2) = \Pr(E_1) \cdot \Pr(E_2|E_1). \tag{2.18}$$

We will elaborate further on Equation 2.18 when we discuss Bayes's theorem. It is easy to see that when E_1 and E_2 are independent, and Equation 2.16 is applied, Equation 2.18 reduces to

$$\Pr(E_1 \cap E_2) = \Pr(E_1) \cdot \Pr(E_2). \tag{2.19}$$

In general, the probability of the joint occurrence of n independent events $E_1, E_2, \ldots, E_n$ is the product of their individual probabilities. That is,

$$\Pr(E_1 \cap E_2 \cap \cdots \cap E_n) = \Pr(E_1) \cdot \Pr(E_2) \cdots \Pr(E_n) = \prod_{i=1}^{n} \Pr(E_i). \tag{2.20}$$

The probability of the joint occurrence of n dependent events $E_1, E_2, \ldots, E_n$ is obtained from

$$\Pr(E_1 \cap E_2 \cap \cdots \cap E_n)$$
$$= \Pr(E_1) \cdot \Pr(E_2|E_1) \cdot \Pr(E_3|E_1 \cap E_2) \cdots \Pr(E_n|E_1 \cap E_2 \cap \cdots \cap E_{n-1}), \quad (2.21)$$

where $\Pr(E_3|E_1 \cap E_2 \cap \cdots)$ denotes the conditional probability of E_3, given the occurrence of both E_1 and E_2, and so on.

Example 2.3

Suppose that Vendor 1 provides 40% and Vendor 2 provides 60% of circuit boards used in a computer. It is further known that 2.5% of Vendor 1's supplies are defective and only 1% of Vendor 2's supplies are defective. What is the probability that a unit is both defective and supplied by Vendor 1? What is the same probability for Vendor 2?

Solution:

E_1 = the event that a circuit board is from Vendor 1
E_2 = the event that a circuit board is from Vendor 2
D = the event that a circuit board is defective
$D|E_1$ = the event that a circuit board is known to be from Vendor 1 and defective
$D|E_2$ = the event that a circuit board is known to be from Vendor 2 and defective

Then

$$\Pr(E_1) = 0.40, \quad \Pr(E_2) = 0.60, \quad \Pr(D|E_1) = 0.025, \quad \text{and} \quad \Pr(D|E_2) = 0.01.$$

From Equation 2.21, the probability that a defective circuit board is from Vendor 1 is

$$\Pr(E_1 \cap D) = \Pr(E_1)\Pr(D|E_1) = (0.4)(0.025) = 0.01.$$

Similarly,

$$\Pr(E_2 \cap D) = 0.006.$$

The evaluation of the probability of the union of two events depends on whether or not these events are mutually exclusive. To illustrate this point, let's consider the 200 electronic parts that we discussed earlier. The union of two events E_1 and E_2 includes those parts that do not meet the length requirement, or the height requirement, or both, that is, a total of $23 + 18 + 7 = 48$. Thus,

$$\Pr(E_1 \cup E_2) = 48/200 = 0.24. \quad (2.22)$$

In other words, 24% of the parts do not meet one or both of the requirements. We can easily see that $\Pr(E_1 \cup E_2) \neq \Pr(E_1) + \Pr(E_2)$, since $0.24 \neq 0.125 + 0.15$. The reason for this inequality is the fact that the two events E_1 and E_2 are not mutually exclusive. In turn, $\Pr(E_1)$ will include the probability of inclusive events $E_1 \cap E_2$, and $\Pr(E_2)$ will also include events $E_1 \cap E_2$. Thus, joint events are counted twice in the expression $\Pr(E_1) + \Pr(E_2)$. Therefore, $\Pr(E_1 \cap E_2)$ must be subtracted from this expression. This description, which can also be seen in a Venn diagram, leads to the following expression for evaluating the probability of the union of two events that are not mutually exclusive:

$$\Pr(E_1 \cup E_2) = \Pr(E_1) + \Pr(E_2) - \Pr(E_1 \cap E_2). \quad (2.23)$$

Since $\Pr(E_1 \cap E_2) = 7/200 = 0.035$, then $\Pr(E_1 \cup E_2) = 0.125 + 0.15 - 0.035 = 0.24$, which is what we expect to get. From Equation 2.23 one can easily infer that if E_1 and E_2 are mutually exclusive, then $\Pr(E_1 \cup E_2) = \Pr(E_1) + \Pr(E_2)$. If events E_1 and E_2 are dependent, then, using Equation 2.18, we can write Equation 2.23 in the following form:

$$\Pr(E_1 \cup E_2) = \Pr(E_1) + \Pr(E_2) - \Pr(E_1) \cdot \Pr(E_2|E_1). \quad (2.24)$$

Equation 2.23 for two events can be logically extended to n events:

$$\begin{aligned}\Pr(E_1 \cup E_2 \cup \cdots \cup E_n) = &\Pr(E_1) + \Pr(E_2) + \cdots + \Pr(E_n)\\ &- [\Pr(E_1 \cap E_2) + \Pr(E_1 \cap E_3) + \cdots + \Pr(E_{n-1} \cap E_n)]\\ &+ [\Pr(E_1 \cap E_2 \cap E_3) + \Pr(E_1 \cap E_2 \cap E_4) + \cdots]\\ &- \cdots - (-1)^{n+1} \Pr(E_1 \cap E_2 \cap \cdots \cap E_n).\end{aligned} \tag{2.25}$$

Equation 2.25 consists of 2^{n-1} terms. If events $E_1, E_2, \ldots, E_n$ are mutually exclusive, then

$$\Pr(E_1 \cup E_2 \cup \cdots \cup E_n) = \Pr(E_1) + \Pr(E_2) + \cdots + \Pr(E_n). \tag{2.26}$$

When events $E_1, E_2, \ldots, E_n$ are not mutually exclusive, a useful method known as a *rare event approximation* can be used. In this approximation, Equation 2.26 is used if all $\Pr(E_i)$ are small, for example, $\Pr(E_i) < 1/(50n)$.

Example 2.4

Determine the maximum error in the right-hand side of Equation 2.25 if Equation 2.26 is used instead of Equation 2.25. Find this error for $n = 2, 3, 4$, and assume $\Pr(E_i) < (50n)^{-1}$.

Solution:

For maximum error assume $\Pr(E_i) = (50n)^{-1}$.

For $n = 2$, using Equation 2.25,

$$\Pr(E_1 \cup E_2) = \frac{2}{50(2)} - \left[\frac{1}{50(2)}\right]^2 = 0.01990.$$

Using Equation 2.26,

$$\Pr(E_1 \cup E_2) = \frac{2}{50(2)} = 0.02000,$$

$$|\text{max \% Error}| = \left|\frac{0.1990 - 0.02000}{0.1990} \times 100\right| = 0.50\%.$$

For $n = 3$, using Equation 2.26,

$$\Pr(E_1 \cup E_2 \cup E_3) = \frac{3}{50(3)} - 3\left[\frac{1}{50(3)}\right]^2 + \left[\frac{1}{50(3)}\right]^3 = 0.01987,$$

$$\Pr(E_1 \cup E_2) = \frac{3}{50(3)} = 0.02000,$$

$$|\text{max \% Error}| = 0.65\%.$$

Similarly, for $n = 4$,

$$|\text{max \% Error}| = 0.76\%.$$

For dependent events, Equation 2.25 can also be expanded to the form of Equation 2.24 by using Equation 2.21. If all events are independent, then, according to Equation 2.20, Equation 2.23 can be further simplified to

$$\Pr(E_1 \cup E_2) = \Pr(E_1) + \Pr(E_2) - \Pr(E_1) \cdot \Pr(E_2). \tag{2.27}$$

Equation 2.27 can be algebraically reformatted to the easier form of

$$\Pr(E_1 \cup E_2) = 1 - [1 - \Pr(E_1)] \cdot [1 - \Pr(E_2)]. \tag{2.28}$$

Equation 2.27 can be expanded in the case of n independent events to

$$\Pr(E_1 \cup E_2 \cup \cdots \cup E_n) = 1 - \prod_{i=1}^{n} [1 - \Pr(E_i)]. \tag{2.29}$$

Example 2.5

A particular type of computer chip is manufactured by three different suppliers. It is known that 5% of chips from Supplier 1, 3% from Supplier 2, and 8% from Supplier 3 are defective. If one chip is selected from each supplier, what is the probability that at least one of the chips is defective?

Solution:

D_1 = the event that a chip from Supplier 1 is defective
D_2 = the event that a chip from Supplier 2 is defective
D_3 = the event that a chip from Supplier 3 is defective

$D_1 \cup D_2 \cup D_3$ is the event that at least one chip from Suppliers 1, 2, or 3 is defective. Since the occurrence of events D_1, D_2, and D_3 is independent, we can use Equation 2.29 to determine the probability of $D_1 \cup D_2 \cup D_3$. Thus,

$$\Pr(D_1 \cup D_2 \cup D_3) = 1 - (1 - 0.05)(1 - 0.03)(1 - 0.08) = 0.152.$$

In probability evaluations, it is sometimes necessary to evaluate the probability of the complement of an event, that is, $\Pr(\bar{E})$. To obtain this value, let's begin with Equation 2.15 and recognize that the probability of all events in the sample space S is 1. The sample space can also be expressed by event E and its complement $\bar{E}$. That is,

$$\Pr(S) = 1 = \Pr(E \cup \bar{E}). \tag{2.30}$$

Since E and $\bar{E}$ are mutually exclusive,

$$\Pr(E \cup \bar{E}) = \Pr(E) + \Pr(\bar{E}). \tag{2.31}$$

Thus, $\Pr(E) + \Pr(\bar{E}) = 1$. By rearrangement, it follows that

$$\Pr(\bar{E}) = 1 - \Pr(E). \tag{2.32}$$

It is important to emphasize the difference between independent events and mutually exclusive events, since these two concepts are sometimes confused. In fact, two events that are mutually exclusive are not independent. Since two mutually exclusive events E_1 and E_2 have no intersection, that is, $E_1 \cap E_2 = \emptyset$, then $\Pr(E_1 \cap E_2) = \Pr(E_1) \cdot \Pr(E_2|E_1) = 0$. This means that $\Pr(E_2|E_1) = 0$, since $\Pr(E_1) \neq 0$. For two independent events, we expect to have $\Pr(E_2|E_1) = \Pr(E_2)$, which is not zero except for the trivial case of $\Pr(E_2) = 0$. This indicates that two mutually exclusive events are indeed dependent.

2.2.3 Bayes's Theorem

An important theorem known as Bayes's theorem follows directly from the concept of conditional probability, a form of which is described in Equation 2.18. For example, three forms of Equation 2.18 for events A and E are

$$\Pr(A \cap E) = \Pr(A) \cdot \Pr(E|A) \tag{2.33}$$

and

$$\Pr(A \cap E) = \Pr(E) \cdot \Pr(A|E), \tag{2.34}$$

and therefore

$$\Pr(A) \cdot \Pr(E|A) = \Pr(E) \cdot \Pr(A|E). \tag{2.35}$$

By solving for $\Pr(A|E)$, it follows that

$$\Pr(A|E) = \frac{\Pr(A) \cdot \Pr(E|A)}{\Pr(E)}. \tag{2.36}$$

This equation is known as Bayes's theorem.

It is easy to prove that if event E depends on some previous events that can occur in one of the n different ways $A_1, A_2, \ldots, A_n$, then Equation 2.36 can be generalized to

$$\Pr(A_j|E) = \frac{\Pr(A_j) \cdot \Pr(E|A_j)}{\sum_{i=1}^{n} \Pr(A_i) \cdot \Pr(E|A_i)}. \tag{2.37}$$

The right-hand side of Bayes's equation consists of two terms: $\Pr(A_j)$, called the *prior probability*, and

$$\frac{\Pr(E|A_j)}{\sum_{i=1}^{n} \Pr(A_i) \cdot \Pr(E|A_i)}, \tag{2.38}$$

the relative likelihood or the factor by which the prior probability is revised based on evidential observations (e.g., limited failure observations). $\Pr(A_j|E)$ is called the posterior probability; that is, given event E, the probability of event A_i can be updated [from prior probability $\Pr(A_j)$]. Clearly, when more evidence (in the form of events E) becomes available, $\Pr(A_j|E)$ can be further updated. Bayes's theorem provides a means of changing one's knowledge about an event in light of new evidence related to the event. We return to this topic and its application in failure data evaluation in Chapter 3. For further studies about Bayes's theorem, refer to Lindley [2].

Example 2.6

Two experts are queried about the expected reliability of a product at the end of its useful life. One expert assesses the reliability as 0.98 and the other gives 0.60. Assume both experts are considered equally credible. That is, there is a 50/50 chance that each expert is correct. Calculate the probability that the experts are correct in the following situations:

a. The product is tested for useful life and does not fail (i.e., the outcome of the test is a success).
b. Two tests are performed and both are successful.
c. One test is performed, and it results in failure.
d. Two tests are performed. One is successful, but the other results in failure.

Solution:

a. Denote
   A_1 = event that expert 1 is correct
   A_2 = event that expert 2 is correct
   B = event that the test is successful

Then

$$\Pr(A_1|B) = \frac{\Pr(A_1) \cdot \Pr(B|A_1)}{\Pr(A_1) \cdot \Pr(B|A_1) + \Pr(A_2) \cdot \Pr(B|A_2)},$$

$$\Pr(A_1|B) = \frac{0.5(0.98)}{0.5(0.98) + 0.5(0.60)} = 0.62,$$

$$\Pr(A_2|B) = 1 - 0.62 = 0.38.$$

It is clear that, because of the successful test, the credibility of expert A_1 rises in light of this test, and the credibility of expert A_2 declines.

b. Denote
B_1 = event that test 1 is successful
B_2 = event that test 2 is successful
Then

$$\Pr(A_1|B_1 \cap B_2) = \frac{\Pr(A_1) \cdot \Pr(B_1 \cap B_2|A_1)}{\Pr(A_1) \cdot \Pr(B_1 \cap B_2|A_1) + \Pr(A_2) \cdot \Pr(B_1 \cap B_2|A_2)},$$

$$\Pr(A_1|B_1 \cap B_2) = \frac{0.5(0.98)^2}{0.5(0.98)^2 + 0.5(0.6)^2} = 0.73,$$

$$\Pr(A_2|B_1 \cap B_2) = 1 - 0.73 = 0.27.$$

As expected, the credibility of expert 1 is further increased.

c. Denote
$\bar{B}$ = event that the test is a failure
Then

$$\Pr(\bar{B}|A_1) = 0.02, \Pr(\bar{B}|A_2) = 0.40,$$

$$\Pr(A_1|\bar{B}) = \frac{\Pr(A_1) \cdot \Pr(\bar{B}|A_1)}{\Pr(A_1) \cdot \Pr(\bar{B}|A_1) + \Pr(A_2) \cdot \Pr(\bar{B}|A_2)},$$

$$\Pr(A_1|\bar{B}) = \frac{0.5(0.02)}{0.5(0.02) + 0.5(0.40)} = 0.05,$$

$$\Pr(A_2|\bar{B}) = 1 - 0.05 = 0.95.$$

The credibility of expert 2's estimation substantially rises due to the originally pessimistic reliability value that he or she provided.

d. Denote
B_1 = event that the first test is successful
$\bar{B}_2$ = event that the second test is a failure
Then

$$\Pr(A_1|B_1 \cap \bar{B}_2) = \frac{\Pr(A_1) \cdot \Pr(B_1 \cap \bar{B}_2|A_1)}{\Pr(A_1) \cdot \Pr(B_1 \cap \bar{B}_2|A_1) + \Pr(A_2) \cdot \Pr(B_1 \cap \bar{B}_2|A_2)},$$

$$\Pr(A_1|B_1 \cap \bar{B}_2) = \frac{0.5(0.98)(0.02)}{0.5(0.98)(0.02) + 0.5(0.6)(0.4)} = 0.07,$$

$$\Pr(A_2|B_1 \cap \bar{B}_2) = 1 - 0.07 = 0.93.$$

Clearly the results fall between the results of (b) and (c).

Example 2.7

Suppose that 70% of an inventory of the memory chips used by a computer manufacturer comes from Vendor 1 and 30% from Vendor 2, and that 99% of the chips from Vendor 1 and 88% of the chips from Vendor 2 are not defective. If a chip from the manufacturer's inventory is selected and is defective, what is the probability that the chip was made by Vendor 1? What is the probability of selecting a defective chip (irrespective of the vendor)?

Solution: Let

A_1 = event that a chip is supplied by Vendor 1,
A_2 = event that a chip is supplied by Vendor 2,
E = event that a chip is defective,
$E|A_1$ = event that a chip known to be made by Vendor 1 is defective,
$A_1|E$ = event that a chip known to be defective is made by Vendor 1.

Thus,
$\Pr(A_1) = 0.7$,
$\Pr(A_2) = 0.3$,
$\Pr(E|A_1) = 1 - 0.99 = 0.01$,
$\Pr(E|A_2) = 1 - 0.88 = 0.12$.

Using Equation 2.37,

$$\Pr(A_1|E) = \frac{0.7(0.01)}{0.7(0.01) + 0.3(0.12)} = 0.163.$$

Thus, the prior probability of 0.7 that Vendor 1 was the supplier is changed to a posterior probability of 0.163 in light of the evidence that the chosen unit is defective. From the denominator of Equation 2.37,

$$\Pr(E) = \sum_{i=1}^{n} \Pr(A_i) \cdot \Pr(E|A_i) = 0.043.$$

Example 2.8

A passenger air bag (PAB) disable switch is used to deactivate the PAB in cases when the passenger seat of a car is not occupied. This saves the PAB from being wasted when the car gets into a frontal collision. The switch itself is an expensive component, so its feasibility needs to be justified based on the probability of a passenger seat being occupied when a collision happens. Available data show that a commercial van driver usually has a passenger 30% of the time. In addition, the expert opinion analysis indicates that the driver is 40% less likely to get into a collision with a passenger than without one. Given a frontal collision has occurred, what is the probability of the passenger seat in a commercial van being occupied?

Solution: Denote

$\Pr(P)$ = probability of a passenger seat being occupied;
$\Pr(\bar{P})$ = probability of a passenger seat not being occupied;
$\Pr(C)$ = probability of getting into a collision;
$\Pr(C|P)$ = probability of getting into a collision, given the passenger seat is occupied;
$\Pr(P|C)$ = probability of the passenger seat being occupied when a collision happens.

Using Bayes's theorem, the probability in question can be found as

$$\Pr(P|C) = \frac{\Pr(P) \cdot \Pr(C|P)}{\Pr(P) \cdot \Pr(C|P) + \Pr(\bar{P}) \cdot \Pr(C|\bar{P})}.$$

According to the statement of the problem, the probabilities of getting into a collision with and without a passenger are related to each other in the following manner:

$$\Pr(C|P) = 0.4 \cdot \Pr(C|\bar{P}).$$

Keeping this in mind, as well as, that $\Pr(P) = 1 - \Pr(\bar{P})$, one finally obtains

$$\Pr(P|C) = \frac{\Pr(P)}{(1/0.4)[1 - \Pr(P)] + \Pr(P)} = \frac{0.3}{(1/0.4)(1 - 0.3) + 0.3} = 0.146.$$

Therefore, in 100 collisions, 15 are expected to have the passenger seat occupied.

2.3 PROBABILITY DISTRIBUTIONS

In this section, we concentrate on basic probability distributions that are used in mathematical theory of reliability and reliability data analysis. In Chapter 3, we also discuss applications of certain probability distributions in reliability analysis. A fundamental aspect in describing probability distributions is the concept of a r.v. We begin with this concept and then continue with the basics of probability distributions applied in reliability analysis.

2.3.1 RANDOM VARIABLE

Let's consider an experiment with a number of possible outcomes. If the occurrence of each outcome is governed by chance (random outcome), then possible outcomes may be assigned numerical values.

An uppercase letter (e.g., X, Y) is used to represent a r.v., and a lowercase letter is used to determine the numerical value that the r.v. can take. For example, if r.v. X represents the number of system breakdowns during a given period of time t (e.g., the number of breakdowns per year) in a process plant, then x_i shows the actual number of observed breakdowns.

Random variables can be divided into two classes, namely, *discrete* and *continuous*. A r.v. is said to be discrete if its sample space is countable, such as the number of system breakdowns in a given period of time. A r.v. is said to be continuous if it can take on a continuum of values. That is, it takes on values from an interval(s) as opposed to a specific countable number. Continuous r.v.s are a result of measured variables as opposed to counted data. For example, the operation of several light bulbs can be modeled by a r.v. T, which takes on a continuous survival time t for each light bulb. Clearly, time t is not countable.

2.3.2 SOME BASIC DISCRETE DISTRIBUTIONS

Consider a discrete r.v., X. The probability distribution for a discrete r.v. is usually denoted by the symbol $\Pr(x_i)$, where x_i is one of the values that r.v. X takes on. Let r.v. X have the sample space S, designating the countable realizations of X, which can be expressed as $S = \{x_1, x_2, \ldots, x_k\}$, where k is a finite or infinite number. The discrete probability distribution for this space is then a function of $\Pr(x_i)$, such that

$$\Pr(x_i) \geq 0, \quad i = 1, 2, \ldots, k \tag{2.39}$$

and

$$\sum_{i=1}^{n} \Pr(x_i) = 1. \tag{2.40}$$

2.3.2.1 Discrete Uniform Distribution

Suppose that all possible k outcomes of an experiment are equally likely. Thus, for the sample space $S = \{x_1, x_2, \ldots, x_k\}$, one can write

$$\Pr(x_i) = p = \frac{1}{k}, \quad i = 1, 2, \ldots, k. \tag{2.41}$$

A traditional model example for this distribution is rolling a die. If r.v. X describes the numbered 1–6 faces, then the discrete number of outcomes is $k = 6$. Thus, $\Pr(x_i) = 1/6$, $x_i = 1, 2, \ldots, 6$.

2.3.2.2 Binomial Distribution

Consider a random trial having two possible outcomes: for instance, success with probability p and failure with probability $1 - p$. Consider a series of n independent trials with these outcomes. Let r.v. X denote the total number of successes. Since the number is a nonnegative integer, the sample space is $S = \{0, 1, 2, \ldots, n\}$. The probability distribution of r.v. X is given by the binomial distribution

$$\Pr(x) = \binom{n}{x} p^x (1 - p)^{n-x}, \quad x = 0, 1, 2, \ldots, n, \tag{2.42}$$

which gives the probability that a known event or outcome occurs exactly x times out of n trials. In Equation 2.42, x is the number of times that a given outcome has occurred. The parameter p indicates the probability that a given outcome will occur. The symbol $\binom{n}{x}$ denotes the total number of ways that a given outcome can occur without regard to the order of occurrence. By definition,

$$\binom{n}{x} = \frac{n!}{x!(n - x)!}, \tag{2.43}$$

where $n! = n(n - 1)(n - 2), \ldots, 1$, and $0! = 1$.

In the following examples, the binomial probability is treated (in the framework of the classical statistical inference approach) as a constant nonrandom quantity. The situations where this probability or other distribution parameters are treated as random are considered in the framework of Bayes's approach, discussed in Chapter 3.

EXAMPLE 2.9

A random sample of 15 valves is observed. From past experience, it is known that the probability of a given failure within 500 h following maintenance is 0.18. Calculate the probability that these valves will experience $0, 1, 2, \ldots, 15$ independent failures within 500 h following their maintenance.

Solution: Here the r.v. X designates the failure of a valve that can take on values of $0, 1, 2, \ldots, 15$ and $p = 0.18$. Using Equation 2.42,

x	$\Pr(x_i)$	x	$\Pr(x_i)$
0	5.10×10^{-2}	8	2.90×10^{-3}
1	1.68×10^{-1}	9	4.66×10^{-4}
2	2.58×10^{-1}	10	5.66×10^{-5}
3	2.45×10^{-1}	11	5.26×10^{-6}
4	1.61×10^{-1}	12	3.15×10^{-7}
5	7.80×10^{-2}	13	1.60×10^{-8}
6	5.40×10^{-2}	14	4.95×10^{-10}
7	1.40×10^{-2}	15	6.75×10^{-12}
	$\Sigma \Pr(x_i) = 1.00$		

EXAMPLE 2.10

In a process plant, there are two identical diesel generators for emergency AC needs. One of these diesels is sufficient to provide the needed emergency AC. Operational history indicates that there is one failure in 100 demands for each of these diesels.

a. What is the probability that at a given time of demand both diesel generators will fail?
b. If, on the average, there are 12 test-related demands per year for emergency AC, what is the probability of at least one failure for diesel A in a given year? (Assume diesel A is demanded first.)
c. What is the probability that for the case described in (b), both diesels A and B will fail on demand simultaneously at least one time in a given year?
d. What is the probability in (c) of exactly one simultaneous failure in a given year?

Solution:

a. $q = 1/100 = 0.01, p = 99/100 = 0.99$
   Assume A and B are independent (see Chapter 7 for dependent failures treatments):

$$\Pr(A \cap B) = \Pr(A) \cdot \Pr(B) = (0.01)(0.01) = 0.0001.$$

   That is, there is a 1/10,000 chance that both A and B will fail on a given demand.
b. Using the binomial distribution, one can find

$$\Pr(X = 0) = \binom{12}{0} (0.01)^0 (0.99)^{12} = 0.886,$$

   which is the probability of observing no failure in 12 trials (demands). Therefore, the probability of at least one failure per diesel generator in a year is

$$\Pr(X \geq 1) = 1 - \Pr(X < 1) = 1 - \Pr(X = 0),$$

$$\Pr(X \geq 1) = 1 - 0.886 = 0.114.$$

c. Using the results obtained in (a), the probabilities of simultaneous failure and nonfailure of generators A and B are $q = 0.0001$ and $p = 0.9999$, respectively. Then, similarly to the previous case,

$$\Pr(Y = 12) = \binom{12}{12} (0.9999)^{12} (0.0001)^0 = 0.9988,$$

   which is the probability of 12 successes for the system composed of both A and B on 12 demands (trials).

continued

Therefore, the probability of at least one failure of both A and B in a year is

$$\Pr(X \geq 1) = 1 - \Pr(X < 1) = 1 - \Pr(X = 0),$$

$$\Pr(X \geq 1) = 1 - \Pr(Y = 12) = 1 - 0.9988 = 0.0012.$$

d. For exactly one simultaneous failure in a given year,

$$\Pr(X = 1) = \binom{12}{1}(0.0001)^1(0.9999)^{11} = 0.00119.$$

2.3.2.3 Hypergeometric Distribution

The hypergeometric distribution is the only distribution associated with a finite population considered in this book. Let us have a finite population of N items, among which there are D items of interest: for example, N identical components, among which D components are defective. The probability of finding x ($x \leq D$) objects of interest within a sample (without replacements) of n ($n \leq N$) items is given by the hypergeometric distribution:

$$\Pr(x; N, D, n) = \frac{\binom{D}{x}\binom{N-D}{n-x}}{\binom{N}{n}}, \tag{2.44}$$

where

$$x = 0, 1, 2, \ldots, n; \quad x \leq D; \quad n - x \leq N - D. \tag{2.45}$$

The hypergeometric distribution is commonly used in statistical quality control and acceptance–rejection test practices. This distribution approaches the binomial one with parameters $p = D/N$ and n, when the ratio n/N becomes small.

Example 2.11

A manufacturer has a stockpile of 286 computer units. It is known that 121 of the units are more reliable than the other units. If a random sample of four computer units is selected without replacement, what is the probability that no units, two units, and all four units are from high-reliability units?

Solution: Use Equation 2.44 with

x = number of high-reliability units in the sample,
n = number of units in the selected sample,
$n - x$ = number of non-high-reliability units in the sample,
N = number of units in the stockpile,
D = number of high-reliability units in the stockpile,
$N - D$ = number of non-high-reliability units in the stockpile.

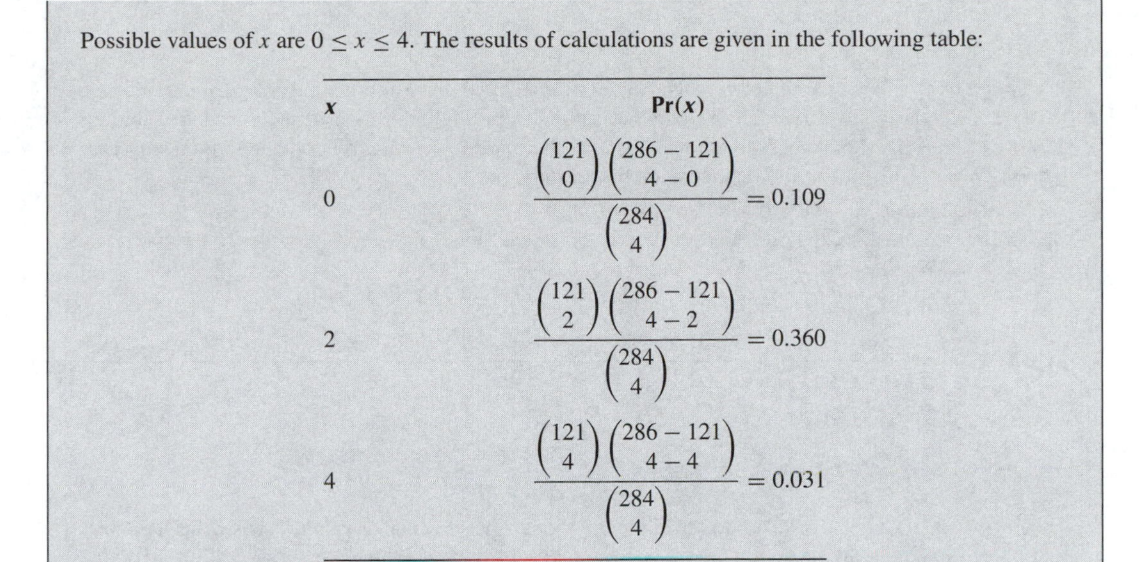

Possible values of x are $0 \le x \le 4$. The results of calculations are given in the following table:

x	$\Pr(x)$
0	$\dfrac{\binom{121}{0}\binom{286-121}{4-0}}{\binom{284}{4}} = 0.109$
2	$\dfrac{\binom{121}{2}\binom{286-121}{4-2}}{\binom{284}{4}} = 0.360$
4	$\dfrac{\binom{121}{4}\binom{286-121}{4-4}}{\binom{284}{4}} = 0.031$

2.3.2.4 Poisson Distribution

This model assumes that objects or events of interest are evenly dispersed at random in a time or space domain, with some constant intensity λ. For example, r.v. X can represent the number of failures observed at a process plant per year (time domain) or the number of buses arriving at a given station per hour (time domain), if they arrive randomly and independently in time. It can also represent the number of cracks or flaws per unit area of a metal sheet (space domain). It is clear that a r.v. X following the Poisson distribution is, in a sense, a number of random events, so that it takes on only integer values. If r.v. X follows the Poisson distribution, then

$$\Pr(x) = \frac{\rho^x \exp(-\rho)}{x!}, \quad \rho > 0, \quad x = 0, 1, 2, \ldots, \tag{2.46}$$

where ρ is the only parameter of the distribution, which is also its mean. For example, if X is a number of events observed in a nonrandom time interval, t, then $\rho = \lambda t$, where λ is the so-called *rate* (time domain) or *intensity* (space domain) *of occurrence* of Poisson events.

Example 2.12

A nuclear plant receives its electric power from a utility grid outside of the plant. From past experience, it is known that loss of grid power occurs at a rate of once a year. What is the probability that over a period of 3 years no power outage will occur? That at least two power outages will occur?

Solution: Denote $\lambda = 1\ \text{year}^{-1}$, $t = 3$ years, $\rho = 1 \times 3 = 3$. Using Equation 2.46 finds

$$\Pr(X = 0) = \frac{3^0 \exp(-3)}{0!} = 0.050,$$

$$\Pr(X = 1) = \frac{3^1 \exp(-3)}{1!} = 0.149,$$

$$\Pr(X \ge 2) = 1 - \Pr(X \le 1) = 1 - \Pr(X = 0) - \Pr(X = 1),$$

$$\Pr(X \ge 2) = 1 - 0.050 - 0.149 = 0.801.$$

Example 2.13

Inspection of a high-pressure pipe reveals that, on average, two corrosion-caused pits per meter of pipe have occurred during its service. If the hypothesis that this pitting intensity is constant and the same for all similar pipes is true, what is the probability that there are fewer than five pits in a 10-m-long pipe branch? What is the probability that there are five or more pits?

Solution: Denote $\lambda = 2$, $t = 10$, $\rho = 2 \times 10 = 20$.

Based on Equation 2.46, the probabilities of interest can be found as

$$\Pr(X < 5) = \Pr(X = 4) + \Pr(X = 3) + \Pr(X = 2) + \Pr(X = 1) + \Pr(X = 0),$$

$$\Pr(X < 5) = \frac{20^4 \exp(-20)}{4!} + \frac{20^3 \exp(-20)}{3!} + \frac{20^2 \exp(-20)}{2!} + \frac{20^1 \exp(-20)}{1!} + \frac{20^0 \exp(-20)}{0!},$$

$$\Pr(X < 5) = 1.69 \times 10^{-5},$$

$$\Pr(X \geq 5) = 1 - 1.69 \times 10^{-5} = 0.99983.$$

The Poisson distribution can be used as an approximation to the binomial distribution when the parameter p of the binomial distribution is small (e.g., when $p \leq 0.1$) and the parameter n is large. In this case, the parameter of the Poisson distribution, ρ, is substituted by np in Equation 2.46. In addition, it should be noted that as ρ increases, the Poisson distribution approaches the normal distribution with a mean and variance of ρ. This asymptotical property is used as the normal approximation for the Poisson distribution.

Example 2.14

A radar system uses 650 similar electronic devices. Each device has a failure rate of 0.00015 per month. If all these devices operate independently, what is the probability that there are no failures in a given year?

Solution: The average number of failures of a device per year is

$$\rho' = 0.00015 \times 12 = 0.0018.$$

The average number of failures of a radar system per year is

$$\rho = n\rho' = 0.0018 \times 650 = 1.17.$$

Finally, the probability of zero failures per year, according to Equation 2.46, is given by

$$\Pr(X = 0) = \frac{1.17^0 \exp(-1.17)}{0!} = 0.31.$$

2.3.2.5 Geometric Distribution

Consider a series of binomial trials with probability of success p. Introduce a r.v., X, equal to the length of a series (number of trials) of successes before the first failure is observed. The distribution of r.v. X is given by the geometrical distribution

$$\Pr(x) = p(1 - p)^{x-1}, \quad x = 1, 2, \ldots. \tag{2.47}$$

The term $(1 - p)^{x-1}$ is the probability that the failure will not occur in the first $(x - 1)$ trials. When multiplied by p, it accounts for the probability of a failure in the xth trial.

Example 2.15

In a nuclear power plant, diesel generators are used to provide emergency electric power to the safety systems. It is known that 1 out of 52 tests performed on a diesel generator results in a failure. What is the probability that the failure occurs at the 10th test?

Solution: Using Equation 2.47 with $x = 10$ and $p = 1/52 = 0.0192$ yields

$$\Pr(x = 10) = (0.0192)(1 - 0.0192)^9 = 0.016.$$

The books by Johnson and Kotz [3], Hahn and Shapiro [4], and Nelson [5] are good references for other discrete probability distributions.

2.3.3 Some Basic Continuous Distributions

In this section, we present certain continuous probability distributions that are fundamental to reliability analysis. A continuous r.v. X has a probability of zero of assuming one of the exact values of its possible outcomes. For example, if r.v. T represents the time interval within which a given emergency action is performed by a pilot, then the probability that a given pilot will perform this emergency action, for example, in *exactly* 2 min is equal to zero, because time is a continuous variable and not countable. In this situation, it is appropriate to introduce the probability associated with a small range of values that the r.v. can take on. For example, one can determine $\Pr(t_1 < T < t_2)$, that is, the probability that the pilot would perform the emergency action sometime between 1.5 and 2.5 min. To define the probability that a r.v. assumes a value less than given, one can introduce the so-called *cumulative distribution function* (cdf), $F(t)$, of a continuous r.v. T as

$$F(x) = \Pr(T \le t). \tag{2.48}$$

Similarly, the cdf of a discrete r.v. X is defined as

$$F(x_i) = \Pr(X \le x_i) = \sum_{\text{all } x_i \le x} \Pr(x_i). \tag{2.49}$$

For a continuous r.v. X, a *probability density function* (pdf), $f(t)$, is defined as

$$f(t) = \frac{\mathrm{d}F(t)}{\mathrm{d}t}. \tag{2.50}$$

It is obvious that the cdf of a r.v. t, $F(t)$, can be expressed in terms of its pdf, $f(t)$, as

$$F(t) = \Pr(T \le t) = \int_0^t f(\xi)\,\mathrm{d}\xi. \tag{2.51}$$

$f(t)\,\mathrm{d}t$ is called the *probability element*, which is the probability associated with a small interval $\mathrm{d}t$ of a continuous r.v. T.

The cdf of any continuous r.v. must satisfy the following conditions:

$$F(-\infty) = 0, \tag{2.52}$$

$$F(\infty) = \int_{-\infty}^{\infty} f(\xi)\,\mathrm{d}\xi = 1, \tag{2.53}$$

$$0 \le F(t) = \int_{-\infty}^{t} f(\xi)\,\mathrm{d}\xi \le 1, \tag{2.54}$$

and

$$F(x_1) = \int_{-\infty}^{x_1} f(\xi)\,\mathrm{d}\xi \le F(x_2) = \int_{-\infty}^{x_2} f(\xi)\,\mathrm{d}\xi, \quad x_1 \le x_2. \tag{2.55}$$

The cdf is used to determine the probability that a r.v., T, falls in an interval (a, b):

$$\Pr(a < T < b) = \int_{-\infty}^{b} f(\xi)\,\mathrm{d}\xi - \int_{-\infty}^{a} f(\xi)\,\mathrm{d}\xi = \int_{a}^{b} f(\xi)\,\mathrm{d}\xi \tag{2.56}$$

or

$$\Pr(a < T < b) = F(b) - F(a). \tag{2.57}$$

For a discrete r.v. X, the cdf is defined in a similar way, so that the analogous equations are

$$\Pr(c < T < d) = \sum_{x_i \le d} p(x_i) - \sum_{x_i \le c} p(x_i) = F(d) - F(c). \tag{2.58}$$

It is obvious that the pdf $f(t)$ of a continuous r.v. T must have the following properties:

$$f(T) \ge 0 \quad \text{for all } t, \int_{-\infty}^{\infty} f(t)\,\mathrm{d}t = 1 \tag{2.59}$$

and

$$\int_{t_1}^{t_2} f(t)\,\mathrm{d}t = \Pr(t_1 < T < t_2). \tag{2.60}$$

Example 2.16

Let the r.v. T have the pdf

$$f(T) = \begin{cases} \dfrac{t^2}{a}, & 0 < t < 6; \\ 0 & \text{otherwise.} \end{cases}$$

What is the value of parameter a? Find $\Pr(1 < T < 3)$.

Solution:

$$\int_{-\infty}^{\infty} f(t)\,\mathrm{d}t = \int_{t_1}^{t_2} f(t)\,\mathrm{d}t = 1,$$

$$\int_{t_1}^{t_2} f(t)\,\mathrm{d}t = \int_{0}^{6} \frac{t^2}{a}\,\mathrm{d}t = .\frac{t^3}{3a}\bigg|_0^6 = \frac{216}{3a} - 0 = 1.$$

Then

$$a = \frac{216}{3} = 72,$$

$$\Pr(1 < T < 3) = \int_{1}^{3} \frac{t^2}{72}\,\mathrm{d}t = \frac{t^3}{216}\bigg|_1^3 = \frac{27}{216} - \frac{1}{216} = 0.12.$$

2.3.3.1 Normal Distribution

Perhaps the best known and important continuous probability distribution is the normal distribution (sometimes called a Gaussian distribution). A normal pdf has the symmetric bell-shaped curve shown in Figure 2.5, called the normal curve.

In 1733, de Moivre developed the mathematical representation of the normal pdf, as follows:

$$f(t) = \frac{1}{\sqrt{2\pi}\sigma} \exp\left[-\frac{1}{2}\left(\frac{t-\mu}{\sigma}\right)^2\right], \quad -\infty < t < \infty, \quad -\infty < \mu < \infty, \tag{2.61}$$

where μ and σ are the parameters of the distribution $\sigma > 0$.

From Equation 2.61 it is evident that once μ and σ are specified, the normal curve can be determined. We will see later, in Chapter 3, that the parameter μ, which is referred to as the mean, and the parameter σ, which is referred to as the standard deviation, have special statistical meanings.

According to Equation 2.61, the probability that the r.v. T takes on a value between abscissas $t = t_1$ and $t = t_2$ is given by

$$\Pr(t_1 < T < t_2) = \frac{1}{\sqrt{2\pi}\sigma} \int_{t_1}^{t_2} \exp\left[-\frac{1}{2}\left(\frac{t-\mu}{\sigma}\right)^2\right] \mathrm{d}t. \tag{2.62}$$

The integral cannot be evaluated in a closed form, so the numerical integration and tabulation of normal cdf are required. However, it would be impractical to provide a separate table for every conceivable value of μ and σ. One way to get around this difficulty is to use the transformation of the normal pdf into the so-called standard normal pdf, which has a mean of zero ($\mu = 0$) and a standard deviation of 1 ($\sigma = 1$). This can be achieved by means of the r.v. transformation Z, such that

$$Z = \frac{T-\mu}{\sigma}. \tag{2.63}$$

That is, whenever r.v. T takes on a value t, the corresponding value of r.v. Z is given by $z = (t - \mu)/\sigma$. Therefore, if T takes on values $t = t_1$ or $t = t_2$, the r.v. Z takes on values $z_1 = (t_1 - \mu)/\sigma$ and

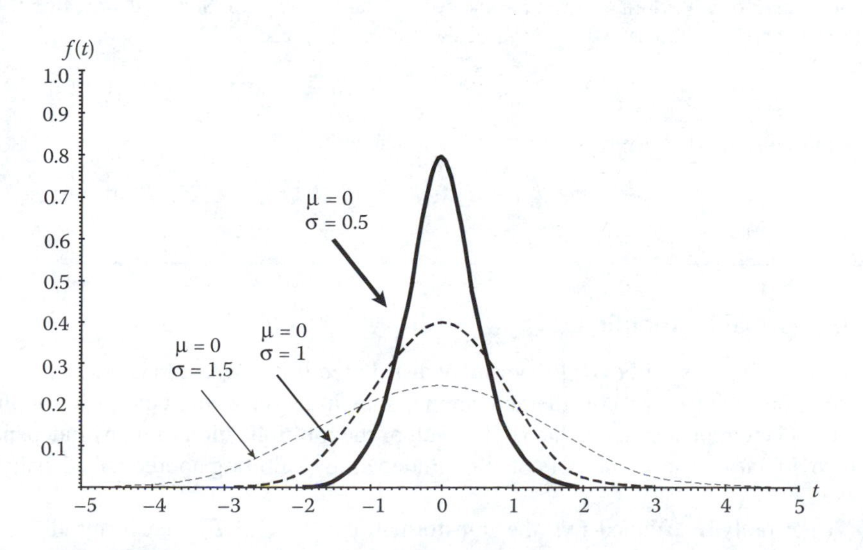

FIGURE 2.5 Normal distribution.

$z_2 = (t_2 - \mu)/\sigma$. Based on this transformation, we can write

$$\Pr(t_1 < T < t_2) = \frac{1}{\sqrt{2\pi}\sigma}\int_{t_1}^{t_2} \exp\left[-\frac{1}{2}\left(\frac{t-\mu}{\sigma}\right)^2\right] \mathrm{d}t,$$

$$\Pr(t_1 < T < t_2) = \frac{1}{\sqrt{2\pi}\sigma}\int_{z_1}^{z_2} \exp\left(-\frac{1}{2}Z^2\right) \mathrm{d}Z, \tag{2.64}$$

$$\Pr(t_1 < T < t_2) = \Pr(z_1 < Z < z_2),$$

where Z, also, has the normal pdf with a mean of zero and a standard deviation of 1. Since the standard normal pdf is characterized by a fixed mean and standard deviation, only one table would be necessary to provide the areas under the normal pdf curves. Table A.1 presents the area under the standard normal curve corresponding to $\Pr(a < Z < \infty)$.

Example 2.17

A manufacturer states that his light bulbs have a mean life of 1700 h and a standard deviation of 280 h. Assuming that the light bulb lives are normally distributed, calculate the probability that a given light bulb will last less than 1000 h.

Solution: First, the corresponding Z value is calculated as

$$Z = \frac{1000 - 1700}{280} = -2.5.$$

Note that the lower tail of a normal pdf goes to $-\infty$, so that the formal solution is given by

$$\Pr(-\infty < T < 1000) = \Pr(-\infty < Z < -2.5) = 0.0062,$$

which can be represented as

$$\Pr(-\infty < T < 1000) = \Pr(-\infty < T \leq 0) + \Pr(0 < T < 1000).$$

The first term on the right side does not have a physical meaning and is negligibly small. A more proper distribution, therefore, would have been a truncated normal distribution. Such a distribution will only exist in $t \geq 1$. This is unnecessary for this particular problem because

$$\Pr(-\infty < T < 0) = \Pr(-\infty < Z = -1700/280 = -6.07) = 6.42 \times 10^{-10},$$

which can be considered as negligible. So, finally, one can write

$$\Pr(-\infty < T < 1000) \approx \Pr(0 < T < 1000) \approx \Pr(-\infty < Z < -2.5) = 0.0062.$$

2.3.3.2 Lognormal Distribution

A positively defined r.v. is said to be lognormally distributed if its logarithm is normally distributed. The lognormal distribution has considerable applications in engineering. One major application of this distribution is to represent a r.v. that is the result of the multiplication of many independent r.v.s, or one that could vary by several orders of magnitude while exhibiting appreciable density over this range.

If T is a normally distributed r.v., the transformation $Y = \exp(T)$ transforms the normal pdf representing r.v. T with mean μ_t and standard deviation σ_t to a lognormal pdf, $f(y)$, which is given by

$$f(y) = \frac{1}{\sqrt{2\pi}\sigma_t y} \exp\left[-\frac{1}{2}\left(\frac{\ln y - \mu_t}{\sigma_t}\right)^2\right], \quad y > 0, \quad -\infty < \mu_t < \infty, \quad \sigma_t > 0. \tag{2.65}$$

Figure 2.6 shows the pdfs of the lognormal distributions for different values of μ_t and σ_t.

The area under the lognormal pdf curve $f(y)$ between two points, y_1 and y_2, which is equal to the probability that r.v. Y takes a value between y_1 and y_2, can be determined using a procedure similar to that outlined for the normal distribution. Since a logarithm is a monotonous function and $\ln y$ is normally distributed, the standard normal r.v. with

$$z_1 = \frac{\ln y_1 - \mu_t}{\sigma_t}, \quad z_2 = \frac{\ln y_2 - \mu_t}{\sigma_t} \tag{2.66}$$

provides the necessary transformations to calculate the probabilities as follows:

$$\begin{aligned} \Pr(y_1 < Y < y_2) &= \Pr(\ln y_1 < \ln Y < \ln y_2), \\ \Pr(y_1 < Y < y_2) &= \Pr(\ln y_1 < T < \ln y_2), \\ \Pr(y_1 < Y < y_2) &= \Pr(z_1 < Z < z_2). \end{aligned} \tag{2.67}$$

If μ_t and σ_t are not known but μ_y and σ_y are known, the following equations can be used to obtain μ_t and σ_t:

$$\mu_t = \ln \frac{\mu_y}{\sqrt{1 + (\sigma_y^2/\mu_y^2)}}, \tag{2.68}$$

$$\sigma_t = \sqrt{\ln\left(1 + \frac{\sigma_y^2}{\mu_y^2}\right)}. \tag{2.69}$$

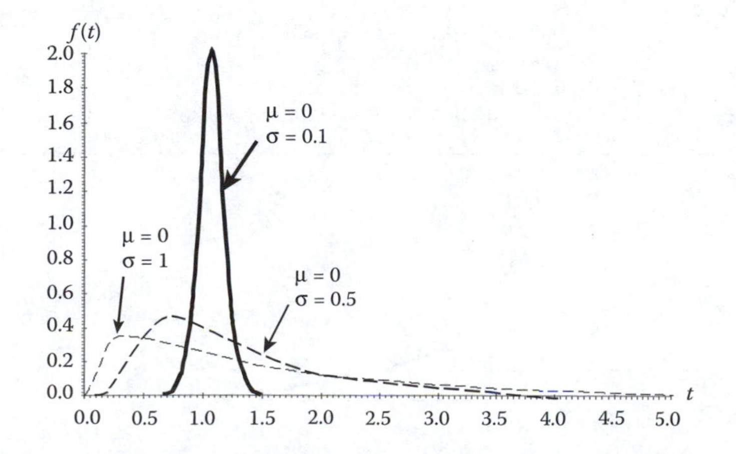

FIGURE 2.6 Lognormal distribution.

From Equations 2.68 and 2.69, μ_y and σ_y can also be determined in terms of μ_t and σ_t:

$$\mu_y = \exp\left(\mu_t + \frac{\sigma_t^2}{2}\right), \tag{2.70}$$

$$\sigma_y = \sqrt{\exp(\sigma_t^2) - 1} \cdot \mu_y. \tag{2.71}$$

2.3.3.3 Exponential Distribution

This distribution was historically the first distribution used as a model of a time-to-failure distribution, and it is still the most widely used in reliability problems. The distribution has a one-parameter pdf given by

$$f(t) = \begin{cases} \lambda \exp(-\lambda t), & \lambda, t > 0; \\ 0, & t \leq 0. \end{cases} \tag{2.72}$$

Figure 2.7 illustrates the exponential pdf. In reliability engineering applications, the parameter λ represents the probability per unit of time that a device fails (referred to as the failure rate of the device).

This notion is introduced in Chapter 3. The exponential distribution is closely associated with the Poisson distribution. Consider the following test. A unit is placed on test at $t = 0$. When the unit fails, it is instantaneously replaced by an identical new one, which, in turn, is instantaneously replaced on its failure by another identical new unit, and so on. The test is terminated at nonrandom time T. It can be shown that if the number of failures during the test is distributed according to the Poisson distribution with a mean of λT, then the time between successive failures (including the time to the first failure) has an exponential distribution with parameter λ. The test considered is an example of the so-called homogeneous Poisson process (HPP). We will elaborate more on the HPP in Chapter 5.

Example 2.18

A system has a constant failure rate of 0.001 failures/h. What is the probability that this system will fail before $t = 1000$ h? Determine the probability that it will work for at least 1000 h.

Solution: Calculate the cdf for the exponential distribution at $t = 1000$ h:

$$\Pr(t < 1000) = \int_0^{1000} \lambda \exp(-\lambda t) dt = -\exp(-\lambda t)\Big|_0^{1000},$$

$$\Pr(t < 1000) = 1 - \exp(-1) = 0.632.$$

Therefore,

$$\Pr(t > 1000) = 1 - \Pr(t < 1000) = 0.368.$$

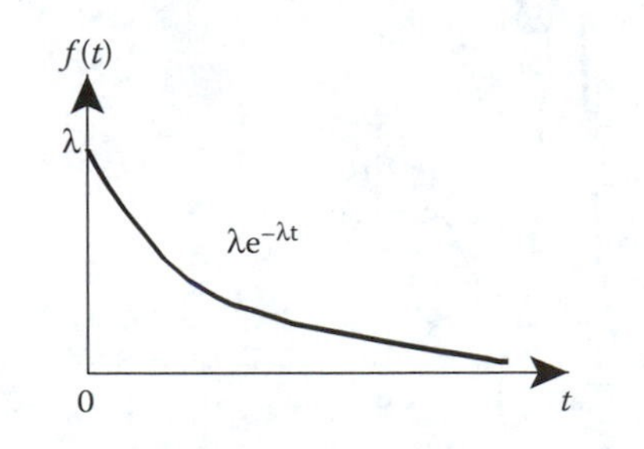

FIGURE 2.7 Exponential distribution.

2.3.3.4 Weibull Distribution

This distribution is widely used to represent the time to failure or life duration of components as well as systems. The continuous r.v. T representing the time to failure follows a Weibull distribution if its pdf is given by

$$f(t) = \begin{cases} \dfrac{\beta t^{\beta-1}}{\alpha^{\beta}} \exp\left[-\left(\dfrac{t}{\alpha}\right)^{\beta}\right], & t, \alpha, \beta > 0; \\ 0 & \text{otherwise.} \end{cases} \tag{2.73}$$

Figure 2.8 shows the Weibull pdfs with various values of parameters α and β.

A careful inspection of these graphs reveals that the parameter β determines the shape of the distribution pdf. Therefore, β is referred to as the *shape parameter*. The parameter α, on the other hand, controls the scale of the distribution. For this reason, α is referred to as the *scale parameter*. In the case when $\beta = 1$, the Weibull distribution is reduced to the exponential distribution with $\lambda = 1/\alpha$, so the exponential distribution is a particular case of the Weibull distribution. For the values of $\beta > 1$, the distribution becomes bell-shaped with some skew. We will elaborate further on this distribution and its use in reliability analysis in Chapter 3.

Example 2.19

For the Weibull distribution with the shape parameter β and the scale parameter α, find the cdf.

Solution:

$$F(t) = \Pr(T \le t) = \int_0^t \left\{\frac{\beta}{\alpha}\left(\frac{\tau}{\alpha}\right)^{\beta-1} \exp\left[-\left(\frac{\tau}{\alpha}\right)^{\beta}\right]\right\} d\tau = 1 - \exp\left[-\left(\frac{t}{\alpha}\right)^{\beta}\right].$$

2.3.3.5 Gamma Distribution

The gamma distribution can be thought of as a generalization of the exponential distribution. For example, if the time T_i between successive failures of a system has an exponential distribution, then a r.v. T such that $T = T_1 + T_2 + \cdots + T_n$ follows a gamma distribution. In the given context, T represents the cumulative time to the nth failure.

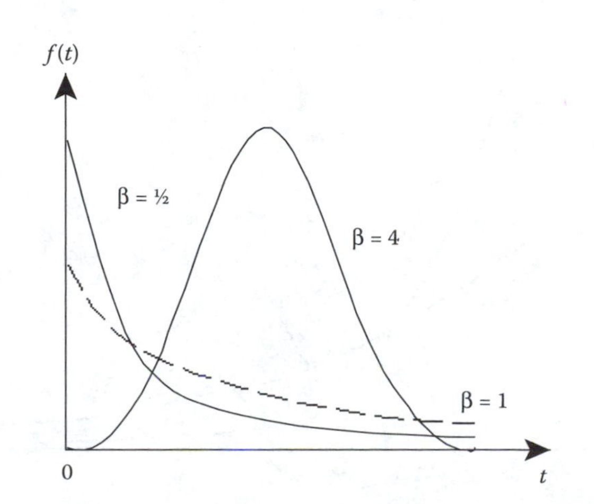

FIGURE 2.8 Weibull distribution.

A different way to interpret this distribution is to consider a situation in which a system is subjected to shocks occurring according to the Poisson process (with parameter λ). If the system fails after receiving n shocks, then the time to failure of such a system follows a gamma distribution.

The pdf of a gamma distribution with parameters β and α is given by

$$f(t) = \frac{1}{\beta^{\alpha}\Gamma(\alpha)} t^{\alpha-1} \exp\left(-\frac{t}{\beta}\right), \quad \alpha, \beta, t \geq 0, \tag{2.74}$$

where $\Gamma(\alpha)$ denotes the so-called gamma function, defined as

$$\Gamma(\alpha) = \int_0^{\infty} x^{\alpha-1} e^{-x} \, dx. \tag{2.75}$$

Note that when α is a positive integer, $\Gamma(\alpha) = (\alpha - 1)!$, but in general the parameter α is not necessarily an integer.

The mean and the variance of the gamma distribution are

$$\begin{aligned} E(t) &= \alpha\beta, \\ \text{var}(t) &= \alpha\beta^2. \end{aligned} \tag{2.76}$$

The parameter α is referred to as the shape parameter and the parameter β is referred to as the scale parameter. It is clear that if $\alpha = 1$, Equation 2.74 is reduced to an exponential distribution. Another important special case of the gamma distribution is the case when $\beta = 2$ and $\alpha = n/2$, where n is a positive integer, referred to as the number of *degrees of freedom*. This one-parameter distribution is known as the χ^2 distribution. This distribution is widely used in reliability data analysis.

Chapter 3 provides more information about applications of the gamma distribution in reliability analysis. Figure 2.9 shows the gamma distribution pdf curves for some values of α and β.

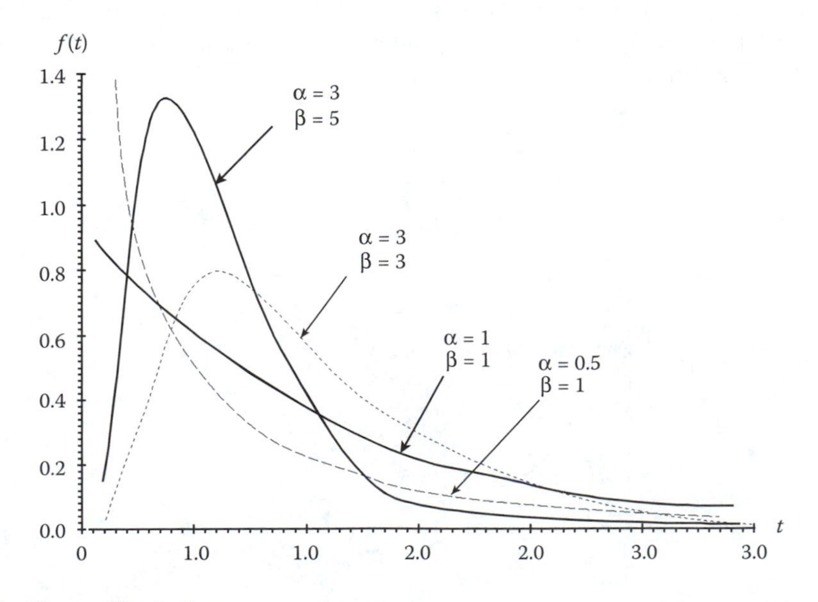

FIGURE 2.9 Gamma distribution.

2.3.3.6 Beta Distribution

The beta distribution is a useful model for r.v.s that are distributed in a finite interval. The pdf of the standard beta distribution is defined over the interval (0, 1) as

$$f(t;\alpha,\beta) = \begin{cases} \dfrac{\Gamma(\alpha+\beta)}{\Gamma(\alpha)\Gamma(\beta)} t^{\alpha-1}(1-t)^{\beta-1}, & 0 \le t \le 1, \alpha > 0, \beta \ge 0; \\ 0 & \text{otherwise.} \end{cases} \quad (2.77)$$

Similar to the gamma distribution, the cdf of the beta distribution cannot be written in closed form. It is expressed in terms of the so-called incomplete beta function, $I_t(\alpha,\beta)$, that is,

$$\mathrm{I}_t(\alpha,\beta) = \begin{cases} \dfrac{\Gamma(\alpha+\beta)}{\Gamma(\alpha)\Gamma(\beta)} \int_0^t x^{\alpha-1}(1-x)^{\beta-1}\,\mathrm{d}x, & 0 \le t \le 1, \alpha > 0, \beta \ge 0; \\ 0, & t < 0; \\ 1, & t > 1. \end{cases} \quad (2.78)$$

The mean value and the variance of the beta distribution are

$$E(t) = \frac{\alpha}{\alpha+\beta},$$
$$\mathrm{var}(t) = \frac{\alpha\beta}{(\alpha+\beta)^2(\alpha+\beta+1)}. \quad (2.79)$$

For the special case of $\alpha = \beta = 1$, the beta distribution reduces to the standard uniform distribution. Practically, the distribution is not used as a time-to-failure distribution. On the other hand, the beta distribution is widely used as an auxiliary distribution in nonparametric classical statistical distribution estimations, as well as a prior distribution in the Bayesian statistical inference, especially when the r.v. can only range between 0 and 1: for example, in a reliability or any other probability estimate. These special applications of beta distributions are discussed in Chapter 3. Figure 2.10 shows the beta distribution pdf curves for some selected values of α and β.

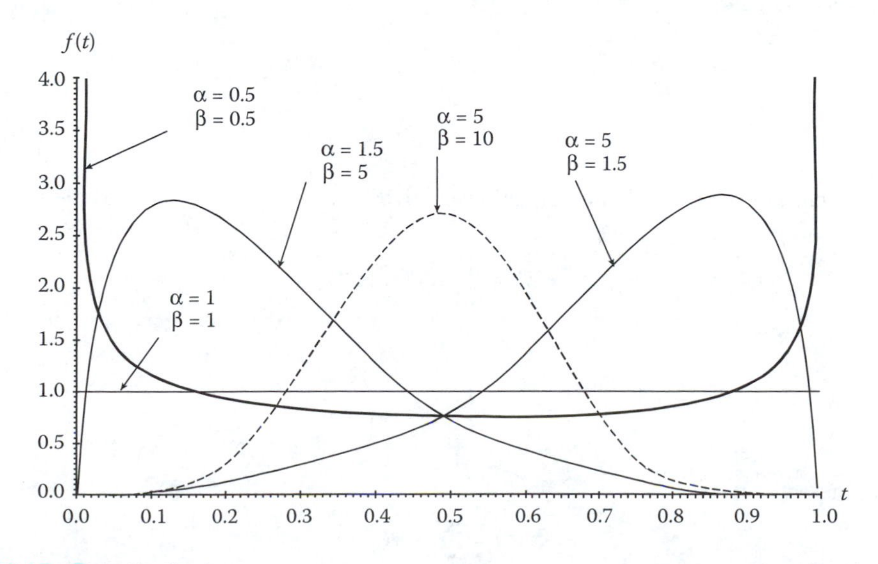

FIGURE 2.10 Beta distribution.

2.3.4 Joint and Marginal Distributions

Thus far, we have discussed distribution functions that are related to one-dimensional sample spaces. There exist, however, situations in which more than one r.v. is simultaneously measured and recorded. For example, in a study of human reliability in a control room situation, one can simultaneously estimate (1) the r.v. T representing time that various operators spend to fulfill an emergency action, and (2) the r.v. E representing the level of training that these various operators have had for performing these emergency actions. Since one expects E and T to have some relationships (e.g., better-trained operators act faster than less-trained ones), a joint distribution of both r.v.s T and E can be used to express their mutual dispersion.

Let X and Y be two r.v.s (not necessarily independent). The pdf for their simultaneous occurrence is denoted by $f(x, y)$ and is called the *joint* pdf of X and Y. If r.v.s X and Y are discrete, the joint pdf can be denoted by $\Pr(X = x, Y = y)$, or simply $\Pr(x, y)$. Thus, $\Pr(x, y)$ gives the probability that the outcomes x and y occur simultaneously. For example, if r.v. X represents the number of circuits of a given type in a process plant and Y represents the number of failures of the circuit in the most recent year, then $\Pr(7, 1)$ is the probability that a randomly selected process plant has seven circuits and that one of them has failed once in the most recent year. The function $f(x, y)$ is a joint pdf of continuous r.v.s X and Y if

1. $f(x, y) \geq 0, \quad -\infty < x, y < \infty.$
2. $\int_{-\infty}^{\infty}\int_{-\infty}^{\infty} f(x, y)\,\mathrm{d}x\,\mathrm{d}y = 1.$

Similarly, the function $\Pr(x, y)$ is a joint probability function of discrete r.v.s X and Y if

1. $\mathrm{P_r}(x, y) \geq 0$ for all values of x and y.
2. $\sum_x \sum_y \mathrm{P_r}(x, y) = 1.$

The probability that two or more joint r.v.s fall within a specified subset of the sample space is given by

$$\Pr(x_1 < X \leq x_2, y_1 < Y \leq y_2) = \int_{x_1}^{x_2}\int_{y_1}^{y_2} f(x, y)\,\mathrm{d}x\,\mathrm{d}y \tag{2.80}$$

for continuous r.v.s, and by

$$\Pr(x_1 < X \leq x_2, y_1 < Y \leq y_2) = \sum_{x_1 < x \leq x_2, y_1 < y \leq y_2} \Pr(x, y) \tag{2.81}$$

for discrete r.v.s.

The *marginal* pdfs of X and Y are defined respectively as

$$g(x) = \int_{-\infty}^{\infty} f(x, y)\,\mathrm{d}y \tag{2.82}$$

and

$$h(y) = \int_{-\infty}^{\infty} f(x, y)\,\mathrm{d}x \tag{2.83}$$

for continuous r.v.s, and as

$$\Pr(x) = \sum_y \Pr(x, y) \tag{2.84}$$

and

$$\Pr(y) = \sum_{x} \Pr(x, y) \tag{2.85}$$

for discrete r.v.s.

Using Equation 2.18, the conditional probability of an event y, given event x, is

$$\Pr(Y = y \mid X = x) = \frac{\Pr(X = x \cap Y = y)}{\Pr(X = x)} = \frac{\Pr(x, y)}{\Pr(x)}, \quad \Pr(x) > 0, \tag{2.86}$$

where X and Y are discrete r.v.s. Similarly, one can extend the same concept to continuous r.v.s X and Y and write

$$f(x \mid y) = \frac{f(x, y)}{h(y)}, \quad h(y) > 0 \tag{2.87}$$

or

$$f(x \mid y) = \frac{f(x, y)}{g(x)}, g(x) > 0, \tag{2.88}$$

where Equations 2.86 through 2.88 are called the *conditional* pdfs of discrete and continuous r.v.s, respectively. The conditional pdfs have the same properties as any other pdf. Similar to Equation 2.18, if r.v.s X and Y are independent, then $f(x \mid y) = f(x)$ for continuous r.v.s, and $\Pr(x \mid y) = \Pr(x)$ for discrete r.v.s. This would lead to the conclusion that for independent r.v.s X and Y,

$$f(x, y) = g(x) \cdot h(y) \tag{2.89}$$

if X and Y are continuous, and

$$\Pr(x, y) = \Pr(x) \cdot \Pr(y) \tag{2.90}$$

if X and Y are discrete.

Equation 2.90 can be expanded to a more general case as

$$f(x_1, x_2, \ldots, x_n) = f_1(x_1) \cdot f_2(x_2) \cdots f_n(x_n), \tag{2.91}$$

where $f(x_1, x_2, \ldots, x_n)$ is a joint pdf of r.v.s $X_1, X_2, \ldots, X_n$ and $f_1(x_1), f_2(x_2), \ldots, f_n(x_n)$ are marginal pdfs of $X_1, X_2, \ldots, X_n$, respectively.

EXAMPLE 2.20

Let r.v. T_1 represent the time (in minutes) that a machinery operator spends to locate and correct a routine problem, and let r.v. T_2 represent the length of time (in minutes) that he needs to spend reading procedures for correcting the problem. If r.v.s T_1 and T_2 are represented by the joint probability function,

$$f(t_1, t_2) = \begin{cases} c(t_1^{1/3} + t_2^{1/5}), & 60 > t_1 > 0, 10 < t_2 > 0; \\ 0 & \text{otherwise.} \end{cases}$$

continued

Find

a. The value of c.
b. The probability that an operator will be able to take care of the problem in less than 10 min. Assume that the operator in this accident should spend less than 2 min to read the necessary procedures.
c. Whether r.v.s X and Y are independent.

Solution:

a. $\Pr(t_1 < 60, t_2 < 10) = \int_0^{t_1=60} \int_0^{t_2=10} c(t_1^{1/3} + t_2^{1/5})\, \mathrm{d}t_1\, \mathrm{d}t_2 = 1, \quad c = 3.92 \times 10^{-4}$.
b. $\Pr(t_1 < 10, t_2 < 2) \int_0^{t_1=10} \int_0^{t_2=2} 3.92 \times 10^{-4}(t_1^{1/3} + t_2^{1/5})(176.17 + t_2^{1/5})\, \mathrm{d}t_2\, \mathrm{d}t_1 = 0.02$.
c. $f(t_2) = \int_0^{t_1=60} f(t_1, t_2)\, \mathrm{d}t_1 = 3.92 \times 10^{-4}(176.17 + 60t_2^{1/5})$.

Similarly,

$$f(t_1) = \int_0^{t_2=10} f(t_1, t_2)\, \mathrm{d}t_2 = 3.92 \times 10^{-4}(10t_2^{1/3} + 13.21).$$

Since $f(t_1, t_2) \neq f(t_1) \times f(t_2)$, t_1 and t_2 are not independent.

2.4 BASIC CHARACTERISTICS OF RANDOM VARIABLES

In this section, we introduce some other basic characteristics of r.v.s that are widely used in reliability engineering.

The *expectation* or *expected value* of r.v. X is a characteristic applied to continuous as well as discrete r.v.s. Consider a discrete r.v. X that takes on values x_i with corresponding probabilities $\Pr(x_i)$. The expected value of X denoted by $E(X)$ is defined as

$$E(X) = \sum_i x_i \Pr(x_i). \tag{2.92}$$

Analogously, if T is a continuous r.v. with a pdf $f(t)$, then the expectation of T is defined as

$$E(T) = \int_{-\infty}^{\infty} tf(t)\, \mathrm{d}t. \tag{2.93}$$

$E(X)$ [or $E(T)$] is a widely used concept in statistics known as the *mean*, or in mechanics known as the center of mass, and sometimes denoted by μ. $E(X)$ is also referred to as the *first moment about the origin*. In general, the kth moment about the origin (ordinary moment) is defined as

$$E(T^k) = \int_{-\infty}^{\infty} t^k f(t)\, \mathrm{d}t \tag{2.94}$$

for all integer $k \geq 1$.

In general, one can obtain the expected value of any real-value function of a r.v. In the case of a discrete distribution, $\Pr(x_i)$, the expected value of function $g(X)$ is defined as

$$E[g(x)] = \sum_{i=1}^{k} g(x_i) \Pr(x_i). \tag{2.95}$$

Similarly, for a continuous r.v. T, the expected value of $g(T)$ is defined as

$$E[g(T)] = \int_{-\infty}^{\infty} g(t)f(t)\, \mathrm{d}t. \tag{2.96}$$

Example 2.21

Determine (a) the first and (b) the second moments about origin for the Poisson distribution.

Solution:

a. $E(X) = \sum_{x=0}^{\infty} \frac{x\exp(-\rho)\rho^x}{x!} = \sum_{x=1}^{\infty} \frac{x\exp(-\rho)\rho^x}{x!} = \rho \sum_{x=1}^{\infty} \frac{\exp(-\rho)\rho^{x-1}}{(x-1)!}.$

Using the substitution $y = x - 1$, the sum can be written as

$$E(X) = \rho \sum_{y=0}^{\infty} \frac{\exp(-\rho)\rho^y}{y!}.$$

According to Equation 2.39,

$$\sum_{y=0}^{\infty} \frac{\exp(-\rho)\rho^y}{y!} = 1.$$

Thus, $E(X) = \rho$.

b. Using Equation 2.95, we can write

$$E(X^2) = \sum_{x=0}^{\infty} \frac{x^2\exp(-\rho)\rho^x}{x!} = \sum_{x=1}^{\infty} \frac{x^2\exp(-\rho)\rho^x}{x!} = \rho \sum_{x=1}^{\infty} \frac{\exp(-\rho)\rho^{x-1}}{(x-1)!}.$$

Let $y = x - 1$ [as in (a)]. Then

$$E(X^2) = \rho \sum_{y=0}^{\infty} (y+1)\frac{\exp(-\rho)\rho^y}{y!} = \rho \sum_{y=0}^{\infty} \frac{y\exp(-\rho)\rho^y}{y!} + \rho \sum_{y=0}^{\infty} \frac{\exp(-\rho)\rho^y}{y!}.$$

Since

$$\sum_{y=0}^{\infty} \frac{y\exp(-\rho)\rho^y}{y!} = \rho$$

and

$$\sum_{y=0}^{\infty} \frac{\exp(-\rho)\rho^y}{y!} = 1,$$

then

$$E(X^2) = \rho^2 + \rho.$$

Example 2.22

Find $E(T)$ and $E(T^2)$ if T is an exponential r.v. with parameter λ.

Solution: Using Equation 2.93, we obtain

$$E(T) = \lambda \int_0^{\infty} t\exp(-\lambda t)\,\mathrm{d}t = -t\exp(-\lambda t)\Big|_0^{\infty} + \int_0^{\infty} \exp(-\lambda t)\,\mathrm{d}t = 0 - \frac{\exp(-\lambda t)}{\lambda}\Big|_0^{\infty} = \frac{1}{\lambda}.$$

Similarly, using Equation 2.96,

$$E(T^2) = \int_0^{\infty} t^2\lambda\exp(-\lambda t)\,\mathrm{d}t = \frac{2}{\lambda^2}.$$

A measure of dispersion or variation of r.v. about its mean is called the variance, and is denoted by var(X) or $\sigma^2(X)$. The variance is also referred to as the *second moment about the mean* (which is analogous to the moment of inertia in mechanics). Sometimes it is referred to as the *central moment* and is defined as

$$\mathrm{var}(X) = \sigma^2(X) = E[(X - \mu)^2], \tag{2.97}$$

where σ is known as the *standard deviation* and σ^2 is known as the variance. In general, the kth moment about the mean is defined (similar to Equation 2.94) as

$$E[(X - \mu)^k] \quad \text{for all integer } k > 0. \tag{2.98}$$

Table 2.2 represents useful simple algebra associated with expectations. The rules given in Table 2.2 can be applied to discrete as well as continuous r.v.s.

One useful method of determining the moments about the origin of a distribution is by using the Laplace transform. Suppose the Laplace transform of pdf $f(t)$ is $F(S)$, then

$$F(S) = \int_0^\infty f(t)\exp(-St)\,\mathrm{d}t \tag{2.99}$$

and

$$\frac{-\mathrm{d}F(S)}{\mathrm{d}S} = \int_0^\infty tf(t)\exp(-St)\,\mathrm{d}t. \tag{2.100}$$

Since for $S = 0$ the right-hand side of Equation 2.100 reduces to the expectation $E(t)$,

$$E(T) = -\left[\frac{\mathrm{d}F(S)}{\mathrm{d}S}\right]_{S=0}. \tag{2.101}$$

In general, it is possible to show that

$$E(T^k) = \left[(-1)^k \frac{\mathrm{d}^k F(S)}{\mathrm{d}S^k}\right]_{S=0}. \tag{2.102}$$

Expression 2.102 is useful to determine moments of pdfs whose Laplace transforms are known or can be easily derived.

TABLE 2.2
Algebra of Expectations

1. $E(aX) = aE(X)$, a = constant
2. $E(a) = a$, a = constant
3. $E[g(X) \pm h(X)] = E[g(X)] \pm E[h(X)]$
4. $E[X \pm Y] = E[X] \pm E[Y]$
5. $E[X \cdot Y] = E[X] \cdot E[Y]$, if X and Y are independent

EXAMPLE 2.23

Using the results of Example 2.22, find the variance var(X) for the exponential distribution.

Solution: From Table 2.2 and Equation 2.97,

$$\mathrm{var}(T) = E(T-\mu)^2 = E(T^2+\mu^2-2\mu T) = E(T^2)+E(\mu^2)-E(2\mu T).$$

Since

$$E(T^2) = 2/\lambda^2$$

and

$$E(\mu^2) = \mu^2 = [E(T)]^2 = 1/\lambda^2,$$

then

$$E(2\mu T) = 2\mu E(T) = 2\mu^2 = 2/\lambda^2$$

and

$$\mathrm{var}(T) = 2/\lambda^2 + 1/\lambda^2 - 2/\lambda^2 = 1/\lambda^2.$$

The concept of expectation equally applies to joint probability distributions. The expectation of a real-value function h of discrete r.v.s $X_1, X_2, \ldots, X_n$ is

$$E[h(X_1,X_2,\ldots,X_n)] = \sum_{x_1}\sum_{x_2}\cdots\sum_{x_n} h(x_1,x_2,\ldots,x_n)\,\mathrm{Pr}(x_1,x_2,\ldots,x_n), \tag{2.103}$$

where $\mathrm{Pr}(x_1,x_2,\ldots,x_n)$ is the discrete joint pdf of r.v.s X_i. When dealing with continuous r.v.s, the summation terms in Equation 2.103 are replaced with integrals

$$E[h(X_1,X_2,\ldots,X_n)] = \int_{-\infty}^{\infty}\int_{-\infty}^{\infty}\cdots\int_{-\infty}^{\infty} h(x_1,x_2,\ldots,x_n)f(x_1,x_2,\ldots,x_n)\,\mathrm{d}x_1\,\mathrm{d}x_2\cdots\mathrm{d}x_n, \tag{2.104}$$

where $f(x_1,x_2,\ldots,x_n)$ is the continuous joint pdf of r.v.s X_i. In the case of a bivariate distribution with two r.v.s X_1 and X_2, the expectation of the function

$$h(X_1,X_2) = [X_1 - E(X_1)][X_2 - E(X_2)] \tag{2.105}$$

is called the *covariance* of r.v.s X_1 and X_2, and is denoted by $\mathrm{cov}(X_1,X_2)$. Using Table 2.2, it is easy to show that

$$\mathrm{cov}(X_1,X_2) = E(X_1\cdot X_2) - E(X_1)E(X_2). \tag{2.106}$$

A common measure of determining the linear relation between two r.v.s is a *correlation coefficient*, which carries information about two aspects of the relationship:

1. Strength, measured on a scale from 0 to 1.
2. Direction, indicated by the plus or minus sign.

Denoted by $\rho(X_1,X_2)$, the correlation coefficient between r.v.s X_1 and X_2 is defined as

$$\rho(X_1,X_2) = \frac{\mathrm{cov}(X_1,X_2)}{\sqrt{\mathrm{var}(X_1)\mathrm{var}(X_2)}}. \tag{2.107}$$

Clearly, if X_1 and X_2 are independent, then from Equation 2.106, $\text{cov}(X_1, X_2) = 0$, and from Equation 2.107, $\rho(X_1, X_2) = 0$. For a linear function of several r.v.s, the expectation and variance are given by

$$E\left(\sum_{i=1}^{n} a_i X_i\right) = \sum_{i=1}^{n} a_i E(X_i), \tag{2.108}$$

$$\text{var}\left(\sum_{i=1}^{n} a_i X_i\right) = \sum_{i=1}^{n} a_i \text{var}(X_i) + 2\sum_{i=1}^{n-1}\sum_{i=1}^{n} a_i a_j \text{cov}(X_i, X_j). \tag{2.109}$$

In cases where r.v.s are independent, Equation 2.109 is simplified to

$$\text{var}\left(\sum_{i=1}^{n} a_i X_i\right) = \sum_{i=1}^{n} a_i^2 \text{var}(X_i). \tag{2.110}$$

Example 2.24

Find the correlation coefficient between r.v.s T_1 and T_2 (see Example 2.20).

$$f(t_1, t_2) = \begin{cases} 3.92 \times 10^{-4}\left(t_1^{1/3} + t_2^{1/5}\right), & 60 > t_1 > 0, 10 < t_2 > 0; \\ 0 & \text{otherwise.} \end{cases}$$

Solution: From Example 2.20, Part (c),

$$f(t_1) = 3.92 \times 10^{-4}\left(10t_1^{1/3} + 13.2\right),$$

$$f(t_2) = 3.92 \times 10^{-4}\left(176.17 + 60t_2^{1/5}\right),$$

$$E[t_1] = \int_0^{60} t_1 f(t_1)\,\mathrm{d}t_1 = 3.92 \times 10^{-4}\left[\frac{10(3)}{7}(60)^{7/3} + 13.2\left(\frac{1}{2}\right)(60)^2\right] = 23.8,$$

$$E[t_2] = \int_0^{10} t_2 f(t_2)\,\mathrm{d}t_2 = 3.92 \times 10^{-4}\left[176.17\left(\frac{1}{2}\right)(10)^2 + 60\left(\frac{5}{11}\right)(10)^{11/5}\right] = 5.1,$$

$$E[t_1 \cdot t_2] = \int_0^{t_2=10}\int_0^{t_1=60} 3.92 \times 10^{-4} t_1 t_2\left(t_1^{1/3} + t_2^{1/5}\right)\mathrm{d}t_1\,\mathrm{d}t_2 = 169,$$

thus

$$\text{cov}(t_1 \cdot t_2) = E[t_1 \cdot t_2] - E[t_1]E[t_2] = 169 - (23.8)(5.1) = 46.6.$$

Similarly,

$$E\left[t_1^2\right] = \int_0^{60} 3.92 \times 10^{-4}\left[10t_1^{7/3} + 13.2t_1^2\right]\mathrm{d}t_1 = 1367.0,$$

$$E\left[t_2^2\right] = \int_0^{10} 3.92 \times 10^{-4}\left[176.17t_2^2 + 60t_1^{11/5}\right]\mathrm{d}t_2 = 34.7,$$

$$\text{var}(t_1) = E\left[t_1^2\right] - \left[E(t_1)\right]^2 = 1367.0 - (23.8)^2 = 800.6,$$

$$\text{var}(t_2) = E\left[t_2^2\right] - \left[E(t_2)\right]^2 = 34.7 - (5.1)^2 = 8.7,$$

$$\rho(t_1, t_2) = \frac{\text{cov}(t_1, t_2)}{\sqrt{\text{var}(t_1)\text{var}(t_2)}} = \frac{46.6}{\sqrt{800.6(8.7)}} = 0.56.$$

This indicates that there is a somewhat strong positive correlation between the time the operator spends to solve a machinery problem and the length of time he or she needs to spend on reading the problem's correction-related procedures.

EXAMPLE 2.25

Two identical pumps are needed in a process plant to provide a sufficient cooling flow and maintain an adequate safety margin. Given many other prevailing conditions, the output flow of each may vary, but it is known to be normally distributed with a mean of 540 gpm and a standard deviation of 65 gpm. Calculate

a. The distribution of the resulting (total) flow from both pumps.
b. The probability that the resulting flow that guarantees a safety margin of 80 gpm is less than 1000 gpm.

Solution:

a. If M_1 and M_2 are the flows from each pump, then the total flow is $M = M_1 + M_2$. Since each of the r.v.s M_1 and M_2 are normally distributed, it can be shown that, if M_1 and M_2 are independent, M is also normally distributed. From Equation 2.108, the mean of r.v. M is given by

$$\mu_M = E(M) = E(M_1) + E(M_2) = 540 + 540 = 1080\,\text{gpm}.$$

Because M_1 and M_2 are assumed to be independent, then using Equation 2.109, the variance of r.v. M is obtained as

$$\text{var}(M) = \text{var}(M_1) + \text{var}(M_2) = (65)^2 + (65)^2 = 8450$$

and

$$\sigma(M) = 91.9\,\text{gpm}.$$

b. Using standard normal distribution transformation (Equation 2.63),

$$z = \frac{1000 - 1080}{91.9} = -0.87.$$

This corresponds to

$$\Pr(M \leq 1000) = \Pr(Z \leq -0.87) = 0.19.$$

2.5 ESTIMATION AND HYPOTHESIS TESTING

Reliability and performance data obtained from special tests, experiments, or practical use of a product provide a basis for performing statistical inference about underlying distribution. Each observed value is considered a *realization* (or *observation*) of some hypothetical r.v.: that is, a value that the r.v., say X, can take on. For example, the number of pump failures following a demand in a large plant can be considered as realization of some r.v.

A set of observations from a distribution is called a *sample*. The number of observations in a sample is called the *sample size*. In the framework of classical statistics, a sample is usually composed of random, independently and identically distributed observations. From a practical point of view, this assumption means that elements of a given sample are obtained independently and under the same conditions.

To check the applicability of a given distribution (e.g., binomial distribution in the pump failure case) and to estimate the parameters of the distribution, one must use special statistical procedures known as hypothesis testing and estimation, which are very briefly considered below.

2.5.1 Point Estimation

Point and *interval estimation* are the two basic kinds of estimation procedures considered in statistics. Point estimation provides a single number obtained on the basis of a data set (a sample) that represents a parameter of the distribution function or other characteristic of the underlying distribution of interest. As opposed to interval estimation, point estimation does not provide any information about its accuracy. Interval estimation is expressed in terms of *confidence intervals*. The confidence interval includes the true value of the parameter with a specified confidence probability.

Suppose we are interested in estimating a single-parameter distribution $f(X, \theta)$ based on a random sample $x_1, \ldots, x_n$. Let $t(x_1, \ldots, x_n)$ be a single-valued (simple) function of $x_1, x_2, \ldots, x_n$. It is obvious that $t(x_1, x_2, \ldots, x_n)$ is also a r.v., which is referred to as a *statistic*. A point estimate is obtained by using an appropriate statistic and calculating its value based on the sample data. The statistic (as a function) is called the *estimator*; meanwhile, its numerical value is called the *estimate*.

Consider the basic properties of point estimators. An estimator $t(x_1, \ldots, x_n)$ is said to be an *unbiased* estimator for θ if its expectation coincides with the value of the parameter of interest θ, that is, $E[t(x_1, x_2, \ldots, x_n)] = \theta$ for any value of θ. Thus, the bias is the difference between the expected value of an estimate and the true parameter value itself. It is obvious that the smaller the bias, the better the estimator is.

Another desirable property of an estimator $t(x_1, x_2, \ldots, x_n)$ is the property of consistency. An estimator t is said to be consistent if, for every $\varepsilon > 0$,

$$\lim_{n \to \infty} P\left[\left|t(x_1, x_1, \ldots, x_n) - \theta\right| < \varepsilon\right] = 1. \tag{2.111}$$

This property implies that as the sample size n increases, the estimator $t(x_1, x_2, \ldots, x_n)$ gets closer to the true value of θ. In some situations several unbiased estimators can be found. A possible procedure for selecting the best one among the unbiased estimators can be based on choosing the one having the least variance. An unbiased estimator t of θ, having minimum variance among all unbiased estimators of θ, is called *efficient*.

Another estimation property is sufficiency. An estimator $t(x_1, x_2, \ldots, x_n)$ is said to be a *sufficient* statistic for the parameter if it contains all the information that is in the sample $x_1, x_2, \ldots, x_n$. In other words the sample $x_1, x_2, \ldots, x_n$ can be replaced by $t(x_1, x_2, \ldots, x_n)$ without loss of any information about the parameter of interest .

Several methods of estimation are considered in mathematical statistics. In the following section, two of the most common methods, that is, the method of moments and the method of maximum likelihood, are briefly discussed.

2.5.1.1 Method of Moments

In the previous section the mean and the variance of a continuous r.v. X were defined as the expected value of X and the expected value of $(X - \mu)^2$, respectively. Quite naturally, one can define the *sample mean* and *sample variance* as the respective expected values of a sample of size n from the distribution of X, namely, $x_1, x_2, \ldots, x_n$, as follows:

$$\bar{x} = \frac{1}{n} \sum_{i=1}^{n} x_i \tag{2.112}$$

and

$$S^2 = \frac{1}{n}\sum_{i=1}^{n}(x_i - \bar{x})^2, \tag{2.113}$$

so that $\bar{x}$ and S^2 can be used as the point estimates of the distribution mean, μ, and variance, σ^2. It should be mentioned that the estimator of variance (Equation 2.113) is biased, since $\bar{x}$ is estimated from the same sample. However, it can be shown that this bias can be removed by multiplying it by $n/(n-1)$:

$$S^2 = \frac{1}{n-1}\sum_{i=1}^{n}(x_i - \bar{x})^2. \tag{2.114}$$

Generalizing the examples considered, it can be said that the method of moments is an estimation procedure based on empirically estimated (or *sample*) moments of the r.v. According to this procedure, the sample moments are equated to the corresponding distribution moments. The solutions of the equations obtained provide the estimators of the distribution parameters.

Example 2.26

A sample of eight manufactured shafts is taken from a plant lot. The diameters of the shafts are 1.01, 1.08, 1.05, 1.01, 1.00, 1.02, 0.99, and 1.02 inches. Find the sample mean and variance.

Solution:

From Equation 2.112, $\bar{x} = 1.0225$.

From Equation 2.114, $S^2 = 0.0085$.

2.5.1.2 Maximum Likelihood Method

This method is one of the most widely used methods of estimation. Consider a continuous r.v. X with pdf $f(X, \theta)$, where θ is a parameter. Let us have a sample $x_1, x_2, \ldots, x_n$ of size n from the distribution of r.v. X. Under the maximum likelihood approach, the estimate of θ is the value of θ that delivers the highest (or most likely) probability density of observing the particular set $x_1, x_2, \ldots, x_n$. The likelihood of obtaining this particular set of sample values is proportional to the joint pdf $f(x, \theta)$ calculated at the sample points $x_1, x_2, \ldots, x_n$. The likelihood function for a continuous distribution is introduced as

$$L(x_1, x_2, \ldots, x_n; \theta) = f(x_1, \theta) f(x_2, \theta) \cdots f(x_n, \theta). \tag{2.115}$$

Generally speaking, the definition of the likelihood function is based on the probability (for a discrete r.v.) or the pdf (for a continuous r.v.) of the joint occurrence of n events, $X = x_1, x_2, \ldots, x_n$. The maximum likelihood estimate (MLE), $\hat{\theta}$, is chosen as the one that maximizes the likelihood function, $L(x_1, x_2, \ldots, x_n; \theta)$, with respect to θ.

The standard way to find a maximum of a parameter is to calculate the first derivative with respect to this parameter and equate it to zero. This yields the equation

$$\frac{\partial L(x_1, x_2, \ldots, x_n; \theta)}{\partial \theta} = 0, \tag{2.116}$$

from which the MLE $\hat{\theta}$ can be obtained.

Due to the multiplicative form of Equation 2.115, it turns out, in many cases, to be more convenient to maximize the logarithm of the likelihood function instead, that is, to solve the following equation:

$$\frac{\partial \log L(x_1, x_2, \ldots, x_n; \theta)}{\partial \theta} = 0. \tag{2.117}$$

Because the logarithm is a monotonous transformation, the estimate of θ_0 obtained from this equation is the same as that obtained from Equation 2.116. For some cases, Equation 2.116 or 2.117 can be solved analytically; for other cases, they have to be solved numerically.

Under some general conditions, the MLEs are consistent, asymptotically efficient, and asymptotically normal.

Example 2.27

Consider a sample $t_1, t_2, \ldots, t_n$ of n times to failure of a component whose time to failure is assumed to be exponentially distributed with parameter λ (the failure rate). Find the maximum likelihood estimator for λ.

Solution: Using Equations 2.115 and 2.117, one can obtain

$$L(t, \lambda) = \prod_{i=1}^{n} \lambda \exp(-\lambda t_i) = \lambda^n \exp\left(-\lambda \sum_{i=1}^{n} t_i\right),$$

$$\ln L = n(\ln \lambda) - \lambda \sum_{i=1}^{n} t_i \quad \text{or} \quad \frac{\partial \ln L}{\partial \lambda} = \frac{n}{\lambda} - \sum_{i=1}^{n} t_i = 0,$$

$$\hat{\lambda} = \frac{n}{\sum_{i=1}^{n} t_i}.$$

Recalling the second-order condition $\partial^2 \ln L / \partial \lambda^2 = -n/\lambda^2 < 0$, it is clear that the estimate $\hat{\lambda}$ is indeed the MLE for the problem considered. Recalling Example 2.26, it is worthy to mention that the estimate can also be obtained using the method of moments.

2.5.2 Interval Estimation and Hypothesis Testing

A two-sided confidence interval for an unknown distribution parameter θ of a continuous r.v. X, based on a sample $x_1, x_2, \ldots, x_n$ of size n from the distribution of X, is introduced in the following way. Consider two statistics $\theta_l(x_1, x_2, \ldots, x_n)$ and $\theta_u(x_1, x_2, \ldots, x_n)$ chosen in such a way that the probability that parameter θ lies in an interval $[\theta_l, \theta_u]$ is

$$\Pr[\theta_l(x_1, x_2, \ldots, x_n) < \theta < \theta_u(x_1, x_2, \ldots, x_n)] = 1 - \alpha. \tag{2.118}$$

The *random* interval $[l, u]$ is called a $100(1-\alpha)\%$ *confidence interval* for the parameter θ. The endpoints l and u are referred to as the $100(1-\alpha)\%$ upper and lower confidence limits of θ; $(1-\alpha)$ is called the *confidence coefficient* or *confidence level*. The most commonly used values for α are 0.10, 0.05, and 0.01. The case when $\theta > \theta_l$ with the probability of $1-\alpha$, θ_l is called the *one-sided lower confidence limit* for θ. The case when $\theta < \theta_u$ with probability of $1-\alpha$, θ_u is called the *one-sided upper confidence limit* for θ. A $100(1-\alpha)\%$ confidence interval for an unknown parameter θ is interpreted as follows: if a series of repetitive experiments (tests) yields random samples from the same distribution, and the same confidence interval is calculated for each sample, then $100(1-\alpha)\%$ of the constructed intervals will, in the long run, contain the true value of θ.

Consider a typical example illustrating the basic idea of confidence limits construction. Consider a procedure for constructing confidence intervals for the mean of a normal distribution with known variance. Let $x_1, x_2, \ldots, x_n$ be a random sample from the normal pdf, $N(\mu, \sigma^2)$, in which μ is unknown and σ^2 is assumed to be known. It can be shown that the sample mean $\bar{X}$ (as a statistic) has the normal distribution $N(\mu, \sigma^2/n)$. Thus, $(X - \mu)/(n)^{1/2}/\sigma$ has the standard normal distribution. Using this distribution, one can write

$$\Pr\left(-z_{1-\alpha/2} \le \frac{\bar{X} - \mu}{\sigma/\sqrt{n}} \le z_{1-\alpha/2}\right) = 1 - \alpha, \tag{2.119}$$

where $z_{1-(\alpha/2)}$ is the $100(1 - \alpha/2)$th percentile of the standard normal distribution, which can be obtained from Table A.1. After simple algebraic transformations, the inequalities inside the parentheses of Equation 2.119 can be rewritten as

$$\Pr\left(\bar{X} - z_{1-\alpha/2}\frac{\sigma}{\sqrt{n}} \le \mu \le \bar{X} + z_{1-\alpha/2}\frac{\sigma}{\sqrt{n}}\right) = 1 - \alpha. \tag{2.120}$$

Equation 2.120 provides the symmetric $(1 - \alpha)$ confidence interval of interest. Generally, a two-sided confidence interval is wider for a higher confidence level $(1 - \alpha)$. As the sample size n increases, the confidence interval becomes shorter for the same confidence coefficient $(1 - \alpha)$.

In the case when σ^2 is unknown and is estimated using Equation 2.114, the respective confidence interval is given by

$$\Pr\left(\bar{x} - t_{\alpha/2}\frac{S}{\sqrt{n}} < \mu < \bar{x} + t_{\alpha/2}\frac{S}{\sqrt{n}}\right) = 1 - \alpha, \tag{2.121}$$

where $t_{\alpha/2}$ is the percentile of t-student distribution with $(n - 1)$ degrees of freedom. Values of t_α for different numbers of degrees of freedom are given in Table A.2. Confidence intervals for σ^2 for a normal distribution can be obtained as

$$\frac{(n-1)S^2}{\chi^2_{1-\alpha/2}(n-1)} < \sigma^2 < \frac{(n-1)S^2}{\chi^2_{\alpha/2}(n-1)}, \tag{2.122}$$

where $\chi^2_{1-\alpha/2}(n-1)$ is the percentile of χ^2 distribution with $(n - 1)$ degrees of freedom, which are given in Table A.3.

2.5.2.1 Hypothesis Testing

Interval estimation and hypothesis testing may be viewed as mutually inverse procedures. Let us consider a r.v. X with a known pdf $f(x;\theta)$. Using a random sample from this distribution, one can obtain a point estimate $\hat{\theta}$ of parameter θ. Let θ have a hypothesized value of $\theta = \theta_0$. Under these quite realistic conditions, the following question can be raised: is the $\hat{\theta}$ estimate compatible with the hypothesized value θ_0? In terms of *statistical hypothesis* testing, the statement $\theta = \theta_0$ is called the *null hypothesis*, which is denoted by H_0. For the case considered, it is written as

$$H_0 : \theta = \theta_0. \tag{2.123}$$

The null hypothesis is always tested against an *alternative hypothesis*, denoted by H_1, which for the case considered might be the statement $\theta \ne \theta_0$, which is written as

$$H_1 : \theta \ne \theta_0. \tag{2.124}$$

The null and alternative hypotheses are also classified as *simple* (or *exact*, when they specify exact parameter values) and *composite* (or *inexact*, when they specify an interval of parameter values). In the considered example, H_0 is simple and H_1 is composite. An example of a simple alternative hypothesis might be

$$H_1 : \theta = \theta^*. \quad (2.125)$$

For testing statistical hypotheses, *test statistics* are used. In many situations the test statistic is the point estimator of the unknown distribution. In this case, as in the case of the interval estimation, one has to obtain the distribution of the test statistic used.

Recall the example considered above. Let $x_1, x_2, \ldots, x_n$, be a random sample from the normal pdf, $N(\mu, \sigma^2)$, in which μ is an unknown parameter and σ^2 is assumed to be known. One has to test the simple null hypothesis

$$H_0 : \mu = \mu^*. \quad (2.126)$$

against the composite alternative

$$H_0 : \mu \neq \mu^*. \quad (2.127)$$

As the test statistic, use the same (Equation 2.112) sample mean, $\bar{x}$, which has the normal distribution, $N(\mu, \sigma^2/n)$. Having the value of the test statistic $\bar{x}$, one can construct the confidence interval using Equation 2.120 and see whether the value of μ^* falls inside the interval. This is the test of the null hypothesis. If the confidence interval includes μ^*, the null hypothesis is not rejected at *significance level* α.

In terms of hypothesis testing, the confidence interval considered is called the *acceptance region*, and the upper and the lower limits of the acceptance region are called the *critical values*, while the significance level α is referred to as a probability of a *Type I error*. In making a decision about whether or not to reject the null hypothesis, it is possible to commit the following errors:

- Reject H_0 when it is true (Type I error).
- Do not reject H_0 when it is false (Type II error).

The probability of the Type II error is designated by β. These situations are traditionally represented by the following table:

	State of Nature (True Situation)	
Decision	**H_0 is True**	**H_0 is False**
Reject H_0	Type I error	No error
Do not reject H_0	No error	Type II error

It is clear that increasing the acceptance region decreases α and simultaneously results in increasing β. The traditional approach to this problem is to keep the probability of Type I errors at a low level (0.01, 0.05, or 0.10) and to minimize the probability of Type II errors as much as possible. The probability of not making of Type II error is referred to as the *power of the test*. Examples of a special class of hypothesis testing are considered in Section 2.7.

2.6 FREQUENCY TABLES AND HISTOGRAMS

When studying distributions, it becomes convenient to start with some preliminary procedures useful for data editing and detecting outliers by constructing *empirical distributions* and *histograms*. Such preliminary data analysis procedures might be useful themselves (the data speak for themselves), as well as being used for other elaborate analyses (the goodness-of-fit testing, for instance).

To illustrate some of these procedures, consider the data set composed of observed times to failure of 100 identical electronic devices that are placed on a life test. The observed time-to-failure data are given in Table 2.3.

The measure of interest here is the probability associated with each interval of time to failure. This can be obtained using Equation 2.15, that is, by dividing each interval frequency by the total number of devices tested. Sometimes it is important in reliability estimation to indicate how well a set of observed data fits a known distribution, that is, to determine whether a hypothesis that the data originate from a known distribution is true. For this purpose, it is necessary to calculate the expected frequencies of failures from the known distribution and to compare them with the observed frequencies. Several methods exist to determine the adequacy of such a fit. We discuss these methods further in Section 2.7.

Example 2.28

Consider the time-to-failure data for an electronic device in Table 2.3. It is believed that the data come from an exponential distribution with parameter $\lambda = 0.005\,\text{h}^{-1}$. Determine the expected frequencies.

Solution: The probability that a r.v. T takes values between 0 and 100 h according to Equation 2.56 is

$$\Pr(0 < T < 100) = \int_0^{100} 5 \times 10^{-3} \exp\left(-5 \times 10^{-3} t\right) \mathrm{d}t,$$

$$\Pr(0 < T < 100) = \left[1 - \exp(-5 \times 10^{-3} t)\right]_0^{100} = 0.393.$$

By multiplying this probability by the total number of devices observed (100), we will be able to determine the expected frequency. The expected frequency here would be $0.393 \times 100 = 39.3$ for the 0–100 interval. The results for the rest of the intervals are shown in Table 2.3.

A comparison of the observed and expected frequencies would reveal differences as great as 4.6. Figure 2.11 illustrates the respective graphic representation and its comparison to the exponential distribution with $\lambda = 5 \times 10^{-3}$.

The graphical representation of empirical data is commonly known as a histogram.

TABLE 2.3
Frequency Table

Class Interval	Observed Frequency	Expected Frequency
0–100	35	39.3
100–200	26	23.8
200–300	11	14.5
300–400	12	8.8
400–500	6	5.3
500–600	3	3.2
600–700	4	2.0
700–800	2	1.2
800–900	0	0.7
900–1000	1	0.4

continued

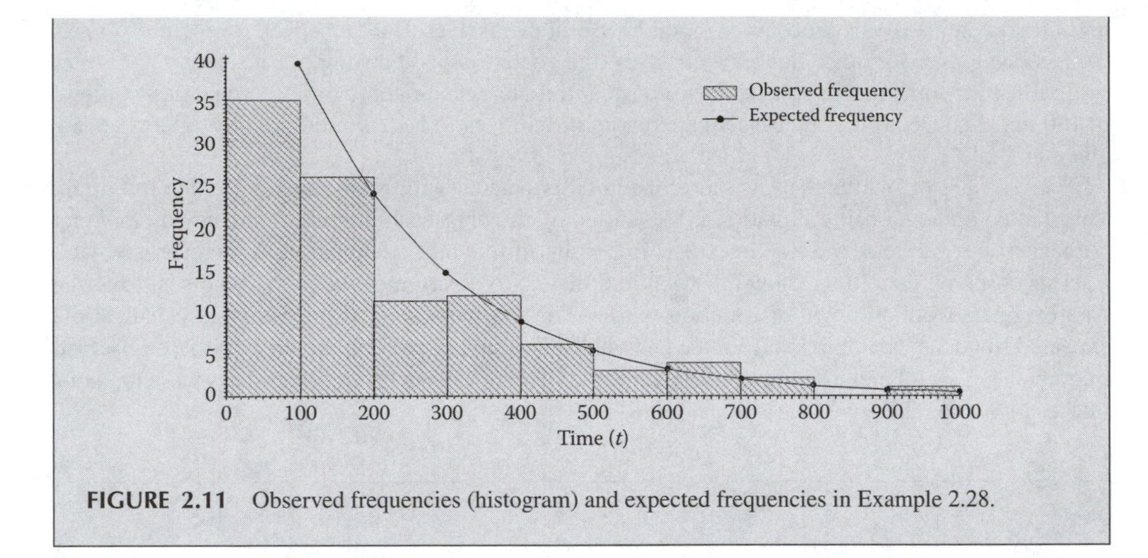

FIGURE 2.11 Observed frequencies (histogram) and expected frequencies in Example 2.28.

2.7 GOODNESS-OF-FIT TESTS

Consider the problem of determining whether a sample belongs to a hypothesized theoretical distribution. For this purpose, we need to perform a test that estimates the adequacy of a fit by determining the difference between the frequency of occurrence of a r.v. characterized by an observed sample and the expected frequencies obtained from the hypothesized distribution. For this purpose, the so-called goodness-of-fit tests are used.

Below we briefly consider two procedures often used as goodness-of-fit tests: the χ^2 and Kolmogorov goodness-of-fit tests.

2.7.1 Chi-Square Test

As the name implies, this test is based on a statistic that has an approximate χ^2 distribution. To perform this test, an observed sample taken from the population representing a r.v. X must be split into k ($k \geq 5$) nonoverlapping intervals (the lower limit for the first interval can be $-\infty$, and the upper limit for the last interval can be $+\infty$). The assumed (hypothesized) distribution model is then used to determine the probabilities p_i that the r.v. X would fall into each interval i ($i = 1, 2, \ldots, k$). This process was described to some extent in Section 2.6. By multiplying p_i by the sample size n, we obtain the expected frequency for each interval. Denote the expected frequency as e_i. It is obvious that $e_i = np_i$. If the observed frequency for each interval i of the sample is denoted by o_i, then the magnitude of differences between e_i and o_i can characterize the adequacy of the fit. The χ^2 test uses the statistic χ^2, which is defined as

$$W = \chi^2 = \sum_{i=1}^{k} \frac{(o_i - e_i)^2}{e_i}. \tag{2.128}$$

The χ^2 statistic approximately follows the χ^2 distribution (mentioned in Section 2.3). If the observed frequencies o_i differ considerably from the expected frequencies e_i, then W will be large and the fit is considered to be poor. A good fit would obviously lead to not rejecting the hypothesized distribution, whereas a poor fit leads to the rejection. It is important to note that the hypothesis can only be rejected, not positively affirmed. Therefore, the hypothesis is either *rejected* or *not rejected* as opposed to accepted or not accepted. The test can be summarized as follows:

Step 1. Choose a hypothesized distribution for the given sample.
Step 2. Select a specified significance level of the test denoted by α.
Step 3. Define the rejection region $R \geq \chi^2_{1-\alpha}(k-m-1)$, where $\chi^2_{1-\alpha}(k-m-1)$ is the $(1-\alpha)100$th percentile of the χ^2 distribution with $k-m-1$ degrees of freedom (the percentiles are given in Table A.3), k is the number of intervals, and m is the number of parameters estimated from the sample. If the parameters of the distribution were estimated without using the given sample, then $m=0$.
Step 4. Calculate the value of the χ^2 statistic, W, from Equation 2.128.
Step 5. If $W > R$, reject the hypothesized distribution; otherwise do not reject the distribution.

It is important at this point to specify the role of α in the χ^2 test. Suppose that the calculated value of W in Equation 2.128 exceeds the 95th percentile, $\chi^2_{0.95}(\cdot)$, given in Table A.3. This indicates that chances are lower than 1 in 20 that the observed data are from the hypothesized distribution. In this case, the model should be rejected (by not rejecting the model, one makes the Type II error discussed above). On the other hand, if the calculated value of W is smaller than $\chi^2_{0.95}(\cdot)$, chances are greater than 1 in 20 that the observed data match the hypothesized distribution model. In this case, the model should not be rejected (by rejecting the model, one makes the Type I error discussed above).

One instructive step in χ^2 testing is to compare the observed data with the expected frequencies to note which classes (intervals) contributed most to the value of W. This can sometimes help to indicate the nature of the deviations.

Example 2.29

The number of parts ordered per week by a maintenance department in a manufacturing plant is believed to follow a Poisson distribution. Use a χ^2 goodness-of-fit test to determine the adequacy of the Poisson distribution. Use the data found in the following table:

No. of Parts per Week (x)	Observed Frequency (o_i)	Expected Frequency(e_i)	χ^2 Statistic ($(o_i-e_i)^2/e_i$)
0	18	15.783	0.311
1	18	18.818	0.036
2	8	11.219	0.923
3	5	4.459	0.066
4	2	1.329	0.339
5	1	0.317	1.472
Total	52	52	3.147

Solution: Since under the Poisson distribution model, events occur at a constant rate, then a natural estimate of ρ is

$$\hat{\rho} = \frac{\text{Number of part used}}{\text{Number of weeks}} = \frac{62}{52} = 1.19 \text{ parts/week.}$$

From the Poisson distribution,

$$\Pr(X = x_i) = \frac{\rho e^{-\rho}}{x_i!}. \tag{2.129}$$

continued

Using $\hat{\rho} = 1.2$, one obtains $\Pr(X = 0) = 0.301$. Therefore, $e_i = 0.301 \times 52 = 15.7$. Other expected frequencies are calculated in the same way since we obtained one parameter (ρ) from the sample, $m = 1$. Therefore, $R = \chi^2_{0.95}(6 - 1 - 1) = 9.49$ from Table A.3. Since $W = 3.147 < R$, there is no reason to reject the hypothesis that the data are from a Poisson distribution (Figure 2.12).

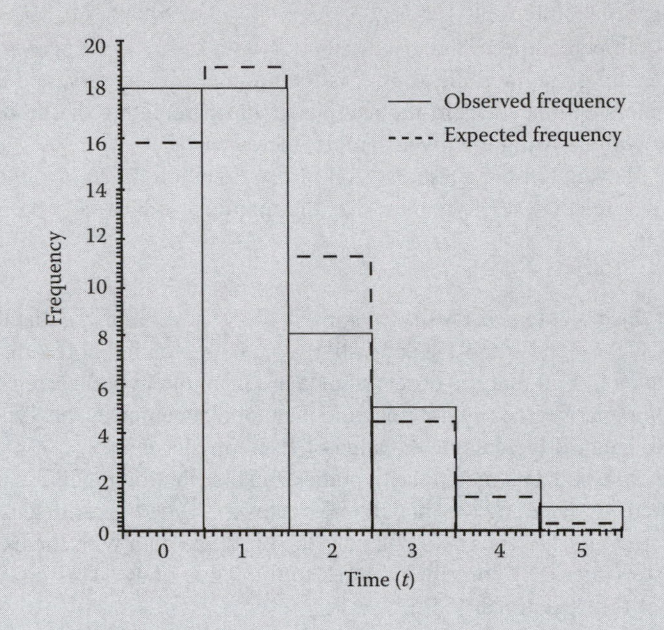

FIGURE 2.12 Observed and expected frequencies in Example 2.29.

Example 2.30

Table 2.4 shows the accumulated mileage for a sample of 100 automobiles after 2 years in service. The mileage accumulation pattern is believed to follow a normal distribution. Use the χ^2 test to check this hypothesis at the 0.05 significance level (Figure 2.13).

TABLE 2.4
Accumulated Mileage of a 100-Passenger Van after 2 Years in Service

32,797	38,071	16,768	26,713	25,754	37,603	39,485	15,261	45,283	41,064
47,119	35,589	43,154	35,390	32,677	26,830	25,056	20,269	16,651	27,812
33,532	44,264	22,418	40,902	29,180	25,210	28,127	14,318	27,300	28,433
55,627	20,588	14,525	22,456	28,185	16,946	29,015	19,938	36,837	36,531
11,538	25,746	52,448	35,138	22,374	30,368	10,539	32,231	21,075	45,554
34,107	28,109	28,968	27,837	41,267	24,571	41,821	44,404	27,836	8734
26,704	29,807	32,628	28,219	33,703	43,665	49,436	32,176	47,590	32,914
9979	16,735	31,388	21,293	36,258	55,269	37,752	42,911	21,248	28,172
10,014	28,688	26,252	31,084	30,935	29,760	43,939	18,318	21,757	26,208
22,159	22,532	31,565	27,037	49,432	17,438	27,322	37,623	17,861	24,993

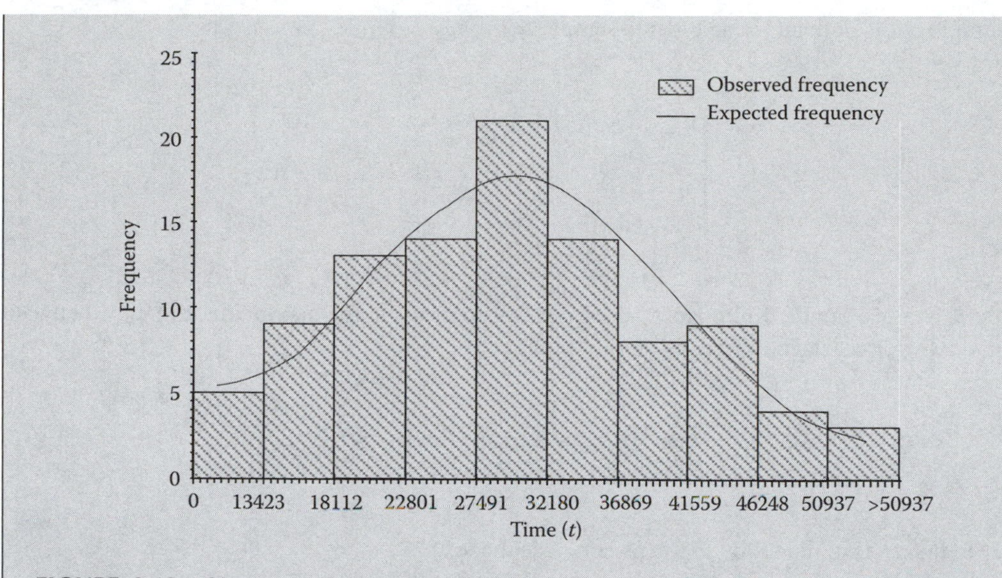

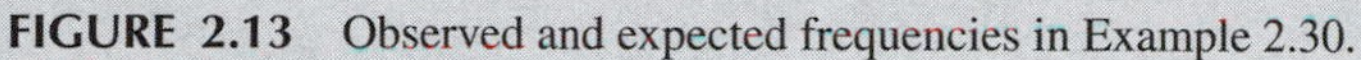
FIGURE 2.13 Observed and expected frequencies in Example 2.30.

Solution: Using Equations 2.112 and 2.114, find that the estimates of mean and standard deviation for the hypothesized normal distribution are equal to 30,011 and 10,472 miles, respectively. Group the data from Table 2.4 to calculate the observed frequencies. Use the equation of normal pdf (Equation 2.61) to find the expected frequencies.

Grouped Data				
Interval		Frequency	Expected	χ^2 Statistic
Start	End	(o_i)	Frequency (e_i)	$((o_i - e_i)^2/e_i)$
1×10^{-8}	13,417	5	5.4479	0.0368
13,417	18,104	9	7.1245	0.4937
18,104	22,791	13	11.7500	0.1330
22,791	27,478	14	15.9143	0.2303
27,478	32,165	20	17.7014	0.2985
32,165	36,852	15	16.1699	0.0846
36,852	41,539	8	12.1305	1.4064
41,539	46,226	9	7.4733	0.3119
46,226	>46,226	7	6.2882	0.0806
Total		100	100.0000	3.0758

Since both of the distribution parameters were estimated from the given sample, then $m = 2$. The critical χ^2 value of the statistic is, therefore, $\chi^2_{0.95}(10 - 2 - 1) = 14.1$. This is higher than the test statistic $W = 3.748$; therefore, there is no reason to reject the hypothesis about the normal distribution at 0.05 significance level.

2.7.2 Kolmogorov–Smirnov Test

In the framework of this test, the individual sample components are treated without clustering them into intervals. Similar to the χ^2 test, a hypothesized cdf is compared with its estimate known as the *empirical* (or *sample*) *cdf*.

A sample cdf is defined for an ordered sample $t_{(1)} < t_{(2)} < t_{(3)} < \cdots < t_{(n)}$ as

$$S_n(t) = \begin{cases} 0, & -\infty < t < t_{(1)}; \\ \frac{i}{n}, & t_{(i)} \le t < t_{(i+1)}, i = 1, 2, \ldots, n-1; \\ 1, & t_{(n)} \le t < \infty. \end{cases} \quad (2.130)$$

Statistic $K - S$ used in the Kolmogorov–Smirnov test to measure the maximum difference between $S_n(t)$ and a hypothesized cdf, $F(t)$, is introduced as

$$K - S = \max_i \left[\left| F(t_i) - S_n(t_i) \right|, \left| F(t_i) - S_n(t_{i-1}) \right| \right]. \quad (2.131)$$

Similar to the χ^2 test, the following steps compose the test:

Step 1. Choose a hypothesized cumulative distribution $F(T)$ for the given sample.
Step 2. Select a specified significance level of the test, α.
Step 3. Define the rejection region $R > D_n(\alpha)$, where $D_n(\alpha)$ can be obtained from Table A.4.
Step 4. If $K - S > D_n(\alpha)$, reject the hypothesized distribution and conclude that $F(t)$ does not fit the data; otherwise, do not reject the hypothesis.

Example 2.31

Time to failure of an electronic device is measured in a life test. The failure times are 254, 586, 809, 862, 1381, 1923, 2542, and 4211 h. Is the exponential distribution with $\lambda = 5 \times 10^{-4}$ an adequate representation of this sample?

Solution:
For an exponential distribution with $\lambda = 5 \times 10^{-4}$, we obtain $F_n(t) = 1 - \exp(-5 \times 10^{-4} t)$. For $\alpha = 0.05$, $D_8(0.05) = 0.457$. Thus, the rejection area is $R > 0.457$.

Time to Failure (t)	i	Empirical cdf $S_n(t_i)$	Empirical cdf $S_n(t_{i-1})$	Fitted cdf ($F_n(t_i)$)	$K - S$ Statistic $\lvert F_n(t_i) - S_n(t_i)\rvert$	$K - S$ Statistic $\lvert F_n(t_i) - S_n(t_{i-1})\rvert$
254	1	0.125	0.000	0.119	0.006	0.119
586	2	0.250	0.125	0.254	0.004	0.129
809	3	0.375	0.250	0.333	0.042	0.083
862	4	0.500	0.375	0.350	0.150	0.025
1381	5	0.625	0.500	0.499	0.126	0.001
1923	6	0.750	0.625	0.618	0.132	0.007
2542	7	0.875	0.750	0.719	**0.156**	0.031
4211	8	1.000	0.875	0.878	0.122	0.003

Since $K - S = 0.156 < 0.457$, we should not reject the hypothesized exponential distribution model (Figure 2.14).

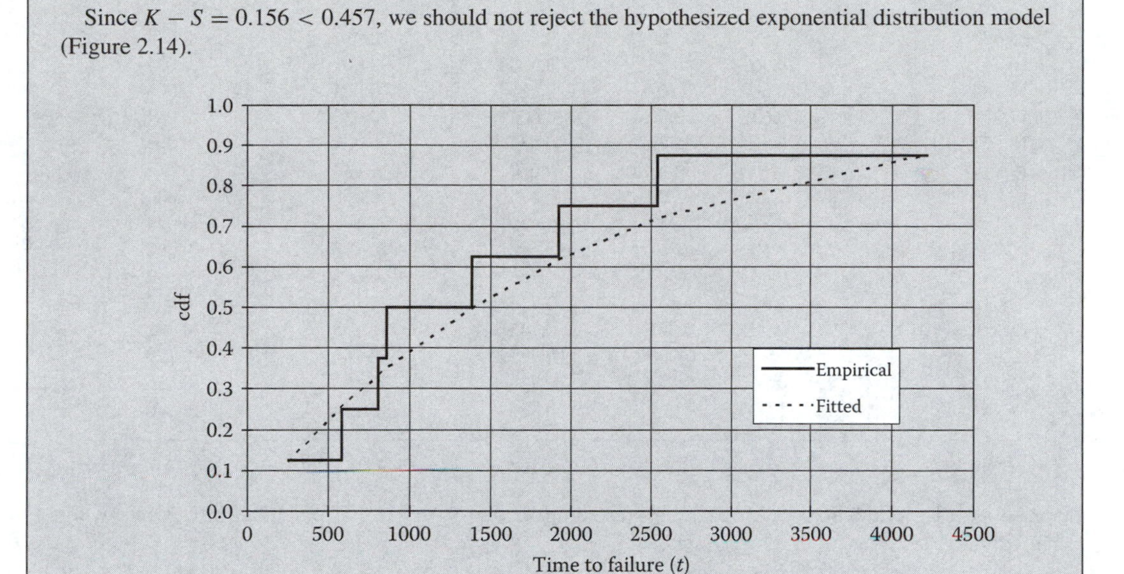

FIGURE 2.14 Empirical and fitted cdf in Example 2.31.

Example 2.32

The wearout time (to failure) of automobile brake pads shown in Table 2.5 is believed to follow a Weibull distribution with parameters $\alpha = 18,400$ miles and $\beta = 1.5$. Use the Kolmogorov–Smirnov test to check this hypothesis at 0.1 and 0.2 significance levels.

TABLE 2.5
Wearout Time of Automobile Brake Pads

		Empirical cdf			$K - S$ Statistic	
Time to Failure (t)	i	$S_n(t_i)$	$S_n(t_{i-1})$	Fitted cdf $F_n(t_i)$	$\lvert F_n(t_i) - S_n(t_i)\rvert$	$\lvert F_n(t_i) - S_n(t_{i-1})\rvert$
1643	1	0.033	0.000	0.028	0.005	0.028
1664	2	0.067	0.033	0.029	0.038	0.004
2083	3	0.100	0.067	0.040	0.060	0.027
3625	4	0.133	0.100	0.088	0.045	0.012
7230	5	0.167	0.133	0.224	0.057	0.091
9095	6	0.200	0.167	0.299	0.099	0.132
9968	7	0.233	0.200	0.334	0.100	0.134
11,689	8	0.267	0.233	0.401	0.135	0.168
12,989	9	0.300	0.267	0.451	0.151	0.184
13,622	10	0.333	0.300	0.474	0.141	0.174
13,953	11	0.367	0.333	0.486	0.119	0.153
14,527	12	0.400	0.367	0.506	0.106	0.140
15,263	13	0.433	0.400	0.532	0.099	0.132
15,428	14	0.467	0.433	0.538	0.071	0.104
15,503	15	0.500	0.467	0.540	0.040	0.073

continued

15,629	16	0.533	0.500	0.544	0.011	0.044
16,342	17	0.567	0.533	0.568	0.001	0.035
16,584	18	0.600	0.567	0.576	0.024	0.009
17,374	19	0.633	0.600	0.601	0.033	0.001
18,571	20	0.667	0.633	0.637	0.030	0.003
19,739	21	0.700	0.667	0.670	0.030	0.003
19,936	22	0.733	0.700	0.675	0.058	0.025
20,102	23	0.767	0.733	0.679	0.087	0.054
20,832	24	0.800	0.767	0.698	0.102	0.068
23,378	25	0.833	0.800	0.758	0.075	0.042
23,612	26	0.867	0.833	0.763	0.103	0.070
23,678	27	0.900	0.867	0.765	0.135	0.102
23,971	28	0.933	0.900	0.771	0.163	0.129
24,341	29	0.967	0.933	0.778	**0.188**	0.155
26,964	30	1.000	0.967	0.826	0.174	0.140

Solution: Use Equations 2.73 and 2.131 to compute the expected and empirical cdf, respectively (Figure 2.15).

The $K - S$ statistic of the given data set is equal to 0.211, which is lower than $D_{30}(0.1) = 0.218$, but higher than $D_{30}(0.2) = 0.190$. This means that we do not reject the null hypothesis at 0.1 significance level but reject it at 0.2 significance level.

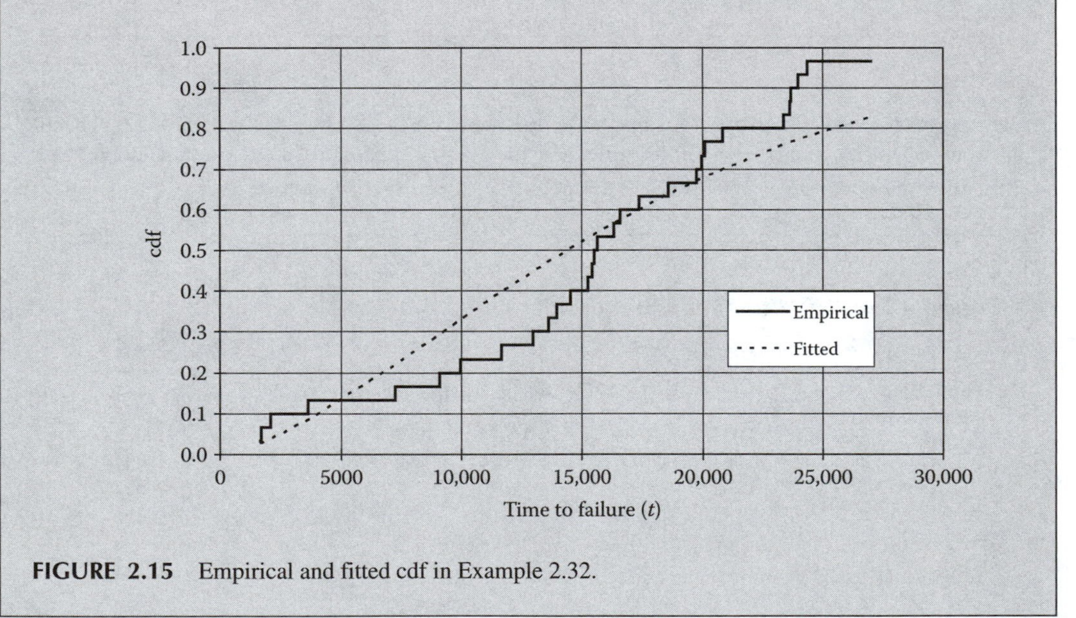

FIGURE 2.15 Empirical and fitted cdf in Example 2.32.

2.8 REGRESSION ANALYSIS

In Section 2.7, we mainly dealt with one or two r.v.s. However, reliability engineering and risk-assessment problems often require relationships among several r.v.s or between random and nonrandom variables. For example, time to failure of an electrical generator can depend on its age, environmental temperature, and power capacity. In this case, we can consider the time to failure as a r.v. Y, which is a function of the variables x_1 (age), x_2 (temperature), and x_3 (power capacity).

In regression analysis one refers to Y as the *dependent variable* and to $x_1, x_2, \ldots, x_k$ as the *independent variables*, *explanatory variables*, or *factors*. Generally speaking, independent variables

$x_1, x_2, \ldots, x_k$ might be random or nonrandom variables whose values are known or chosen by the experimenter [in the case of the so-called *design of experiments* (DoE)]. The conditional expectation of Y for any given values of $x_1, x_2, \ldots, x_k$, $E(Y \mid x_1, x_2, \ldots, x_k)$ is known as the *regression* of Y on $x_1, x_2, \ldots, x_k$. In other words, regression analysis estimates the average value for the dependent variable corresponding to each value of the independent variable.

In the case when the regression of Y is a linear function with respect to the independent variables $x_1, x_2, \ldots, x_k$, it can be written in the form

$$E(Y \mid x_1, x_2, \ldots, x_k) = \beta_0 + \beta_1 x_1 + \beta_2 x_2 + \cdots + \beta_k x_k. \quad (2.132)$$

The coefficients $\beta_0, \beta_1, \beta_2, \ldots, \beta_k$ are called *regression coefficients* or *parameters*. When the expectation of Y is nonrandom, the relationship (Equation 2.132) is a deterministic one. The corresponding regression model for the r.v. Y can be written in the following form:

$$Y = \beta_0 + \beta_1 x_1 + \beta_2 x_2 + \cdots + \beta_k x_k + \varepsilon, \quad (2.133)$$

where ε is the *random error*, assumed to be independent (for all combinations of x considered) and r.v. is distributed with mean $E(\varepsilon) = 0$ and finite variance σ^2. If it is normally distributed, one deals with the *normal* regression.

2.8.1 Simple Linear Regression

Consider the regression model for the simple deterministic relationship:

$$Y = \beta_0 + \beta_1 x_1. \quad (2.134)$$

Let us have n pairs of observations $(x_1, y_1), (x_2, y_2), \ldots, (x_n, y_n)$. Also, assume that for any given value x, the dependent variable Y is related to the value of x by

$$Y = \beta_0 + \beta_1 x + \varepsilon, \quad (2.135)$$

where ε is normally distributed with mean zero and variance σ^2. The r.v. Y has, for a given x, normal distribution with mean $\beta_0 + \beta_1 x$ and variance σ^2. Also, suppose that for any given values $x_1, x_2, \ldots, x_n$, r.v.s $Y_1, Y_2, \ldots, Y_n$ are independent. For the above n pairs of observations, the joint pdf of $y_1, y_2, \ldots, y_n$ is given by

$$f_n(y \mid x, \beta_0, \beta_1, \sigma^2) = \frac{1}{\sqrt{(2\pi)^n \sigma^n}} \exp\left[-\frac{1}{2\sigma^2}\sum_{i=1}^{n}(y_i - \beta_0 - \beta_1 x_i)^2\right]. \quad (2.136)$$

Equation 2.136 is the likelihood function (discussed in Section 2.5) for the parameters β_0 and β_1. Maximizing this function with respect to β_0 and β_1 reduces the problem to minimizing the sum of squares

$$S(\beta_0, \beta_1) = \sum_{i=1}^{n}(y_i - \beta_0 - \beta_1 x_i)^2 \quad (2.137)$$

with respect to β_0 and β_1.

Thus, the maximum likelihood estimation of the parameters β_0 and β_1 is the estimation by the *method of least squares*. The values of β_0 and β_1 minimizing $S(\beta_0, \beta_1)$ are those for which the derivatives

$$\frac{\partial S(\beta_0, \beta_1)}{\partial \beta_0} = 0, \quad \frac{\partial S(\beta_0, \beta_1)}{\partial \beta_1} = 0. \quad (2.138)$$

The solution of the above equations yields the least-squares estimates of the parameters β_0 and β_1 (denoted $\hat{\beta}_0$ and $\hat{\beta}_1$) as

$$\hat{\beta}_0 = \bar{y} - \hat{\beta}_1 \bar{x}, \quad \hat{\beta}_1 = \frac{\sum_{i=1}^{n}(x_i - \bar{x})(y_i - \bar{y})}{\sum_{i=1}^{n}(x_i - \bar{x})^2}, \tag{2.139}$$

where

$$\bar{y} = \frac{1}{n}\sum_{i=1}^{n} y_i, \quad \bar{x} = \frac{1}{n}\sum_{i=1}^{n} x_i. \tag{2.140}$$

Note that the estimates are linear functions of the observations y_i; they are also unbiased and have the minimum variance among all unbiased estimates.

The estimate of the dependent variable variance σ^2 can be found as

$$S^2 = \frac{\sum_{i=1}^{n}(y_i - \hat{y}_i)^2}{n-2}. \tag{2.141}$$

where

$$\hat{Y}_i = \hat{\beta}_0 + \hat{\beta}_1 x_i, \tag{2.142}$$

is predicted by the regression model values for the dependent variable and $(n - 2)$ is the number of degrees of freedom (2 is the number of the estimated parameters of the model). The estimate of variance of Y (Equation 2.141) is also called the *residual variance* and it is used as a measure of accuracy of model fitting as well. The positive square root of S^2 in Equation 2.141 is called the *standard error of the estimate* of Y, and the numerator in Equation 2.141 is called the *residual sum of squares*. For a more detailed discussion on the reliability applications of regression analysis, see [6].

EXAMPLE 2.33

An electronic device was tested under the elevated temperatures of 50°C, 60°C, and 70°C. The test results as times to failure for samples of 10 items in hours are given in Table 2.6 (see Figure 2.16).

This is an example of *accelerated life testing* discussed in Chapter 7. Assuming that the logarithm of time to failure t follows the normal distribution with the mean given by the Arrhenius model—that is,

$$E(\ln t) = A + \frac{B}{T}, \tag{2.143}$$

where $T = t°\text{C} + 273$ is the absolute temperature—find the estimates of parameters A and B.

Solution:

The equation above can be easily transformed to the simple linear regression (Equation 2.134),

$$Y = \beta_0 + \beta_1 x, \tag{2.144}$$

TABLE 2.6
Times to Failure of an Electronic Component under Different Temperatures

Time to Failure (t)			$\ln(t)$		
50°C	**60°C**	**70°C**	**50°C**	**60°C**	**70°C**
1950	607	44	7.5756	6.4085	3.7842
3418	644	53	8.1368	6.4677	3.9703
4750	675	82	8.4659	6.5147	4.4067
5090	758	88	8.5350	6.6307	4.4773
7588	1047	123	8.9343	6.9537	4.8122
10,890	1330	189	9.2956	7.1929	5.2417
11,601	1369	204	9.3588	7.2218	5.3181
15,288	1884	243	9.6348	7.5412	5.4931
19,024	2068	317	9.8535	7.6343	5.7589
22,700	2931	322	10.0301	7.9831	5.7746
		$E[\ln(t)]$	8.9820	7.0549	4.9037

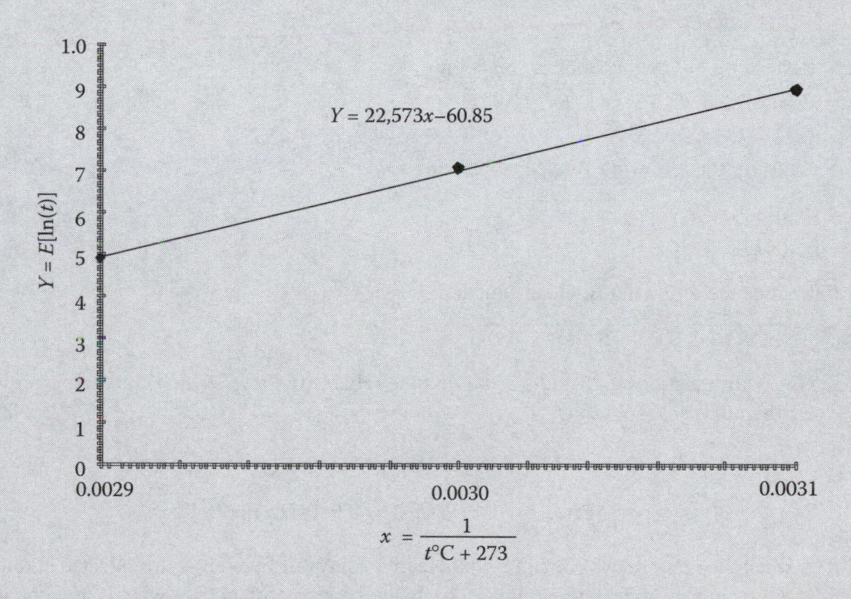

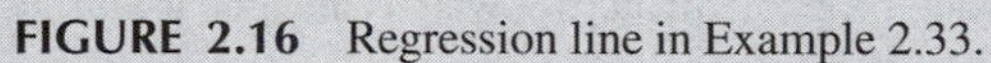

FIGURE 2.16 Regression line in Example 2.33.

using transformations $Y = \ln t$, $x = 1/T$, $\beta_0 = \text{A}$, and $\beta_1 = \text{B}$. Accordingly, from the data in Table 2.6:

$Y = E[\ln(t)]$	$t(°\text{C})$	$x = (t°\text{C} + 273)^{-1}$
8.9820	50	0.0031
7.0549	60	0.0030
4.9037	70	0.0029

continued

Using the transformed data above and Equation 2.139, one can find the estimates of parameters A and B as

$$\hat{A} = \exp(-60.85) = 3.76 \times 10^{-27}\,\text{h}^{-1},$$
$$\hat{B} = 22,573^{\circ}\text{K}. \tag{2.145}$$

EXERCISES

2.1 Simplify the following Boolean functions:

a. $\overline{\overline{(A \cap B \cup C)} \cap \overline{B}}$
b. $(A \cup B) \cap (\overline{A} \cup \overline{B} \cap \overline{A})$
c. $\overline{A \cap B \cap B \cap C} \cap \overline{B}$.

2.2 Reduce the following Boolean function:
$A \cap B \cap \overline{(C \cup (\overline{C} \cup A) \cup \overline{B})}$.

2.3 Simplify the following Boolean expressions:

a. $\overline{\overline{[(A \cap B) \cup C]} \cap \overline{B}}$
b. $[(A \cup B) \cap \overline{A}] \cup (\overline{B} \cap \overline{A})$.

2.4 Reduce the following Boolean function:
$G = (A \cup B \cup C) \cap \overline{(A \cap \overline{B} \cap \overline{C})} \cap \overline{C}$.
If $\Pr(A) = \Pr(B) = \Pr(C) = 0.9$, what is $\Pr(G)$?

2.5 Simplify the following Boolean equations:

a. $(A \cup B \cup C) \cap \overline{(A \cap \overline{B} \cap \overline{C})} \cap \overline{C}$
b. $(A \cup B) \cap \overline{B}$.

2.6 Reduce the following Boolean equation:

$$\overline{(A \cup (B \cap C))} \cap \overline{(B \cup (D \cap A))}.$$

2.7 Use both Equations 2.25 and 2.29 to find the reliability $\Pr(s)$. Which equation is preferred for the numerical solution?

$$\Pr(s) = \Pr(E_1 \cup E_2 \cup E_3),$$
$$\Pr(E_1) = 0.8, \Pr(E_2) = 0.9, \Pr(E_3) = 0.95.$$

2.8 A stockpile of 40 relays contain eight defective relays. If five relays are selected at random, and the number of defective relays is known to be greater than two, what is the probability that exactly four relays are defective?

2.9 Given that $P = 0.006$ is the probability of an engine failure on a flight between two cities, find the probability of

a. No engine failure in 1000 flights.
c. At least one failure in 1000 flights.
d. At least two failures in 1000 flights.

2.10 A random sample of 10 resistors is to be tested. From past experience, it is known that the probability of a given resistor being defective is 0.08. Let X be the number of defective resistors.

a. What kind of distribution function would be recommended for modeling the r.v. X?
b. According to the distribution function in (a), what is the probability that in the sample of 10 resistors, there are more than 1 defective resistors in the sample?

2.11 How many different license plates can be made if each consists of three numbers and three letters, and no number or letter can appear more than once on a single plate?

2.12 The consumption of maneuvering jet fuel in a satellite is known to be normally distributed with a mean of 10,000 h and a standard deviation of 1000 h. What is the probability of being able to maneuver the satellite for the duration of a 1-year mission?

2.13 Suppose a process produces electronic components, 20% of which are defective. Find the distribution of x, the number of defective components, in a sample size of five. Given that the sample contains at least three defective components, find the probability that four components are defective.

2.14 If the heights of 300 students are normally distributed, with a mean of 68 inches and a standard deviation of 3 inches, how many students have

a. Heights of more than 70 inches?
b. Heights between 67 and 68 inches?

2.15 Assume that 1% of a certain type of resistor is bad when purchased. What is the probability that a circuit with 10 resistors has exactly 1 bad resistor?

2.16 Between the hours of 2 and 4 p.m., the average number of phone calls per minute coming into an office is two and one-half. Find the probability that during a particular minute, there will be more than five phone calls.

2.17 A guard works between 5 p.m. and 12 midnight; he sleeps an average of 1 h before 9 p.m., and 1.5 h between 9 and 12. If an inspector finds him asleep, what is the probability that this happens before 9 p.m.?

2.18 The number of system breakdowns occurring with a constant rate in a given length of time has a mean value of two breakdowns. What is the probability that in the same length of time, two breakdowns will occur?

2.19 An electronic assembly consists of two subsystems, A and B. Each assembly is given one preliminary checkout test. Records on 100 preliminary checkout tests show that subsystem A failed 10 times. Subsystem B alone failed 15 times. Both subsystems A and B failed together five times.

a. What is the probability of A failing, given that B has failed?
b. What is the probability that A alone fails?

2.20 A presidential election poll shows one candidate leading with 60% of the vote. If the poll is taken from 200 random voters throughout the United States, what is the probability that the candidate will get less than 50% of the votes in the election? (Assume that the 200 voters sampled are true representatives of the voting profile.)

2.21 A newspaper article reports that a New York medical team has introduced a new male contraceptive method. The effectiveness of this method was tested using a number of couples over a period of 5 years. The following statistics are obtained:

Year	Total Number of Times the Method was Employed	Number of Unwanted Pregnancies
1	8200	19
2	10,100	18
3	2120	1
4	6120	9
5	18,130	30

a. Estimate the mean probability of an unwanted pregnancy per use. What is the standard deviation of the estimate?
b. What are the 95% upper and lower confidence limits of the mean and standard deviation?

2.22 Suppose that the lengths of the individual links of a chain distribute themselves with a uniform distribution, as shown below.

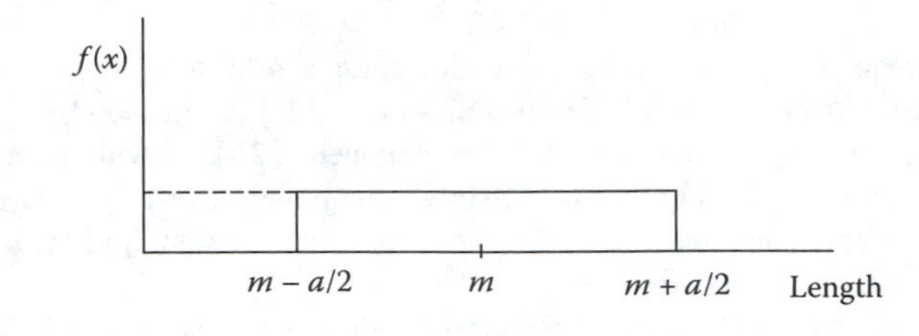

a. What is the height of the rectangle?
b. Find the cumulative pdf for the above distribution. Make a sketch of the distribution and label the axes.
c. If numerous chains are made from two such links hooked together, what is the pdf of two link chains?
d. Consider a 100-link chain. What is the probability that the length of the chain will be less than 100.5 m if $a = 0.1$ m?

2.23 If $f(x, y) = 1/2xy^2 + 1/2yx^2$, and $0 < x < 1$, and $0 < y < 2$:

a. Show that $f(x, y)$ is a joint pdf.
b. Find $\Pr(x > y)$, $\Pr(y > x)$, and $\Pr(x = y)$.

2.24 A company is studying the feasibility of buying an elevator for a building under construction. One proposal is a 10-passenger elevator that, on average, would arrive in the lobby once per minute. The company rejects this proposal because it expects an average of five passengers per minute to use the elevator.

a. Support the proposal by calculating the probability that in any given minute, the elevator does not show up, and 10 or more passengers arrive.
b. Determine the probability that the elevator arrives only once in a 5-minute period.

2.25 The frequency distribution of time to establish the root causes of a failure by a group of experts is observed and given below:

Time (h)	Frequency
45–55	7
55–65	18
55–65	35
75–85	28
85–95	12

Test whether a normal distribution with known $\sigma = 10$ is an appropriate model for these data.

2.26 A random-number generator yields the following sample of 50 digits:

Digit	0	1	2	3	4	5	6	7	8	9
Frequency	4	8	8	4	10	3	2	2	4	5

Is there any reason to doubt that the digits are uniformly distributed? (Use the χ^2 goodness-of-fit test.)

2.27 A set of 40 high-efficiency pumps is tested; all of the pumps fail ($F = 40$) after 400 pump-hours ($T = 400$). It is believed that the time to failure of the pumps follows an exponential distribution. Using the following table and the goodness-of-fit method, determine if the exponential distribution is a good choice.

Time Interval (h)	**Number of Observed Failures**
0–2	6
2–6	12
6–10	7
10–15	6
15–25	7
25–100	2
	Total = 40

2.28 Use Equation 2.102 to calculate the mean and variance of a Weibull distribution.

2.29 Consider the following repair times:

Repair Time (y)	0–4	4–24	24–72	72–300	300–5400
Number of Observed Frequency	17	41	12	7	9

Use the χ^2 goodness-of-fit test to determine the adequacy of a lognormal distribution:

a. For 5% level of significance.

b. For 1% level of significance.

2.30 Consider the following time-to-failure data with the ranked value of t_i. Test the hypothesis that the data fit a normal distribution. (Use the Kolmogorov test for this purpose.)

Event	1	2	3	4	5	6	7	8	9	10
Time to Failure (h)	10.3	12.4	13.7	13.9	14.1	14.2	14.4	15.0	15.9	16.1

2.31 If a device has a cycle to failure, t, which follows an exponential distribution with $\lambda = 0.003$ failures/cycle,

a. Determine the mean cycle to failure for this device.

b. If the device is used in a space experiment and is known to have survived for 300 cycles, what is the probability that it will fail sometimes after 1000 cycles?

2.32 A highly stressed machine part's time to failure is normally distributed with 90% of the failures occurring between 200 and 270 h of use (i.e., 5% below 200 h and 5% above 270 h).

a. Find the mean and standard deviation of time to failure.

b. Determine the part's life if no more than 1% replacement probability can be tolerated.

c. Compute the probability that life time of the part exceeds 50 h of use, if the part has been operating for 200 h without failure.

2.33 Fifty compressors were stress tested for 200 days, and the following failure times (grouped) were regarded.

Interval (days)	**Number of Failures**
$0 < T < 50$	9
$50 < T < 75$	11
$75 < T < 100$	9
$100 < T < 150$	12

2.34 Perform a χ^2 goodness-of-fit test at the 10% significance level on the following grouped data to fit a uniform pdf of time to failure between the two limits of $a = 0$ and $b = 1000$ h.

Interval	Observed Number of Failures
0–200	16
200–400	18
400–600	25
600–800	23
800–1000	24

2.35 A turbine in a generator can cause a great deal of damage if it fails critically. The manager of the plant can choose either to replace the old turbine with a new one, or to leave it in place.

The state of a turbine is categorized as either "good," "acceptable," or "poor." The probability of critical failure per quarter depends on the state of the turbine:

$$P(\text{failure}|\text{good}) = 0.0001,$$

$$P(\text{failure}|\text{acceptable}) = 0.001,$$

$$P(\text{failure}|\text{poor}) = 0.01.$$

The technical department has made a model for the degradation of the turbine. In this model it is assumed that, if the state is "good" or "acceptable," then at the end of the quarter it stays the same with probability 0.95 or degrades with probability 0.05 (i.e., good becomes "acceptable" and "acceptable" becomes "poor"). If the state is "poor," then it stays "poor." Determine the probability that a turbine that is "good" becomes "good," "acceptable," and "poor" in the next three quarters and does not fail.

BIBLIOGRAPHY

1. Cox, R. T., Probability, frequency and reasonable expectation, *Am. J. Phys.*, 14, 1, 1946.
2. Lindley, D. V., *Introduction to Probability and Statistics from a Bayesian Viewpoint*, 2 Volumes, Cambridge University Press, Cambridge, 1965.
3. Johnson, N. L. and S. Kotz, *Distribution in Statistics*, 2 Volumes, Wiley, New York, 1969 and 1970.
4. Hahn, G. J. and S. S. Shapiro, *Statistical Models in Engineering*, Wiley, New York, 1967.
5. Nelson, W., *Applied Life Data Analysis*, Wiley, New York, 1982.
6. Lawless, J. F., *Statistical Models and Methods for Life Time Data*, Wiley, New York, 1982.
7. Hill, H. E. and J. W. Prane, *Applied Techniques in Statistics for Selected Industries: Coatings, Paints and Pigments*, Wiley, New York, 1984.
8. Mann, N., R. E. Schafer, and N. D. Singpurwalla, *Methods for Statistical Analysis of Reliability and Life Data*, Wiley, New York, 1974.

3 Elements of Component Reliability

In this chapter, we discuss the basic elements of component reliability estimation. The discussion centers around the classical frequency approach to component reliability. However, we also present some aspects of component reliability analysis based on the subjectivist or Bayesian approach.

We start with a formal definition of reliability and define commonly used terms and metrics. These formal definitions are not necessarily limited to reliability of an actual component; rather, they encompass a broad group of physical items (i.e., components, subsystems, systems, etc.), which are components in the framework of reliability formalism. We then focus on some important aspects of component reliability analysis in the rest of the chapter.

3.1 CONCEPT OF RELIABILITY

Reliability has many connotations. In general, it refers to an item's ability to successfully perform an intended function during a mission. The longer the item performs its intended function, the more reliable it is. Formally, reliability is viewed as both an engineering and a probabilistic notion. Indeed, both of these views form the fundamental basis for reliability studies. In engineering, reliability deals with those design and analysis activities that extend an item's life by controlling or eliminating its potential failure modes. Examples include designing stronger and more durable elements, reducing harmful environmental conditions, minimizing loads and stresses applied to an item during its use, and providing a preventive maintenance program to minimize the occurrence of failures.

To quantitatively measure the reliability of an item, we use a probabilistic metric, which treats reliability as a probability of the successful achievement of an item's intended function. The formal probabilistic definition of reliability, given in Section 1.5, is its mathematical representation. The right-hand side of Equation 1.1 denotes the probability that a specified failure time T exceeds a specified mission time $T' = t$ given that stress conditions $c_1, c_2, \ldots$ are met.

Practically, r.v. T represents the time to failure of an item, and stress conditions $c_1, c_2, \ldots$ represent conditions (e.g., design-related conditions) that are specified, *a priori*, for successful performance of the item. Other representations of r.v. T include the number of cycles to failure or miles to failure and so on. In the remainder of this book, we consider only time-to-failure representation, although the same treatment equally applies to other representations. Conditions $c_1, c_2, \ldots$ are often implicitly considered; therefore, Equation 1.1 is written in a simplified form of Equation 1.2. We use Equation 1.2 in the remainder of this book except for the Section 7.1 in Chapter 7.

3.1.1 Reliability Function

Let us start with the formal definition given by expression 1.1. Furthermore, let $f(t)$ denote a pdf representing the r.v. T. According to Equation 2.48, the probability of failure of the item as a function of time is defined by

$$\Pr(T \leq t) = \int_0^t f(\theta)\,\mathrm{d}\theta = F(t) \quad \text{for } t \geq 0, \tag{3.1}$$

where $F(t)$ denotes the probability that the item will fail sometime up to time t. According to our formalism expressed in Equation 1.2, Equation 3.1 is the *unreliability* of the item. Formally, we can call $F(t)$ (which is the time-to-failure cdf) the *unreliability function*. Conversely, we can define the *reliability function* (a.k.a., *the survivor* or *survivorship function*) as

$$R(t) = 1 - F(t) = \int_t^{\infty} f(\tau)\,\mathrm{d}\tau. \tag{3.2}$$

The *p-level quantile* of a continuous r.v. T with cdf $F(T)$ is defined as the value t_p, such that

$$F(t_p) = p, \quad 0 < p < 1.$$

The *median* is defined as the quantile of the level of $p = 0.5$. Similar to the mean, it is used as a location parameter. A quantile is often referred to as the "100p percent point," or the "100pth *percentile*." In reliability studies, the 100pth percentile of time to failure is the point at which the probability of an item failure is equal to p. For example, the so-called B_{10} life of mechanical components, frequently quoted by manufacturers, is the time at which 10% of the components are expected to fail. The most popular percentiles used in reliability are the 1st, 5th, 10th, and 50th percentiles.

Provided we have the pdf, $f(t)$, we can obtain $R(t)$. The basic characteristics of time-to-failure distribution and basic reliability measures can be expressed in terms of pdf, $f(t)$, cdf, $F(t)$, or reliability function, $R(t)$. The *mean time to failure* (MTTF), for example, illustrates the expected time during which the item will perform its function successfully (sometimes called *expected life*). According to Equation 2.93,

$$\mathrm{MTTF} = E(t) = \int_0^{\infty} t f(t)\,\mathrm{d}t. \tag{3.3}$$

If $\lim_{t\to\infty} t f(t) = 0$, then, integrating by parts, it is easy to obtain another form of Equation 3.3 given by

$$E(t) = \int_0^{\infty} R(t)\,\mathrm{d}t. \tag{3.4}$$

It is important to make a distinction, at this point, between MTTF and the *mean time between failures* (MTBF). Obviously, the former metric is associated with nonrepairable components, whereas the latter is related to the repairable components. In the case of MTBF, the pdf in Equation 3.3 can be the pdf of time between the first failure and the second failure, the second failure and the third failure, and so on. If we have surveillance, and the item is completely renewed through replacement, maintenance, or repair, the MTTF coincides with MTBF. Theoretically, this means that the renewal process is assumed to be perfect. That is, the item that goes through repair or maintenance is assumed to exhibit the characteristics of a new item. In practice this may not be true. In this case, one needs to determine the MTBF for the item for each renewal cycle (each ith time-between-failures interval). However, the approach based on the *as good as new* assumption can be quite adequate for many reliability considerations. The topic of MTTF and MTBF will be revisited in Chapter 5.

Let $R(t)$ be the reliability function of an item at time t. The probability that the item will survive for time τ, given that it has survived for time t, is called the *conditional reliability function* and is given by

$$R(\tau|t) = \frac{R(t+\tau)}{R(t)}. \tag{3.5}$$

Therefore, the conditional probability of failure during the same interval is

$$F(\tau|t) = 1 - R(\tau|t). \tag{3.6}$$

3.1.2 Failure Rate

The *failure rate*, or *hazard rate*, $h(t)$, is introduced as

$$h(t) = \lim_{\tau \to 0} \frac{1}{\tau} \frac{F(t+\tau) - F(t)}{R(t)} = \frac{f(t)}{R(t)}, \tag{3.7}$$

so it is evident that $h(t)$ is the time-to-failure conditional pdf. The failure rate can also be expressed in terms of the reliability function as

$$h(t) = -\frac{\mathrm{d}}{\mathrm{d}t}[\ln R(t)], \tag{3.8}$$

so that

$$R(t) = \exp\left[-\int_0^t h(x)\,\mathrm{d}x\right]. \tag{3.9}$$

The integral of the failure rate in the exponent is known as the *cumulative hazard function*, $H(t)$:

$$H(t) = \int_0^t h(x)\,\mathrm{d}x. \tag{3.10}$$

As mentioned above, the failure rate can be defined as the conditional pdf of the component time to failure, given the component has survived to time t. The expected value associated with such pdf is referred to as the *residual* MTTF.

Example 3.1

A device time to failure follows the exponential distribution. If the device has survived up to time t, determine its residual MTTF.

Solution: According to Equation 3.3,

$$\mathrm{MTTF} = \frac{\int_0^\infty \tau f(t+\tau)\,\mathrm{d}\tau}{\int_t^\infty f(\tau)\,\mathrm{d}\tau} = \frac{\int_0^\infty \tau\lambda \exp[-\lambda(t+\tau)]\,\mathrm{d}\tau}{\int_t^\infty \lambda \exp(-\lambda\tau)\,\mathrm{d}\tau} = \frac{\exp(-\lambda t)\int_0^\infty \tau\lambda \exp(-\lambda\tau)\,\mathrm{d}\tau}{\exp(-\lambda t)} = \frac{1}{\lambda}.$$

Let us introduce another useful reliability measure related to failure rate. For a given time interval, t, the average failure rate, $\langle h(t)\rangle$, is given by

$$\langle h(t)\rangle = \frac{1}{t}\int_0^t h(x)\,\mathrm{d}x \tag{3.11}$$

or

$$\langle h(t)\rangle = -\frac{\ln R(t)}{t}, \tag{3.12}$$

therefore

$$\langle h(t)\rangle = \frac{H(t)}{t}. \tag{3.13}$$

If the time interval, t, is equal to a given percentile, t_p, then

$$\langle h(t_p)\rangle = -\frac{\ln(1-p)}{t_p}. \tag{3.14}$$

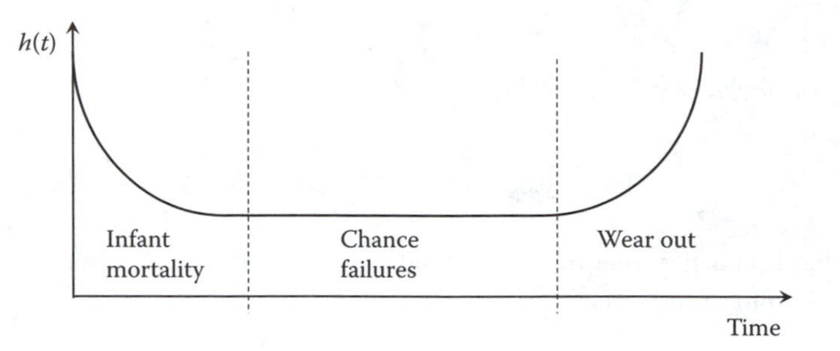

FIGURE 3.1 Typical bathtub curve.

Hazard rate is an important function in reliability analysis since it shows changes in the probability of failure over the lifetime of a component. In practice, $h(t)$ often exhibits a bathtub shape and is referred to as a *bathtub curve*. A bathtub curve is shown in Figure 3.1.

Generally, a bathtub curve can be divided into three regions. The *burn-in* early failure region exhibits a *decreasing failure rate* (DFR), characterized by early failures attributable to defects in design, manufacturing, or construction. Most components do not experience the early failure characteristic, so this part of the curve is representative of the population and not individual units. A time-to-failure distribution having a DFR is referred to as a distribution belonging to the class of DFR distribution.

Analogously, a time-to-failure distribution having a decreasing average failure rate is referred to as a distribution belonging to the class of decreasing failure rate average (DFRA) distribution.

The *chance failure region* of the bathtub curve exhibits a reasonably *constant failure rate*, characterized by random failures of the component. In this period, many mechanisms of failure due to complex underlying physical, chemical, or nuclear phenomena give rise to this approximately constant failure rate. The third region, called the *wearout region*, which exhibits an *increasing failure rate* (IFR), is characterized mainly by complex aging phenomena. Here, the component deteriorates (e.g., due to accumulated fatigue) and is more vulnerable to outside shocks. It is helpful to note that these three regions can be radically different for different types of components. Figures 3.2 and 3.3 show typical bathtub curves for mechanical and electrical devices, respectively.

These figures demonstrate that electrical devices can exhibit a relatively large chance-failure period. Figure 3.4 shows the effect of various levels of stress on a device.

As stress level increases, the chance-failure region decreases and premature wearout occurs. Therefore, it is important to minimize stress factors, such as a harsh operating environment, to maximize reliability.

Similar to the DFR and DFRA distributions, the IFR and *increasing failure rate average* (IFRA) distributions are considered in the framework of a mathematical theory of reliability [1].

Table 3.1 lists the cdfs (unreliability functions) and hazard rate functions for important pdfs.

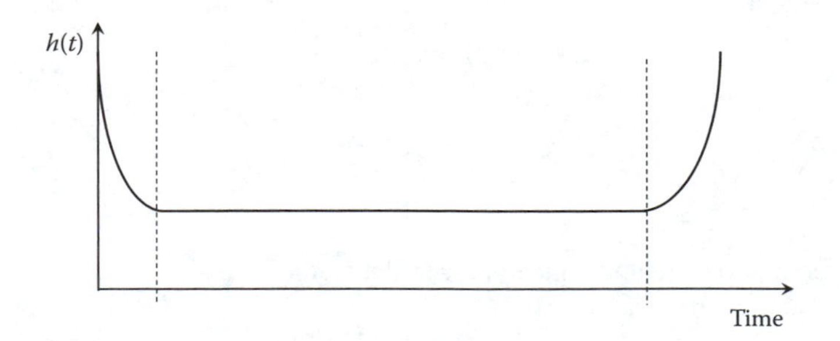

FIGURE 3.2 A typical bathtub curve for electrical devices.

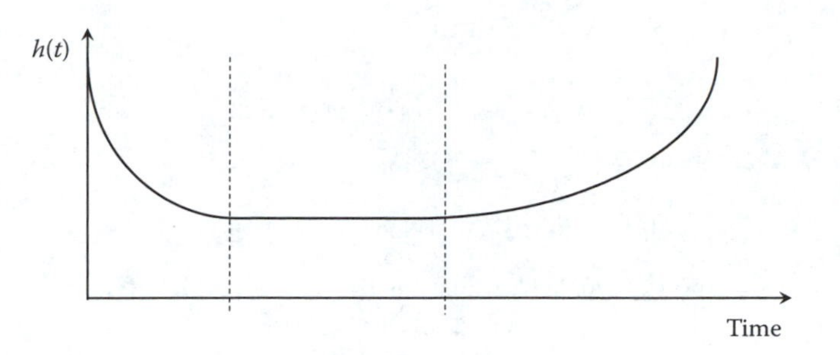

FIGURE 3.3 A typical bathtub curve for mechanical devices.

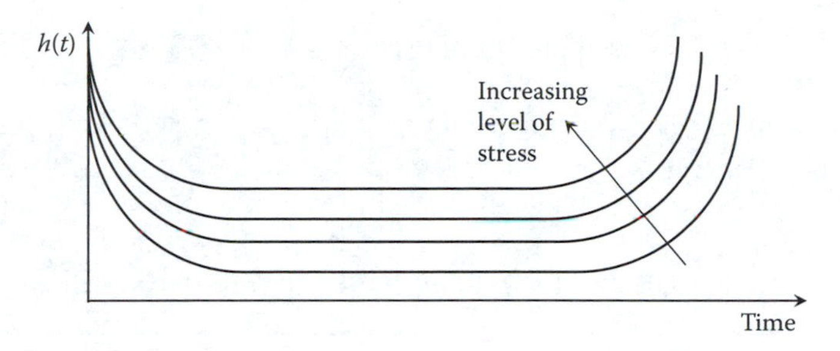

FIGURE 3.4 Effect of stress in a typical bathtub curve.

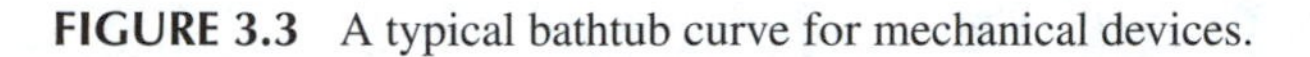

Example 3.2

Failure rate $h(t)$ of a device is approximated by

$$h(t) = \begin{cases} 0.1 - 0.001t, & 0 \leq t \leq 100; \\ -0.1 + 0.001t, & t > 100; \end{cases}$$

as shown in the figure below. Find the pdf and the reliability function for $t \leq 200$.

Solution:

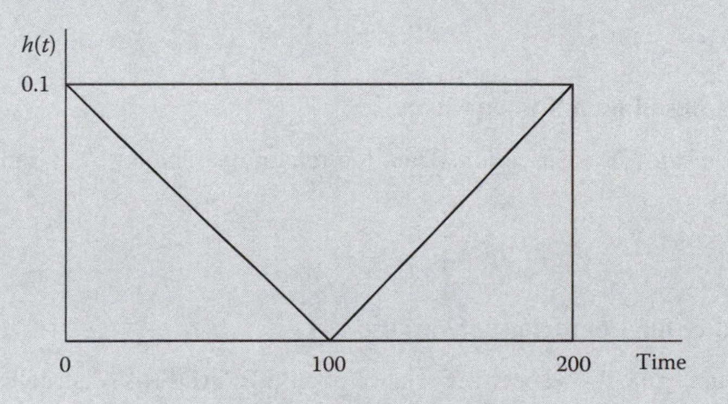

For $0 \leq t \leq 100$,

$$\int_0^t h(\theta)\,d\theta = \int_0^t (0.1 - 0.001\theta)\,d\theta = \left(0.1\theta - \frac{0.001}{2}\theta^2\right)_0^t = 0.1t - 0.0005t^2$$

continued

thus

$$R(t) = \exp(-0.1t + 0.0005t^2).$$

Using Equation 3.7, one obtains

$$f(t) = (0.1 - 0.001t)\exp(-0.1t + 0.0005t^2).$$

Note that $R(100) = \exp(-5) = 0.0067$, so the solution of the problem for $t > 100$ is of academic interest only.

For $t > 100$,

$$h(t) = -0.1 + 0.001t.$$

Accordingly,

$$R(t) = R(100)\exp\left[\int_{100}^{t}(0.1 - 0.001\theta)\,d\theta\right]$$
$$= 0.0067\exp(0.1t - 0.0005t^2 - 5),$$
$$f(t) = 0.0067(-0.1 + 0.001t)\exp\left[\int_{100}^{t}(0.1 - 0.001\theta)\,d\theta\right]$$
$$= 0.0067(-0.1 + 0.001t)R(100)\exp(0.1t - 0.005t^2 - 5).$$

3.1.3 Some Useful Bounds and Inequalities for IFR (DFR)

The following simple bounds and inequalities [1] are given in terms of moments and quantiles (percentiles) of IFR (DFR) distributions. It is worth recalling that any IFR (DFR) distribution belongs to the class of IFRA (DFRA) distributions.

3.1.3.1 Bounds Based on a Known Quantile

Let t_p be pth quantile* or 100pth percentile of an IFRA (DFRA) distribution. Then the reliability function $R(t)$ satisfies the following inequality:

$$R(t)\begin{cases} \geq (\leq)e^{-\alpha t} & \text{for } 0 \leq t \leq t_p, \\ \leq (\geq)e^{-\alpha t} & \text{for } t \geq t_p, \end{cases} \tag{3.15}$$

where $\alpha = \frac{-\ln(1-p)}{t_p}$.

3.1.3.2 Bounds Based on a Known Mean

Let μ be a mean of an IFR distribution. Then the reliability function $R(t)$ satisfies the following inequality:

$$R(t)\begin{cases} \geq e^{-t/\mu} & \text{for } t < \mu, \\ \geq 0 & \text{for } t \geq \mu. \end{cases} \tag{3.16}$$

3.1.3.3 Inequality for Coefficient of Variation

Let μ be a mean and σ^2 be the respective variance of an IFRA (DFRA) distribution. In this case, the coefficient of variation σ/μ satisfies the following inequality:

$$\frac{\sigma}{\mu} \leq (\geq)1. \tag{3.17}$$

It is worth recalling that, for exponential distribution, the coefficient of variation is equal to 1.

*By definition, if t_p is pth quantile, then $F(t_p) = 1 - R(t_p) = p$.

TABLE 3.1
Important Time-to-Failure Distributions and their Characteristics

Distribution Characteristic	Exponential Distribution	Normal Distribution	Lognormal Distribution
pdf, $f(t)$	$\lambda \exp(-\lambda t)$	$\frac{1}{\sigma\sqrt{2\pi}} \exp\left[-\frac{1}{2}\left(\frac{t-\mu}{\sigma}\right)^2\right]$	$\frac{1}{\sigma_t t\sqrt{2\pi}} \exp\left[-\frac{1}{2\sigma_t^2}(\ln t - \mu_t)^2\right]$
cdf, $F(t)$	$1-\exp(-\lambda t)$	$\frac{1}{\sigma\sqrt{2\pi}} \int_0^t \exp\left[-\frac{(\theta-\mu)^2}{2\sigma^2}\right] d\theta$	$\frac{1}{\sigma_t\sqrt{2\pi}} \int_0^t \frac{1}{\theta} \exp\left[-\frac{1}{2}\frac{(\ln\theta-\mu_t)^2}{2\sigma_t^2}\right] d\theta$
Instantaneous failure rate, $h(t)$	λ	$\frac{f(t)}{1-F(t)}$	$\frac{f(t)}{1-F(t)}$
MTTF	$1/\lambda$	μ	$\exp\left(\mu_t + \frac{1}{2}\sigma_t^2\right)$
Major applications in component reliability	– Life distribution of complex nonrepairable systems – Life distribution "burn-in" of some components	– Life distribution of high stress components – Stress–strength analysis – Tolerance analysis	– Size distribution of breaks (in pipes, etc.) – Life distribution of some transistors – Prior parameter distribution in Bayesian analysis

Distribution Characteristic	Weibull Distribution	Gamma Distribution	Smallest Extreme Value Distribution
pdf, $f(t)$	$\frac{\beta(t)^{\beta-1}}{\alpha^\beta} \exp\left[-\left(\frac{t}{\alpha}\right)^\beta\right]$	$\frac{1}{\beta^\alpha\Gamma(\alpha)} t^{\alpha-1} \exp\left(-\frac{t}{\beta}\right)$	$\frac{1}{\delta}\exp\left[\frac{1}{\delta}(t-\lambda) - \exp\left(-\frac{t-\lambda}{\delta}\right)\right]$
cdf, $F(t)$	$1-\exp\left[-\left(\frac{t}{\alpha}\right)^\beta\right]$	$\frac{1}{\beta^\alpha\Gamma(\alpha)} \int_0^t y^{\alpha-1} \exp\left(-\frac{y}{\beta}\right) dy$	$1-\exp\left[-\exp\left(-\frac{t-\lambda}{\delta}\right)\right]$
Instantaneous failure rate, $h(t)$	$\frac{\beta}{\alpha}\left(\frac{t}{\alpha}\right)^{\beta-1}$	$\frac{t^{\alpha-1}\exp\left(-\frac{t}{\beta}\right)}{\beta^\alpha\left[\Gamma(\alpha) - \int_0^t y^{\alpha-1}\exp\left(-\frac{y}{\beta}\right) dy\right]}$	$\frac{1}{\delta}\exp\left(-\frac{t-\lambda}{\delta}\right)$
MTTF	$\alpha\Gamma\left(\frac{1+\beta}{\beta}\right)$	$\beta\alpha$	$\lambda - 0.5776$
Major applications in component reliability	– Corrosion resistance – Life distribution of many basic components, such as capacitors, relays ball bearings, and certain motors	– Distributions of time between recalibration or maintenance of components – Time to failure of system with standby components – Prior distribution in Bayes's estimation	– Distribution of breaking strength of some components – Breakdown voltage of capacitors – Extreme natural phenomena, such as temperature and rainfall minima

3.2 COMMON DISTRIBUTIONS IN COMPONENT RELIABILITY

Table 3.1 displays some basic reliability characteristics of exponential, normal, lognormal, Weibull, gamma, and smallest extreme-value distributions that are commonly used as time-to-failure distribution models for components. Some other characteristics of each of these distributions are further discussed in this section.

3.2.1 EXPONENTIAL DISTRIBUTION

The exponential distribution is one of the most commonly used distributions in reliability analysis. This can be attributed primarily to its simplicity and to the fact that it gives a simple, constant hazard rate model corresponding to a situation that is often realistic. In the context of the bathtub curve, this distribution can simply represent the chance-failure region. It is evident that for components whose chance-failure region is long, in comparison with the other two regions, this distribution might be adequate. This is often the case for electronic components and mechanical components, especially in certain applications, when new components are screened and only those that are determined to have passed (the burn-in period) are used. For such components, exponential distribution is a reasonable choice. In general, exponential distribution is considered as a good model for representing systems and complex, nonredundant components consisting of many interacting parts.

In Section 2.3, we noted that the exponential distribution can be introduced using the HPP. Now let us assume that each failure in this process is caused by a random shock, and the number of shocks occurring in a time interval of length t is described by a Poisson distribution with the mean number of shocks equal to λt. Then, the random number of shocks, n, occurring in the interval $[0, t]$ is given by

$$\Pr[X = n] = \frac{\exp(-\lambda t)(\lambda t)^n}{n!}, \quad n = 0, 1, 2, \ldots, \lambda, \ t > 0, \tag{3.18}$$

where λ is the rate at which the shocks occur. Since based on this model, the first shock causes component failure, then the component is functioning only when no shocks occur, that is, $n = 0$. Thus, one can write

$$R(t; \lambda) = \Pr[X = 0] = \exp(-\lambda t). \tag{3.19}$$

Using relationship 3.2, the exponential pdf can be obtained as

$$f(t) = \lambda \exp(-\lambda t). \tag{3.20}$$

Let us now consider one of the most interesting properties of the exponential distribution: *a failure process represented by the exponential distribution has no memory*. Consider the law of conditional probability and assume that an item has survived after operating for a time t. The probability that the item will fail sometime between t and $t + \Delta t$ is

$$\Pr(t \le T \le t + \Delta t | T > t) = \frac{\exp(-\lambda t) - \exp[-\lambda(t + \Delta t)]}{\exp(-\lambda t)} = 1 - \exp(-\lambda \Delta t), \tag{3.21}$$

which is independent of t. In other words, the component that has worked up to time t has no memory of its past. This property can also be easily described by the shock model. That is, at any point along time t, the rate at which fatal shocks occur is the same regardless whether any shock has occurred up to time t.

3.2.2 WEIBULL DISTRIBUTION

The Weibull distribution has a wide range of applications in reliability analysis. This distribution covers a variety of shapes. Due to its flexibility for describing hazard rates, all three regions of the bathtub curve can be represented by the Weibull distribution. It is possible to show that the Weibull distribution is appropriate for a system or complex component composed of a number of components or parts whose failure is governed by the most severe defect or vulnerable of its components or parts (the *weakest link model*). The pdf of the Weibull distribution is given by

$$f(t) = \frac{\beta(t)^{\beta-1}}{\alpha^{\beta}} \exp\left[-\left(\frac{t}{\alpha}\right)^{\beta}\right], \quad \alpha, \beta > 0,\ t > 0. \tag{3.22}$$

Using Equation 3.7, the failure rate, $h(t)$, can be derived as

$$h(t) = \frac{\beta}{\alpha}\left(\frac{t}{\alpha}\right)^{\beta-1}, \quad \alpha, \beta > 0,\ t > 0. \tag{3.23}$$

Sometimes the transformation $\lambda = 1/\alpha^{\beta}$ is used. In this case Equation 3.23 is transformed to $h(t) = \lambda\beta t^{\beta-1}$. This form will be used later in Chapter 5.

Parameters α and β of the Weibull distribution are referred to as the *scale* and *shape* parameters, respectively. If $0 < \beta < 1$ in Equation 3.23, the Weibull distribution is a DFR distribution that can be used to describe burn-in (early) failure behavior. For $\beta = 1$, the Weibull distribution reduces to the exponential distribution. If $\beta > 1$, the Weibull distribution can be used as a model for the wearout region of the bathtub curve (as an IFR distribution). Main applications of the Weibull distribution include

- Corrosion resistance studies.
- Time to failure of many types of hardware, including capacitors, relays, electron tubes, germanium transistors, photoconductive cells, ball bearings, and certain motors.
- Time to failure of basic elements of a system (components, parts, etc.), although the time to failure of the system itself can be better represented by the exponential distribution.

In some cases, a parameter, called the *location parameter*, is used in the Weibull distribution to account for a period of guaranteed (failure-free) life. The failure rate might be represented by

$$h(t) = \frac{\beta}{\alpha}\left(\frac{t-\theta}{\alpha}\right)^{\beta-1}, \quad \beta, \alpha > 0,\ 0 < \theta \le t < \infty. \tag{3.24}$$

Accordingly, the pdf and reliability function become

$$f(t) = \frac{\beta}{\alpha}\left(\frac{t-\theta}{\alpha}\right)^{\beta-1} \exp\left[-\left(\frac{t-\theta}{\alpha}\right)^{\beta}\right], \quad t \ge \theta \tag{3.25}$$

and

$$R(t) = \exp\left[-\left(\frac{t-\theta}{\alpha}\right)^{\beta}\right], \quad t \ge \theta. \tag{3.26}$$

3.2.3 Gamma Distribution

The gamma distribution was introduced in Section 2.3 as a generalization of the exponential distribution. Recalling the simple shock model considered in Section 3.2.1, one can expand this model for the case when a component fails after being subjected to k successive random shocks arriving according to the HPP. Time-to-failure distribution of the component in this case follows the gamma distribution.

Examples of its application include the distribution of times between recalibration of an instrument that needs recalibration after k uses; time between maintenance of items that require maintenance after k uses; and time to failure of a system with standby components, having the same exponential time-to-failure distribution.

The pdf of the gamma distribution has two parameters, α and β, and it was given in Chapter 2 by Equation 2.74:

$$f(t) = \frac{1}{\beta^\alpha \Gamma(\alpha)} t^{\alpha-1} \exp\left(-\frac{t}{\beta}\right), \quad \alpha, \beta, t \geq 0.$$

The mean value and the variance of the gamma distribution are, respectively,

$$\begin{aligned} E(T) &= \alpha\beta, \\ \sigma^2(T) &= \alpha\beta^2. \end{aligned} \tag{3.27}$$

The gamma cdf and reliability function, in the general case, do not have closed forms. In the case when the shape parameter α is an integer, the gamma distribution is known as the *Erlangian* distribution. In this case, the reliability and failure rate functions can be expressed in terms of Poisson distribution as

$$R(t) = \sum_{k=0}^{\alpha-1} \frac{(t/\beta)^k \exp(-(t/\beta))}{k!} \tag{3.28}$$

and

$$h(t) = \frac{t^{\alpha-1}}{\beta^\alpha \Gamma(\alpha) \sum_{k=0}^{\alpha-1}((t/\beta)^k/k!)}. \tag{3.29}$$

Accordingly, α shows the number of shocks required before a failure occurs and β represents the mean time to occurrence of a shock.

The gamma distribution is a DFR distribution for $\alpha < 1$, a constant failure rate for $\alpha = 1$, and an IFR distribution for $\alpha > 1$. Thus, the gamma distribution can represent each of three regions of the bathtub curve.

EXAMPLE 3.3

The mean time to adjustment of an engine in a fighter plane is $M = 100$ h (assume time to adjustment follows the exponential distribution). Suppose there is a rule to replace certain parts of the engine after three consecutive adjustments.

a. What is the distribution of the time to replace?
b. What is the probability that a given engine does not require part replacement for at least 200 h?
c. What is the mean time to replace?

Solution:

a. Use gamma distribution for T with $\alpha = 3$, $\beta = 100$.
b.

$$\begin{aligned} R(t) &= \sum_{k=0}^{2} \frac{(t/100)^k \exp(-(t/100))}{k!} \\ &= \frac{(200/100)^0 \exp(-2)}{0!} + \frac{(2)^1 \exp(-2)}{1!} + \frac{(2)^2 \exp(-2)}{2!} \\ &= 0.135 + 0.271 + 0.271 = 0.677. \end{aligned}$$

c. Mean time to replace $= E(T) = \alpha\beta = 3(100) = 300$ h.

3.2.4 Normal Distribution

The normal distribution is a basic distribution of statistics. The popularity of this distribution in reliability engineering can be explained by the *central limit theorem*. In engineering terms, according to this theorem, the sum of the large number, n, of independent r.v.s approaches the normal distribution. This distribution is an appropriate model for many practical engineering situations; for example, it can be used as the distribution of diameters of manufactured shafts. Since a normally distributed r.v. can take on a value from the $(-\infty, +\infty)$ range, it has limited applications in reliability problems that involve time-to-failure estimations because time cannot take on negative values. However, for cases where the mean μ is positive and is larger than σ by several folds, the probability that the r.v. T takes negative values can be negligible. For those cases where the probability that r.v. T takes negative values is not negligible, the respective truncated normal distribution can be used (see [2]).

The normal pdf was introduced in Chapter 2 by Equation 2.61 as

$$f(t) = \frac{1}{\sigma\sqrt{2\pi}} \exp\left[-\frac{1}{2}\left(\frac{t-\mu}{\sigma}\right)^2\right], \quad -\infty < t < \infty, \ -\infty < \mu < \infty,$$

where μ is the MTTF and σ is the standard deviation-of-failure time. The normal distribution failure rate is always a monotonically increasing function of time, t, so the normal distribution is an IFR distribution. Thus, the normal distribution can be used as a model representing the wearout region of the bathtub curve. The normal distribution is also a widely used model representing stress and/or strength in the framework of the *stress–strength* reliability models, which are time-independent reliability models (see Section 6.1.1 in Chapter 6).

3.2.5 Lognormal Distribution

The lognormal distribution is widely used in reliability engineering. The lognormal distribution represents the distribution of a r.v. whose logarithm follows the normal distribution. This model is particularly suitable for failure processes that are the result of many small multiplicative errors. Specific applications of this distribution include time to failure of components due to fatigue cracks [3,4]. Other applications of the lognormal distribution are associated with failures attributed to maintenance activities and distribution of cracks initiated and grown by mechanical fatigue. The distribution is also used as a model representing the distribution of particle sizes observed in breakage processes and the life distribution of some electronic components. In Bayesian reliability analysis the lognormal distribution is a popular model to represent the so-called prior distributions. We discuss this topic further in Section 3.6.

The lognormal distribution is a two-parameter distribution. For a r.v. T, the lognormal pdf is

$$f(t) = \frac{1}{\sigma_t t\sqrt{2\pi}} \exp\left[-\frac{1}{2\sigma_t^2}(\ln t - \mu_t)^2\right], \quad 0 < t < \infty, \ -\infty < \mu_t < \infty, \ \sigma_t > 0, \tag{3.30}$$

where $\mu_t = E(\ln t)$ and $\sigma_t^2 = \mathrm{Var}(\ln t)$. The failure rate for the lognormal distribution initially increases over time and then decreases. The rate of increase and decrease depends on the values of the parameters μ_t and σ_t. In general, this distribution is appropriate for representing time to failure for a component whose early failures (or processes resulting in failures) dominate its overall failure behavior.

3.2.6 Extreme Value Distributions

Extreme value distributions are considered in the framework of extreme value theory. Basic applications of this theory are associated with distributions of extreme loads in structural and maritime engineering (distributions of extreme winds, earthquakes, floods, ocean waves, etc.), and other reliability engineering problems, as well as in environmental contamination studies.

3.2.6.1 Some Basic Concepts and Definitions

Let $x_1, x_2, \ldots, x_n$ be a sample of independently and identically distributed r.v.s (e.g., representing environmental contamination concentration values) with cdf, $F(X)$. In extreme value theory, the distribution $F(X)$ is called *parent distribution*. When we rearrange the sample in increasing value order, so that $x_{1:n} < x_{2:n} < \cdots < x_{n:n}$, the statistics $x_{i:n}$ $(i = 1, 2, \ldots, n)$ obtained are called the *order statistics*.

It can be shown that the cdf, $F_{r:n}(x)$, of rth order statistic, $X_{r:n}$, can be expressed in terms of the binomial distribution as

$$F_{r:n}(X) = \sum_{k=r}^{n} \binom{n}{k} F^k(X)[1 - F(X)]^{n-k}. \tag{3.31}$$

Clearly, the maximum statistic of a sample of size n is the last order statistic ($r = n$), so its cdf can be written, using Equation 3.31, as

$$F_{r:n}(X) = F^n(X). \tag{3.32}$$

The distribution obtained is called the *distribution of maxima*. Getting ahead of our discussion, we will mention that this is (if X is the time to failure) the time-to-failure distribution of a parallel system (discussed in Chapter 4) composed of n identical components.

The distribution of minima of a random sample of size n can be obtained from Equation 3.31, as a particular case, when $r = 1$. That is,

$$F_{1:n}(X) = 1 - [1 - F(X)]^n. \tag{3.33}$$

Similar to the case considered above, this is the time-to-failure distribution of a series system (considered in Chapter 4) composed of n identical components.

Example 3.4 [5]

One of the main engineering concerns in the design of a nuclear power plant is the estimation of the probability distribution of the distances of possible earthquakes to the tentative location of the plant. Due to the presence of a fault in the area surrounding a potential location, it has been established that the epicenter of an earthquake can occur, with equal likelihood, at any point within the 50 km radius of the fault. If the plant location is aligned with the fault, and its closest extreme is 200 km away, then the distance between the epicenter and the plant can be assumed to be distributed uniformly between 200 and 250 km. Find the probability of having minimum distances of less than 210 km.

Solution: The cdf of the distance to the closest earthquake in series of 5 and 10 earthquakes is given by

$$F_{1:n}(X) = 1 - \left(1 - \frac{x - 200}{50}\right)^n, \quad n = 5, 10.$$

The probability of having minimum distances less than 210 km, which is equivalent to the standard uniform distribution $U(0, 1)$ value of (210–200)/50, are 0.672 and 0.893 for the series of 5 and 10 earthquakes, respectively.

3.2.6.2 Order Statistics from Samples of Random Size

The statistics discussed above have been associated with a fixed sample size. However, there are many practical situations in which one may be interested in the extreme statistics where the sample size is random. For example, the number of defects having random sizes in a material can be itself random.

Denote the distribution of sample size, n, by

$$p_i = \Pr[n = n_i]. \tag{3.34}$$

Let the statistic of interest, x, have, for a fixed sample size n, pdf $f(x, n)$ and cdf $F(x, n)$. Using the total probability rule, the pdf $g(x)$ and cdf $G(x)$ of the statistic x can be written as

$$\begin{aligned} g(x) &= \sum_i p_i f(x, n_x), \\ G(x) &= \sum_i p_i F(x, n_x). \end{aligned} \tag{3.35}$$

An example of this model's application to the problem of probability estimation of nuclear power plant core damage is considered in Chapter 8.

When the sample size n goes to infinity, the distributions of maxima and minima have the following limits:

$$\lim_{n\to\infty} F^n(x) = 0, \quad F(x) < 1 \tag{3.36}$$

and

$$\lim_{n\to\infty} \{1 - [1 - F(x)]^n\} = 1, \quad F(x) \leq 1. \tag{3.37}$$

These limits show that the limit distributions take on only values 0 and 1. It is said that such distribution *degenerates*. To avoid this degeneration, the following linear transformation is used:

$$Y = a_n + b_n x, \tag{3.38}$$

where a_n and b_n are the constants chosen to get the following *limit distributions*:

For maxima,

$$\lim_{n\to\infty} H_n(a_n + b_n x) = \lim_{n\to\infty} F^n(a_n + b_n x) = H(x) \quad \text{for all } x. \tag{3.39}$$

For minima,

$$\lim_{n\to\infty} L_n(c_n + c_n x) = \lim_{n\to\infty} [1 - F(c_n + c_n x)]^n = L(x) \quad \text{for all } x, \tag{3.40}$$

which do not degenerate.

3.2.6.3 Asymptotic Distributions of Maxima and Minima

We discussed how to get the distribution of maxima and minima from a given parent distribution in the case of finite samples of fixed and random sample sizes. The *asymptotic* distributions of extreme values are used in the following situations [5]:

1. The sample size increases to infinity.
2. The parent distribution is unknown.
3. The sample size is large but unknown.

3.2.6.4 Three Types of Limit Distributions

The fundamental result of extreme value theory is the existence of only three feasible types of limit distributions for maxima, H, and three similar feasible types of limit distribution for minima, L [6–8]. This result is given by the following theorems.

There are only three types of nondegenerated distributions for maxima, $H(x)$, satisfying the condition 3.39 (the cdfs are given in the standard forms, that is, with scale parameter equal to 1):

Type I (the Gumbel distribution):

$$H_1(x) = \exp\left[-\exp\left(-\frac{x}{\delta}\right)\right], \quad \delta > 0,\ -\infty < x < \infty. \tag{3.41}$$

Type II (the Freshet distribution):

$$H_2(x) = \begin{cases} \exp\left[-\left(\frac{x}{\delta}\right)^{-\gamma}\right], & \delta > 0,\ \gamma > 0,\ x > 0; \\ 0, & x \le 0. \end{cases} \tag{3.42}$$

Type III (the Weibull distribution):

$$H_3(x) = \begin{cases} 1, & x > 0; \\ \exp\left[-\left(\frac{x}{\delta}\right)^{\gamma}\right], & \delta > 0,\ \gamma > 0,\ x \le 0. \end{cases} \tag{3.43}$$

A similar theorem for the distributions of minima states that there are only three types of nondegenerated distributions for minima, $L(x)$, satisfying the condition 3.40. These are as follows:

Type I (the Gumbel distribution):

$$L_1(x) = 1 - \exp\left[-\exp\left(\frac{x}{\delta}\right)\right], \quad \delta > 0,\ -\infty < x < \infty. \tag{3.44}$$

Type II (the Freshet distribution):

$$L_2(x) = \begin{cases} 1 - \exp\left[-\left(\frac{x}{\delta}\right)^{-\gamma}\right], & \delta > 0,\ \gamma > 0,\ x < 0; \\ 1, & x \ge 0. \end{cases} \tag{3.45}$$

Type III (the Weibull distribution):

$$L_3(x) = \begin{cases} 0, & x < 0; \\ 1 - \exp\left[-\left(\frac{x}{\delta}\right)^{\gamma}\right], & \delta > 0,\ \gamma > 0,\ x \ge 0. \end{cases} \tag{3.46}$$

Currently, the following three extreme value distributions are widely used in reliability engineering: the Weibull distribution for minima (discussed in Section 3.2.2), the Gumbel distribution for minima, and the Gumbel distribution for maxima. The last two distributions are also referred to as *the smallest extreme value distribution* and *the largest extreme value distribution* [9].

Similar to the three-parameter Weibull distribution (Equation 3.25), the smallest extreme value distribution and the largest extreme value distribution are sometimes used as two-parameter distributions. In this case, their pdfs take on the following forms.

The pdf of the smallest extreme value distribution is given by

$$f(x) = \frac{1}{\delta}\exp\left[\frac{1}{\delta}(t-\lambda) - \exp\left(\frac{t-\lambda}{\delta}\right)\right], \quad -\infty < \lambda < \infty,\ \delta > 0,\ -\infty < t < \infty. \tag{3.47}$$

The parameter λ is called the *location* parameter and can take on any value. The parameter δ is called the *scale* parameter and is always positive. The failure rate for the smallest extreme value distribution is

$$h(x) = \frac{1}{\delta}\exp\left(\frac{t-\lambda}{\delta}\right), \tag{3.48}$$

which is an increasing function of time, so that the smallest extreme value distribution is the IFR distribution that can be used as a model for component failures due to aging. In this model, the component's wearout period is characterized by an exponentially IFR. Clearly, negative values of t are not meaningful when it is representing time to failure.

The Weibull distribution and the smallest extreme value distribution are closely related to each other. If a r.v. X follows the Weibull distribution with pdf (Equation 3.22), the transformed r.v. $T = \ln(X)$ follows the smallest extreme value distribution with parameters

$$\lambda = \ln(\alpha), \quad \delta = \frac{1}{\beta}. \tag{3.49}$$

The two-parameter largest extreme value pdf is given by

$$f(x) = \frac{1}{\delta}\exp\left[-\frac{1}{\delta}(t-\lambda) - \exp\left(\frac{t-\lambda}{\delta}\right)\right], \quad -\infty < \lambda < \infty,\ \delta > 0,\ -\infty < t < \infty. \tag{3.50}$$

The largest extreme value distribution, although not very useful for component failure behavior modeling, is useful for estimating natural extreme phenomena.

For further discussions regarding extreme value distributions, see [2,5,10].

Example 3.5

The maximum demand for electric power at any given time during a year is directly related to extreme weather conditions. An electric utility has determined that the distribution of maximum power demands can be presented by the largest extreme value distribution with $\lambda = 1200$ (MW) and $\delta = 480$ (MW). Determine the probability (per year) that the demand will exceed the utility's maximum installed power of 3000 (MW).

Solution: Since this is the largest extreme value distribution, we should integrate Equation 3.50 from 3000 to $+\infty$:

$$\Pr(t > 3000) = \int_{3000}^{\infty} f(t)\,dt = 1 - \exp\left[-\exp\left(-\frac{t-\lambda}{\delta}\right)\right].$$

Since

$$\frac{t-\lambda}{\delta} = \frac{3000-1200}{480} = 3.75,$$

then

$$\Pr(t > 3000) = 0.023.$$

3.3 COMPONENT RELIABILITY MODEL SELECTION

In the previous section, we discussed several distribution models useful for reliability analysis of components. A *probability model* is referred to as a mathematical expression that describes in terms of probabilities how a r.v. is spread over its range. It is necessary at this point to discuss how field and test data can support the selection of a probability model for reliability analysis. In this section, we consider several procedures for selecting and estimating the models using observed failure data. These methods can be divided into two groups: nonparametric methods (that do not need a particular distribution function) and parametric methods (that are based on a selected distribution function). We discuss each of these methods in more detail. In addition, some graphic-based exploratory data analysis procedures are considered in Section 7.3.

3.3.1 Graphical Nonparametric Procedures

The nonparametric approach, in principle, attempts to directly estimate the reliability characteristic of an item (e.g., the pdf, reliability, and hazard rates) from a sample. The shape of these functions, however, is often used as an indication of the most appropriate parametric distribution representation. Thus, such procedures can be considered as tools for exploratory (preliminary) data analysis. It is important to mention that failure data from a maintained item can be used as the sample only if a maintained component is assumed to be *as good as new* following maintenance. Then each failure time can be considered a sample observation independent of the previously observed failure times. Therefore, n observed times to failure of such a maintained component is equivalent to putting n independent new components under test.

3.3.1.1 Small Samples

Suppose n times to failure makes a small sample (e.g., $n < 25$). Let the data be ordered such that $t_1 \le t_2 \le \cdots \le t_n$. Blom [11] introduced the following nonparametric estimators for the reliability functions of interest:

$$\hat{h}(t_i) = \frac{1}{(n - i + 0.625)(t_{i+1} - t_i)}, \quad i = 1, 2, \ldots, n - 1, \tag{3.51}$$

$$\hat{R}(t_i) = \frac{n - i + 0.625}{n + 0.25}, \quad i = 1, 2, \ldots, n - 1, \tag{3.52}$$

and

$$\hat{f}(t_i) = \frac{1}{(n + 0.25)(t_{i+1} - t_i)}, \quad i = 1, 2, \ldots, n - 1. \tag{3.53}$$

Although there are other estimators besides those above, [12] concludes that estimators (Equations 3.51 through 3.53) have good properties and recommends their use. One should keep in mind that 0.625 and 0.25 are correction terms of a minor importance, which result in a small bias and a small mean square error for the Weibull distribution estimation [13].

Example 3.6

A high-pressure pump in a process plant has the failure times t_i (shown in the following table). Plot $\hat{h}(t_i)$, $\hat{R}(t_i)$, $\hat{f}(t_i)$ and discuss the results.

Solution:

i	t_i	$t_{i+1} - t_i$	$\hat{h}(t)$	$\hat{R}(t)$	$\hat{f}(t)$
1	0.20	0.60	0.25	0.91	0.23
2	0.80	0.30	0.59	0.78	0.46
3	1.10	0.41	0.53	0.64	0.34
4	1.51	0.32	0.86	0.50	0.43
5	1.83	0.69	0.55	0.36	0.20
6	2.52	0.42	1.34	0.22	0.30
7	2.98	—	—	0.09	—

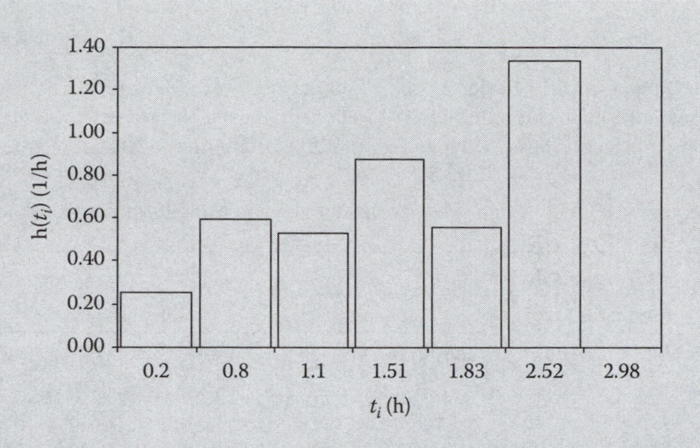

From the above histogram, one can conclude that the failure rate is somewhat constant over the operating period of the component, with an increase toward the end. However, as a point of caution, although a constant hazard rate might be concluded, several other tests and additional observations may be needed to support this conclusion. Additionally, the histogram is only a representative of the case under study. An extension of the result to future times or other cases (e.g., other high-pressure pumps) may not be accurate.

3.3.1.2 Large Samples

Suppose n times to failure make a large sample. Suppose further that the sample is grouped into a number of equal time-to-failure increments, Δt. According to the definition of reliability, a nonparametric estimate of the reliability function is

$$\hat{R}(t_i) = \frac{N_s(t_i)}{N}, \tag{3.54}$$

where $N_s(t_i)$ represents the number of surviving components. Note that the estimator (Equation 3.54) is absolutely compatible with the empirical (sample) cdf (Equation 2.130) introduced in Section 2.5. Time t_i is usually taken to be the upper endpoint of each interval. Similarly, the pdf is estimated by

$$\hat{f}(t_i) = \frac{N_f(t_i)}{N\Delta t}, \tag{3.55}$$

where $N_f(t_i)$ is the number of failures observed in the interval $(t_i,\ t_i + \Delta t)$. Finally, using Equations 3.54 and 3.55, one obtains

$$\hat{h}(t_i) = \frac{N_f(t_i)}{N_s(t_i)\Delta t}, \tag{3.56}$$

It is clear that for $i = 1, N_s(t_i) = N$, and for $i > 1, N_s(t_i) = N_s(t_{i-1}) - N_f(t_i)$. Equation 3.56 gives an estimate of average failure rate during the interval $(t_i, t_i + \Delta t)$. When $N_s(t_i) \to \infty$ and $\Delta t \to 0$, estimate 3.56 approaches the true hazard rate $h(t)$.

In Equation 3.56, $N_f(t_i)/N_s(t_i)$ is the estimate of probability that the component will fail in the interval $(t_i, t_i + \Delta t)$, since $N_s(t_i)$ represents the number of components functioning at t_i. Dividing this quantity by Δt, the estimate of failure rate (probability of failure per unit of time for interval Δt) is obtained. It should be noted that the accuracy of this estimate depends on Δt. Therefore, if smaller ts are used, we would, theoretically, expect to obtain a better estimation. However, the drawback of using smaller Δts is the decrease in the amount of data for each interval to estimate $\hat{R}(t_i)$ and $\hat{f}(t_i)$. Therefore, selecting Δt requires consideration of both of these opposing factors.

Example 3.7

Times to failure for an electrical device are obtained during three stages of the component's life. The first stage represents the infant mortality of the component; the second stage represents chance failures; and the third stage represents the wearout period. Plot the failure rate for this component using the data provided below.

Solution: Use Equations 3.54 through 3.56 to calculate the empirical hazard rate, reliability, and pdf.

Given Data				Calculated Data		
Interval (t_i)		Frequency				
Beginning	End	$N_f(t_i)$	$N_s(t_i)$	$h(t_i)$	$R(t_i)$	$f(t_i)$
Infant Mortality Stage						
0	20	79	71	0.02633	1.00000	0.02633
20	40	37	34	0.02606	0.47333	0.01233
40	60	15	19	0.02206	0.22667	0.00500
60	80	6	13	0.01579	0.12667	0.00200
80	100	2	11	0.00769	0.08667	0.00067
100	120	1	10	0.00455	0.07333	0.00033
120	>120	10	0	0.05000	0.06667	0.00333
		Total 150				
Chance Failure Stage						
0	2000	211	289	0.00021	1.00000	0.00021
2000	4000	142	147	0.00025	0.57800	0.00014
4000	6000	67	80	0.00023	0.29400	0.00007
6000	8000	28	52	0.00018	0.16000	0.00003
8000	10,000	21	31	0.00020	0.10400	0.00002
10,000	$>10,000$	31	0	0.00050	0.06200	0.00003
		Total 500				
Wearout Stage						
0	100	34	266	0.00113	1.00000	0.00113
100	200	74	192	0.00278	0.88667	0.00247
200	300	110	82	0.00573	0.64000	0.00367
300	>300	82	0	0.01000	0.27333	0.00273
		Total 300				

The graph below plots the estimated hazard rate functions for the three observation periods. (Please note that the three periods, each having a different scale, are combined on the same x-axis. Chance failure represents the majority of the device's life.)

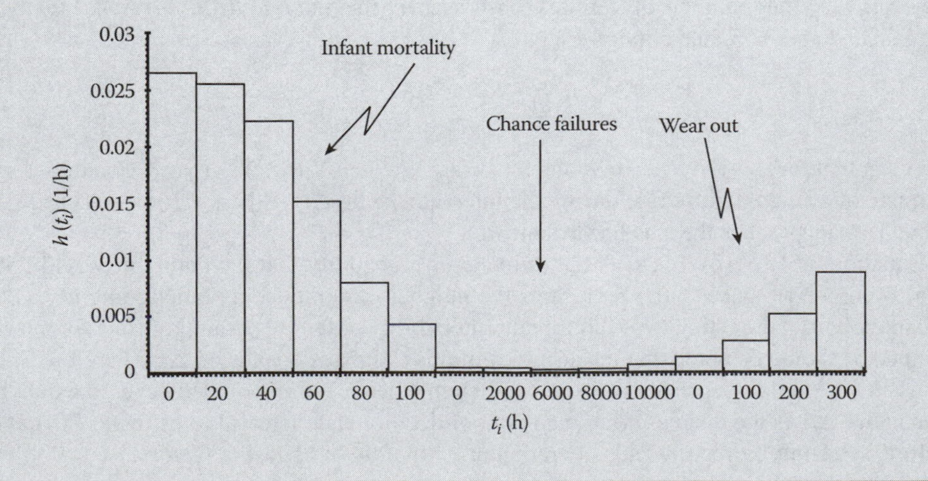

3.3.2 Probability Plotting

Probability plotting is a simple graphical method of displaying and analyzing observed data. The data are plotted on special probability papers similar to the way that a transformed cdf would plot as a straight line. Each type of distribution has its own probability paper. If a set of data is hypothesized to originate from a known distribution, the graph can be used to conclude whether or not the hypothesis might be rejected. From the plotted line, one can also roughly estimate the parameters of the hypothesized distribution. Probability plotting is often used in reliability analysis to test the appropriateness of using known distributions to present a set of observed data. This method is used because it provides simple, visual representation of the data. This approach is an informal, qualitative decision-making method. In contrast, the goodness-of-fit tests discussed in Chapter 2 are more formal, quantitative decision-making methods. Therefore, the plotting method should be used with care and preferably as an exploratory data analysis procedure. In this section, we will discuss some factors that should be considered when probability plotting is used.

It should be noted that with the present level of computer support in reliability data analysis, the term "probability paper" is used only to refer to the graphical method of parameter estimation, where the paper itself has become obsolete. It takes a simple electronic spreadsheet to program the above equations of cdf linearization and estimate the distribution parameters by the least-squares method. Modern reliability engineers no longer have to use a ruler and eyeball judgment to analyze reliability data.

We will briefly discuss the probability papers for the basic distributions considered in this book. See [9,14,15], and [16] for further discussion regarding other distributions and various plotting techniques.

3.3.2.1 Exponential Distribution Probability Plotting

Taking the logarithm of the expression for the reliability function of the exponential distribution (Equation 3.19), one obtains

$$\ln R(t) = -\lambda t. \tag{3.57}$$

If $R(t)$ is plotted as a function of time, t, on semilogarithmic plotting paper, according to Equation 3.57, the resulting plot will be a straight line with a slope of $(-\lambda)$.

Consider the following n times to failure observed from a life test: $t_1 \leq t_2 \leq \cdots \leq t_n$. According to Equation 3.57, an estimate of the reliability $R(t_i)$ can be made for each t_i. A crude, nonparametric estimate of $R(t_i)$ is $1 - i/n$ (recall Equations 3.54 and 2.130). However, as noted in Section 3.3.1, the statistic Equation 3.53 provides a better estimation for $R(t)$ for the Weibull distribution (recall that the exponential distribution is a particular case of the Weibull distribution).

Graphically, the y-axis shows $R(t_i)$ and the x-axis shows t_i. The resulting points should reasonably fall on a straight line if these data can be described by the exponential distribution. Since the slope of $\ln R(t)$ versus t is negative, it is also possible to plot $\ln(1/R(t))$ versus t in which the slope is positive. Other appropriate estimators of $R(t_i)$ include the so-called mean rank, $(n - i + 1)/(n + 1)$ and the median rank, $(n - i + 0.7)/(n + 0.4)$ [13].

It is also possible to estimate the MTTF from the plotted graph. For this purpose, at the level of $R = 0.368$ (or $1/R = e \approx 2.718$), a line parallel to the x-axis is drawn. At the intersection of this line and the fitted line, another line vertical to the x-axis is drawn. The value of t on the x-axis is an estimate of MTTF, and its inverse is $\hat{\lambda}$. The exponential plot is a particular case of the Weibull distribution, and therefore, the Weibull paper may be used to determine whether or not the exponential distribution is a good fit.

Example 3.8

Nine times to failure of a diesel generator are recorded as 31.3, 45.9, 78.3, 22.1, 2.3, 4.8, 8.1, 11.3, and 17.3 days. If the diesel is restored to "as good as new" after each failure, determine whether the data represent the exponential distribution. That is, find $\hat{\lambda}$ and $\hat{R}$ (193 h).

Solution: First arrange the data in increasing order and then calculate the corresponding $\hat{R}(t_i)$.

i	t	$\frac{n-i+0.625}{n+0.25}$	$\frac{n+0.25}{n-i+0.625}$
1	2.3	0.93	1.07
2	4.8	0.82	1.21
3	8.1	0.72	1.40
4	11.3	0.61	1.64
5	17.3	0.50	2.00
6	22.1	0.39	2.55
7	31.3	0.28	3.53
8	45.9	0.18	5.69
9	78.3	0.07	14.80

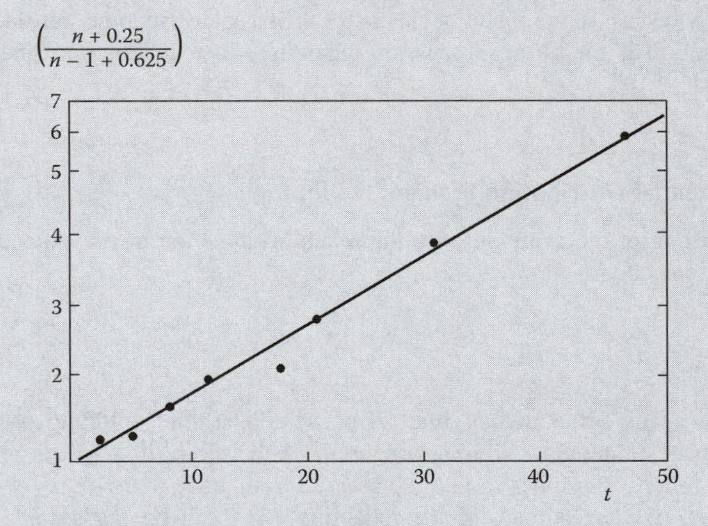

FIGURE 3.5 Exponential probability plot in Example 3.8.

Figure 3.5 shows a plot of the above data on logarithmic paper.

$$\bar{\lambda} = \frac{7-3}{48.6-28} = 0.194\,\text{failures/day} = 8.1 \times 10^{-3}\,\text{failures/day}$$

$$\hat{R}(193) = \exp[-8.1 \times 10^{-3}(193)] = 0.21.$$

3.3.2.2 Weibull Distribution Probability Plotting

Similar to plots of the exponential distribution, plots of the Weibull distribution require special probability papers. If the observed data form a reasonably straight line, the Weibull distribution can be considered as a competing model. Recalling the expression for the Weibull cdf from Table 3.1

or from Example 2.19, one can obtain the following relationships for the respective reliability functions:

$$\frac{1}{R(t)} = \exp\left[\left(\frac{t}{\alpha}\right)^{\beta}\right],$$
$$\ln\left[\ln\left(\frac{1}{R(t)}\right)\right] = \beta \ln t - \beta \ln \alpha. \tag{3.58}$$

This linear relationship (in $\ln t$) provides the basis for the Weibull plots. It is evident that $\ln\{\ln[1/R(t)]\}$ plots as a straight line against $\ln t$ with slope β and y-intercept of $(-\beta \ln \alpha)$. Accordingly, the values of the Weibull parameters α and β can be obtained from the y-intercept and the slope of the graph, respectively.

As was mentioned earlier, several estimators of $R(t)$ can be used. The most recommended estimator is Equation 3.53 (identical to that used in exponential plots). The corresponding plotting procedure is simple. On the special Weibull paper (Figure 3.6), t_i is plotted in the logarithmic x-axis, and the estimate of $F(t) = (i - 0.375)/(n + 0.25) \times 100$ is plotted on the y-axis (often labeled % failure). The third scale shown in Figure 3.6 is for $\ln\{\ln[1/R(t)]\}$, but it is more convenient to use the estimate of $F(t)$.

The degree to which the plotted data fall on a straight line determines the conformance of the data to the Weibull distribution. If the data plot reasonably well as a straight line, the Weibull distribution is a reasonable fit, and the shape parameter and the scale parameter β can be roughly estimated. If a line is drawn parallel to the plotted straight line from the center of a small circle (sometimes called the *origin*) until it crosses the *small beta estimator* axis or scale, the value of β can be obtained. To find α, draw a horizontal line from the 63.2% cdf level until it intersects the fitted straight line. From this point, draw a vertical line until it intersects with the x-axis, and read the value of parameter α at this intersection.

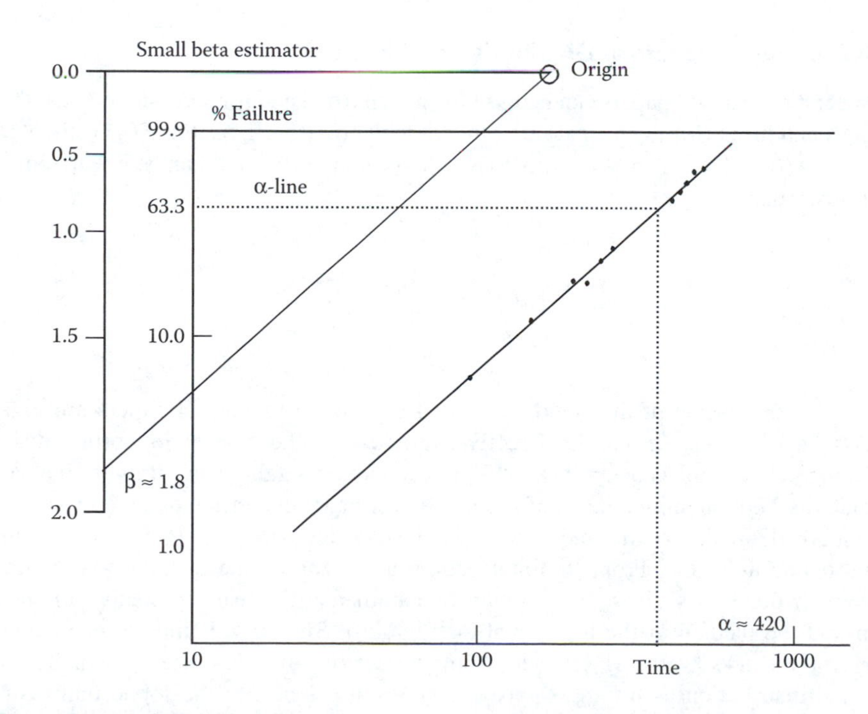

FIGURE 3.6 Weibull probability plot.

EXAMPLE 3.9

Time to failure of a device is assumed to follow the Weibull distribution. Ten of these devices undergo a reliability test. The times to failure (in hours) are 89, 132, 202, 263, 321, 362, 421, 473, 575, and 663. If the Weibull distribution is the correct choice to model the data, what are the parameters of this distribution? What is the reliability of the device at 1000 h?

Solution:

i	1	2	3	4	5	6	7	8	9	10
t_i	89	132	202	263	321	362	421	473	575	663
$\frac{i-0.375}{n+0.25} \times 100$	6.10	15.85	25.61	35.37	45.12	54.88	64.46	74.39	84.15	93.90

Figure 3.6 shows the fitted line on the Weibull probability paper. The fitting is reasonably good. The graphical estimate of β is approximately 1.8, and the estimate of α is approximately 420 h. Percent failure at 1000 h is about 99.1%; the reliability is about 0.9% [$R(t = 1000) = 0.009$].

In cases where the data do not fall on a straight line but are concave or convex in shape, it is possible to find a *location parameter* θ (i.e., to try using the three-parameter Weibull distribution (Equation 3.25) introduced in Section 3.2.2) that might "straighten out" these points. For this procedure, see [14] and [16].

If the failure data are grouped, the class midpoints t_i' (rather than t_i) should be used for plotting, where $t_i' = (t_{i-1} + t_i)/2$. One can also use class endpoints instead of midpoints. Recent studies suggest that the Weibull parameters obtained by using class endpoints in the plots are better matched with those of the maximum likelihood estimation method.

3.3.2.3 Normal and Lognormal Distribution Probability Plotting

The same special probability papers can be used for both normal and lognormal plots. On the x-axis, t_i (for normal) and $\ln(t_i)$ (for lognormal) are plotted, while on the y-axis, the $F(t_i)$ estimated by the same $(i - 0.375)/(n + 0.25)$ is plotted. It is easy to show that normal cdf can be linearized using the following transformation:

$$\Phi^{-1}[F(t_i)] = \frac{1}{\sigma}t_i - \frac{\mu}{\sigma}, \tag{3.59}$$

where $\Phi^{-1}(\cdot)$ is the inverse of the standard normal cdf. Some lognormal papers are logarithmic on the x-axis, in which case t_i can be directly expressed. In the case of lognormal distribution, t_i in Equation 3.59 is replaced by $\ln(t_i)$. If the plotted data fall on a straight line, a normal or lognormal distribution might be conformed. To estimate the mean parameter μ, the value of 50% is marked on the x-axis and a line parallel to the y-axis is drawn until it intersects with the plotted straight line. From the intersection, a horizontal line to the x-axis is drawn. Its intersection with the y-axis gives the estimate of parameter μ (mean or median of the normal distribution and the median of the lognormal distribution). Similarly, if the corresponding y-axis intersection for the 84% value is selected from the x-axis, the parameter σ can be estimated for the normal distribution as $\sigma \approx t_{84\%} - t_{50\%}$ or $\sigma \approx t_{84\%} - \mu$. For the lognormal distribution, $\sigma \approx \ln t_{84\%} - \mu_t$.

EXAMPLE 3.10

The time it takes for a thermocouple to drift upward or downward to an unacceptable level is measured and recorded in a process plant (see the following table.). Determine whether the drifting time can be modeled by a normal distribution.

i	t_i (months)	$\frac{i-0.375}{n+0.25} \times 100$	i	t_i (months)	$\frac{i-0.375}{n+0.25} \times 100$
1	11.2	4.39	8	17.2	53.50
2	12.8	11.40	9	18.1	60.53
3	14.4	18.42	10	18.9	67.54
4	15.1	25.44	11	19.3	74.56
5	16.2	32.46	12	20.0	81.58
6	16.3	39.47	13	21.8	88.60
7	17.0	46.49	14	22.7	95.61

Solution: Figure 3.7 shows that the data conform the normal distribution, with $\mu = t_{50\%} = 17.25$ months and $t_{84\%} = 20.75$ months. Therefore, the estimate of $\sigma \approx 20.75 - 17.25 = 3.5$ months.

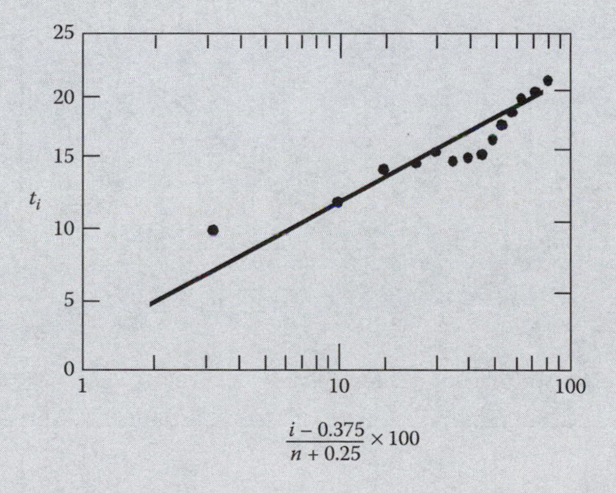

FIGURE 3.7 Normal distribution plot in Example 3.10.

EXAMPLE 3.11

Five components undergo low-cycle fatigue crack tests. The failure cycles are given in the table below. Determine the conformity of the data with a lognormal distribution. Estimate the parameters of the lognormal distribution.

continued

Solution:

i	t_i (cycles)	$\frac{i-0.375}{n+0.25} \times 100$
1	363	11.90
2	1115	30.95
3	1982	50.00
4	4241	69.05
5	9738	88.10

From the probability plot in Figure 3.8,

$$\hat{\mu}_t = \ln(\hat{\mu}_y) = \ln(2000) = 7.61,$$

$$\sigma_t \approx \ln(t_{84\%}) - \hat{\mu}_t$$

$$= \ln(6700) - 7.61 \approx 1.12.$$

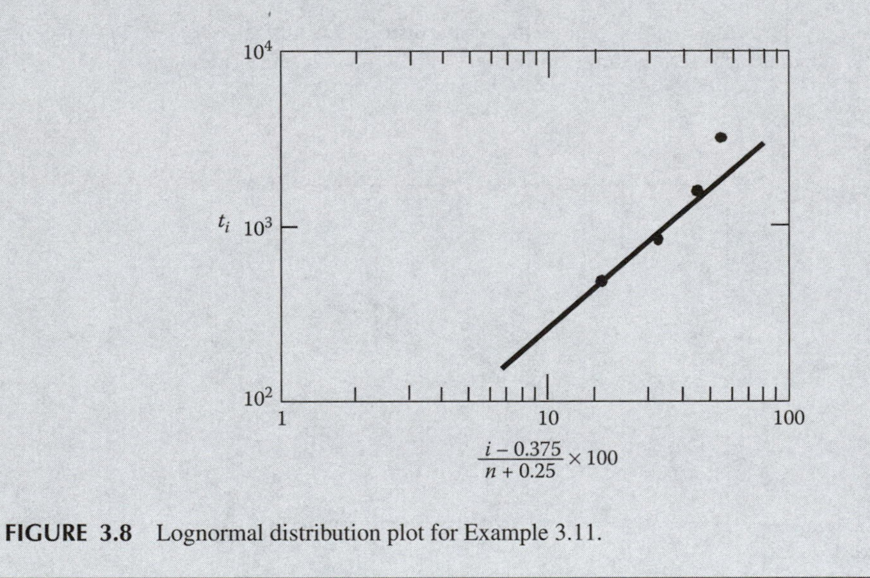

FIGURE 3.8 Lognormal distribution plot for Example 3.11.

3.3.3 Total-Time-on-Test Plots

The total-time-on-test plot is a graphical procedure that helps to determine whether the underlying distribution exhibits an IFR, a constant failure rate (the distribution is exponential), or a DFR. The procedure for this type of plot is discussed detail in [17] and [18]. Additionally, [19] discusses the use of this method for optimal replacement policy problems. Although it is possible to treat grouped failure data using this method, we discuss its use for ungrouped failure data only. Consider the observed failure times of n components such that $t_1 \leq t_2 \leq \cdots \leq t_n$. If the number of survivors to time t_i is denoted by $N_s(t_i)$, then the survival probability (the reliability function) can be estimated as

$$R(t_i) = \frac{N_s(t_i)}{n}. \tag{3.60}$$

It is clear that $N_s(t_i)$ is a stepwise function. The total time on test to age t_i, denoted by $T(t_i)$, is obtained from

$$T(t_i) = \int_0^{t_i} N_s(t)\,dt. \quad (3.61)$$

Equation 3.61 can be expressed in a more tractable form:

$$\int_0^{t_i} N_s(t)\,dt = nt_1 + (n-1)(t_2 - t_1) + \cdots + (n-i+1)(t_i - t_{i-1}), \quad (3.62)$$

since n components have survived up to time t_1 (time to the first failure), $(n-1)$ of the components survived during the period between t_1 and t_2, and so on. The so-called *scaled total time on test* at time t_i is defined as

$$\tilde{T}(t_i) = \frac{\int_0^{t_i} N_s(t)\,dt}{\int_0^{t_n} N_s(t)\,dt}. \quad (3.63)$$

For the exponential distribution, $\tilde{T}(t_i) = i/n$. A graphical representation of total-time-on-test is formed by plotting i/n on the x-axis and $\tilde{T}(t_i)$ on the y-axis. Its deviation from the reference line $\tilde{T}(t_i) = i/n$ is then assessed. If the data fall on the $\tilde{T}(t_i) = i/n$ line, the exponential distribution can be assumed. If the plot is concave over most of the graph, there is a possibility of an IFR. If the plot is convex over most of the graph, there is a possibility of a DFR. If the plot does reasonably fall on a straight line and does not reveal concavity or convexity, one can assume the exponential distribution.

The total-time-on-test plot, similar to other graphical procedures, should not be used as the only test for determining model adequacy. This is of a particular importance when only a small sample is available. The total-time-on-test plots are simple to carry out and provide a good alternative to more elaborate hazard and probability plots. These plots are scale invariant and, unlike probability plots, no special plotting papers are needed.

EXAMPLE 3.12

In a nuclear power plant, the times to failure (in hours) of certain pumps are recorded and given in the table below. Use the total-time-on-test plot to draw a conclusion about the time dependence of the failure rate function.

Solution: Using Equations 3.61 through 3.63, the following results are obtained:

i	t_i	$(n-i+1)\cdot(t_i - t_{i-1})$	$T(t_i)$	i/n	$\tilde{T}(t_i)$
1	1400	14,000	14,000	0.1	0.14
2	3500	18,900	32,900	0.2	0.33
3	5900	19,200	52,100	0.3	0.53
4	7600	11,900	64,000	0.4	0.65
5	8600	6000	70,000	0.5	0.71
6	9000	2000	72,000	0.6	0.73
7	11,600	10,400	82,400	0.7	0.83
8	12,000	1200	83,600	0.8	0.84
9	19,100	14,200	97,800	0.9	0.99
10	20,400	1300	99,100	1.0	1.00

continued

The graph indicates a mild tendency toward an IFR over the observation period (Figure 3.9).

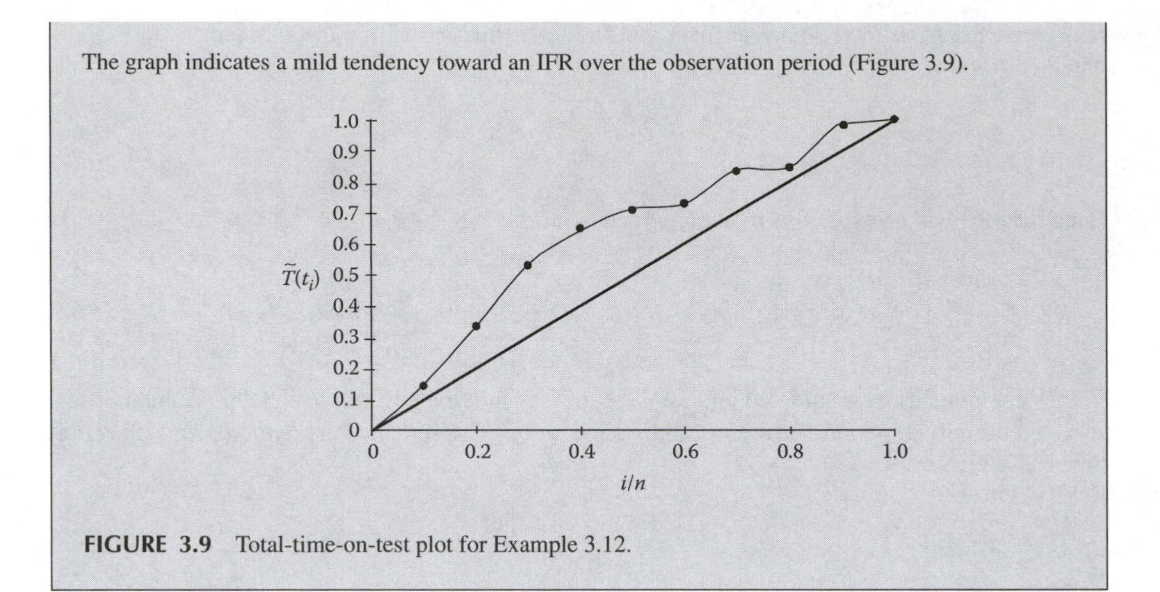

FIGURE 3.9 Total-time-on-test plot for Example 3.12.

3.4 MAXIMUM LIKELIHOOD ESTIMATION OF RELIABILITY DISTRIBUTION PARAMETERS

This section deals with statistical methods for estimating reliability distribution model parameters, such as parameter λ of the exponential distribution, μ and σ of the normal and lognormal distribution, p of the binomial distribution, and α and β of the Weibull distribution. The objective is to find a point estimate and a confidence interval for the parameters of interest. We briefly discussed this topic in Chapter 2 for estimating parameters and confidence intervals associated with a normal distribution. In this section, we expand the discussion to include estimation of parameters of other distributions useful to reliability analysis.

It is important to appreciate why we need to consider confidence intervals in this estimation process. In essence, we have a limited amount of information (e.g., on times to failure), and thus we cannot state our estimation with certainty. Therefore, the confidence interval is highly influenced by the amount of data available. Of course other factors, such as diversity and accuracy of the data sources and adequacy of the selected model, can also influence the state of our uncertainty regarding the estimated parameters. When discussing the goodness-of-fit tests, we dealt with uncertainty due to the adequacy of the model by using the concept of levels of significance. However, uncertainty due to the diversity and accuracy of the data sources is a much more difficult issue to deal with.

The methods of parameter estimation discussed in this section are more formal and accurate methods of determining distribution parameters than the methods described previously (such as the plotting methods).

Field data and data obtained from special life (reliability) tests can be used to estimate parameters of time-to-failure or failure-on-demand distribution. In life testing, a sample of components from a hypothesized population of such components is tested using the environment in which the components are expected to function, and their times to failure are recorded. In general, two major types of tests are performed. The first is testing *with replacement* of the failed items, and the second is testing *without replacement* of the failed items. The test with replacement is sometimes called *monitored testing*.

Samples of times to failure or times between failures (later, the term *time to failure* will be used wherever it does not result in a loss of generality) are seldom *complete* samples. A complete sample is one in which all items have failed during a test for a given observation period, and all the failure times are known (distinct). So far in this chapter, we have referred to complete data sets only. The likelihood function for a complete sample was introduced in Section 2.5. In the following sections, the likelihood functions for some types of censoring are discussed. Modern products are usually so reliable that a complete sample is a rarity, even in accelerated life testing. Thus, as a rule, reliability data are almost always incomplete.

3.4.1 Censored Data

3.4.1.1 Left and Right Censoring

Let N be the number of items in a sample, and assume that all units of the sample are tested simultaneously. If during the test period, T, only r units have failed, the failure times are known, and when the failed items are not replaced, the sample is called *singly censored on the right at* T. In this case, the only information we have about $N - r$ unfailed units is that their failure times are greater than the duration of the test T. Formally, an observation is called *right censored at* T, if the exact value of the observation is not known, but it is known that it is greater than or equal to T [20].

If a distinct failure time for an item is not known, but is known to be less than a given value, the failure time is called *left censored*. This type of censoring practically never appears in reliability data collection, and so it is not discussed. If the only information available is that an item is failed in an interval (e.g., between successive inspections), the respective data are called *grouped* or *interval* data. Such data were considered in Section 3.3.

It is important to understand the way—or *mechanism* by which—censored data are obtained. The basic discrimination is associated with *random* and *nonrandom* censoring, the simplest cases of which are discussed below.

3.4.1.2 Type I Censoring

Consider the situation of right censoring. If the test is terminated at a given nonrandom time, T, the number of failures, r, observed during the test period will be a r.v. These censored data are *Type I* or *time right singly censored* data, and the corresponding test is sometimes called *time-terminated*. For the general case, Type I censoring is considered under the following scheme of observations.

Let each unit in a sample of n units be observed during different periods of time $L_1, L_2, \ldots, L_n$. The time to failure of an individual unit, t_i, is a distinct value if it is less than the corresponding time period, that is, if $t_i < L_i$. Otherwise, t_i is the *time to censoring*, which indicates that the time to failure of the ith item is greater than L_i. This is the case of *Type I multiply censored* data; the case considered above is its particular case, when $L_1 = L_2 = \cdots = L_n = T$. Type I multiply censored data are quite common in reliability testing. For example, a test may start with a sample size of n but at some given times $L_1, L_2, \ldots, L_k$ $(k < n)$ the prescribed number of units can be deleted from (or placed on) the test.

Another example of multiply censored data is the miles-to-failure information on a fleet of vehicles that is observed for a given period of time. The mileage corresponding to failures of a particular component (e.g., the alternator) would be the distinct miles to failure. At the same time, the failure mileage of other components (e.g., the battery, connectors, or power distribution box) may be considered the miles to censoring.

3.4.1.3 Type II Censoring

A test may also be terminated when a nonrandom number of failures (say, r), specified in advance, occurs. In this case, the duration of the test number is a r.v. This situation is known as *Type II right censoring*.

Under Type II censoring, only the r smallest times to failure $t_{(1)} < t_{(2)} < \cdots < t_{(r)}$ out of the sample of N times to failure are distinct times to failure. The times to failure $t_{(i)}$ $(i = 1, 2, \ldots, r)$ are identically distributed r.v.'s (as in the previous case of the Type I censoring).

3.4.1.4 Types of Reliability Tests

When n components are placed on a life test, whether with replacement or not, it is sometimes necessary, due to the long life of certain components, to terminate the test and perform the reliability analysis based on the observed data up to the time of termination. There are two basic types of possible life test terminations. The first type is *time-terminated* (which results in Type I right censored data), and the second is *failure-terminated* (resulting in Type II right censored data). In the time-terminated life test, n units are placed on a test and the test is terminated after a predetermined time has elapsed. The number of components that failed during the test time and the corresponding time to failure of each component are recorded. In the failure-terminated life tests, n units are placed on test and the test is terminated when a predetermined number of component failures have occurred. The time to failure of each failed component, including the time of the last failure, is recorded.

Type I and Type II life tests can be performed with replacement or without replacement. Therefore, four types of life test experiments are possible. Each of these is discussed in more detail in the remainder of this section.

3.4.1.5 Random Censoring

Random censoring is typically used in reliability data analysis when there are several failure modes that must be estimated separately. The times to failure due to each failure mode are considered r.v.s having different distributions, while the whole system is called a *competing risks* (or *series*) system.

An example of this type of censoring is a car dealer collecting failure data during the warranty service of a given car model. The dealer and the car producer are interested in reliability estimations of the car model. A car in the sample under service can fail not only due to a generic failure cause, but also due to accidents or driver errors. In the latter cases, time-to-accident or time-to-human error must be treated as time to censoring. Note that the situation might be opposite. For example, some organizations (say, an insurance company) might be interested in studying the psychology of drivers of the same car model, so they would need to estimate an accident rate for the given model. In this case, from the data analysis point of view, the time to generic failure becomes time to censoring; meanwhile the time to accident becomes time to failure. Such dualism of censored data is also important for reliability databank development. Reliability estimation for this type of censoring is discussed in Section 3.5.

3.4.2 Exponential Distribution Point Estimation

3.4.2.1 Type I Life Test with Replacement

Suppose n components are placed on test with replacement (i.e., monitored), and the test is terminated after a specified time t_0. The accumulated (by both failed and unfailed components) time on test, T, (in hours or other time units), is given by

$$T = n\,t_0. \tag{3.64}$$

Time T is also called the *total time on test*. Equation 3.64 shows that at each time instant from the beginning of the test up to time t_0, exactly n components have been on test. Accordingly, if r failures have been observed up to time t_0, then assuming the exponential distribution, the maximum likelihood point estimate of the failure rate of the component can be found similar to that in the method used in Example 2.27, that is, as

$$\hat{\lambda} = \frac{r}{T}. \tag{3.65}$$

The corresponding MLE of the MTTF is given by

$$\widehat{\text{MTTF}} = \frac{T}{r}. \tag{3.66}$$

The number of units tested during the test, n', is

$$n' = n + r. \tag{3.67}$$

3.4.2.2 Type I Life Test without Replacement

Suppose n components are placed on test without replacement, and the test is terminated after a specified time t_0 during which r failures have occurred. The total time on test, T, for the failed and survived components is

$$T = \sum_{i=1}^{r} t_i + (n - r)t_0, \tag{3.68}$$

where $\sum_{i=1}^{r} t_i$ represents the accumulated time on test of the r failed components (r is random here), and $(n - r)t_0$ is the accumulated time on test of the surviving components. Using Equation 3.68, the failure rate and MTTF estimates can be obtained from Equations 3.65 and 3.66, respectively. Since no replacement has taken place, the total number of components tested during the test is $n' = n$.

3.4.2.3 Type II Life Test with Replacement

Consider a situation in which n components are being tested with replacement (i.e., monitored), and a component is replaced with an identical component as soon as it fails (except for the last failure). If the test is terminated after a time t_r when the rth failure has occurred (i.e., r is specified [nonrandom] but t_r is random), then the total-time-on-test, T, associated with failed and unfailed components, is given by

$$T = nt_r. \tag{3.69}$$

Note that t_r, unlike t_0, is a variable in this case. If the time to failure follows the exponential distribution, the maximum likelihood estimation of λ is

$$\hat{\lambda} = \frac{r}{T} \tag{3.70}$$

and the respective estimate of MTTF is

$$\widehat{\text{MTTF}} = \frac{T}{r}. \tag{3.71}$$

The total number of units tested, n', is

$$n' = n + r - 1, \tag{3.72}$$

where $(r - 1)$ is the total number of failed *and* replaced components. All failed components are replaced except the last one, because the test is terminated when the last component fails (i.e., the rth failure has been observed).

3.4.2.4 Type II Life Test without Replacement

Consider another situation when n components are being tested without replacement, that is, when a failure occurs, the failed component is not replaced by a new one. The test is terminated at time t_r when the rth failure has occurred (i.e., r is specified but t_r is random). The total time on test of both failed and unfailed components is obtained from

$$T = \sum_{i=1}^{r} t_i + (n - r)t_r, \tag{3.73}$$

where $\sum_{i=1}^{r} t_i$ is the accumulated time contribution from the failed components, and $(n-r)t_r$ is the accumulated time contribution from the survived components. Accordingly, the failure rate and MTTF estimates for the exponentially distributed time to failure can be obtained using Equations 3.73, 3.70, and 3.71. It should also be noted that the total number of units in this test is

$$n' = n, \tag{3.74}$$

since no components are being replaced.

EXAMPLE 3.13

Ten light bulbs are placed under life test. The test is terminated at $t_0 = 850$ h. Eight components fail before 850 h have elapsed. Determine the accumulated component hours and an estimate of the failure rate and MTTF for the following situations:

a. The components are replaced when they fail.
b. The components are not replaced when they fail.
c. Repeat a. and b., assuming the test is terminated when the eighth component fails.

The failure times obtained are 183, 318, 412, 432, 553, 680, 689, and 748.

Solution:

a. *Type I test*
Using Equation 3.64, $T = 10(850) = 8500$ component hours.
$\hat{\lambda} = 8/8500 = 9.4 \times 10^{-4}\,\text{h}^{-1}$, from Equation 3.58, $\widehat{\text{MTTF}} = 1062.5$ h.
b. *Type I test*
Using Equation 3.68, $\sum_{i=1}^{r} t_i = 4015$, $(n - r)t_0 = (10 - 8)850 = 1700$.
Thus, $T = 4015 + 1700 = 5715$ component hours.
$\hat{\lambda} = 8/5715 = 1.4 \times 10^{-3}\,\text{h}^{-1}$.
From Equation 3.66, $\widehat{\text{MTTF}} = 714.4$ h.
c. *Type II test*
Here, t_r is the time to the eighth failure, which is 748.
Using Equation 3.69, $T = 10(748) = 7480$ component hours.
From Equation 3.70, $\hat{\lambda} = 8/7480 = 1.1 \times 10^{-3}\,\text{h}^{-1}$.
From Equation 3.71, $\widehat{\text{MTTF}} = 935$ h.
Using Equation 3.73, $\sum_{i=1}^{r} T_i = 4015$, $(n - r)T_r = (10 - 8)748 = 1496$.
Thus, $T = 4015 + 1496 = 5511$ component hours.
From Equation 3.70, $\hat{\lambda} = 8/5511 = 1.5 \times 10^{-3}\,\text{h}^{-1}$.
From Equation 3.71, $\widehat{\text{MTTF}} = 688.8$ h.

A simple comparison of the results shows that although the same set of data is used, the effect of the type of the test and of the replacement of the failed units may be significant.

If there are censored units, a term should be added to Equations 3.64, 3.68, 3.69, and 3.73 to account for test time accumulated by the censored (or suspended) units. All other equations and terms remain the same.

3.4.3 EXPONENTIAL DISTRIBUTION INTERVAL ESTIMATION

In Example 2.27 and in Section 3.4.2, we discussed the maximum likelihood estimator for the parameter λ (failure rate or hazard rate) of the exponential distribution. This point estimator is $\hat{\lambda} = r/T$, where r is the number of failures observed and T is the total time on test. Epstein [21] has shown that if the time to failure is exponentially distributed with parameter λ, the quantity $2r\lambda/\hat{\lambda} = 2\lambda T$ has the χ^2 distribution with $2r$ degrees of freedom for the Type II censored data (failure-terminated test). Based on this information, one can construct the corresponding confidence intervals. Because uncensored data can be considered as the particular case of the Type II right censored data (when $r = n$), the same procedure is applicable to the complete (uncensored) sample.

Using the distribution of $2r\lambda/\hat{\lambda}$, one can write

$$\Pr\left[\chi^2_{\alpha/2}(2r) \le \frac{2r\lambda}{\hat{\lambda}} \le \chi^2_{1-\alpha/2}(2r)\right] = 1 - \alpha. \tag{3.75}$$

By rearranging and using $\hat{\lambda} = r/T$, the two-sided confidence interval for the true value of λ can be obtained:

$$\Pr\left[\frac{\chi^2_{\alpha/2}(2r)}{2T} \le \lambda \le \frac{\chi^2_{1-\alpha/2}(2r)}{2T}\right] = 1 - \alpha. \tag{3.76}$$

The corresponding upper confidence limit (the one-sided confidence interval) is

$$\Pr\left[0 \le \lambda \le \frac{\chi^2_{1-\alpha/2}(2r)}{2T}\right] = 1 - \alpha. \tag{3.77}$$

Accordingly, confidence intervals for MTTF and $R(t)$ at a time $t = t_0$ can also be obtained as one-sided and two-sided confidence intervals from Equations 3.76 and 3.77. The results are summarized in Table 3.2 [15].

TABLE 3.2
100(1 − α)% Confidence Limits on λ, MTTF, and $R(t_0)$

	Type I (Time Terminated Test)			
	One-Sided Confidence Limits		Two-Sided Confidence Limits	
Parameter	**Lower Limit**	**Upper Limit**	**Lower Limit**	**Upper Limit**
λ	0	$\frac{\chi^2_{1-\alpha}(2r+2)}{2T}$	$\frac{\chi^2_{\alpha/2}(2r)}{2T}$	$\frac{\chi^2_{1-\alpha/2}(2r+2)}{2T}$
MTTF	$\frac{2T}{\chi^2_{1-\alpha}(2r+2)}$	∞	$\frac{2T}{\chi^2_{1-\alpha/2}(2r+2)}$	$\frac{2T}{\chi^2_{\alpha/2}(2r)}$
$R(t_0)$	$\exp\left[-\frac{\chi^2_{1-\alpha}(2r+2)}{2T}t_0\right]$	1	$\exp\left[-\frac{\chi^2_{1-\alpha/2}(2r+2)}{2T}t_0\right]$	$\exp\left[-\frac{\chi^2_{\alpha/2}(2r)}{2T}t_0\right]$

	Type II (Failure Terminated Test)			
	One-Sided Confidence Limits		Two-Sided Confidence Limits	
Parameter	**Lower Limit**	**Upper Limit**	**Lower Limit**	**Upper Limit**
λ	0	$\frac{\chi^2_{1-\alpha}(2r)}{2T}$	$\frac{\chi^2_{\alpha/2}(2r)}{2T}$	$\frac{\chi^2_{1-\alpha/2}(2r)}{2T}$
MTTF	$\frac{2T}{\chi^2_{1-\alpha}(2r)}$	∞	$\frac{2T}{\chi^2_{1-\alpha/2}(2r)}$	$\frac{2T}{\chi^2_{\alpha/2}(2r)}$
$R(t_0)$	$\exp\left[-\frac{\chi^2_{1-\alpha}(2r)}{2T}t_0\right]$	1	$\exp\left[-\frac{\chi^2_{1-\alpha/2}(2r)}{2T}t_0\right]$	$\exp\left[-\frac{\chi^2_{\alpha/2}(2r)}{2T}t_0\right]$

As opposed to Type II censored data, the corresponding exact confidence limits for Type I censored data are not available. The approximate two-sided confidence interval for failure rate λ for Type I (time-terminated test) data usually is constructed as

$$\Pr\left[\frac{\chi^2_{\alpha/2}(2r)}{2T} \leq \lambda \leq \frac{\chi^2_{1-\alpha/2}(2r+2)}{2T}\right] = 1 - \alpha. \tag{3.78}$$

The respective upper confidence limit (a one-sided confidence interval) is given by

$$\Pr\left[0 \leq \lambda \leq \frac{\chi^2_{1-\alpha/2}(2r+2)}{2T}\right] = 1 - \alpha. \tag{3.79}$$

If no failure is observed during a test, the formal estimation gives $\hat{\lambda} = 0$, or MTTF $= \infty$. This cannot realistically be true, since we may have had a small or limited test. Had the test been continued, eventually a failure would have been observed. An upper confidence estimate for λ can be obtained for $r = 0$. However, the lower confidence limit cannot be obtained with $r = 0$. It is possible to relax this limitation by conservatively assuming that a failure occurs exactly at the end of the observation period. Then $r = 1$ can be used to evaluate the lower limit for the two-sided confidence interval. This conservative modification, although sometimes used to allow a complete statistical analysis, lacks firm statistical basis. Welker and Lipow [22] have investigated methods to determine approximate nonzero point estimates in these cases.

Example 3.14

Twenty-five units are placed on a reliability test that lasts 500 h. In this test, eight failures occur at 75, 115, 192, 258, 312, 389, 410, and 496 h. The failed units are replaced. Find $\hat{\lambda}$, one-sided and two-sided confidence intervals for λ, and MTTF at the 90% confidence level; and one-sided and two-sided 90% confidence intervals for reliability at $t_0 = 1000$ h.

Solution: This is a Type I test. The accumulated time T is obtained from Equation 3.64

$$T = 25(500) = 12{,}500\text{ h}.$$

The point estimate of failure rate is

$$\hat{\lambda} = 8/12{,}500 = 6.4 \times 10^{-4}\text{ h}^{-1}.$$

One-sided confidence interval for λ is

$$0 \leq \lambda \leq \frac{\chi^2(2 \times 8 + 2)}{2 \times 12{,}500}.$$

From Table A.3,

$$\chi^2_{0.9}(18) = 25.99, \quad 0 \leq \lambda \leq 1.04 \times 10^{-3}\text{ h}^{-1}.$$

Two-sided confidence interval for λ is

$$\frac{\chi^2_{0.05}(2 \times 8)}{2 \times 12{,}500} \leq \lambda \leq \frac{\chi^2_{0.95}(2 \times 8 + 2)}{2 \times 12{,}500}.$$

From Table A.3,

$$\chi^2_{0.05}(16) = 7.96 \quad \text{and} \quad \chi^2_{0.95}(18) = 28.87.$$

Thus,

$$3.18 \times 10^{-4}\text{ h}^{-1} \leq \lambda \leq 1.15 \times 10^{-3}\text{ h}^{-1}.$$

One-sided 90% confidence interval for $R(1000)$ is

$$\exp[(-1.04 \times 10^{-3})(1000)] \le R(1000) \le 1$$

or

$$0.35 \le R(1000) \le 1.$$

Two-sided 90% confidence interval for $R(t)$ is

$$\exp[(-1.15 \times 10^{-3})(1000)] \le R(1000) \le \exp[(-3.18 \times 10^{-4})(1000)]$$

or

$$0.32 \le R(1000) \le 0.73.$$

3.4.4 Lognormal Distribution

The lognormal distribution is commonly used to represent the occurrence of certain events in time or space whose values span more than one order of magnitude. For example, a r.v. representing the length of time interval required for repair of hardware follows a lognormal distribution. Because the lognormal distribution has two parameters, parameter estimation poses a more challenging problem than for the exponential distribution. Taking the natural logarithm of the data, the analysis is reduced to the case of the normal distribution, so that the point estimates for the two parameters of the lognormal distribution for a complete sample of size n can be obtained from

$$\hat{\mu}_t = \sum_{i=1}^{n} \frac{\ln t_i}{n}, \tag{3.80}$$

$$\hat{\sigma}_t^2 = \frac{\sum_{i=1}^{n} (\ln t_i - \hat{\mu}_t)^2}{n-1}. \tag{3.81}$$

The confidence interval for μ_t is given by

$$\Pr\left[\hat{\mu}_t - \frac{\hat{\sigma}_t t_{\alpha/2}}{\sqrt{n}} \le \mu_t \le \hat{\mu}_t + \frac{\hat{\sigma}_t t_{\alpha/2}}{\sqrt{n}}\right] = 1 - \alpha. \tag{3.82}$$

The respective confidence interval for σ_t^2 is:

$$\Pr\left[\hat{\mu}_t - \frac{\hat{\sigma}_t^2 (n-1)}{\chi^2_{1-\alpha/2}(n-1)} \le \sigma_t^2 \le \hat{\mu}_t + \frac{\hat{\sigma}_t^2 (n-1)}{\chi^2_{\alpha/2}(n-1)}\right] = 1 - \alpha. \tag{3.83}$$

In the case of censored data, the corresponding statistical estimation turns out to be much more complicated, see [9] and [20] for a comprehensive treatment of these cases. Below we give a cursory overview of cases involving censored data.

If right censored data exist, then the maximum likelihood estimation of parameters and are obtained by maximizing the likelihood functions as

$$l = \prod_{i=1}^{r} f(t_i, \Theta) \prod_{j=1}^{n-r} [1 - F(t_j, \Theta)], \tag{3.84}$$

where Θ is the vector of unknown parameters to be estimated, n is the total number of units, and r is the number of units that failed. So the number of censored unit is $(n - r) \cdot f(\cdot)$ is the pdf of time to failure and $F(\cdot)$ is the unreliability or cumulative probability of failure. For lognormal distribution, the log likelihood function and its derivatives take the following forms:

$$L = \ln l = \sum_{i=1}^{n} \ln f(t_i, \Theta) - \sum_{j=1}^{n-r} \ln F(t_j, \Theta), \tag{3.85}$$

$$\frac{1}{\hat{\sigma}_t^2} \sum_{i=1}^{r} [\ln(t_i) - \hat{\mu}_t] + \frac{1}{\hat{\sigma}_t} \sum_{j=1}^{n-r} \frac{\phi[\ln(t_j) - \hat{\mu}_t/\hat{\sigma}_t]}{1 - \Phi[\ln(t_j) - \mu_t/\sigma_t]} = 0, \tag{3.86}$$

$$\sum_{i=1}^{r} \left\{ \frac{[\ln(t_i) - \hat{\mu}_t]^2}{\hat{\sigma}_t^3} - \frac{1}{\hat{\sigma}_t} \right\} + \frac{1}{\hat{\sigma}_t} \sum_{j=1}^{n-r} \frac{[\ln(t_j) - \hat{\mu}_t/\hat{\sigma}_t]\phi[\ln(t_j - \hat{\mu}_t)/\hat{\sigma}_t]}{1 - \Phi[\ln(t_j - \hat{\mu}_t)/\hat{\sigma}_t]} = 0, \tag{3.87}$$

where n is the total number of units observed under test or from the field, r is the number of failed units (i.e., $n - r$ is the number of censored units), t_i is the time of failure, t_j is the time of right censoring, and

$$\phi(x) = \frac{1}{2\sqrt{2\pi}} \mathrm{e}^{-(1/2)x^2}, \tag{3.88}$$

$$\Phi(x) = \frac{1}{\sqrt{2\pi}} \int_{-\infty}^{x} \mathrm{e}^{-(1/2)x^2}\, \mathrm{d}t. \tag{3.89}$$

Solving the system of two equations with two unknowns $\hat{\mu}_t$ and $\hat{\sigma}_t$ in Equations 3.86 and 3.87 requires a numerical solution. For estimation of confidence intervals, refer to Equations (2.120 through 2.128).

3.4.5 Weibull Distribution

The Weibull distribution can be used for data that are assumed to be from IFR, DFR, or constant failure rate distributions. Similar to the lognormal distribution, the Weibull distribution is a two-parameter distribution, and its estimation, even in the case of complete (uncensored) data, is not a trivial problem.

It can be shown that, in the special situation when r units fail out of n units placed on test or under observation for time t_c, the MLEs of β and α parameters of the Weibull distribution (Equation 3.22) can be obtained as a solution for the following system of nonlinear equations if censored data exist:

$$\frac{\sum_{i=1}^{r} (t_i)^{\hat{\beta}} \ln t_i + (n-r) t_c^{\hat{\beta}} \ln t_c}{\sum_{i=1}^{r} (t_i)^{\hat{\beta}} + (n-r) t_c^{\hat{\beta}}} - \frac{1}{\hat{\beta}} = \frac{1}{r} \sum_{i=1}^{r} \ln t_i \tag{3.90}$$

and

$$\hat{\alpha} = \left(\frac{\sum_{i=1}^{r} (t_i)^{\hat{\beta}}}{n} + (n-r) t_c^{\hat{\beta}} \right)^{1/\hat{\beta}}. \tag{3.91}$$

This system can be solved using an appropriate numerical procedure. The corresponding confidence estimation is also complicated. See [3,9,23,24] for further discussions. In Chapter 5, we will discuss another form of estimating the Weibull distribution parameters.

If right censored data exists, then the MLEs of parameters α and β are obtained by solving the following two maximum likelihood derivative functions:

$$\frac{r}{\hat{\beta}} + \sum_{i=1}^{r} \ln\left(\frac{t_i}{\hat{\alpha}}\right) - \sum_{i=1}^{r}\left(\frac{t}{\hat{\alpha}}\right)^{\hat{\beta}} \ln\left(\frac{t_i}{\hat{\alpha}}\right) - \sum_{j=1}^{n-r}\left(\frac{t_j}{\hat{\alpha}}\right)^{\hat{\beta}} \ln\left(\frac{t_j}{\hat{\alpha}}\right) = 0, \quad (3.92)$$

$$\frac{r\hat{\beta}}{\hat{\alpha}} + \frac{\hat{\beta}}{2}\sum_{i=1}^{r}\left(\frac{t_i}{\hat{\alpha}}\right)^{\hat{\beta}} \ln\left(\frac{t_i}{\hat{\alpha}}\right) + \frac{\hat{\beta}}{\alpha}\sum_{j=1}^{n-r}\left(\frac{t_j}{\hat{\alpha}}\right)^{\hat{\beta}} = 0, \quad (3.93)$$

where n is the total number of units observed under test or from the field, and r is the number of failed units. Thus, $n - r$ would be the number of censored units, t_i the time that unit i failed, and t_j the time that unit j was right censored. Two Equations 3.92 and 3.93 can be solved for the two unknowns $\hat{\alpha}$ and $\hat{\beta}$.

The confidence interval of the $\hat{\alpha}, \hat{\beta}$ parameters of the distributions, especially for censored data, may be obtained from the so-called Fisher Information Matrix as follows:

$$\hat{\beta}e^{-K_\alpha\sqrt{\text{var}(\hat{\beta})}/\hat{\beta}} \le \beta \le \hat{\beta}e^{K_\alpha\sqrt{\text{var}(\hat{\beta})}/\hat{\beta}}, \quad (3.94)$$

$$\hat{\alpha} - K_\alpha\sqrt{\text{var}(\hat{\alpha})} \le \alpha \le \hat{\alpha} + K_\alpha\sqrt{\text{var}(\hat{\alpha})}, \quad (3.95)$$

where $\text{var}(\hat{\beta})$ and $\text{var}(\hat{\alpha})$ can be found from the Fisher Information Matrix below, and $K_\alpha = 1/\sqrt{2\pi}\int_{K_\alpha}^{\infty} e^{-t^2/2}\,dt$ is the desirable confidence level.

$$\begin{bmatrix} \text{var}(\hat{\beta}) & \text{cov}(\hat{\alpha}, \hat{\beta}) \\ \text{cov}(\hat{\beta}, \hat{\alpha}) & \text{var}(\hat{\alpha}) \end{bmatrix} = \begin{bmatrix} -\dfrac{\partial^2 L}{\partial \beta^2} & -\dfrac{\partial^2 L}{\partial \alpha \partial \beta} \\ -\dfrac{\partial^2 L}{\partial \beta \partial \alpha} & -\dfrac{\partial^2 L}{\partial \alpha^2} \end{bmatrix}^{-1}. \quad (3.96)$$

Note that L is the log likelihood function as described in Equation 3.85 with the pdf and cdf having the form of the Weibull distribution, $f(t_i; \alpha, \beta)$ and $F(t_i; \alpha, \beta)$, respectively.

Example 3.15

Estimate the parameters of a lognormal distribution and a Weibull distribution.

a. Estimate the two parameters of the lognormal distribution for the following time-to-failure data (with no censored units). Now assume that at $t = 3000$, five units were still operating without failure (i.e., they are censored). Repeat the estimation of the two parameters and compare the difference.
Solution:

Unit No.	1	2	3	4	5	6
Time of Failure	144	385	747	1144	1576	2612

Using Equations 3.86 and 3.87

$$\hat{\mu}_t = 6.635, \quad \hat{\sigma}_t = 0.953.$$

Accordingly, using Equations 2.70 and 2.71, the actual mean and standard deviation is $\hat{\mu} = 1198.8$ and $\hat{\sigma} = 1458.4$ h, respectively.

continued

Repeating the use of Equations 3.86 and 3.87 with five censored units at 3000 h, the new estimates would be:

$$\hat{\mu}_t = 7.800, \quad \hat{\sigma}_t = 0.158.$$

The solution requires iterative numerical calculations such as a Monte Carlo simulation.

b. Using the data given in Example 3.9, obtain the maximum likelihood estimators for the parameters of a Weibull distribution. Then assume that three units have been censored at 192, 323, and 685 h. What are the new estimates for the parameters $\hat{\alpha}$ and $\hat{\beta}$ and their 90% confidence intervals?

Solution: Using the numerical procedure, systems 3.92 and 3.93 are solved, which results in estimates of $\hat{\beta} = 2.092$, $\hat{\alpha} = 395.919$ h. A comparison of these results with the plot from Example 3.9 is reasonable, but it illustrates the approximate nature of these data analysis methods and demonstrates the importance of using more than one of the methods discussed. For the second part, the MLE Equations 3.92 and 3.93 need to be solved again considering the three censored data in addition to the failures. The iterative numerical solution results in estimates of $\hat{\alpha}$ and $\hat{\beta}$ as well as their 90% confidence intervals as follows:

Parameter	**5%**	**Mean**	**95%**
$\hat{\alpha}$	355.203	463.296	604.282
$\hat{\beta}$	1.296	1.969	2.992

3.4.6 Binomial Distribution

When the data are in the form of failures occurring on demand, that is, x failures observed in n trials, there is a constant probability of failure (or success), and the binomial distribution can be used as an appropriate model. This is often the situation for standby components (or systems). For instance, a redundant pump is demanded for operation n times in a given period of test or observation.

The best estimator for p is given by the formula

$$\hat{p} = \frac{x}{n}. \tag{3.97}$$

The lower and the upper confidence limits for p can be found, using the so-called Clopper–Pearson procedure see [9]:

$$p_l = \{1 + (n - x + 1)x^{-1}F_{1-\alpha/2}[2n - 2x + 2; 2x]\}^{-1}, \tag{3.98}$$

$$p_u = \{1 + (n - x)\{(x + 1)F_{1-\alpha/2}[2x + 2; 2n - 2x]\}^{-1}\}^{-1}, \tag{3.99}$$

where $F_{(1-\alpha/2)}(f_1; f_2)$ is the $(1 - \alpha/2)$ quantile [or the $100(1 - \alpha/2)$ percentiles] of the F-distribution with f_1 degrees of freedom for the numerator, and f_2 degrees of freedom for the denominator. Table A.5 contains some percentiles of the F-distribution. As was mentioned in Chapter 2, the Poisson distribution can be used as an approximation for the binomial distribution when the parameter, p, of the binomial distribution is small and the parameter n is large, for example, $x < n/10$, which means that approximate confidence limits can be constructed using Equation 3.78 with $r = x$ and $T = n$.

EXAMPLE 3.16

An emergency pump in a nuclear power plant is in a standby mode. There have been 563 start tests for the pump, and only three failures have been observed. No degradation in the pump's physical characteristics or changes in operating environment are observed. Find the 90% confidence interval for the probability of failure per demand.

Solution: Denote $n = 563$, $x = 3$. Using Equations 3.85 through 3.87, find $\hat{p} = 3/563 = 0.0053$.

$$p_l = \{1 + (563 - 3 + 1)/3F_{0.95}(2 \times 563 - 2 \times 3 + 2; 2 \times 3)\}^{-1} = 0.0014,$$

where $F_{0.95}(1122; 6) = 3.67$ from Table A.5.

Similarly,

$$p_u = \{1 + (563 - 3)\{(3 + 1)F_{0.95}(2 \times 3 + 2; 2 \times 563 - 2 \times 3\}^{-1}\}^{-1} = 0.0137.$$

Therefore,

$$\Pr(0.0014 \leq p \leq 0.0137) = 90\%.$$

EXAMPLE 3.17

In a commercial nuclear plant, the performance of the emergency diesel generators has been observed for about 5 years. During this time, there have been 35 real demands with four observed failures. Find the 90% confidence limits and point estimate for the probability of failure per demand. What would the error be if we used Equation 3.78 instead of Equations 3.98 and 3.99 to solve this problem?

Solution: Here, $x = 4$ and $n = 35$. Using Equation 3.97,

$$\hat{p} = \frac{4}{35} = 0.114.$$

To find lower and upper limits, use Equations 3.98 and 3.99. Thus,

$$p_l = \{1 + (35 - 4 + 1)/4F_{0.95}(2 \times 35 - 2 \times 4 + 2; 2 \times 4)\}^{-1} = 0.04.$$

$$p_u = \{1 + (35 - 4)\{(4 + 1)F_{0.95}(2 \times 4 + 2; 2 \times 35 - 2 \times 4\}^{-1}\}^{-1} = 0.243.$$

If we used Equation 3.78,

$$p_l = \frac{\chi^2_{0.05}(8)}{2 \times 35} = \frac{2.733}{70} = 0.039.$$

$$p_u = \frac{\chi^2_{0.95}(10)}{2 \times 35} = \frac{18.31}{70} = 0.262.$$

The error due to this approximation is

$$\text{Lower limit error} = \frac{|0.04 - 0.039|}{0.04} \times 100 = 2.5\%.$$

$$\text{Upper limit error} = \frac{|0.243 - 0.262|}{0.243} \times 100 = 7.8\%.$$

Note that this is not a negligible error, and Equation 3.78 should not be used. Since $x > n/10$, Equation 3.78 is not a good approximation.

3.5 CLASSICAL NONPARAMETRIC DISTRIBUTION ESTIMATION

Based on Section 3.1, any reliability measure or index can be expressed in terms of time-to-failure cdf or reliability function. Thus, the problem of estimating these functions is of great importance. The commonly used estimate of cdf is the *empirical (or sample) distribution function* (edf) introduced for uncensored data in Chapter 2 (see Equation 2.130). In this section, we consider some other nonparametric point and confidence distribution estimation procedures applicable for censored data.

3.5.1 CONFIDENCE INTERVALS FOR cdf AND RELIABILITY FUNCTION FOR COMPLETE AND SINGLY CENSORED DATA

The construction of an edf requires a complete sample but can also be constructed for the right censored samples for the failure times, which are less than the last time to failure observed ($t < t_{(r)}$). The edf is a random function, since it depends on the sample units. For any given point, t, the edf, $S_n(t)$, is the fraction of sample items that failed before t.

The edf is, in a sense, the estimate of the probability, p, in a binomial trial, and this probability is $p = F(t)$. Note that it is easy to show that the maximum likelihood estimator of the binomial parameter p coincides with $S_n(t)$, and $S_n(t)$ is a consistent estimator of the cdf, $F(t)$.

Using relationship 3.2 between the cdf, the reliability function, and the edf (Equation 2.130), one can obtain the respective estimate of the reliability function. This estimate, called the *empirical (or sample) reliability function,* is given by

$$R_n(t) = \begin{cases} 1, & 0 < t < t_{(1)}; \\ 1 - \frac{i}{n}, & t_{(i)} \le t < t_{(i+1)},\ i = 1, 2, \ldots, n-1; \\ 0, & t_{(n)} \le t < \infty; \end{cases} \tag{3.100}$$

here $t_{(1)} < t_{(2)} < \cdots < t_{(n)}$ are the ordered sample data (the so-called, order statistics).

The mean number of failures observed during time, t, is $E(r) = pn = F(t)n$, and so the mean value of the fraction of sample items failed before t is $E(r/n) = p = F(t)$. The variance of this fraction is given by

$$\mathrm{var}\left(\frac{r}{n}\right) = \frac{p(1-p)}{n} = \frac{F(t)[1-F(t)]}{n}. \tag{3.101}$$

For practical problems, Equation 3.101 is used, replacing $F(t)$ with $S_n(t)$. As the sample size, n, increases, the binomial distribution can be approximated by a normal distribution (consistent with the discussion in Chapter 2) with the same mean and variance [i.e., $\mu = np, \sigma^2 = np(1-p)$], which provides reasonable results if both np and $n(1-p)$ are ≥ 5. Using this approximation, the following $100(1-\alpha)\%$ confidence interval for the unknown cdf, $F(t)$, at any given point t can be constructed as:

$$S_n(t) - z_{1-\alpha/2}\sqrt{\frac{S_n(t)[1-S_n(t)]}{n}} \le F(t) \le S_n(t) + z_{1-\alpha/2}\sqrt{\frac{S_n(t)[1-S_n(t)]}{n}}, \tag{3.102}$$

where z is the quantile of level α of the standard normal distribution.

The corresponding estimates for the reliability (survivor) function can be obtained using Equation 3.2:

$$R_n(t) = 1 - S_n(t).$$

Example 3.18

Using the data from Example 3.9, find the point nonparametric estimate and the 95% confidence interval for the cdf, $F(t)$, for $t = 350$ h.

Solution: Using Equation 2.130 the point estimate for $F(350)$ is $S_n(350) = 5/10$ (note that we have five observations out of 10 that are <350 h).

The respective approximate 95% confidence interval based on Equation 3.102 is

$$0.5 - 1.96\sqrt{\frac{0.5(1-0.5)}{10}} \leq F(350) \leq 0.5 + 1.96\sqrt{\frac{0.5(1-0.5)}{10}}.$$

Therefore,

$$\Pr(0.1900 < F(350) < 0.6099) = 0.95.$$

Using a complete or right censored sample from an unknown cdf, one can also obtain the strict *confidence intervals for the unknown cdf, $F(t)$*. This can be done using the same Clopper–Pearson procedure for constructing the confidence intervals for a binomial parameter p, that is, using Equations 3.98 and 3.99. These limits can also be expressed in more compact form in terms of the so-called incomplete beta function as follows.

The lower confidence limit, $F_l(t)$, at the point t where $S_n(t) = r/n\,(r = 0, 1, 2, \ldots, n)$, is the largest value of p that satisfies the following inequality

$$I_p(r, n - r + 1) \leq \frac{\alpha}{2} \tag{3.103}$$

and the upper confidence limit, $Fu(t)$, at the same point is the smallest p satisfying the inequality

$$I_{1-p}(n - r, r + 1) \leq \frac{\alpha}{2}, \tag{3.104}$$

where $I_t(\alpha, \beta)$ is the incomplete beta function, which was introduced in Chapter 2 as the cdf of the beta distribution (Equation 2.77). The incomplete beta function is difficult to tabulate; however, its numerical approximation is available in the literature (e.g., see http://functions.wolfram.com/).

Example 3.19

For the data from Example 3.18 find the strict 95% confidence interval for the cdf, $F(t)$, for $t = 350$ h, using Equations 3.103 and 3.104.

Solution: Using Equation 3.103, the lower confidence limit is found from

$$I_p(5, 10 - 5 + 1) \leq 0.025 \text{ as } 0.1871$$

and, using Equation 3.104, the upper confidence limit is found from

$$I_{1-p}(5, 6) \leq 0.025 \text{ as } 0.8131.$$

Therefore, the strict confidence interval is

$$\Pr(0.1871 < F(350) < 0.8131) = 0.95,$$

which is reasonably close to the approximate interval obtained in the previous example.

Another typical reliability estimation problem that can be solved using this nonparametric approach is the estimation of the lower confidence limit for the reliability function using the same type of data. This can be accomplished using Equation 3.104, in which $1 - p = 1 - F(t)$ is replaced by the reliability function, $R(t)$. Accordingly, one obtains

$$I_R(n - r, r + 1) \leq \alpha. \tag{3.105}$$

This procedure is illustrated by the following example.

EXAMPLE 3.20

Twenty-two identical components were tested over 1000 h, and no failure was observed. Find the lower confidence limit, $R_1(t)$, for the reliability function at $t = 1000$ h and for confidence probability $1 - \alpha = 0.90$.

Solution: We need to find the largest value of R satisfying Equation 3.105. For the problem considered $\alpha = 0.1$, $r = 0$, and $n = 22$; therefore,

$$I_R(22, 1) \leq 0.1.$$

Solving for R, one obtains $R_l \approx 0.90$.

Another possible application of Equation 3.105 is for reliability demonstration tests when the lower $1 - \alpha$ confidence limit for reliability, R_1, is given together with acceptable number of failures, r, during the test duration. The problem is to find the sample size, n, to be tested with the number of failures not exceeding r, so that quantities n, r, and R_1 would satisfy Equation 3.105. It should be noted that for given R_1, r, and α, relationship 3.105 cannot be satisfied for any sample size n. For given values of R_1 and α, the minimum sample size for which Equation 3.105 can be satisfied corresponds to $r = 0$. This is illustrated by the following example.

EXAMPLE 3.21

The reliability test on an automotive component must demonstrate the lower limit of reliability of 86% with 90% confidence. Assuming that no failure is tolerated, what sample size has to be tested to satisfy the above requirements?

Solution: When $r = 0$, Equation 3.105 can be rewritten in the following simple form:

$$(R_1)^n = \alpha,$$

from which the required sample size is obtained as follows.

$$n = \frac{\ln(\alpha)}{\ln R_1} = \frac{\ln(0.1)}{\ln(0.86)} = 15.$$

3.5.2 CONFIDENCE INTERVALS FOR cdf AND RELIABILITY FUNCTION FOR MULTIPLY CENSORED DATA

The point and confidence estimates considered so far do not apply to multiply censored data. For such samples, the *Kaplan–Meier* or *product-limit* estimate, which is the MLE of the cdf, can be used.

Suppose we have a sample of n times to failure, among which only k failure times are distinct times to failure. Denote these ordered times (order statistics) as $t_{(1)} \leq t_{(2)} \leq \cdots \leq t_{(k)}$, and let $t_{(0)}$ be equal to zero; that is, $t_{(0)} \equiv 0$. Let n_j be the number of items under observation just before $t_{(j)}$. Assume that the time to failure pdf is continuous, so that there is only one failure at every $t_{(i)}$. Then, $n_{j+1} = n_j - 1$. Under these conditions, the product limit estimate is given by

$$S_n(t) = 1 - R_n(t) = \begin{cases} 0, & 0 \leq t < t_{(1)}, \\ 1 - \prod_{j=1}^{i} \frac{n_j - 1}{n_j}, & t_{(i)} \leq t < t_{(i+1)}, \quad i = 1, 2, \ldots, m-1, \\ 1, & t \geq t_{(m)}, \end{cases} \tag{3.106}$$

where integer $m = k$, if $k < n$, and $m = n$, if $k = n$.

It is possible to show that for uncensored (complete) data samples, the product limit estimate coincides with the edf given by Equation 2.130. In the general case (including discrete distribution, censored or grouped data), the Kaplan–Meier estimate is given by

$$S_n(t) = 1 - R_n(t) = \begin{cases} 0, & 0 \leq t < t_{(1)}; \\ 1 - \prod_{j=1}^{i} \frac{n_j - d_j}{n_j}, & t_{(i)} \leq t < t_{(i+1)}, \quad i = 1, 2, \ldots, m-1; \\ 1, & t \geq t_{(m)}, \end{cases} \tag{3.107}$$

where d_j is the number of failures at $t_{(j)}$.

For estimation of variance of S_n (or R_n), Greenwood's formula [20] is used:

$$\text{var}[\hat{S}_n(t)] = \text{var}[\hat{R}_n(t)] = \sum_{j:t_{(j)}<t} \frac{d_j}{n_j(n_j - d_j)}. \tag{3.108}$$

Combining the product-limit estimate and Greenwood's formula into equations similar to Equation 3.102, one can construct the corresponding approximate confidence limits for the reliability or the cdf of interest.

If the data are shown in intervals, then it is possible to use an adjusted number of units at risk, assuming that censored time occurs uniformly over the interval. If F_i = number of failures in the ith interval, C_i = number of censored units in the ith interval, and H_i = number of units at risk at $t_{i-1} = H_{i-1} - F_{i-1} - C_{i-1}$, then the unconditional probability of survival is $H_i' = H_i - (C_i/2)$. Similarly, the conditional probability of a unit failure in the ith interval given survival to time t_{i-1} is (F_i/H_i'), and conditional probability of survival would be $p_i = 1 - (F_i/H_i')$. As such, the unconditional probability of survival is $\hat{R}_i = p_i \times \hat{R}_{i-1}$ with $R_0 = 1$.

EXAMPLE 3.22

The table below shows censored test data of a mechanical component (censored data points are shown with *). Find the Kaplan–Meier estimate of the time-to-failure cdf. Plot the data on a Weibull paper and estimate the parameters of the distribution. Use Equations 3.92 and 3.93 to estimate the parameters through maximum likelihood estimation.

continued

i	Ordered Failure Time, $t_{(i)}$, or Time to Censoring, $t^*_{(i)}$	$S_n[t_{(i)}]$
0	0	0
1	32	$1 - 15/16 = 0.06255$
2	41	$1 - (15/16)(14/15) = 0.125$
3	58	$1 - (15/16)(14/15)(13/14) = 0.187$
4	64*	0.187 (65.0 is time to censoring)
5	66	$1 - (13/16)[(13 - 1 - 1)/12] = 1 - 0.745 = 0.255$
6	72	$1 - 0.745(10/11) = 0.323$
7	74*	0.323 (because 75 is time to censoring)
8	76*	0.323
9	83*	0.323
10	88*	0.323
11	92*	0.323
12	100*	0.323
13	104	0.492
14	108*	0.492
15	109	0.746
16	121*	0.746

Solution: The second column contains the product-limit estimate of the cdf as a function of time, calculated using Equation 3.106. To make the calculations clear, the detailed calculations for the first seven lines of the table are also given. The MLE of the parameters is then calculated using the probability plot of the data.

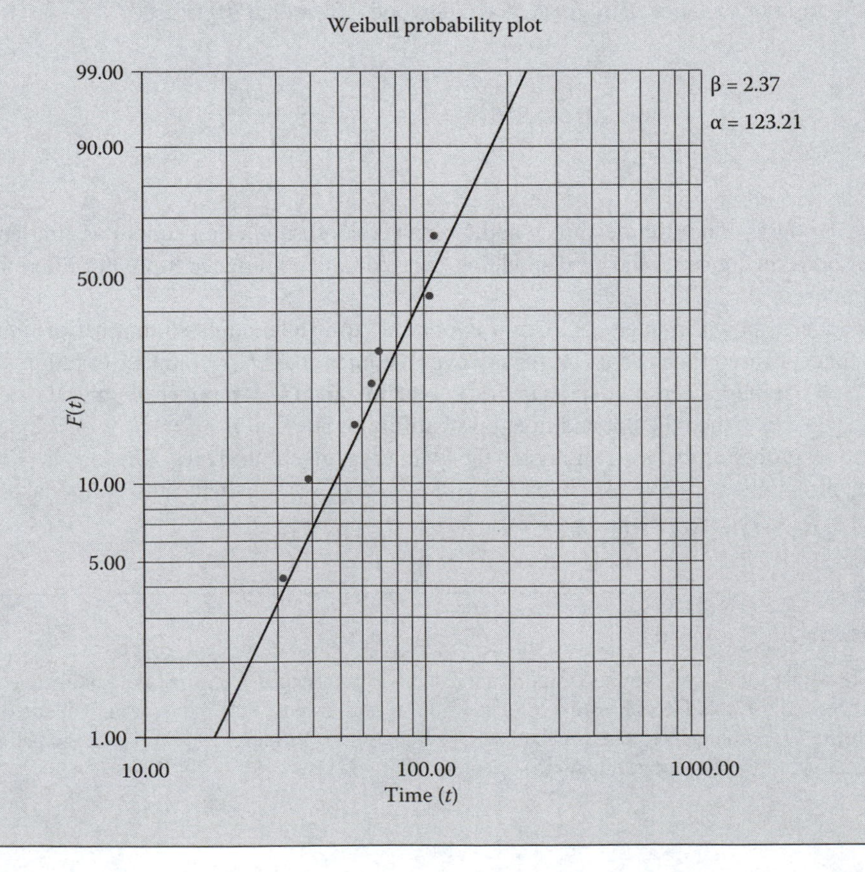

Another useful method for estimating cdf and reliability function for multiply censored data is the method of rank adjustment. In this case, if n units are observed, then the rank adjustment is computed using

$$i_{t_i} = i_{t_{i-1}} + \text{rank increment}.$$

In this case, rank increment is computed from $(n+1) - i_{t_{i-1}}/1 + m$, where m is the number of units survived, including and beyond the ith unit. A plot of the rank order against any plotting position point, such as the median rank or $(i_{t_{i-1}} - 0.375/n + 0.25)$, provides the basis for estimating the reliability function, cdf, and their parameters.

Example 3.23

Consider the failure cycles in a low-cycle fatigue test shown in the table below. Using the rank adjustment method, find the corresponding two-parameter Weibull cdf of cycles to failure and associated parameters.

i	1	2	3	4	5	6	7	8	9	10
c_i (cycles)	150	340*	560	800	1130*	1720	2470*	4210*	5230	6890

Solution:
Evaluation of the data and plot of $i_{c_{i-1}}$ versus $(i_{c_{i-1}} - 0.375/n + 0.25)$ are shown below.

i	c_i (cycles)	Rank Increments	$i_{c_{i-1}}$	Rank Order $\frac{i_{c_{i-1}} - 0.375}{n + 0.25}$
1	150	1	1	6.1
2	340*	—	—	—
3	560	$(11-1)/(1+8) = 1.111$	$1 + 1.111 = 2.111$	16.9
4	800	$(11-2.111)/(1+7) = 1.111$	$2.111 + 1.111 = 3.222$	27.8
5	1130*	—	—	—
6	1720	$(11-3.222)/(1+5) = 1.2963$	$3.222 + 1.2963 = 4.512$	40.4
7	2470*	—	—	—
8	4210*	—	—	—
9	5230	$(11-4.518)/(1+2) = 2.16$	$4.518 + 2.160 = 6.679$	61.5
10	6890	$(11-6.679)/(1+1) = 2.16$	$6.679 + 2.160 = 8.839$	82.6

3.6 BAYESIAN ESTIMATION PROCEDURES

In Section 3.5, we discussed the importance of quantifying uncertainties associated with the estimates of various distribution parameters. In the preceding sections, we also discussed formal statistical methods for quantifying the uncertainties. Namely, we discussed the concept of confidence intervals, which, in essence, are a statistical treatment of available information. The more data available, the more confident and accurate the statistical inference is. For this reason, the statistical approach is sometimes called the *frequentist* method of treatment. In the framework of the Bayesian approach, the parameters of interest are treated as r.v.s, the true values of which are unknown. Thus, a distribution can be assigned to represent the parameter; the mean (or, for some cases, the median) of the distribution can be used as an estimate of the parameter of interest. The pdf of a parameter in Bayesian terms can be obtained from a prior and posterior pdf.

In practice, however, the prior pdf is used to represent the relevant prior knowledge, including subjective judgment regarding the characteristics of the parameter and its distribution. When the prior knowledge is combined with other relevant information (often statistics obtained from tests and observations), a posterior distribution is obtained, which better represents the parameter of interest. Since the selection of the prior and the determination of the posterior often involve subjective judgments, the Bayesian estimation is sometimes called the subjectivist approach to parameter estimation.

The basic concept of Bayes's theorem was discussed in Chapter 2. In essence, to estimate parameters of interest, such as parameters of a pdf model, this theorem can be written in one of three forms: discrete, continuous, or mixed. Martz and Waller [15] have significantly elaborated on the concept of the Bayesian technique and its application to reliability analysis. The discrete form of Bayes's theorem was discussed in Section 2.2.3. The continuous and mixed forms, which are the common forms used for parameter estimation in reliability and risk analysis, are briefly discussed below.

Let θ be a parameter of interest. It can be a parameter of a time-to-failure distribution or a reliability index, such as MTTF, failure rate, and so on. Suppose parameter θ is a continuous r.v., so that the prior and posterior distributions of θ can be represented by continuous pdf's. Let $h(\theta)$ be a continuous prior pdf of θ, and let $l(\theta|t)$ be the likelihood function based on sample data, t. Then the posterior pdf of θ is given by

$$f(\theta|t) = \frac{h(\theta)l(t|\theta)}{\int_{-\infty}^{\infty} h(\theta)l(t|\theta)\,\mathrm{d}\theta}. \tag{3.109}$$

Relationship 3.109 is the Bayes theorem for a continuous r.v. The Bayesian inference process includes the following three stages:

- Constructing the likelihood function based on the distribution of interest and type of data available (complete samples, censored data, grouped data, etc.).
- Quantifying the prior information about the parameter of interest in the form of a prior distribution.
- Choosing a loss function, $L(\theta, \hat{\theta})$, which is a measure of discrepancy between the true value of the parameter θ and its estimate $\hat{\theta}$. The most popular loss function is the quadratic (squared-error) loss function

  $$L(\theta, \hat{\theta}) = (\theta - \hat{\theta})^2, \tag{3.110}$$

  which is equal to the square of the distance between θ and its estimate $\hat{\theta}$. Also, the loss function of the following form can be used and includes the following steps:

  $$L(\theta, \hat{\theta}) = |\theta - \hat{\theta}|. \tag{3.111}$$

- Calculating the posterior distribution using Bayes's theorem.
- Obtaining point and interval estimates.

A point Bayesian estimate is the estimate that minimizes the so-called Bayesian risk, $G(\hat{\theta})$, which is introduced as the expected (mean) value of the loss function with respect to the posterior distribution; that is,

$$G(\hat{\theta}) = \int_{-\infty}^{\infty} L(\theta, \hat{\theta}) f(\theta|t)\,\mathrm{d}\theta. \tag{3.112}$$

The estimate $\hat{\theta}$ is chosen as one minimizing the mean of the loss function 3.112; that is,

$$\hat{\theta} = \arg \min_{-\infty<\theta<\infty} G(\theta). \tag{3.113}$$

If loss function 3.110 is used, the corresponding Bayes point estimate is the posterior mean of θ; that is,

$$\hat{\theta} = \int_{-\infty}^{\infty} \theta f(\theta|t)\, d\theta. \tag{3.114}$$

If loss function 3.111 is chosen, the corresponding Bayes point estimate is the median of the posterior distribution of θ.

The Bayes analog of the classical confidence interval is known as the Bayes's probability interval. For constructing Bayes's probability interval, the following relationship based on the posterior distribution is used:

$$\Pr(\theta_l < \theta \leq \theta_u) = 1 - \alpha. \tag{3.115}$$

Similar to the classical estimation, the Bayesian estimation procedures can be divided into parametric and nonparametric ones. The following are examples of the parametric Bayesian estimation.

3.6.1 Estimation of the Parameter of Exponential Distribution

Consider a test of n units which results in r distinct times to failure $t_{(1)} < t_{(2)} < \cdots < t_{(r)}$ and $n - r$ times to censoring $t_{c1}, t_{c2}, \ldots, t_{c(n-r)}$, so that the total time on test, T, is

$$T = \sum_{i=1}^{r} t_{(i)} + \sum_{i=1}^{n-r} t_{ci}. \tag{3.116}$$

A time to failure is supposed to have the exponential distribution. The problem is to estimate the parameter λ of the exponential distribution.

Suppose a gamma distribution is used as the prior distribution of parameter λ. This distribution was already discussed in the present chapter as well as in Chapter 2. The pdf of the gamma distribution is given by Equation 2.74 as a time-to-failure distribution. Rewrite the pdf as a function of λ, which is now being considered as a r.v.:

$$h(\lambda; \delta, \rho) = \frac{1}{\Gamma(\delta)} \rho^{\delta} \lambda^{\delta-1} \exp(-\rho\lambda), \quad \lambda > 0, \rho \geq 0, \delta \geq 0. \tag{3.117}$$

In the Bayesian context, the parameters δ and ρ, as the parameters of prior distribution, are sometimes called *hyperparameters*. Selection of the hyperparameters is discussed later; but for the time being, suppose that these parameters are known. Also, suppose that the quadratic loss function 3.110 is used.

For the available data and the exponential time-to-failure distribution, using Equation 3.84 the likelihood function can be written as

$$\begin{aligned} l(\lambda|t) &= \prod_{i=1}^{r} \lambda e^{-\lambda t_{(i)}} \prod_{i=1}^{n-r} e^{-\lambda t_{ci}} \\ &= \lambda^r e^{-\lambda T}, \end{aligned} \tag{3.118}$$

where T is the total time on test given by Equation 3.116.

Using the prior distribution Equation 3.117, the likelihood function 3.102 and Bayes's theorem in the form Equation 3.109, one can find the posterior pdf of the parameter λ as

$$f(\lambda|T) = \frac{e^{-\lambda(T+\rho)} \lambda^{r+\delta-1}}{\int_0^{\infty} e^{-\lambda(T+\rho)} \lambda^{r+\delta-1}\, d\lambda}. \tag{3.119}$$

Recalling the definition of the gamma function (Equation 2.75), it is easy to show that the integral in the denominator of Equation 3.119 is

$$\int_0^\infty \lambda^{r+\delta-1} e^{-\lambda(T+\rho)}\, d\lambda = \frac{\Gamma(\delta+r)}{(\rho+T)^{\delta+r}}. \tag{3.120}$$

Finally, the posterior pdf of λ can be written as

$$f(\lambda|T) = \frac{(\rho+T)^{\delta+r}}{\Gamma(\delta+r)} \lambda^{r+\delta-1} e^{-\lambda(T+\rho)}. \tag{3.121}$$

A comparison with the prior pdf (Equation 3.117) shows that the posterior pdf (Equation 3.121) is also a gamma distribution with parameters $\rho' = r + \delta$ and $\lambda' = T + \rho$. Prior distributions that result in posterior distributions of the same family are referred to as *conjugate prior distributions.*

Since the quadratic loss function is used, the point Bayesian estimate of λ is the mean of the posterior gamma distribution with parameters ρ' and λ', so that the point Bayesian estimate, B, is

$$\lambda_B = \frac{\rho'}{\lambda'} = \frac{r+\delta}{T+\rho}. \tag{3.122}$$

The corresponding probability intervals can be obtained using Equation 3.115. For example, the $100(1-\alpha)$-level upper one-sided Bayes's probability interval for λ can be obtained from the following equation based on the posterior distribution Equation 3.121:

$$\Pr(\lambda < \lambda_u) = 1 - \alpha. \tag{3.123}$$

The same upper one-sided probability interval for λ can be expressed in a more convenient form similar to the classical confidence interval, that is, in terms of the χ^2 distribution, as

$$\Pr\{2\lambda(\rho+T) < \chi^2_{1-\alpha}[2(\delta+r)]\} = 1 - \alpha \tag{3.124}$$

and

$$\lambda_u = \frac{\chi^2_{1-\alpha}[2(\delta+r)]}{2(\rho+T)}. \tag{3.125}$$

Note that, contrary to the classical estimation, the number of degrees of freedom, $2(\delta + r)$, for the Bayes confidence limits are not necessarily integers.

The gamma distribution was chosen as the prior distribution for the purpose of simplicity and performance. Now let us consider the reliability interpretation of the Bayesian estimate obtained.

3.6.1.1 Selecting Parameters of Prior Distribution

The point Bayesian estimate λ_B, Equation 3.122, can be interpreted as follows. The parameter δ is a prior (fictitious) number of failures observed during a prior (fictitious) test, having ρ as the total time on test. So, intuitively, one would choose the prior estimate of λ as the ratio δ/ρ, which coincides with the *mean value* (recall Equation 3.27) of the prior gamma distribution used Equation 3.117.

The corresponding practical situation is quite the opposite—usually one has a prior estimate of λ; meanwhile, the parameters δ and ρ must be found. Given the prior point estimate λ_p, one can only estimate the ratio $\delta/\rho = \lambda_p$. To estimate these parameters separately, additional information about the degree of belief or accuracy of this prior estimate is required. Since the variance of the gamma distribution is δ/ρ^2, the coefficient of variation of the prior distribution is $1/\delta^{1/2}$ (the coefficient of

TABLE 3.3
Parameters and Coefficients of Variation of Gamma Prior Distribution with Mean Equal to 0.01

Shape Parameter, δ [Prior (Fictitious) Number of Failures]	Scale Parameter, ρ [Prior (Fictitious) Total Time on Test]	Coefficient of Variation (%)
1	100	100
5	500	45
10	1000	32
100	10,000	10

variation is the ratio of standard deviation to mean). The coefficient of variation can be used as a measure of relative accuracy of the prior point estimate of λ_p.

Thus, given the prior point estimate, λ_p, and the relative error of this estimate, one can estimate the corresponding parameters of the prior gamma distribution. To get a sense of the scale of these errors, consider the following numerical example. Let the prior point estimate λ_p be 0.01 (in some arbitrary units). The corresponding values of the coefficient of variation, expressed in percentages, for different values of the parameters δ and ρ are given in Table 3.3.

This approach is a simple and convenient way of expressing prior information (e.g., knowledge or degree of belief expressed by the subject matter experts) in terms of the gamma prior distribution. Another approach to selecting parameters of the prior gamma distribution is based on the quantile (or percentile) estimation of the prior distribution [13,15]). Let $H(\lambda, \rho, \delta)$ be the cdf of the prior gamma distribution. Recalling the definition of a quantile of level p, λ_p, one can write

$$H(\lambda_p, \rho, \delta) = \int_0^{\lambda_p} h(x)\, dx = p, \tag{3.126}$$

where $h(\lambda)$ is given by Equation 3.123.

Given a pair of quantiles, say, λ_{p1} and λ_{p2} of levels p_1 and p_2, and using Equation 3.126, one obtains two equations with two unknowns. These equations uniquely determine the parameters of gamma distribution. Practically, the procedure is as follows:

1. An expert specifies the values p_1, λ_{p1}, p_2, and λ_{p2} such that

$$H(\lambda_{p1}, \rho, \delta) = \Pr(\lambda < \lambda_{p1}) = p_1,$$
$$H(\lambda_{p2}, \rho, \delta) = \Pr(\lambda < \lambda_{p2}) = p_2.$$

   For example, she/he specifies that there is 90% probability that the parameter is <0.1 and 50% probability that λ is <0.01.
2. A numerical procedure is used to solve the system of equations in 1. to find the values of the parameters δ and ρ.

Usually the value of p_1 is chosen as 0.90 or 0.95, and the value of p_2 as 0.5 (the median value). In [15], a special table and graphs are provided for solving the system of equations above.

The procedure discussed above is not limited by gamma distribution. It is known as *the method of quantiles* and is applied as an estimation method of distribution parameters, not necessarily in a Bayesian context.

Example 3.24

A sample of identical units was tested. Six failures were observed during the test. The total time on test is 1440 h. The time-to-failure distribution is assumed to be exponential. The gamma distribution with a mean of 0.01 h^{-1} and with a coefficient of variation of 30% was selected as a prior distribution to represent the parameter of interest, λ. Find the posterior point estimate and the upper 90% probability limit for λ. See Figure 3.10 for the prior and the posterior distribution of λ.

Solution: The respective parameters of the distribution are $\delta = 11.1$ and $\rho = 1100$ h. Using Equation 3.122, the point posterior estimate (the mean of the posterior distribution) of the hazard rate is evaluated as

$$\lambda_B = \frac{11.1 + 6}{1110 + 1440} = 6.71 \times 10^{-3}\,\text{h}^{-1}.$$

Using Equation 3.125, calculate the 90% upper limit of one-sided Bayes's probability interval for λ as

$$\lambda_u = \frac{\chi^2_{0.9}(2 \times 17.1)}{2(2550)} = 8.80 \times 10^{-3}\,\text{h}^{-1}.$$

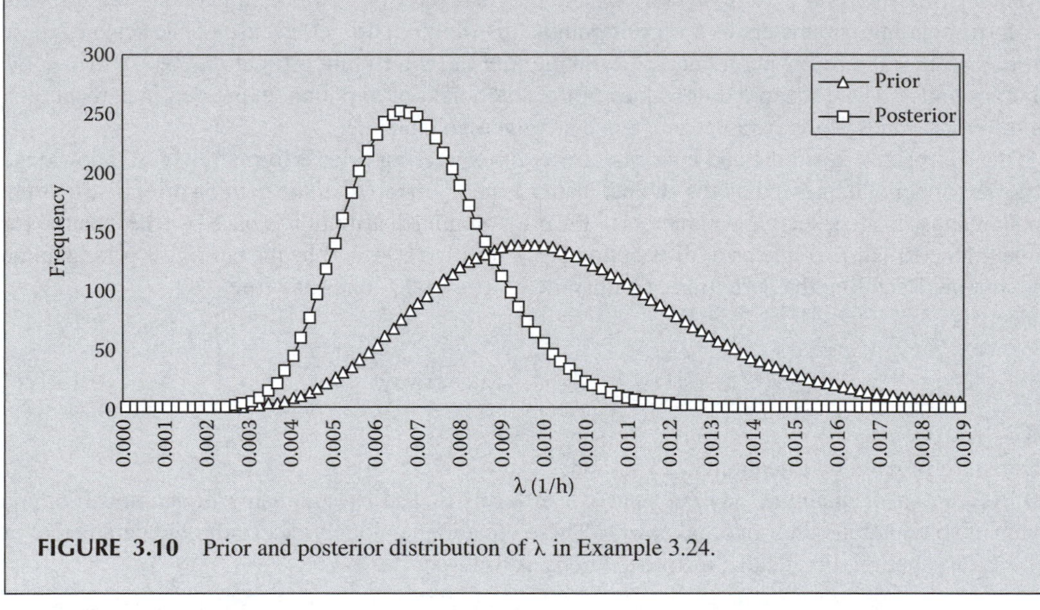

FIGURE 3.10 Prior and posterior distribution of λ in Example 3.24.

3.6.1.2 Uniform Prior Distribution

This prior distribution has very simple form, so it is convenient to use as an expression of prior information. Consider the prior pdf in the form

$$h(\lambda; a, b) = \begin{cases} \dfrac{1}{b-a}, & a < \lambda < b; \\ 0 & \text{otherwise.} \end{cases} \tag{3.127}$$

Using the likelihood function, Equation 3.118, the posterior pdf can be written as

$$f(\lambda|t) = \frac{\lambda^r e^{-\lambda T}}{\int_a^b \lambda^r e^{-\lambda T}\, d\lambda}, \quad a < \lambda < b. \tag{3.128}$$

Substituting $y = \lambda T$ in the denominator of Equation 3.128, one obtains

$$\int_a^b \lambda^r \mathrm{e}^{-\lambda T}\, \mathrm{d}\lambda = \int_{aT}^{bT} \frac{y^r \mathrm{e}^{-y}\, \mathrm{d}y}{T^{r+1}}. \tag{3.129}$$

Recall the definition of the incomplete gamma function,

$$\Gamma(c, z) = \int_0^z y^{c-1} \mathrm{e}^{-y}\, \mathrm{d}y, \quad c > 0, \tag{3.130}$$

in which $\Gamma(c, \infty) = \Gamma(c)$, where $\Gamma(c)$ is the ("complete") gamma function (see Equation 2.75 in Chapter 2). Accordingly, the denominator in Equation 3.128 can be written as

$$\int_a^b \lambda^r \mathrm{e}^{-\lambda T}\, \mathrm{d}\lambda = \frac{1}{T^{r+1}}[\Gamma(r+1, bT) - \Gamma(r+1, aT)] \tag{3.131}$$

and the posterior pdf (Equation 3.128) takes on the form

$$f(\lambda|t) = \frac{t^{r+1}\lambda^r \mathrm{e}^{-\lambda T}}{\Gamma(r+1, bT) - \Gamma(r+1, aT)}, \quad a < \lambda < b. \tag{3.132}$$

Note that the classical maximum likelihood point estimate, r/T, is the mode of the posterior pdf (Equation 3.132). The corresponding mean value (which is the Bayesian point estimator of λ for the case of the squared-error loss function) is given by [15]:

$$\lambda_{\mathrm{B}} = \frac{\Gamma(r+2, bT) - \Gamma(r+2, aT)}{\Gamma(r+1, bT) - \Gamma(r+1, aT)}. \tag{3.133}$$

The two-sided Bayesian probability interval can now be found in the standard way, that is, by integrating Equation 3.132, which results in solving the following equations with respect to λ_{l} and λ_{u}:

$$\begin{aligned}
\Pr(\lambda < \lambda_{\mathrm{l}}) &= \int_a^{\lambda_{\mathrm{l}}} \frac{t^{r+1}\lambda^r \mathrm{e}^{-\lambda T}\, \mathrm{d}\lambda}{\Gamma(r+1, bT) - \Gamma(r+1, aT)} \\
&= \frac{\Gamma(r+1, \lambda_{\mathrm{l}} T) - \Gamma(r+2, aT)}{\Gamma(r+1, bT) - \Gamma(r+1, aT)} = \frac{\alpha}{2}. \\
\Pr(\lambda < \lambda_{\mathrm{l}}) &= \frac{\Gamma(r+1, bT) - \Gamma(r+2, \lambda_{\mathrm{u}} T)}{\Gamma(r+1, bT) - \Gamma(r+1, aT)} = \frac{\alpha}{2}
\end{aligned} \tag{3.134}$$

EXAMPLE 3.25

An electronic component has the exponential time-to-failure distribution. The uniform prior distribution of λ is given by $a = 1 \times 10^{-6}$ and $b = 5 \times 10^{-6}\,\mathrm{h}^{-1}$ in Equation 3.127. A life test of the component results in $r = 30$ failures in total time on test of $T = 10^7$ h. Find the point estimate (mean) and 90% two-sided Bayesian probability interval for the failure rate λ (Figure 3.11).

Solution: Using Equations 3.133 and 3.134, the point estimate $\lambda_{\mathrm{B}} = 3.1 \times 10^{-6}\,\mathrm{h}^{-1}$ and the 90% two-sided Bayesian probability interval is $(2.24 \times 10^{-6} < \lambda_{\mathrm{B}} < 4.04 \times 10^{-6})\,\mathrm{h}^{-1}$.

continued

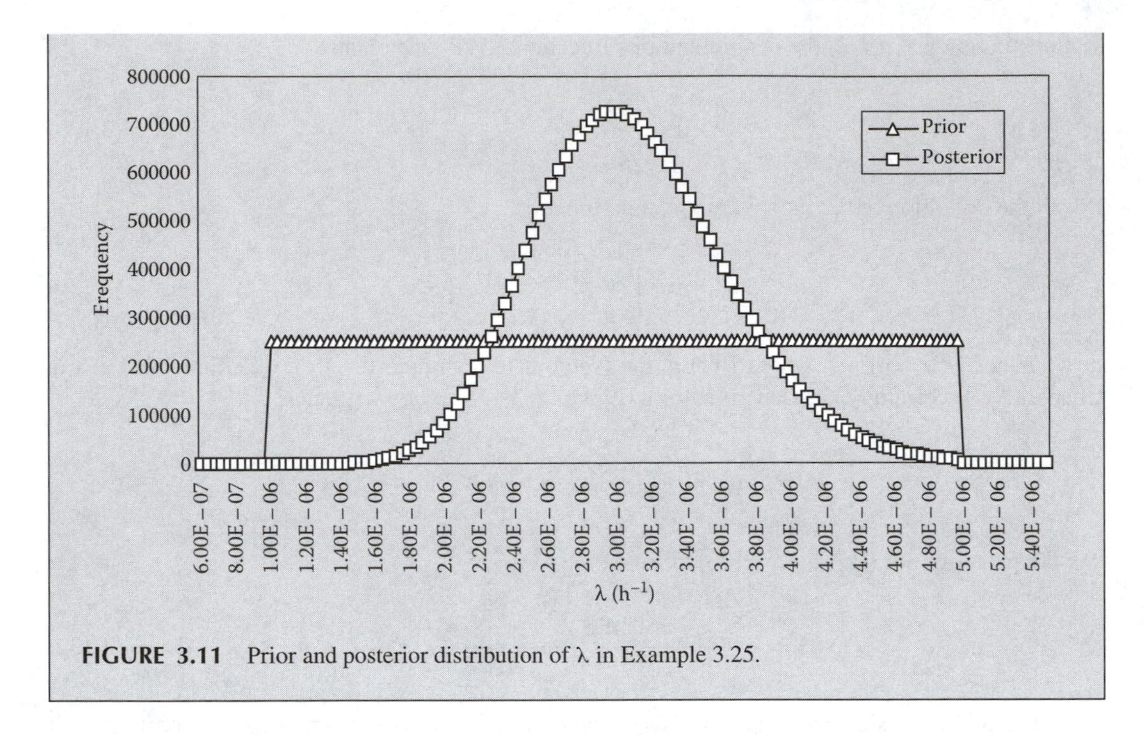

FIGURE 3.11 Prior and posterior distribution of λ in Example 3.25.

3.6.2 Bayesian Estimation of the Parameter of Binomial Distribution

The binomial distribution plays an important role in reliability. Suppose that n identical units have been placed on test (without replacement of the failed units) for a specified time, t, and that the test yields r failures. The number of failures, r, can be considered as a discrete r.v. having the binomial distribution with parameters n and $p(t)$, where $p(t)$ is the probability of failure of a single unit during time t. As discussed in Section 3.5, $p(t)$, as a function of time, is the time-to-failure cdf, and $1 - p(t)$ is the reliability or survivor function. A straightforward application of the binomial distribution is the modeling of a number of failures to start on demand for a redundant unit. The probability of failure in this case might be considered as time independent. Thus, one should keep in mind two possible applications of the binomial distribution:

1. The survivor (reliability) function or time-to-failure cdf.
2. The binomial distribution itself.

The MLE of the parameter p is the ratio r/n, which is widely used as the classical estimate. To obtain a Bayesian estimation procedure for the reliability (survivor) function, let us consider p as the survivor probability in a single Bernoulli trial (so, now "success" means surviving). If the number of units placed on test, n, is fixed in advance, the probability distribution of the number, x, of unfailed units during the test (i.e., the number of successes) is given by the binomial distribution with parameters n and x (see Equation 2.42):

$$f(x : n, p) = \frac{n!}{(n-x)!x!} p^x (1-p)^{n-x}. \tag{3.135}$$

The corresponding likelihood function can be written as

$$l(p|x) = cp^x (1-p)^{n-x}, \tag{3.136}$$

where c is a constant that does not depend on the parameter of interest, p. For any continuous prior distribution with pdf $h(p)$, the corresponding posterior pdf can be written as

$$f(p|x) = \frac{p^x(1-p)^{n-x}h(p)}{\int_{-\infty}^{\infty} p^x(1-p)^{n-x}h(p)\,\mathrm{d}p}. \tag{3.137}$$

3.6.2.1 Standard Uniform Prior Distribution

Consider the particular case of uniform distribution, $U(a,b)$, which in the Bayesian context represents a state of total ignorance. While this seems to have little practical importance, it is interesting, nevertheless, from a methodological point of view. For this case, one can write

$$h(p) = \begin{cases} 1, & 0 < p \leq 1; \\ 0 & \text{otherwise.} \end{cases} \tag{3.138}$$

and

$$f(p|x) = \frac{p^{(x+1)-1}(1-p)^{(n-x+1)-1}}{\int_0^1 p^{(x+1)-1}(1-p)^{(n-x+1)-1}\,\mathrm{d}p}. \tag{3.139}$$

The integral in the denominator can be expressed as

$$\int_0^1 p^{(x+1)-1}(1-p)^{(n-x+1)-1}\,\mathrm{d}p = \frac{\Gamma(x+1)\Gamma(n-x+1)}{\Gamma(n+2)}. \tag{3.140}$$

So, the posterior cdf can be easily recognized as the pdf of the beta distribution, $f(p; x+1, n-x+1)$, which was introduced in Chapter 2. Recalling the expression for the mean value of the beta distribution (Equation 2.78), the point Bayesian estimate of p can be written as

$$p_{\mathrm{B}} = \frac{x+1}{n+2}. \tag{3.141}$$

Note that the estimate is different from the respective classical estimate (x/n), but when the sample size increases the estimates are closer to each other.

Because the cdf of the beta distribution is expressed in terms of the incomplete beta function (see Equation 3.105), the $100(1-\alpha)\%$ two-sided Bayesian probability interval for p can be obtained by solving the following equations:

$$\begin{aligned} \Pr(p < p_{\mathrm{l}}) &= I_{p_{\mathrm{l}}}(x+1, n-x+1) = \frac{\alpha}{2}, \\ \Pr(p > p_{\mathrm{u}}) &= I_{p_{\mathrm{u}}}(x+1, n-x+1) = 1 - \frac{\alpha}{2}. \end{aligned} \tag{3.142}$$

The probability intervals above are very similar to the corresponding classical confidence intervals Equations 3.103 and 3.104.

Example 3.26

Calculate the point estimate and the 95% two-sided Bayesian probability interval for the reliability of a new component based on a life test of 300 components, out of which four have failed. Suppose that for this component no historical information is available. Accordingly, its prior reliability estimate may be assumed to be uniformly distributed between 0 and 1.

continued

Solution: Using Equation 3.141, find

$$R = 1 - p_{\mathrm{B}} = 1 - \frac{4+1}{300+2} = 0.9834.$$

Using Equation 3.142, the 95% upper and lower limits are evaluated as 0.9663 and 0.9946, respectively.

It is interesting to compare the above results with the classical results. The point estimate of the reliability is $R = 1 - p = 1 - 4/300 = 0.9867$, and the 95% upper and lower limits, according to Equations 3.101 and 3.102, are 0.9662 and 0.9964, respectively.

3.6.2.2 Truncated Standard Uniform Prior Distribution

Consider the following prior pdf of p:

$$h(p|p_1, p_0) = \begin{cases} \dfrac{1}{p_1 - p_0}, & 0 \le p_0 < p < p_1 \le 1; \\ 0 & \text{otherwise.} \end{cases} \tag{3.143}$$

The corresponding posterior pdf cannot be expressed in a closed form, but it can be written in terms of the incomplete beta function [15] as

$$f(p|x) = \frac{(\Gamma(n+2)/\Gamma(x+1)\Gamma(n-x+1))p^{(x+1)-1}(1-p)^{(n-x+1)-1}}{I_{p_1}(x+1, n-x+1) - I_{p_0}(x+1, n-x+1)} \quad 0 \le p_0 < p \le 1, \tag{3.144}$$

where the numerator is in the form of a beta pdf with parameters $x+1$ and $n-x+1$. The same posterior pdf can be written in a simpler form, which can be more convenient for straightforward point estimate calculations:

$$f(p|x) = \frac{p^{(x+1)-1}(1-p)^{(n-x+1)-1}}{\int_{p_2}^{p_1} p^{(x+1)-1}(1-p)^{(n-x+1)-1}\,\mathrm{d}p}. \tag{3.145}$$

The posterior mean can be obtained in terms of the incomplete beta function as

$$p_{\mathrm{B}} = \frac{x+1}{n+2}\left[\frac{I_{p_1}(x+2, n-x+1) - I_{p_0}(x+2, n-x+1)}{I_{p_1}(x+1, n-x+1) - I_{p_0}(x+1, n-x+1)}\right], \tag{3.146}$$

where the first multiplier coincides with the corresponding estimate for the case of the standard uniform prior, and the second multiplier can be considered as a correction term associated with the truncated uniform prior distribution. The same estimate can be written in terms of the posterior pdf (Equation 3.145) as

$$p_{\mathrm{B}} = \frac{\int_{p_2}^{p_1} p^{(x+1)}(1-p)^{(n-x+1)-1}\,\mathrm{d}p}{\int_{p_2}^{p_1} p^{(x+1)-1}(1-p)^{(n-x+1)-1}\,\mathrm{d}p}. \tag{3.147}$$

Using the posterior pdf, the $100(1-\alpha)\%$ two-sided Bayesian probability interval for p can be obtained as solutions of the following equations:

$$\begin{aligned} I_{p_l}(x+1, n-x+1) &= \left(1 - \frac{\alpha}{2}\right) I_{p_0}(x+1, n-x+1) + \frac{\alpha}{2} I_{p_1}(x+1, n-x+1), \\ I_{p_u}(x+1, n-x+1) &= \left(1 - \frac{\alpha}{2}\right) I_{p_1}(x+1, n-x+1) + \frac{\alpha}{2} I_{p_0}(x+1, n-x+1). \end{aligned} \tag{3.148}$$

EXAMPLE 3.27

A new sensor installed on 500 vehicles was observed for 12 months in service (MIS), and four failures were recorded. The reliability of a similar sensor at 12 MIS has been known not to exceed 0.985, which can be expressed in terms of the uniform prior distribution with $p_0 = 0.985$, and $p_1 = 1$. Find the point estimate (posterior mean) and the 90% one-sided lower Bayesian probability interval for the reliability of the new sensor (Figure 3.12).

Solution: According to Equation 3.146, the posterior mean is 0.9926, and the 90% posterior lower limit from Equation 3.148 is calculated as 0.9856.

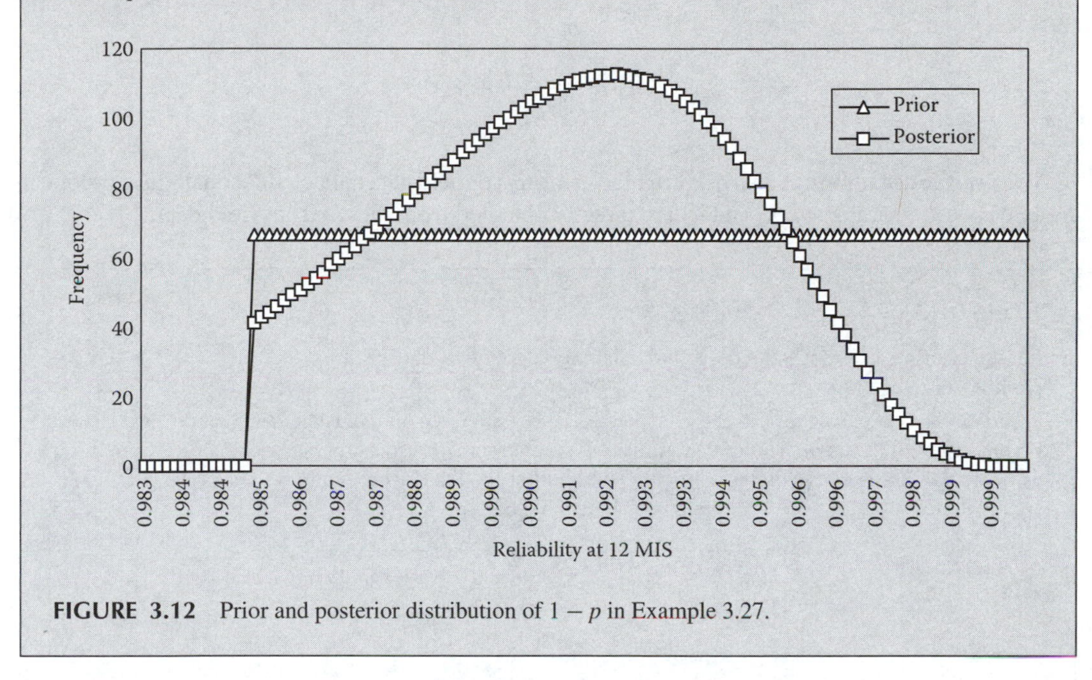

FIGURE 3.12 Prior and posterior distribution of $1 - p$ in Example 3.27.

3.6.2.3 Beta Prior Distribution

The most widely used prior distribution for the parameter, p, of the binomial distribution is the beta distribution, which was introduced in Chapter 2. The pdf of the distribution can be written in the following convenient form:

$$h(p; x_0, n_0) = \begin{cases} \dfrac{\Gamma(n_0)}{\Gamma(x_0)\Gamma(n_0 - x_0)} p^{x_0 - 1}(1 - p)^{n_0 - x_0 - 1}, & 0 \le p \le 1; \\ 0 & \text{otherwise}, \end{cases} \quad (3.149)$$

where $n_0 > x_0 \ge 0$.

The pdf provides a great variety of different shapes. It is important to note that the standard uniform distribution is a particular case of the beta distribution. When x_0 is equal to one and n_0 is equal to two, Equation 3.149 reduces to the standard uniform distribution. Moreover, the beta prior distribution turns out to be a conjugate prior distribution for the estimation of the parameter p of the binomial distribution of interest.

Given the expression for the mean value of the beta distribution (Equation 2.78), the prior mean is x_0/n_0, so that the parameters of the prior, x_0 and n_0, can be interpreted as a pseudonumber of identical units that survive (or fail) a pseudotest of no units during pseudotime t. Thus, while selecting the parameters of the prior distribution an expert can express her/his knowledge in terms of the pseudotest considered (i.e., in terms of x_0 and n_0). On the other hand, an expert can evaluate the prior mean, that is, the ratio, x_0/n_0, and her/his degree of belief in terms of standard deviation or coefficient of variation of the prior distribution. For example, if the coefficient of variation is used, it can be treated as a measure of uncertainty (relative error) of prior assessment.

Let p_{pr} be the prior mean and k be the coefficient of variation of the prior beta distribution. The corresponding parameters x_0 and n_0 can be found as a solution of the following equation system:

$$\begin{aligned} p_{pr} &= \frac{x_0}{n_0}, \\ n_0 &= \frac{1 - p_{pr}}{k^2 p_{pr}} - 1. \end{aligned} \tag{3.150}$$

The prior distribution can also be estimated using test or field data collected for analogous products. In this case, the parameters x_0 and n_0 are directly obtained from these test or field data.

Example 3.28

Let the prior mean (point estimate) of the reliability function be chosen as $p_{pr} = x_0/n_0 = 0.9$. Select the parameters x_0 and n_0.

Solution: The choice of the parameters x_0 and n_0 can be (similar to one considered in Section 3.6.1) based on the values of the coefficient of variation used as a measure of dispersion (accuracy) of the prior point estimate p_{pr}. Some values of the coefficient of variation and the corresponding values of the parameters x_0 and n_0 for $p_{pr} = x_0/n_0 = 0.9$ are given in the table below.

n_0	x_0	Coefficient of Variation (%)
1	0.9	23.6
10	9	10.0
100	90	3.3
1000	900	1.0

The posterior pdf is

$$f(p|x) = \frac{\Gamma(n + n_0)}{\Gamma(x + x_0)\Gamma(n + n_0 - x - x_0)} p^{(x+x_0)-1}(1-p)^{(n+n_0-x-x_0)-1}, \tag{3.151}$$

which is also a beta distribution pdf. The corresponding posterior mean is given by

$$p_B = \frac{x + x_0}{n + n_0}. \tag{3.152}$$

Note that as n approaches infinity, the Bayesian estimate approaches the MLE, x/n, Equation 3.97. In other words, the classical inference tends to dominate the Bayesian inference as the amount of data increases.

One should also keep in mind that the prior distribution parameters can also be estimated based on prior data (data collected on similar equipment, for example), using the respective sample size, n_0, and the number of failures observed, x_0.

The corresponding $100(1-\alpha)\%$ two-sided Bayesian probability interval for p can be obtained as the solutions of the following equations:

$$\Pr(p < p_\mathrm{l}) = I_{p_\mathrm{l}}(x + x_0, n + n_0 - x - x_0) = \frac{\alpha}{2},$$
$$\Pr(p > p_\mathrm{u}) = I_{p_\mathrm{u}}(x + x_0, n + n_0 - x - x_0) = 1 - \frac{\alpha}{2}. \tag{3.153}$$

Example 3.29

A design engineer assesses the reliability of a new component at the end of its useful life ($T = 10,000$ h) as 0.75 ± 0.19. A sample of 100 new components has been tested for 10,000 h, and 29 failures have been recorded. Given the test results, find the posterior mean and the 90% Bayesian probability interval for the component reliability, if the prior distribution of the component reliability is assumed to be a beta distribution.

Solution: The prior mean is 0.75, and the coefficient of variation is $0.19/0.75 = 0.25$. Using Equation 3.150, the parameters of the prior distribution are evaluated as $x_0 = 3.15$ and $n_0 = 4.19$. Thus, according to Equation 3.152, the posterior point estimate of the new component reliability is $R(10,000) = (3.15 + 71)/(4.19 + 100) = 0.712$. According to Equation 3.153, the 90% lower and upper confidence limits are 0.637 and 0.782, respectively. Figure 3.13 shows the prior and posterior distributions of $R = 1 - p$.

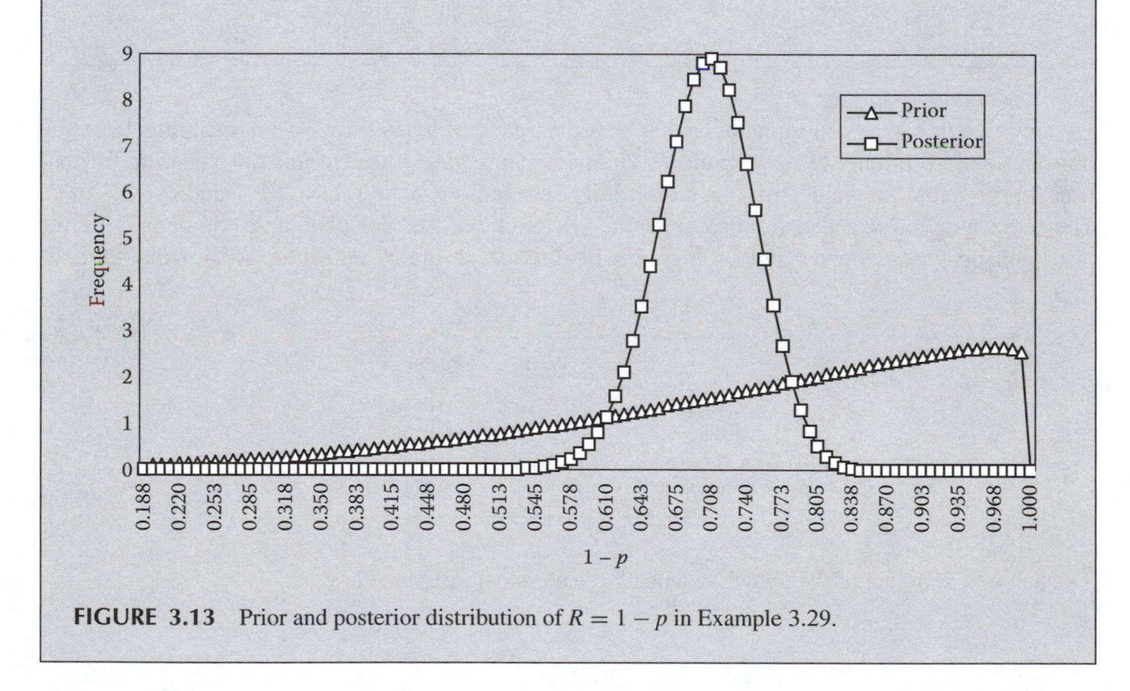

FIGURE 3.13 Prior and posterior distribution of $R = 1 - p$ in Example 3.29.

3.6.2.4 Lognormal Prior Distribution

The following example illustrates the case when the prior distribution and the likelihood function do not result in a conjugate posterior distribution, and the posterior distribution obtained cannot be expressed in terms of standard function. This is the case when a numerical integration is required.

EXAMPLE 3.30

The number of failures to start a diesel generator on demand has a binomial distribution with parameter p. The prior data on the performance of the similar diesel are obtained from field data, and p is assumed to follow the lognormal distribution with known parameters $\mu_y = 0.0516$ and $\sigma_y = 0.0421$ (the respective values of μ_t and σ_t are 3.22 and 0.714). A limited test of the diesel generators of interest shows that eight failures are observed in 582 demands. Calculate the Bayesian point estimate of p (mean and median) and the 90th percentiles of p. Compare these results with the corresponding values for the prior distribution.

Solution: Since we are dealing with a demand failure, a binomial distribution best represents the observed data. The likelihood function is given by

$$\Pr(X|p) = \binom{582}{8} p^8 (1-p)^{574}$$

and the prior pdf is

$$f(p) = \frac{1}{\sigma_t p \sqrt{2\pi}} \exp\left[-\frac{1}{2}\left(\frac{\ln p - \mu_t}{\sigma_t}\right)^2\right], \quad p > 0.$$

Using the initial data, the posterior pdf becomes

$$f(p|X) = \frac{p^8 (1-p)^{574} \exp\left[-\frac{1}{2}\left(\frac{\ln p - 3.22}{0.714}\right)^2\right]}{\int_0^1 p^8 (1-p)^{574} \exp\left[-\frac{1}{2}\left(\frac{\ln p - 3.22}{0.714}\right)^2\right] dp}.$$

It is evident that the denominator cannot be expressed in a closed form, so a numerical integration must be applied. Table 3.4 shows results of a numerical integration used to find the posterior distribution. In this table the values of p_1 are arbitrarily selected between 1.23×10^{-8} and 1.78×10^{-1}. Then, the numerator and denominator of the posterior pdf are calculated. A comparison of the prior and posterior is given below. Figure 3.14 displays the prior and the posterior distributions of p.

	Prior	**Posterior**
Mean	0.0516	0.0170
Median	0.0399	0.0158
5th percentile	0.0123	0.0089
95th percentile	0.1293	0.0267

The point estimate of the actual data using the classical inference is

$$\hat{p} = \frac{8}{582} = 0.0137.$$

3.6.3 BAYESIAN ESTIMATION OF THE WEIBULL DISTRIBUTION

The Bayesian procedures considered in the previous sections are related to the binomial and exponential distributions, which are very important in statistical reliability engineering. At this point, it is important to note that these distributions are one-parameter distributions. It should be noted, as well, that in the framework of the Bayesian approach, both the exponential and binomial

TABLE 3.4
Results of a Numerical Integration in Example 3.30

i	Probability p_i	Prior pdf	Likelihood Function	Prior* Likelihood	Posterior pfd	Posterior cdf
0	1.23×10^{-8}	0.00×10^{0}	1.54×10^{-28}	0.00×10^{0}	0.00×10^{0}	0.00×10^{0}
1	1.78×10^{-3}	3.68×10^{-3}	3.47×10^{-3}	1.28×10^{-5}	1.01×10^{-3}	1.80×10^{-6}
2	3.55×10^{-3}	1.94×10^{-1}	3.98×10^{-2}	7.72×10^{-3}	6.13×10^{-1}	1.09×10^{-3}
3	5.33×10^{-3}	1.15×10^{0}	1.08×10^{-1}	1.24×10^{-1}	9.85×10^{0}	1.86×10^{-2}
4	7.11×10^{-3}	3.05×10^{0}	1.62×10^{-1}	4.93×10^{-1}	3.91×10^{-1}	8.81×10^{-2}
5	8.89×10^{-3}	5.53×10^{0}	1.76×10^{-1}	9.72×10^{-1}	7.71×10^{-1}	2.25×10^{-1}
6	1.07×10^{-2}	8.18×10^{0}	1.55×10^{-1}	1.27×10^{0}	1.01×10^{2}	4.04×10^{-1}
7	1.24×10^{-2}	1.07×10^{1}	1.19×10^{-1}	1.27×10^{0}	1.01×10^{2}	5.83×10^{-1}
8	1.42×10^{-2}	1.28×10^{1}	8.20×10^{-2}	1.05×10^{0}	8.35×10^{1}	7.32×10^{-1}
9	1.60×10^{-2}	1.46×10^{1}	5.22×10^{-2}	7.62×10^{-1}	6.05×10^{1}	8.39×10^{-1}
10	1.78×10^{-2}	1.60×10^{1}	3.11×10^{-2}	4.97×10^{-1}	3.94×10^{1}	9.09×10^{-1}
11	1.95×10^{-2}	1.69×10^{1}	1.76×10^{-2}	2.99×10^{-1}	2.37×10^{1}	9.51×10^{-1}
12	2.13×10^{-2}	1.76×10^{1}	9.56×10^{-3}	1.68×10^{-1}	1.33×10^{1}	9.75×10^{-1}
13	2.31×10^{-2}	1.79×10^{1}	5.00×10^{-3}	8.96×10^{-2}	7.11×10^{0}	9.88×10^{-1}
14	2.49×10^{-2}	1.80×10^{1}	2.53×10^{-3}	4.57×10^{-2}	3.63×10^{0}	9.94×10^{-1}
15	2.67×10^{-2}	1.79×10^{1}	1.25×10^{-3}	2.24×10^{-2}	1.78×10^{0}	9.97×10^{-1}
16	2.84×10^{-2}	1.77×10^{1}	6.01×10^{-4}	1.06×10^{-2}	8.44×10^{-1}	9.99×10^{-1}
17	3.02×10^{-2}	1.73×10^{1}	2.83×10^{-4}	4.90×10^{-3}	3.89×10^{-1}	9.99×10^{-1}
18	3.20×10^{-2}	1.69×10^{1}	1.31×10^{-4}	2.21×10^{-3}	1.75×10^{-1}	1.00×10^{0}
19	3.38×10^{-2}	1.64×10^{1}	5.93×10^{-5}	9.70×10^{-4}	7.70×10^{-2}	1.00×10^{0}
20	3.55×10^{-2}	1.58×10^{1}	2.65×10^{-5}	4.18×10^{-4}	3.32×10^{-2}	1.00×10^{0}
21	3.73×10^{-2}	1.52×10^{1}	1.17×10^{-5}	1.77×10^{-4}	1.41×10^{-2}	1.00×10^{0}
22	3.91×10^{-2}	1.46×10^{1}	5.07×10^{-6}	7.39×10^{-5}	5.87×10^{-3}	1.00×10^{0}
23	4.09×10^{-2}	1.40×10^{1}	2.17×10^{-6}	3.04×10^{-5}	2.41×10^{-3}	1.00×10^{0}
24	4.27×10^{-2}	1.34×10^{1}	9.22×10^{-7}	1.23×10^{-5}	9.77×10^{-4}	1.00×10^{0}
25	4.44×10^{-2}	1.27×10^{1}	3.87×10^{-7}	4.93×10^{-6}	3.91×10^{-4}	1.00×10^{0}
...	...	...	...	...	...	...
91	1.62×10^{-1}	5.19×10^{-1}	3.77×10^{-37}	1.96×10^{-37}	1.55×10^{-35}	1.00×10^{0}
92	1.63×10^{-1}	4.98×10^{-1}	1.17×10^{-37}	5.83×10^{-38}	4.63×10^{-36}	1.00×10^{0}
93	1.65×10^{-1}	4.78×10^{-1}	3.62×10^{-38}	1.73×10^{-38}	1.37×10^{-36}	1.00×10^{0}
94	1.67×10^{-1}	4.59×10^{-1}	1.12×10^{-38}	5.13×10^{-39}	4.07×10^{-37}	1.00×10^{0}
95	1.69×10^{-1}	4.41×10^{-1}	3.43×10^{-39}	1.51×10^{-39}	1.20×10^{-37}	1.00×10^{0}
96	1.71×10^{-1}	4.23×10^{-1}	1.05×10^{-39}	4.46×10^{-40}	3.54×10^{-38}	1.00×10^{0}
97	1.72×10^{-1}	4.07×10^{-1}	3.22×10^{-40}	1.31×10^{-40}	1.04×10^{-38}	1.00×10^{0}
98	1.74×10^{-1}	3.91×10^{-1}	9.79×10^{-40}	3.83×10^{-41}	3.04×10^{-39}	1.00×10^{0}
99	1.76×10^{-1}	3.76×10^{-1}	2.97×10^{-41}	1.12×10^{-41}	8.86×10^{-40}	1.00×10^{0}
100	1.78×10^{-1}	3.61×10^{-1}	8.99×10^{-42}	3.25×10^{-42}	2.58×10^{-40}	1.00×10^{0}
			Sum 7.09×10^{0}			

distributions have their respective conjugate prior distributions, which makes their practical use more convenient.

In the case of the two-parameter distribution, the respective Bayesian estimation procedures become more complicated because they require the two-dimensional prior distributions. In this and the following section, these procedures are developed for another common reliability engineering distribution—the two-parameter Weibull distribution. It should be mentioned that the procedures considered below, generally speaking, can be applied to any two-parameter distribution for which classical probability papers are applicable.

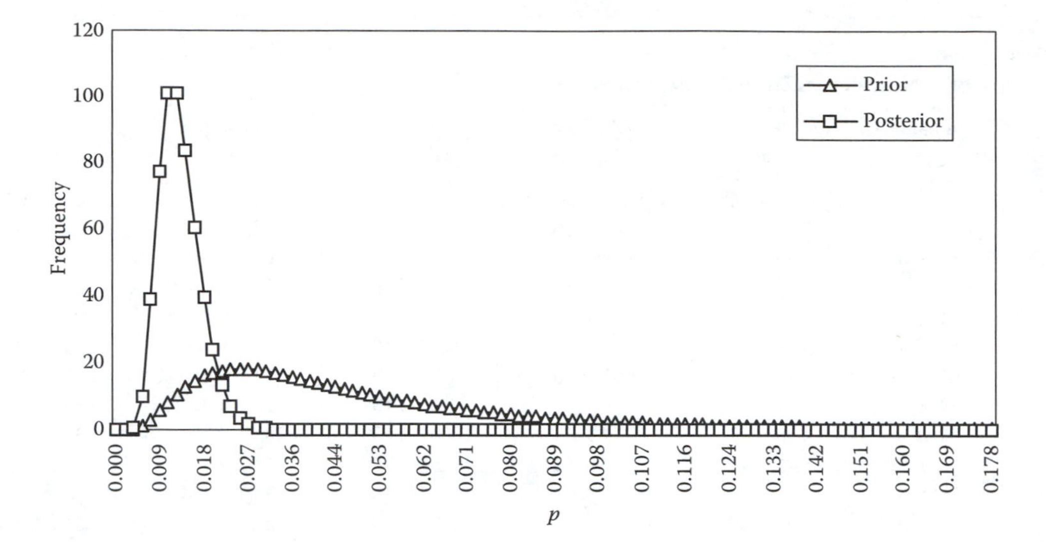

FIGURE 3.14 Prior and posterior distribution of p in Example 3.30.

The fundamental result related to the most realistic case, when both parameters of the Weibull distribution are r.v.s, obtained by Soland [25], states that the Weibull distribution does not have a conjugate continuous joint prior distribution. Later, Soland [26] proved that a conjugate continuous-discrete joint prior distribution exists for the Weibull distribution parameters. The continuous component of this distribution is related to the scale parameter of the Weibull distribution (similar to the Bayesian estimation of the exponential distribution, for which gamma is used as the conjugate prior distribution), and the discrete component to the shape parameter. Although of great academic interest, this result is rather challenging to apply to real-life problems due to the difficulties of evaluating the prior information needed (mostly related to the shape parameter). A reliability application example of this approach can be found in [15]. Below, we consider a new procedure for the Bayesian estimation of the Weibull distribution [27], which allows constructing the continuous joint prior distribution of the Weibull parameters as well as the posterior estimates of the mean and standard deviation of the estimated cdf (or the reliability function) at any given value of the exposure variable (e.g., time).

Consider the prior information, which is available at two values (t_1 and t_2, $t_2 > t_1$) of the exposure variable as the estimates of cdf, $\hat{F}(t_k)$ and its standard deviation, $\hat{\sigma}\{\hat{F}(t_k)\}$; see Figure 3.15. This prior information can be divided into the following two cases. In the first case, it comes in the traditional form as life data sufficient for the Kaplan–Meier and Greenwood estimators to be used. It also might be attributed to the empirical Bayesian inference. In the second case, the prior information is obtained using the expert elicitation methods [28].

Based on these prior data, we are interested in obtaining the following:

- Joint *prior* distribution of the Weibull distribution parameters
- Prior point estimates of the Weibull parameters
- Prior distribution of the estimated cdf at any fixed exposure t_k
- Joint *posterior* distribution of the Weibull distribution parameters (based on data observed at t_3)
- Posterior point estimates of the Weibull parameters as the location of the *highest posterior probability density* (mode) of their joint distribution, similar to the highest posterior density (HPD) estimates considered in [29]
- *Posterior* distribution of the estimated cdf at exposure t_3 (based on data observed at t_3).

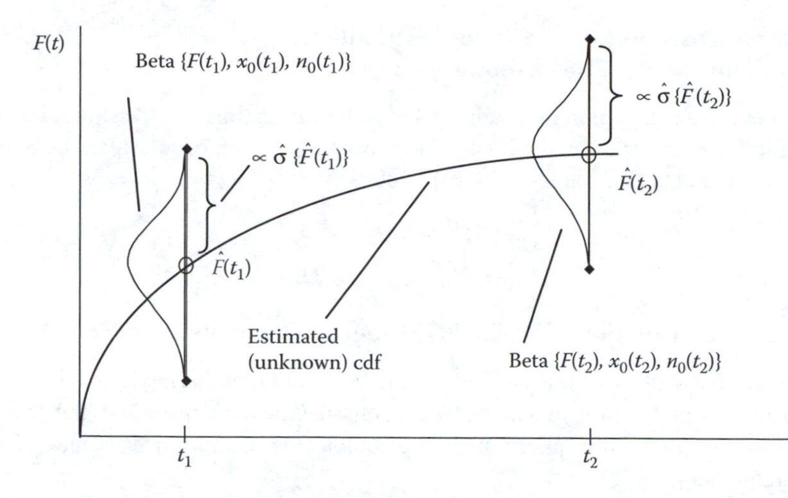

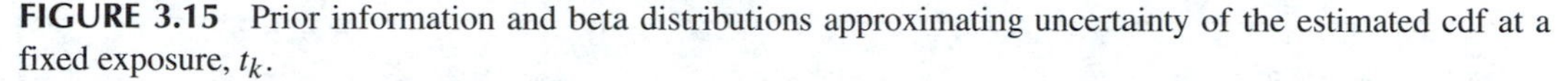

FIGURE 3.15 Prior information and beta distributions approximating uncertainty of the estimated cdf at a fixed exposure, t_k.

3.6.3.1 Prior Distribution

The beta prior distribution with the following pdf is assumed to characterize uncertainty of the estimated cdf at a fixed exposure:

$$f(p) = \frac{\Gamma(n_0)}{\Gamma(x_0)\Gamma(n_0 - x_0)} p^{x_0-1}(1-p)^{n_0-x_0-1}, \quad n_0 > 0, n_0 - x_0 > 0, \tag{3.154}$$

where p is the r.v. representing the estimated cdf at a fixed exposure, $F(t_k)$; n_0, x_0 are the parameters of the beta distribution, both positive quantities; and $\Gamma(\cdot)$ is the gamma function.

Based on the available estimates of the *prior* cdf at a fixed exposure ($t =$ constant), $\hat{F}(t_k)$ and its standard deviation, $\hat{\sigma}\{\hat{F}(t_k)\}$, the parameters of the respective *prior* beta distribution can be obtained through the method of moments [30] as

$$x_0(t) = \frac{\hat{F}(t)^2[1 - \hat{F}(t)]}{\hat{\sigma}^2\{\hat{F}(t)\}} - \hat{F}(t), \quad n_0(t) = \frac{x_0(t)}{\hat{F}(t)}. \tag{3.155}$$

By random sampling from the estimated beta distributions, one can obtain pairs of random realizations of the estimated cdf at exposures t_1 and t_2: $\{F(t_1), F(t_2)\}$, $i = 1, 2, \ldots, n$. For any such pair, as long as $F(t_1) < F(t_2)$, the shape and the scale parameters of the respective Weibull cdf can be obtained as

$$\begin{aligned} \beta &= \frac{F^*(t_2) - F^*(t_1)}{Ln(t_2/t_1)}, \\ \alpha &= \exp\left(Ln(t_1) - \frac{F^*(t_1)}{\beta}\right). \end{aligned} \tag{3.156}$$

Here, $F^*(t_k) = Ln(Ln(1 - F(t_k))^{-1})$, $k = 1, 2$.

It is obvious that n simulated pairs of $\{F_i(t_1), F_i(t_2)\}$ would define n pairs of Weibull distribution parameters $\{\alpha_i, \beta_i\}$, which are used to construct their joint prior distribution.

Further, the obtained realizations of the Weibull cdf, $F_i(t; \alpha_i, \beta_i)$, can be extrapolated to a given exposure, say, t_3, to obtain the distribution of the cdf at t_3. Using Equation 3.155, this distribution can again be approximated by the beta *prior* distribution with parameters $x_0(t_3)$ and $n_0(t_3)$.

3.6.3.2 Posterior Distributions (Posterior Distribution of Weibull cdf at a Fixed Exposure)

The estimated beta prior at t_3 is now available for Bayesian estimation. Assume that observation data are available for exposure t_3 in the binary form of r failures out of N trials. Then, the *posterior* distribution is also the beta distribution with parameters

$$\begin{aligned} x_0(t_3)^* &= x_0(t_3) + r, \\ n_0(t_3)^* &= n_0(t_3) + N. \end{aligned} \tag{3.157}$$

3.6.3.3 Posterior Distribution of Weibull Distribution Parameters

The joint posterior distribution of the Weibull parameters can be numerically obtained by sampling random realizations from the estimated beta distributions at exposures t_1, t_2, and (the newly obtained) t_3, and then using the probability paper method to obtain the sample of estimates of the Weibull distribution parameters.

EXAMPLE 3.31

Expert estimates of the failure probability of a mechanical component at $t_1 = 1$ and $t_2 = 3$ years of operation are $\hat{F}(t_1) = 0.01$ and $\hat{F}(t_2) = 0.05$, with 10% error. Treating the error as the coefficient of variation, estimates of the standard deviation of the failure probability at the two exposures are

$$\hat{\sigma}\{\hat{F}(t_1)\} = 0.001$$

and

$$\hat{\sigma}\{\hat{F}(t_2)\} = 0.005,$$

respectively. Using Equation 3.155, the estimates of the beta distribution parameters at the two exposures are

$$\{x_0(t_1) = 99,\ n_0(t_1) = 9899\}$$

and

$$\{x_0(t_2) = 95,\ n_0(t_2) = 1899\},$$

respectively. By sampling random realizations ($n = 10{,}000$) of the cdf from the two estimated beta distributions, and then using Equation 3.156, one obtains the joint prior distribution of Weibull parameters shown in Figure 3.16. The prior estimates of the Weibull parameters corresponding to the mode of their joint distribution are

$$\{\beta_{pr} = 1.51,\ \alpha_{pr} = 22.95\}.$$

The failure probability and its standard deviation at a future exposure of $t_3 = 5$ years of operation are estimated as 0.105 and 0.016, respectively, which, using Equation 3.141, can be translated into respective beta distribution parameters at t_3:

$$\{x_0(t_3) = 40,\ n_0(t_3) = 386\}.$$

Further, consider the data observed at t_3, which are one failure out of 12 components observed. The parameters of the posterior beta distribution parameters at t_3, according to Equation 3.157, are

$$\{x_0(t_3)^* = 41, n_0(t_3)^* = 398\}.$$

Figure 3.17 shows the joint posterior distribution of the Weibull distribution parameters obtained by sampling random realizations ($n = 10{,}000$) from the estimated beta distributions at exposures t_2 and t_3, and then using Equation 3.156. The posterior HPD estimates of the Weibull parameters are

$$\{\beta_{post} = 1.70,\ \alpha_{post} = 16.55\}.$$

The posterior estimates of the failure probability (cdf), and its standard deviation at t_3 are obtained as 0.104 and 0.069, respectively. The estimation results are summarized in Table 3.5.

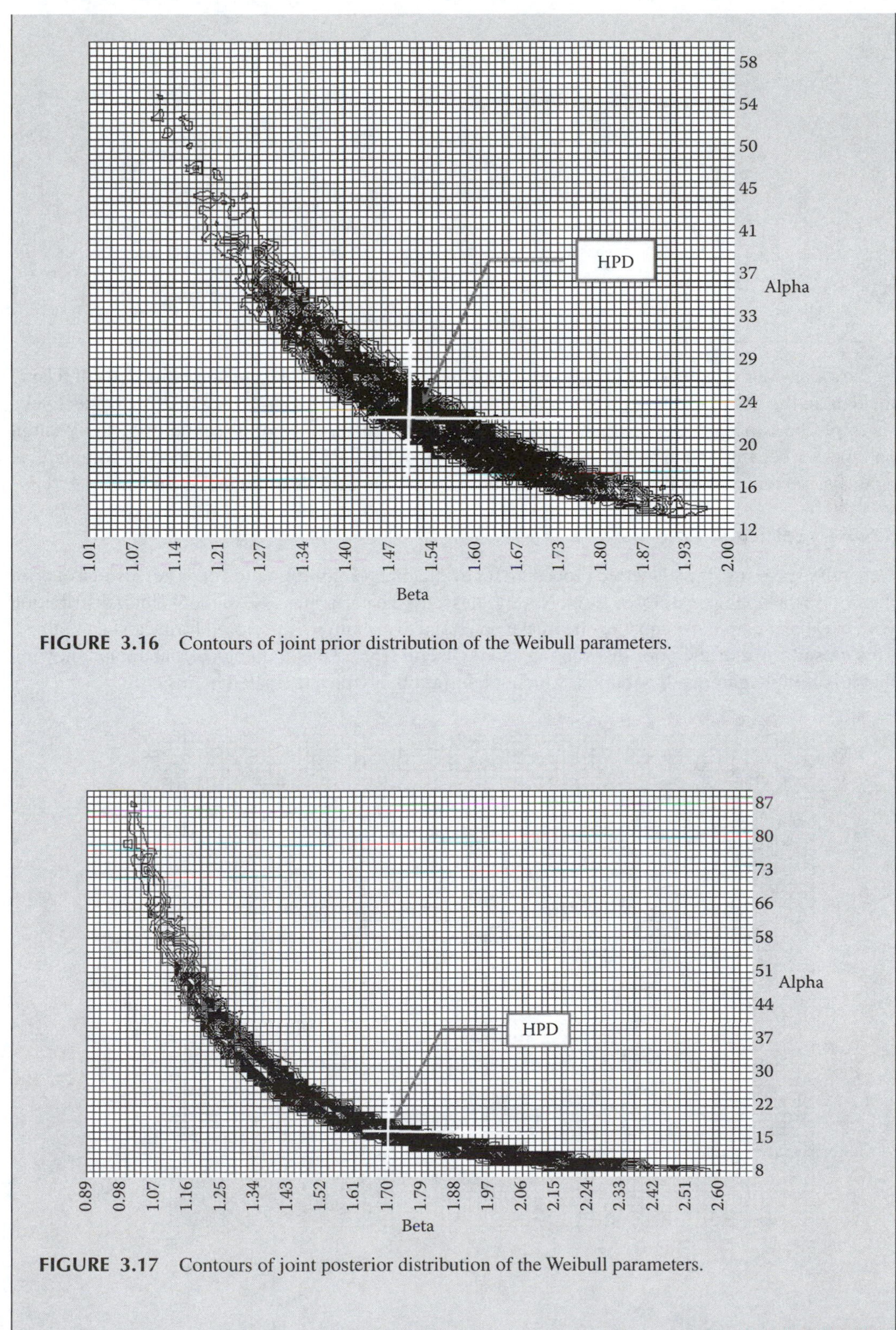

FIGURE 3.16 Contours of joint prior distribution of the Weibull parameters.

FIGURE 3.17 Contours of joint posterior distribution of the Weibull parameters.

continued

TABLE 3.5
Prior and Posterior Estimates

	Prior			Posterior
	$t_1 = 1$ (given)	$t_2 = 3$ (given)	$t_3 = 5$ (inferred)	$t_3 = 5$ (inferred)
$F(t)$	0.010	0.050	0.105	0.104
$\sigma\{F(t)\}$	0.001	0.005	0.016	0.069
α		22.95		16.55
β		1.51		1.70

Now consider sensitivity of the posterior distribution to the initial data on the estimated cdf. Using the data in the above example, we increased the cdf error at t_1 and t_2 to 50% and 20%, respectively. As might be expected, the contours of the prior and the posterior joint distribution of the Weibull parameters became much wider (see Figures 3.17 and 3.18). For the convenience of comparative analysis, we maintained the same scales in Figures 3.18 and 3.19 as in Figures 3.16 and 3.19.

3.6.3.4 Particular Case

Generally speaking, the suggested procedure for evaluating prior information does not result in a prior distribution available in a closed form. Nevertheless, for the particular case of the Weibull distribution with the shape parameter equal to one (the exponential distribution), the closed form can be obtained. This closed form of the prior distribution can be useful from a methodological standpoint showing that it is *not* the gamma distribution, which might intuitively be anticipated in this case.

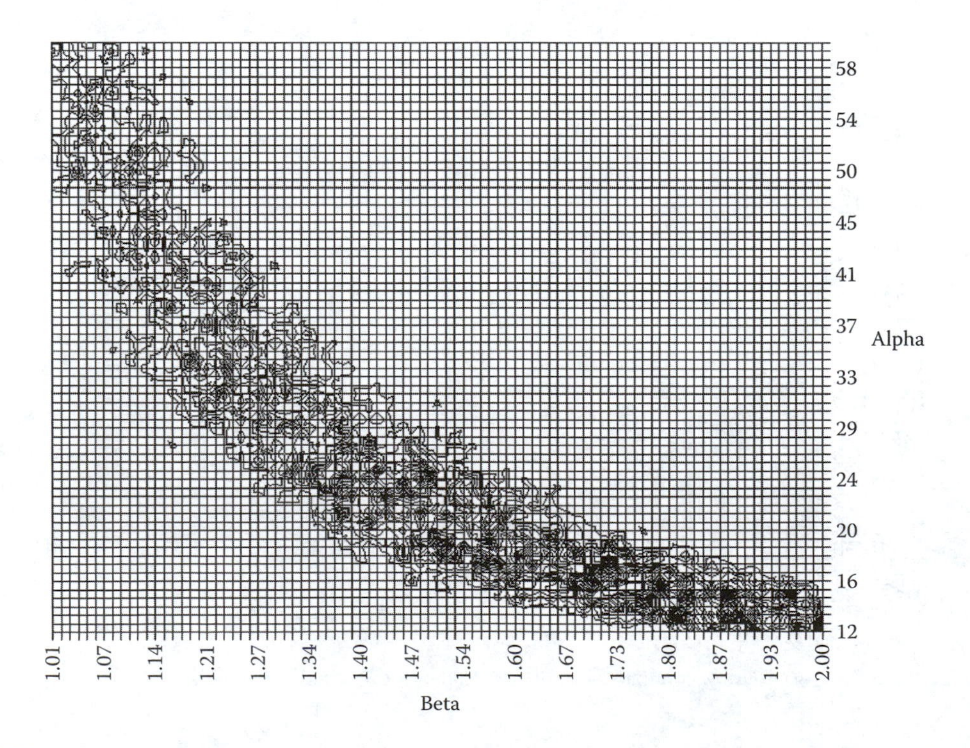

FIGURE 3.18 Contours of joint prior distribution of the Weibull parameters with increased coefficient of variation (CoV) of prior data.

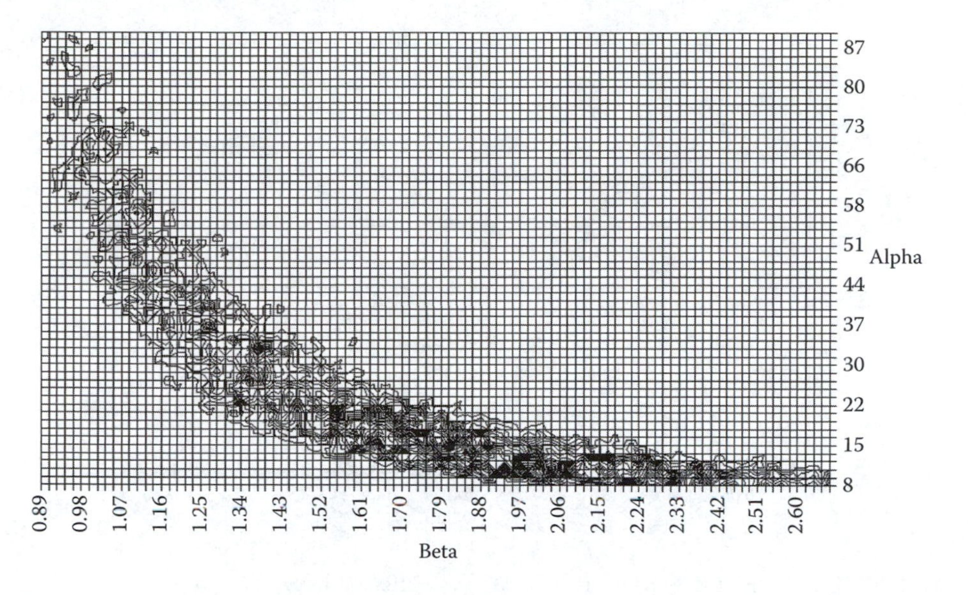

FIGURE 3.19 Contours of joint posterior distribution of the Weibull parameters with increased *CoV* of prior data.

In this case, the prior information can be reduced to the estimates of the cdf and its standard deviation at just *one* value of the nonrandom exposure, t, so that the r.v. of interest, λ, can be represented as

$$\lambda = \frac{-\ln(1 - F(t))}{t}, \tag{3.158}$$

where $F(t)$ is distributed according to the beta distribution with pdf given in Equation 3.154.

It can be shown that the pdf of Equation 3.158 is given by

$$f(\lambda) = t\frac{\Gamma(n_0)}{\Gamma(x_0)\Gamma(n_0 - x_0)}[z^{n_0 - x_0}(1 - z)^{x_0 - 1}], \tag{3.159}$$

where $z = e^{-t\lambda}$ and $\lambda > 0$. Figure 3.20 shows the pdf of the introduced distribution for various values of its parameters. The mode of Equation 3.158, the HPD, happens to be available in a closed form as

$$\lambda = \frac{1}{t}Ln\left(\frac{n_0 - 1}{n_0 - x_0}\right), \quad n_0 > 1,\ n_0 - x_0 > 0. \tag{3.160}$$

The mean of Equation 3.158 can be obtained as

$$E(\lambda) = t\frac{\Gamma(n_0)}{\Gamma(x_0)\Gamma(n_0 - x_0)}\int_0^{\infty} \lambda[e^{-(n_0 - x_0)\lambda t}(1 - e^{-\lambda t})^{x_0 - 1}]\,d\lambda. \tag{3.161}$$

Using the same variable $z = e^{-t\lambda}$, the above integral can be written in the form

$$E(\lambda) = \frac{1}{t}\left(\frac{\Gamma(\delta + \rho)}{\Gamma(\delta)\Gamma(\rho)}\right)\int_0^1 \ln\left(\frac{1}{z}\right)[z^{n_0 - x_0 - 1}(1 - z)^{x_0 - 1}]\,dz. \tag{3.162}$$

From the Tables of Integrals [31], one finds

$$\int_0^1 \left(\ln\frac{1}{z}\right)^h z^{q-1}(1 - z^b)^{(c/b)-1}\,dz = h!\left\{\frac{1}{q^{h+1}} + \sum_{k=1}^{\infty}\frac{(kb - c)[(k-1)b - c]\cdots(b - c)}{k!b^k(q + bk)^{h+1}}\right\}, \tag{3.163}$$

where $q > 0$ and $h + c/b > 0$.

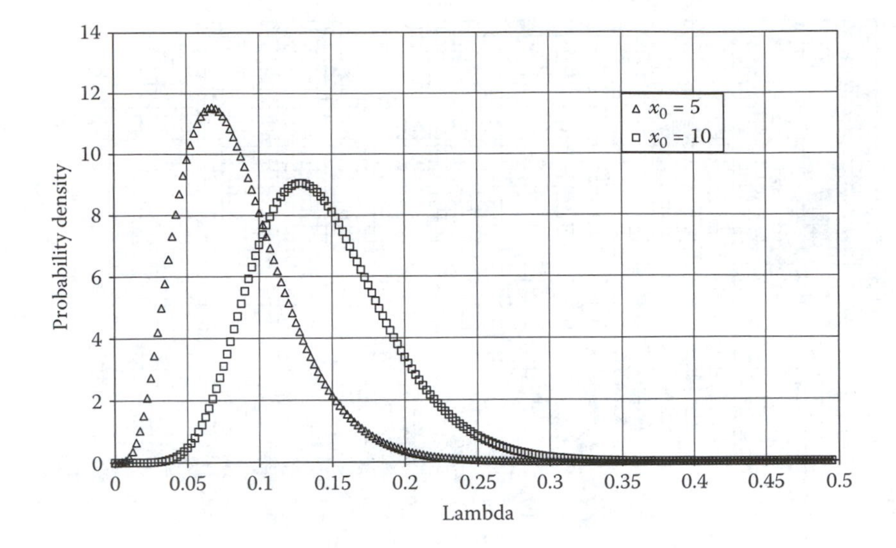

FIGURE 3.20 Pdf of Equation 3.157 with $t = 5$, $n_0 - x_0 = 10$, and x_0 of 5 and 10.

Applying Equation 3.163 to the integral in Equation 3.162, one finally obtains

$$\int_0^1 \ln\left(\frac{1}{z}\right) [z^{n_0-x_0-1}(1-z)^{x_0-1}]\,dz = \left\{\frac{1}{(n_0-x_0)^2} + \sum_{k=1}^{\infty} \frac{(k-x_0)[(k-1)-x_0]\cdots(1-x_0)}{k!(n_0-x_0+k)^2}\right\}. \quad (3.164)$$

3.6.4 Bayesian Probability Papers

In this section another simple Bayesian approach to estimation of the two-parameter reliability distributions is discussed [32]. This approach is similar to the widely used probability papers (discussed in detail in Section 3.3.2), which can be considered as the respective classical analog.

The traditional probability paper technique is applied to the distributions whose cdfs or reliability functions can be linearized in such a way that the distribution parameters are estimated through the simple linear regression model $y = ax + b$. The family of such distributions includes such common distributions as the Weibull, exponential, log-normal, Pareto, and log-logistic distributions. The estimates obtained using the probability papers are the initial estimates (for subsequent nonlinear estimation), but in reliability engineering practice, they often turn out to be the final ones as well.

In this section, we will discuss the basic assumptions related to the simple normal linear regression model in the framework of the probability paper procedures. Analogously, the probability paper procedures are considered from the standpoint of applying the Bayesian simple linear regression model. It will be shown that the Bayesian simple regression model can be applied to the probability paper procedures with approximately the same number of violations of the respective Bayesian assumptions as the classical probability paper procedures have with respect to the classical simple regression model. The discussion below is limited to the respective point estimation procedure.

The linearized cdfs and reliability functions discussed above are applicable to those lifetime (time-to-failure) distributions for which some transform of lifetime has a location-scale parameter distribution. The location-scale distribution for a r.v. t is defined as the distribution having the pdf, which can be written in the following form [33]:

$$f(t) = \frac{1}{b} f_0\left(\frac{t-u}{b}\right) \quad -\infty < t < \infty, \quad (3.165)$$

where $u(-\infty < u < \infty)$ and $b > 0$ are location and scale parameters, and $f_0(x)$ is a specified pdf on $(-\infty, \infty)$.

3.6.4.1 Classical Simple Linear Regression

Consider the basic assumptions associated with the simple, normal linear regression model. Let us assume that a random *response variable* y fluctuates about an unknown nonrandom *response function* $\eta(x)$ of nonrandom known *explanatory variable* x. That is, $y = \eta(x) + \varepsilon$, where ε is the random fluctuation or error. In the following, we consider $\eta(x)$ in the simple linear form, so that it can be written as

$$y(x) = \beta_0 + \beta_1 x + \varepsilon \tag{3.166}$$

or as

$$y(x) = \beta_0 x_0 + \beta_1 x_1 + \varepsilon, \tag{3.167}$$

where β_0 and β_1 are unknown parameters to be estimated, and $x_0 \equiv 1$, and $x_1 \equiv x$. The data related to model 3.167 are the pairs composed of observations $y_i(x_i)$ and the respective values x_i $(i = 1, 2, \ldots, n)$, $n \geq 2$. For these observations, it is assumed that

$$y_i(x_i) = \beta_0 x_0 + \beta_1 x_i + \varepsilon_i, \tag{3.168}$$

where errors ε_i $(i = 1, 2, \ldots, n)$ are independent and normally distributed with mean 0 and variance σ^2. In other words, the observations $y_i(x_i)$ are independently and normally distributed with mean $\beta_0 x_0 + \beta_1 x_i$ and variance σ^2.

For the following discussion let us consider Equation 3.168 in its matrix form, which is given by

$$Y = XB + \varepsilon, \tag{3.169}$$

where

$$Y = \begin{bmatrix} y_1 \\ y_2 \\ \cdot \\ \cdot \\ \cdot \\ y_n \end{bmatrix}, \quad X = \begin{bmatrix} 1 & x_1 \\ 1 & x_2 \\ \cdot & \cdot \\ \cdot & \cdot \\ \cdot & \cdot \\ 1 & x_n \end{bmatrix}, \quad B = \begin{bmatrix} \beta_0 \\ \beta_1 \end{bmatrix}, \quad \varepsilon = \begin{bmatrix} \varepsilon_1 \\ \varepsilon_2 \\ \cdot \\ \cdot \\ \cdot \\ \varepsilon_n \end{bmatrix},$$

where ε is the vector of errors with zero means and the following matrix of variances $\text{Var}(\varepsilon) = \sigma^2 I$, and I is the $(n \times n)$ unit (identity) matrix.

The matrix form (Equation 3.169) is used below in Example 3.32 and in further discussion below. The estimates of parameters β_0 and β_1 are (see Chapter 2):

$$\hat{\beta}_0 = \bar{y} - \hat{\beta}_1 \bar{x} \tag{3.170}$$

and

$$\hat{\beta}_1 = \frac{\Sigma(x_i - \bar{x})(y_i - \bar{y})}{\Sigma(x_i - \bar{x})^2}, \tag{3.171}$$

where $\bar{y} = n^{-1}\Sigma y_i$ and $\bar{x} = n^{-1}\Sigma x_i$.

The estimates (Equations 3.170 and 3.171) can be written in the matrix form as

$$\hat{B} = \begin{bmatrix} \hat{\beta}_0 \\ \hat{\beta}_1 \end{bmatrix} = (X'X)^{-1}X'Y, \tag{3.172}$$

where X' is the transpose of matrix X, and $(X'X)^{-1}$ is the inverse of the matrix product of X' and X.

A more general case of model 3.168 is the so-called *weighted linear regression*, when errors ε_i are still independent but have different variances σ_i^2 $(i = 1, 2, \ldots, n)$. This model will be discussed in the following, so we need to write it here as

$$y_i(x_i) = \beta_0 x_0 + \beta_1 x_i + \varepsilon_i, \tag{3.173}$$

where errors ε_i are independent normally distributed with mean 0 and different variances $\sigma_i^2(x_i)$ $(i = 1, 2, \ldots, n)$. In the matrix form, the model 3.173 can be written as

$$Y = XB + \varepsilon, \tag{3.174}$$

where ε is the vector of independent errors with zero means and (opposite of Equation 3.169) the variance matrix is the following symmetric positively defined diagonal $(n \times n)$ matrix $\mathrm{Var}(\varepsilon) = \Sigma$,

$$\Sigma = \begin{bmatrix} \sigma_1^2 & 0 & 0 & 0 \\ 0 & \sigma_2^2 & 0 & 0 \\ \cdot & \cdot & \cdot & \cdot \\ 0 & 0 & 0 & \sigma_n^2 \end{bmatrix}. \tag{3.175}$$

The above variance matrix Σ can be represented as

$$\Sigma = \sigma^2 \begin{bmatrix} 1/w_1 & 0 & 0 & 0 \\ 0 & 1/w_2 & 0 & 0 \\ \cdot & \cdot & \cdot & \cdot \\ 0 & 0 & 0 & 1/w_n \end{bmatrix}, \tag{3.176}$$

where $w_1, w_2, \ldots, w_n$ are the so-called weights. It is obvious that the greater the variance, the smaller the respective weight. The matrix of weights is defined as

$$\Sigma^{-1} = \sigma^{-2} \begin{bmatrix} w_1 & 0 & 0 & 0 \\ 0 & w_2 & 0 & 0 \\ \cdot & \cdot & \cdot & \cdot \\ 0 & 0 & 0 & w_n \end{bmatrix}. \tag{3.177}$$

The estimates of parameters β_0 and β_1 for model 3.173 are

$$\hat{B} = \begin{bmatrix} \hat{\beta}_0 \\ \hat{\beta}_1 \end{bmatrix} = (X'\Sigma^{-1}X)^{-1}X'\Sigma^{-1}Y. \tag{3.178}$$

3.6.4.2 Classical Probability Papers

Without loss of generality, consider the Weibull probability paper estimation procedure, which is common in life data analysis. Let the cdf of the Weibull time-to-failure distribution $F(t)$ be given in the following form:

$$F(t) = 1 - \exp\left(-\left(\frac{t}{\alpha}\right)^{\beta}\right), \tag{3.179}$$

where t is the time to failure, and α and β are the scale and shape parameters, respectively. Applying the logarithmic transformation twice, the above cdf is transformed to the following expression:

$$\ln(-\ln(1 - F(t))) = \beta \ln t - \beta \ln \alpha. \tag{3.180}$$

Introducing the following notation $y(t) = \ln(-\ln(1 - F(t)), \ln t = x$, $\beta_0 = \beta \ln(\alpha)$, Equation 3.180 takes on the simple linear response function:

$$y(x) = \beta_0 x_0 + \beta_1 x_1 + \varepsilon. \quad (3.181)$$

It should be noted that there is no guarantee that the errors ε are independent and normally distributed with mean 0 and variance σ^2 anymore. Nevertheless, the simple linear regression technique is widely applied to Equation 3.180, which is known as Weibull probability plotting. The corresponding procedure also includes the estimation of cdf $F(t)$ using the order statistic, which is illustrated in the framework of the following example.

EXAMPLE 3.32

One hundred identical components whose life is believed to follow a Weibull distribution were put on a life test. The test data are Type II censored: The test was terminated at the time of the fifth failure. Failure times t_i (in hours) of the five failed components were 11.96, 39.10, 71.52, 74.90, and 123.14. Estimate the parameters of the distribution using the Bayesian plotting approach.

Solution: The traditional estimates $\hat{F}(t_i)$ of cdf $F(t)$, used in the Weibull probability papers, is given by the following formula [33]:

$$\hat{F}(t_i) = \frac{i - 0.5}{n},$$

where t_i ($i = 1, 2, \ldots, r$; and $r \leq n$) are the ordered failure times. In our example, $r = 5$ and $n = 100$.

Calculating these estimates for our data and applying double logarithmic transformation (Equation 3.181) results in the following table (vector):

$$Y = \begin{bmatrix} -5.29581 \\ -4.19216 \\ -3.67625 \\ -3.33465 \\ -3.07816 \end{bmatrix}.$$

The explanatory variable x_1 is the logarithm of the failure times, so our explanatory variable matrix X is

$$X = \begin{bmatrix} 1 & 2.48196 \\ 1 & 3.66605 \\ 1 & 4.27004 \\ 1 & 4.31620 \\ 1 & 4.81332 \end{bmatrix}.$$

Now we can find the estimates of the parameters β and $\beta_0 = \beta \ln(\alpha)$ using Equations 3.170 and 3.172 or Equation 3.172 as $\hat{\beta} = 0.971$ (the estimate of the shape parameter) and $\hat{\beta}_0 = -7.712$, so that the estimate of the scale parameter is $\hat{\alpha} = 2809.852$. It should be noted that the test data of this example are the simulated data from the Weibull distribution with the scale parameter $\alpha = 1000$ and the shape parameter $\beta = 1.5$. It is clear that the estimates obtained are rather biased.

3.6.4.3 Bayesian Simple Linear Regression and Bayesian Probability Papers

Let us begin with the Bayesian interpretation of classical simple linear regression. Consider the simple normal linear regression (Equation 3.168). In the Bayesian context, it is assumed that the parameters of model β_0, β_1, and $\log \sigma$ are uniformly and independently distributed. That is,

$$p(\beta_0, \beta_1, \sigma) \propto \frac{1}{\sigma}. \quad (3.182)$$

Note that this is an extra assumption; that is, the assumptions about the observations $y_i(x_i)$ are not changed.

Assumption (Equation 3.182) is a convenient form of the *noninformative prior distribution.*

It can be shown [34] that under the given assumptions, the conditional posterior pdf for β_0 and β_1 has the *bivariate normal* form with mean $(\hat{\beta}_0, \hat{\beta}_1)$, which is given by

$$\hat{\beta}_0 = \bar{y} - \hat{\beta}_1 \bar{x}, \qquad \hat{\beta}_1 = \frac{\Sigma(x_i - \bar{x})(y_i - \bar{y})}{\Sigma(x_i - \bar{x})^2}, \tag{3.183}$$

where $\bar{y} = n^{-1}\Sigma y_i$ and $\bar{x} = n^{-1}\Sigma x_i$. The above expressions for $\hat{\beta}_0$ and $\hat{\beta}_1$ are the easily recognizable classical least-squares estimates 3.170 and 3.171 for the simple linear regression 3.168.

3.6.4.4 Including Prior Information about Model Parameters

In the framework of Bayesian linear regression analysis, the prior information can be added to one or to several regression parameters. Let us begin with including prior information about a single regression parameter, say β_1. It is supposed that the information can be expressed as the normal distribution with known mean $\beta_{1\text{pr}}$ and variance $\sigma^2_{\beta 1\text{pr}}$ [35]; that is, $\beta_{1\text{pr}} \sim N(\beta_1, \sigma^2_{\beta 1\text{pr}})$.

Note that this prior distribution is similar to classical *assumptions* about observations $y_i(x_i)$ $(i = 1, 2, \ldots, n)$ introduced above.

Based on this similarity, the prior information on parameter β_1 is interpreted as an additional (pseudo) data point in the regression data set, and the posterior point estimates are calculated using the same Equations 3.171 or 3.183. For the case considered, this "observed" value of y corresponds to $x_0 = 0$ and $x_1 = 1$.

Including prior information about a set of regression parameters is performed in a similar way. For example, the prior information about the other regression parameter, β_0, is included as a data point having the prior $\beta_{0\text{pr}} \sim N(\beta_0, \sigma^2_{\beta 0\text{pr}})$. This "observed" value of y corresponds to $x_0 = 1$ and $x_1 = 0$.

Because, for the time being, we are considering the case of independent observations with equal variances, we are expanding this assumption to the priors; in other words, it is assumed that the priors are independently and normally distributed with equal variances. That is,

$$\sigma^2 \beta_{0\text{pr}} = \sigma^2 \beta_{1\text{pr}} = \sigma^2. \tag{3.184}$$

The following example illustrates the issues discussed in this section.

EXAMPLE 3.33

Use the data from Example 3.32 and find the posterior point estimates of the parameters. (a) Use the perfect prior (the original parameters of the generating distribution). (b) Use prior information with negligible value (weight). (c) Use prior information with relatively large weights. (d) Use a prior information with moderate (equal) weight.

Solution: The prior information about the unknown parameters is incorporated as follows.

Part (a)

The shape parameter of the Weibull distribution $\beta_{\text{pr}} = 1.5$ and the scale parameter $\alpha_{\text{pr}} = 1000$. Note that we use the true values of the parameters of the Weibull distribution, from which the data were generated, so that, to an extent, our prior information is ideal.

In terms of the regression model 3.181, parameter β_1 as an additional (pseudo) data point is 1.5 with corresponding $x_0 = 0$ and $x_1 = 1$. The parameter β_0 as another pseudopoint is $\beta \ln(\alpha) = -10.36$ with the corresponding $x_0 = 1$ and $x_1 = 0$. The table (vector) of observations Y with these two new points now is

$$Y = \begin{vmatrix} -5.29581 \\ -4.19216 \\ -3.67625 \\ -3.33465 \\ -3.07816 \\ -10.3616 \\ 1.5000 \end{vmatrix}.$$

The respective explanatory variable matrix X is now

$$X = \begin{vmatrix} 1 & 2.48196 \\ 1 & 3.66605 \\ 1 & 4.27004 \\ 1 & 4.31620 \\ 1 & 4.81332 \\ 1 & 0.00000 \\ 0 & 1.00000 \end{vmatrix}.$$

As in Example 3.32, the estimates of the posterior estimates of parameters β and $\beta_0 = \beta \ln(\alpha)$ are calculated using Equations 3.170 and 3.171 or Equation 3.172, which gives $\hat{\beta}_{1\text{post}} = 1.512$ (the estimate of the shape parameter) and $\hat{\beta}_0 = -9.915$, so that the posterior estimate of the scale parameter is $\hat{\alpha}_{\text{post}} = 705.294$. See Figure 3.21 for a graphical interpretation of Example 3.33.

The above example represents a case when the pseudo-data points are assumed to have the same variances as the real data points (observations). In the framework of the weighted regression, this case corresponds to the equal weights situation. From a Bayesian standpoint, this is the situation when the prior information has as much value as the real data.

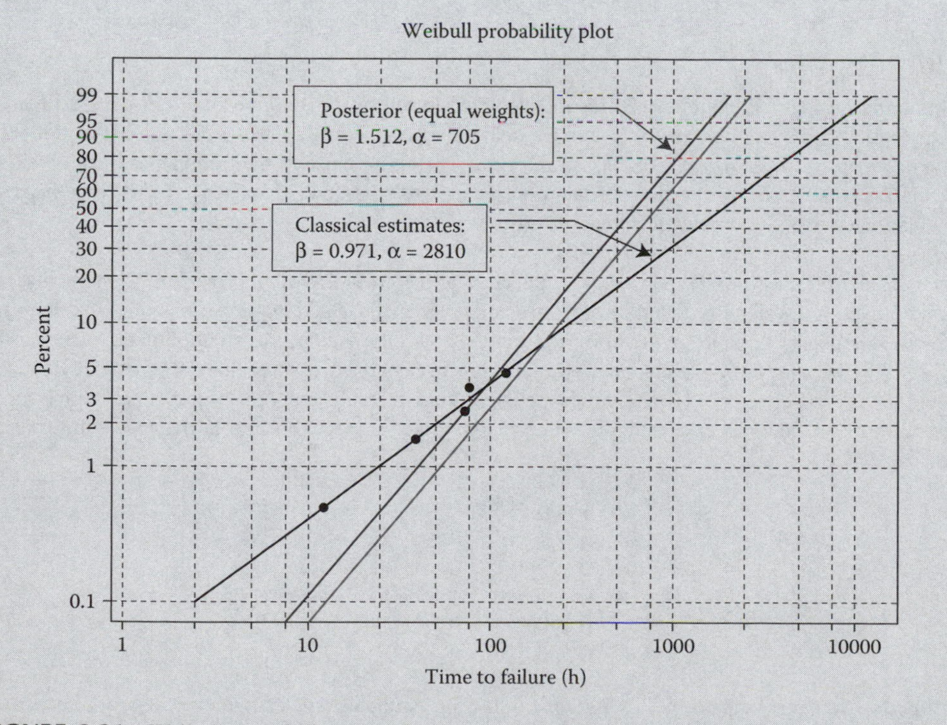

FIGURE 3.21 Weibull probability plot of prior and posterior distributions.

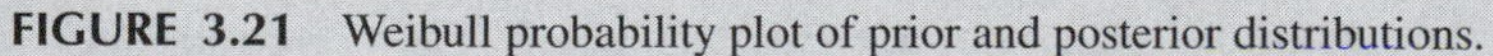

continued

Now consider the following two extreme cases.

Part (b)

In the first case, the prior information has a negligible value. This case can be realized using very small weights (large variances) related to the pseudo-data points on the Weibull plot. Let us consider the data of Example 2.32 with the following variance matrix:

$$\Sigma = \begin{bmatrix} 1 & 0 & 0 & 0 & 0 & 0 & 0 \\ 0 & 1 & 0 & 0 & 0 & 0 & 0 \\ 0 & 0 & 1 & 0 & 0 & 0 & 0 \\ 0 & 0 & 0 & 1 & 0 & 0 & 0 \\ 0 & 0 & 0 & 0 & 1 & 0 & 0 \\ 1 & 0 & 0 & 0 & 0 & 1000 & 0 \\ 0 & 0 & 0 & 0 & 0 & 0 & 1000 \end{bmatrix}.$$

Applying Equation 3.178 for the weighted linear regression results in $\hat{\beta}_{1\text{post}} = 0.975$ (the estimate of the shape parameter) and $\hat{\beta}_0 = -7.726$, so that the estimate of the scale parameter of the Weibull distribution is $\hat{\alpha}_{\text{post}} = 2772.408$. It is clear that the posterior estimates are close to the classical estimates (see Example 3.32). The result shows that in this case, the prior information does not play a significant role in the estimation.

Part (c)

Now consider the opposite case. Let us select very large weights (very small variances) related to the pseudo-data points on the Weibull plot. Let us consider the data of Example 3.33 with the following variance matrix:

$$\Sigma = \begin{bmatrix} 1 & 0 & 0 & 0 & 0 & 0 & 0 \\ 0 & 1 & 0 & 0 & 0 & 0 & 0 \\ 0 & 0 & 1 & 0 & 0 & 0 & 0 \\ 0 & 0 & 0 & 1 & 0 & 0 & 0 \\ 0 & 0 & 0 & 0 & 1 & 0 & 0 \\ 1 & 0 & 0 & 0 & 0 & 0.001 & 0 \\ 0 & 0 & 0 & 0 & 0 & 0 & 0.001 \end{bmatrix}.$$

Applying the same Equation 3.160 for the weighted linear regression gives the following estimates: $\hat{\beta}_{1\text{post}} = 1.509$ (the estimate of the shape parameter), and $\hat{\beta}_0 = -10.359$, so that the estimate of the scale parameter is $\hat{\alpha}_{\text{post}} = 958.276$. The posterior estimates are close to the true values of the Weibull distribution parameters, from which the data were generated. This result reveals that in this case, the prior information does play a dominant role in estimation.

Part (d)

Now consider the case close to a practical application of the given Bayesian procedure. Assume that the data points on the Weibull probability plot have equal variances (standard deviations) and that they are equal to 1. A degree of belief in the prior information about the Weibull distribution parameters can be expressed in the same terms of standard deviations. It is reasonable to assume that the standard deviations related to the respective pseudo-data points are larger than the real data points, for example, three times larger. Let us consider this case using the same example. The respective variance matrix for this case is

$$\Sigma = \begin{bmatrix} 1 & 0 & 0 & 0 & 0 & 0 & 0 \\ 0 & 1 & 0 & 0 & 0 & 0 & 0 \\ 0 & 0 & 1 & 0 & 0 & 0 & 0 \\ 0 & 0 & 0 & 1 & 0 & 0 & 0 \\ 0 & 0 & 0 & 0 & 1 & 0 & 0 \\ 1 & 0 & 0 & 0 & 0 & 9.000 & 0 \\ 0 & 0 & 0 & 0 & 0 & 0 & 9.000 \end{bmatrix}.$$

Applying the same Equation 3.178 gives the following estimates: The estimate of the shape parameter is $\hat{\beta}_{1\text{post}} = 1.934$, and the estimate of the scale parameter is $\hat{\alpha} = 1356.920$. The posterior estimates are based on both types of data: the real observations and the prior information.

TABLE 3.6
Classical and Bayesian Probability Paper Estimation Summary

Estimation Procedure and Data	Estimate of α	Estimate of β
Classical procedure. Real data only	2810	0.971
Bayes's procedure with equal weights based on real data and ideal prior estimates, that is, $\alpha_{pr} = 1000$, $\beta_{pr} = 1.5$	705	1.512
Bayes's procedure with negligible prior information, that is, prior estimates have very small weights (large variances)	2772	0.975
Bayes's procedure with prior information strongly dominating real data, that is, prior estimates have very large weights (small variances)	958	1.509
Bayes's procedure with prior information comparable with real data information.	1357	1.934

Note: True values of the Weibull distribution parameters are $\alpha = 1000$ and $\beta = 1.5$.

3.6.4.5 Including Prior Information about the Reliability or cdf

Prior information about the reliability function or cdf can be included in the data set using a similar approach, that is, by treating the prior knowledge about the reliability function at some given times as additional data points, and expressing the degree of belief in terms of standard deviations of prior reliability function estimates, which can be obtained using either expert opinion or appropriate data (e.g., data on the predecessor product or alpha version testing) (Table 3.6).

EXAMPLE 3.34 REAL ACCELERATED FATIGUE TEST DATA

A sample of 12 identical induction-hardened steel ball-joints underwent an accelerated fatigue life test with the following cycles to failure (in thousands): 150, 170, 180, 200, 200, 215, 220, 220, 250, 260, 265,

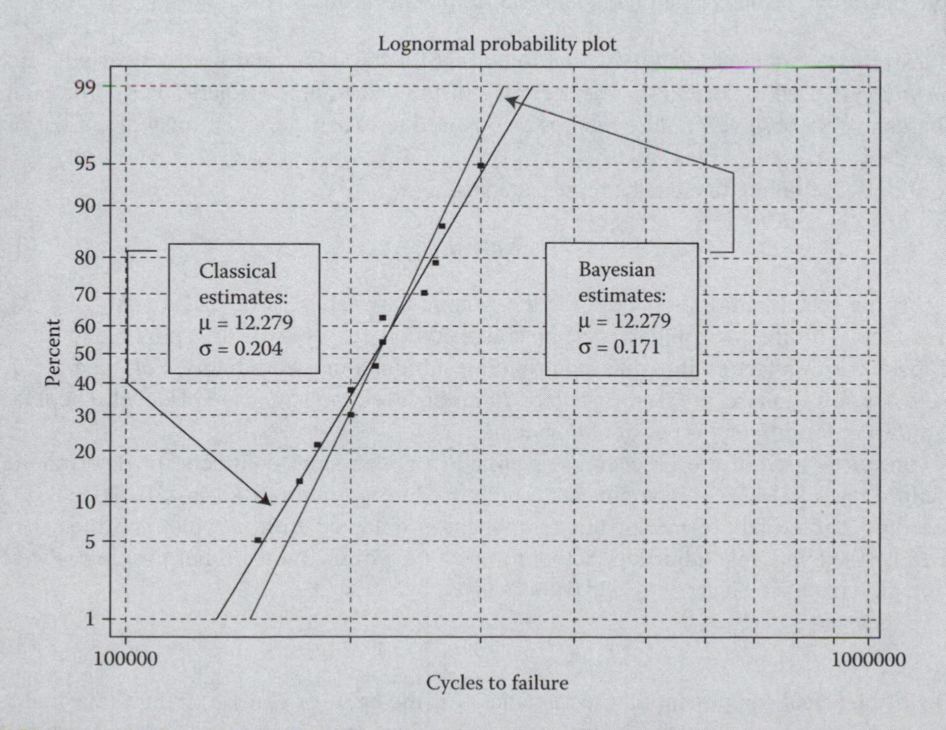

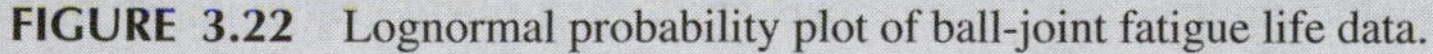
FIGURE 3.22 Lognormal probability plot of ball-joint fatigue life data.

continued

and 300. Based on long-term history of such tests, the underlying life distribution was assumed to be lognormal. The cdf of the lognormal distribution with location parameter μ, and scale parameter σ is linearized using the following simple transformation:

$$\Phi^{-1}[F(t)] = \frac{1}{\sigma}\ln(t) - \frac{\mu}{\sigma},$$

where $\Phi^{-1}[\cdot]$ is the inverse of the standard normal cdf. The classical least-squares estimates of the location and scale parameters in this case are 12.279 and 0.204, respectively. Historical data suggests that the scale parameter should be 0.160. Using a procedure similar to that outlined in part (a) of Example 3.33 (equal weights), the Bayesian posterior estimate of the scale parameter was found to be 0.171. The analysis is graphically summarized in Figure 3.22.

3.7 METHODS OF GENERIC FAILURE RATE DETERMINATION

Due to the lack of observed data, component reliability determination may require the use of generic failure data adjusted for the various factors that influence the failure rate for the component under analysis. Generally, these factors are

1. *Environmental Factors*—These factors affect the failure rate due to extreme mechanical, electrical, nuclear, and chemical environments. For example, a high-vibration environment would lead to high stresses that promote failure of components.
2. *Design Factors*—These factors affect the failure rate due to the quality of material used or workmanship, material composition, functional requirements, geometry, and complexity.
3. *Operating Factors*—These factors affect the failure rate due to the applied stresses resulting from operation, testing, repair and maintenance practices, and so on.

To a lesser extent, the *age factor* is used to correct for the infant and wearout periods, and the *original factor* is used to correct for the accuracy of the data source (generic data). For example, obtaining data from observed failure records as opposed to expert judgment may affect the failure rate dependability.

Accordingly, the failure rate can be represented as

$$\lambda_{\mathrm{a}} = \lambda_{\mathrm{g}} K_{\mathrm{E}} K_{\mathrm{D}} K_{\mathrm{O}}, \ldots, \tag{3.185}$$

where λ_{a} is the actual failure rate and λ_{g} is the generic base failure rate, and K_{E}, K_{D}, and K_{O} are correction factors for the environment, design, and operation, respectively. It is possible to subdivide each of the correction factors into their contributing subfunctions accordingly. For example, $K_{\mathrm{E}} = f(k_a, k_b, \ldots)$, when k_a and k_b are factors such as vibration level, moisture, and pH level. These factors may be different for different types of components.

This concept is used in the procedures specified in government contracts for determining the actual failure rate of electronic components. The procedure is summarized in MIL-HDBK-217. In this procedure, a base failure rate of the component is obtained from a table, and then they are multiplied by the applicable adjusting factors for each type of component. For example, the actual failure rate of a tantalum electrolytic capacitor is given by

$$\lambda_{\mathrm{p}} = \lambda_{\mathrm{b}} \left(\pi_{\mathrm{E}} \cdot \pi_{\mathrm{SR}} \cdot \pi_{\mathrm{Q}} \cdot \pi_{\mathrm{CV}}\right), \tag{3.186}$$

where λ_{p} is the actual component failure rate and λ_{b} is the base (or generic) failure rate, and the π factors are adjusting factors for the environment, series resistance, quality, and capacitance factors.

Values of λ_b and the factors are given in MIL-HDBK-217 for many types of electronic components. Generally, λ_b is obtained from an empirical model called the Arrhenius model:

$$\lambda_b = K \exp(E/kT), \tag{3.187}$$

where E = activation energy for the process, $k = 1.38 \times 10^{-23}\,\text{J} \cdot \text{K}^{-1}$, T = absolute temperature (°K), and K = a constant. The readers are referred to 217Plus (http://quanterion.com/RIAC/) for the most updated generic data on electronics components. Further, readers interested in the extension of the physics-based generic failure data for electronic components should refer to Salemi [36].

The Arrhenius model forms the basis for a large portion of electronic components described in MIL-HDBK-217. However, care must be applied in using this database, especially because the data in this handbook are derived from repairable systems (and hence, apply to such systems). Also, application of the various adjusting factors can drastically affect the actual failure rates. Therefore, proper care must be applied to ensure correct use of the factors and to verify the adequacy of the factors suggested [37]. Also, the appropriateness of the Arrhenius model has been debated many times in the literature. The statistical procedures for fitting the Arrhenius model and other reliability models with explanatory factors are considered in the accelerated life testing section of this book (see Chapter 7).

For other types of components, many different generic sources of data are available. Among them are IEEE-500 [38], Guidelines for Process Equipment Data [39], IEC/TR 62380 Reliability data handbook [40], Nuclear Power Plant, and Probability Risk Assessment (PRA) data sources. For example, Table B.1 (in Appendix B) shows a set of data obtained from NUREG/CR-4550 [41]. Finally the readers are referred to Keller and Amendola [42] for a more detailed discussion of generic reliability engineering databases.

EXERCISES

3.1 For a gamma distribution with a scale parameter of 400, and a shape parameter of 3.8, determine $\Pr(x < 200)$.

3.2 Time to failure of a relay follows a Weibull distribution with $\alpha = 10$ years, $\beta = 0.5$. Find the following:

a. Pr(failure after 1 year).
b. Pr(failure after 10 years).
c. The MTTF.

3.3 The hazard rate of a device is $h(t) = 1/\sqrt{t}$. Find the following:

a. pdf
b. Reliability function
c. MTTF
d. Variance.

3.4 Assume that 100 components are placed on test for 1000 h. From previous testing, we believe that the hazard rate is constant, and the MTTF = 500 h. Estimate the number of components that will fail in the time interval of 100–200 h. How many components will fail if it is known that 15 components failed in $T < 100$ h?

3.5 Assume that t, the r.v. that denotes life in hours of a specified component, has a cdf of

$$F(t) = \begin{cases} 1 - \dfrac{100}{t}, & t \geq 100; \\ 0, & t < 100. \end{cases}$$

Determine the following:

a. pdf $f(t)$
b. Reliability function $R(t)$
c. MTTF.

3.6 Show whether a uniform distribution represents an IFR, DFR, or constant failure rate.

3.7 For the following Rayleigh distribution,

$$f(t) = \frac{t}{\alpha^2} \exp\left(\frac{-t^2}{2\alpha^2}\right), \quad t \geq 0,\ \alpha > 0.$$

a. Find the hazard rate $h(t)$ corresponding to this distribution.

b. Find the reliability function $R(t)$.

c. Find the MTTF. Note that $\int_0^\infty e^{-ax^2} = \frac{1}{2}\sqrt{\frac{\pi}{a}}$.

d. For which part of the bathtub curve is this distribution adequate?

3.8 Due to the aging process, the failure rate of a nonrepairable item is increasing according to $\lambda(t) = \lambda\beta t^{\beta-1}$. Assume that the values of λ and β are estimated as $\hat{\beta} = 1.62$ and $\hat{\lambda} = 1.2 \times 10^{-5}$ h. Determine the probability that the item will fail sometime between 100 and 200 h. Assume an operation beginning immediately after the onset of aging.

3.9 Suppose r.v. X has the exponential pdf $f(x) = \lambda \exp[-\lambda x]$ for $x > 0$, and $f(x) = 0$ for $x \leq 0$. Find $\Pr(x > a + b|x > a)$, given a, $b > 0$.

3.10 The following time-to-failure data are found when 158 transformer units are put under test. Use a nonparametric method to estimate $f(t)$, $h(t)$, and $R(t)$ of the transformers. No failures are observed prior to 1750 h.

Age Range (h)	Number of Failures
1750	17
2250	54
2750	27
3250	17
3750	19
4250	24

3.11 A test was run on 10 electric motors under high temperature. The test was run for 60 h, during which six motors failed. The failures occurred at the following times: 37.5, 46.0, 48.0, 51.5, 53.0, and 54.5 h. We do not know whether an exponential distribution or a Weibull distribution model is better for representing these data. Use the plotting method as the main tool to discuss the appropriateness of these two models.

3.12 A test of 25 integrated circuits over 500 h yields the following data:

Time Interval	Number of Failures in Each Interval
0–100	10
100	7
200	3
300	3
400	2

Plot the pdf, hazard rate, and reliability function for each interval of these integrated circuits using a nonparametric method.

3.13 Total test time of a device is 50,000 h. The test is terminated after the first failure. If the pdf of the device time to failure is known to be exponentially distributed, what is the probability that the estimated failure rate is not greater than 4.6×10^{-5} per hour?

3.14 A manufacturer uses exponential distribution to model the "cycle-to-failure" number of its products. In this case, r.v. T in the exponential pdf represents the number of cycles to failure $\lambda = 0.003$ failure/cycle.

a. What is the mean number of cycles to failure for this product?
b. If a component survives for 300 cycles, what is the probability that it will fail sometime after 500 cycles? Accordingly, if 1000 components have survived 300 cycles, how many would be expected to fail after 500 cycles?

3.15 The shaft diameters in a sample of 25 shafts are measured. The sample mean of diameter is 0.102 cm, with a standard deviation of 0.005 cm. What is the upper 95% confidence limit on the mean diameter of all shafts produced by this process, assuming that the distribution of shaft diameters is normal?

3.16 The sample mean life of 10 car batteries is of 102.5 months, with a standard deviation of 9.45 months. What are the 80% confidence limits for the mean and standard deviation of a pdf that represents these batteries?

3.17 The breaking strength X of five specimens of a rope of 1/4-inch diameter are 660, 460, 540, 580, and 550 lb. Estimate the following:

a. The mean breaking strength by a 95% confidence level assuming normally distributed strength.
b. The point estimate of strength value at which only 5% of such specimens would be expected to break if μ is assumed to be an unbiased estimate of the true mean and σ is assumed to be the true standard deviation. (Assume x is normally distributed)
c. The 90% confidence interval of the estimate of the standard deviation.

3.18 One-hundred and twenty-four devices are placed on an overstress test with failures occurring at the following times.

Time (h)	0.4	1	2	5	8	12	25
Total number of failures	1	3	5	15	20	30	50

a. Plot the data on Weibull probability paper.
b. Estimate the shape parameter.
c. Estimate the scale parameter.
d. What other distributions may also represent these failure data?

3.19 Seven pumps have failure times (in months) of 15.1, 10.7, 8.8, 11.3, 12.6, 14.4, and 8.7. (Assume an exponential distribution.)

a. Find a point estimate of the MTTF.
b. Estimate the reliability of a pump for $t = 12$ months.
c. Calculate the 95% two-sided interval of λ.

3.20 The average life of a certain type of small motor is 10 years, with a standard deviation of 2 years. The manufacturer replaces free of charge all motors that fail while under warranty. If the manufacturer is willing to replace only 3% of the motors that fail, what warranty period should be offered? Assume that the time to failure of the motors follows a normal distribution.

3.21 A manufacturer claims that certain machine parts will have a mean diameter of 4 cm, with a standard deviation of 0.01 mm. The diameters of five parts are measured and found to be (in mm) 39.98, 40.01, 39.96, 40.03, and 40.02. Would you accept this claim with a 90% confidence level?

3.22 You are to design a life test experiment to estimate the failure rate of a new device. Your boss asks you to make sure that the 80% upper and lower limits of the estimate interval (two-sided) do not differ by more than a factor of 2. Due to cost constraints, the components will be tested until they fail. Determine how many components should be put on test.

3.23 For an experiment, 25 relays are allowed to run until the first failure occurs at $t = 15$ h. At this point, the experimenters decide to continue the test for another 5 h. No failures occur during this extended period, and the test is terminated. Using the 90% confidence level, determine the following:

a. Point estimate of MTTF.
b. Two-sided confidence interval for MTTF.
c. Two-sided confidence interval for reliability at $t = 25$ h.

3.24 A locomotive control system fails 15 times out of the 96 times it is activated to function. Determine the following:

a. A point estimate for failure probability of the system.
b. 95% two-sided confidence intervals for the probability of failure. (Assume that after each failure, the system is repaired and put back in an as-good-as-new state.)

3.25 A sample of 10 measurements of a sphere diameter gives a mean of 4.38 inches, with a standard deviation of 0.06 inch. Find the 99% confidence limits of the actual mean and standard deviation.

3.26 A sample of measurements, which is assumed to follow a normal distribution, is taken from a study of an industrial process. For this sample, the 95% confidence error on estimating the mean (μ) is 2.2. What sample size should be taken if we want the 99% confidence error to be 1.5, assuming the same sample variance?

3.27 Suppose the generic failure rate of a component corresponding to an exponential time-to-failure model is $\lambda_g = 10 - 3\,\text{h}^{-1}$ with a standard deviation of $\lambda_g/2$. Assume that 10 components are closely observed for 1500 h and one failure is observed. Using the Bayesian method, calculate the mean and variance of λ from the posterior distribution. Calculate the 90% lower confidence limit.

3.28 In the reactor safety study, the failure rate of a diesel generator can be described as having a lognormal distribution with the upper and lower 90% bounds of 3×10^{-2} and 3×10^{-4}, respectively. If a given nuclear plant experiences two failures in 8760 h of operation, determine the upper and lower 90% bounds given this plant experience. (Consider the reactor safety study values as prior information.)

3.29 Five measurements of the breaking strength of a computer board were recorded as 0.28, 0.30, 0.27, 0.33, and 0.31 kgf. Find the point estimate and the 99% confidence intervals for the actual mean breaking strength, assuming the breaking strength is distributed exponentially.

3.30 The number of days in a 50-day period during which x failures of an assembly line are recorded as follows. Use a χ^2 goodness-of-fit test t_p to determine whether a Poisson distribution is a good fit to these data. Perform the test at a 5% significance level.

Number of failures (x)	0	1	2	3	4
Number of days × Failures observed	21	18	7	3	1

3.31 Fifty identical units of a manufactured product are tested for 300 h. Only one failure is observed (the failed unit is replaced with a good one).

a. Find an estimate of the failure rate of this unit.
b. Find the 90% confidence interval (two-sided) for the actual failure rate.

3.32 A mechanical life test of 18 circuit breakers of a new design was run to estimate the percentage of failures over 10,000 cycles of operation. Breakers were inspected on a schedule, and it is known that failures occurred between certain inspections as shown below:

Cycles (×1000)	10–15	15–17.5	17.5–20	20–25	25–30	30+
Number of failures	2	3	1	1	2	9 survived

a. Make a Weibull plot of these data. Is this a good fit?
b. Graphically estimate percentage of failures over 10,000 cycles.
c. Graphically estimate the Weibull parameters.

3.33 Fifty-eight fans in service are supposed to have an exponential life distribution with an MTTF of 28,700 h. Assuming that a failed fan is replaced with a new that does not fail, predict the number of such fans that will fail in 2000 h.

3.34 A manufacturer tests 125 high-performance contacts and finds that three are defective.

a. Calculate the probability that a random contact is defective.
b. What is the 90% confidence interval for the estimated probability in (a)?

3.35 If the time-to-failure pdf of a component follows a linear model as follows:

$$f(t) = \begin{cases} ct, & 0 < t < 10{,}000, \\ 0 & \text{otherwise.} \end{cases}$$

determine

a. Reliability function.
b. Failure rate function.

3.36 The cycle to failure T for a certain kind of component has the instantaneous failure rate $\lambda(t) = 2.5 \times 10^{-5}\ t^2 \geq 0$ per cycles. Find the MCTF (mean cycle to failure), and the reliability of this component at 100 cycles.

3.37 The following data were collected by Frank Proschan in 1983. Operating hours to the first failure of an engine cooling part in 13 aircrafts are

Aircraft	1	2	3	4	5	6	7	8	9	10	11	12	13
Hours	194	413	90	74	55	23	97	50	359	50	130	487	102

a. Would these data support an IFR, DFR, or constant failure rate assumption?
b. Based on a graphic nonparametric analysis of these data, confirm the results obtained in part (a).

3.38 The following times to failure in hours were observed in an experiment where 14 units were tested until eight of them have failed:

$$80, 310, 350, 470, 650, 900, 1100, 1530.$$

Assuming that the units have a constant failure rate, calculate a point estimate of the failure rate. Also, calculate a 95% one-sided confidence interval of the failure rate.

3.39 A life test of 10 small motors with a newly designed insulator has been performed. The following data are obtained:

Motor no.	1	2	3	4	5	6	7	8	9	10
Failure time (h)	1175	1200	1400	1450	1580	1870	1930	2120	2180	2430

a. Make a Weibull plot of these data and estimate the parameters.
b. Estimate the motor reliability after 6 months of continuous operation.

3.40 Use the data in problem 3.39 to perform a total-time-on-test plot.

3.41 A company redesigns one of its compressors and wants to estimate reliability of the new product. Using its past experience, the company believes that the reliability of the new compressor will be higher than 0.5 (for a given mission time). The company's testing of one new compressor showed that the product successfully achieved its mission.

a. Assuming a uniform prior distribution for the above reliability estimate, find the posterior estimate of reliability based on the test data.
b. If the company conducted another test, which resulted in another mission success, what would be the new estimate of the product reliability?

3.42 The specification calls for a power transistor to have a reliability of 0.95 at 2000 h. Five hundred transistors are placed on test for 2000 h with 15 failures observed. The failure units are substituted upon failure. Has this specification been met? What is the chance that the specification is not met?

3.43 The following failure times, in days, were obtained by placing 30 units on a 100-day life test: 8, 12, 22, 51, 73, and 85 days.

a. Perform a Weibull plot and show that a two-parameter Weibull pdf is a good fit. What are the estimates of the pdfs parameters?
b. Use the nonparametric method to find an empirical hazard rate.
c. Compare the results of (a) and (b).

3.44 The time to failure of a solid-state power unit has a hazard function in the form of $h(t) = 1.34 \times 10^{-4} t^{1/2}$ for $t \geq 0$.

a. Compute reliability at 50 h.
b. Determine design life if a reliability of 0.99 is desired.
c. Compute MTTF.
d. If the unit has operated for 50 h, what is the probability that the unit will operate for another 50 h?
e. What is the mean residual time to failure at $t = 50$ (hour)? Compare it with part (c) results.

3.45 The following censored data reflect failure times in months, of a new laser printer. Censored times result from removal of the printer due to upgrades. Determine the reliability of this printer over its 2-year warranty period.

$$8, 33, 15+, 27, 18, 24+, 13+, 12, 37, 29+, 25, 30.$$

3.46 Consider a r.v. with cdf $F(t) = 2/t, 0 < t \leq 2$. Do the following:

a. Derive expressions for the corresponding time-to-failure pdf, hazard function, and reliability function.
b. Compute $\Pr(0.1 < T \leq 0.2 | T > 0.1)$.

3.47 A life test on silicon photodiode detectors in which 28 detectors were tested at very high stress condition is performed. Failures were found after the inspections at 2500 (1 failure), 3000 (1 failure), 3500 (2 failures), 3600 (1 failure), 3700 (1 failure), and 3800 (1 failure). The other 21 detectors had not failed after 3800 h of operation. Use these data to estimate the life distribution of such photodiode detectors running at the test conditions:

a. Compute and plot a nonparametric estimate of the cdf for time to failure at the test conditions.
b. Compute hazard rate.
c. Compute reliability function.

3.48 Data on minutes to breakdown for an insulating fluid were obtained for 11 tests at 30 kV. After 100 min, there were 7 breakdowns (failures) at the following times (in minutes): 7.74, 17.05, 20.46, 21.02, 22.66, 43.40, and 47.30. The other four units had not failed.

a. Make a Weibull probability plot.
b. What do you think about the suggestion of using an exponential distribution to model the data?
c. Assuming an exponential distribution, obtain the maximum likelihood point estimate of failure rate.
d. Compute an approximate 95% confidence interval for λ.
e. Compute the MLE of the approximate reliability at 95%, one-sided confidence interval of reliability at $t = 50$ min.

3.49 A manufacturer of clocks claims that a certain model will last 5 years on the average with a standard deviation of 1.2 years. A random sample of 6 such clocks lasted 6, 5.5, 4, 5.2, 5, and 4.3 years. Discuss the adequacy of such a claim. Use any method to support your statistical or probabilistic conclusion.

3.50 A sample of 100 specimens of a titanium alloy was subjected to a strain-life fatigue test to determine time to crack initiation. The test was run up to a limit of 100,000 cycles. The observed times of crack initiation (in units of 1000 of cycles) were 18, 32, 39, 53, 59, 68, 77, 78, and 93. No crack had initiated in any of the other 91 specimens.

a. Compute a nonparametric estimate of the pdf and reliability function.
b. Plot the data on Weibull paper. Use the plot to obtain an estimate of the Weibull shape parameter β and α, and comment on the adequacy of the Weibull distribution.

3.51 The reliability of a turbine blade can be represented by the following expression:

$$R(t) = \left(1 - \frac{t}{t_0}\right)^2, \quad 0 \leq t \leq t_0,$$

where t_0 is the maximum life of the blade.

a. Show that the blades are experiencing wearout.
b. Compute MTTF for a maximum life of 3000 h.

3.52 The following censored data reflect failure times in months, of a new computer scanner. Determine the reliability of this scanner over its 2-year warranty period.

$$9, 35+, 17, 28, 25+, 37+, 28+, 27.$$

3.53 Use Rank Increment rank statistic to plot a Weibull pdf using the following multiple censored data. Which part of the bathtub curve are these data most likely coming from?

Test No.	1	2	3	4	5	6	7	8	9	10
Time to failure	309+	386	180	167+	122	229	104+	217+	168	138

"+" = censoring.

3.54 The failure time (in hour) of a component is lognormally distributed with parameters $\mu_t = 4$ and $\sigma_t = 0.9$.

a. What is the MTTF for this component?
b. When should the component be replaced, if the minimum required reliability for the component is 0.95?
c. What is the value of the hazard function at this time (i.e., for the time calculated in part (b))?
d. What is the mean residual life at this time if the component has already survived to this time (i.e., for the time calculated in part (b))?

3.55 On surface of a highly stressed motor shaft there are five crack initiation points observed after the shaft is tested at 6000 rpm for 5 h.

a. Calculate the point estimate of a crack initiation per cycle (per turn of the shaft).
b. Determine the 90% two-sided confidence interval of the probability of a crack initiation per cycle.
c. What is the point estimate of the mean-cycles-to-crack initiation?
d. What is the probability of no crack initiation in a test of 2000 rpm for 10 h?

3.56 A household appliance is advertised as having more than a 10-year life. If the following is its time-to-failure pdf, determine the reliability of a new appliance at 10 year:

$$f(t) = 0.1(1 + 0.05t)^3 \quad t \geq 0.$$

What is its MTTF when it is new, and what is its MTTF when a 1-year warranty period ends, assuming it still works?

3.57 Assume exponential pdf model with the failure rate λ describes the time to failure of a component. Suppose we have observed failures of n such components in a life test as: $t_1, t_2, \ldots, t_n$.

a. Write the likelihood function representing this set of observed failure data.
b. Given the failure data observed, write the equation for the posterior pdf of λ, if the prior pdf of λ can be expressed by a uniform distribution between λ_1 and λ_2.

c. If $n = 2, t_1 = 100, t_2 = 150, \lambda_1 = 0.05$, and $\lambda_2 = 0.001$, calculate the posterior pdf, mean, and 95% bounds of λ.

$$\left[\text{Hint: } \int \lambda^2 e^{\alpha\lambda}\, d\lambda = e^{\alpha\lambda}\left(\frac{\lambda^2}{\alpha} - \frac{2\lambda}{\alpha^2} + \frac{2}{\alpha^3}\right).\right]$$

3.58 Consider the following reliability test data obtained for a given component whose time-to-failure distribution is known to follow exponential distribution.

a. Write the likelihood function used to estimate λ.
b. What is the MLE of the failure rate, λ?
c. Use Kaplan–Meier order statistic to see whether exponential pdf is an appropriate model. What would be the corresponding estimate of λ?

Test No.	1	2	3	4	5	6	7	8
Time to failure	823	900+	1120	2301	3000+	3000+	3211	4000+

"+" = Right censored data.

3.59 a. The time to failure of a component follows an exponential distribution, and a right censored life test of 10 components with replacement yielded three failures in 8760 h; calculate the point estimate of the unreliability associated with the output from the system after 1 year of operation.

b. What is the 90% confidence interval of unreliability of each component?

3.60 Calculate the conditional reliability of a unit at $t = 1000$ h if it has survived up to $t = 500$ h. Assume the time to failure of the unit follow a Weibull distribution with $\alpha = 1800$ (hour) and $\beta = 1.8$.

3.61 What is the reliability function of a component if its hazard rate is expressed as $h(t) = 0.01/(t + 1)$.

3.62 The following multiply censored data reflect failure times, in months, of a new laser printer. Censored times result from removals of the printer due to upgrades. Determine the reliability of this printer over its 2-year warranty period. Use the adjusted rank method, the Kaplan–Meier method.

8	33	15^+	27	18	24^+	13^+	12	37	29^+	25	30

"+" = Right censored data.

3.63 In a reliability test under normal stress load, 500 components have been exposed to a given stress profile under constant environmental condition. After completion of the test, 12 components were failed. Determine:

a. An estimate of the reliability of this component to endure the given stress profile.
b. The upper 90% confidence limit for the expected number of reliable components under the same stress profile, if we use 1000 such components in a plant.

3.64 The pdf of the cycle to failure of a component is expressed by $f(N) = (c/N^{\alpha})$, where c and α are constants and $1 < N < 10^6$.

a. N^{α} determine the value of constant c.
b. If we have gathered the following cycles to failure data, set up the maximum likelihood function and a point estimate of α: $N1 = 12{,}920, N2 = 29{,}805, N3 = 76{,}560, N4 = 96{,}723$, and $N5 = 102{,}902$.
c. Estimate parameter α using probability plotting.

3.65 Life data were gathered in a right censored test that ended in 6 months:

Time Interval in Months	Number of Units Failed	Number of Units Censored
$1 < T < 2$	1	5
$2 < T < 3$	3	2
$3 < T < 4$	5	7
$4 < T < 5$	7	5
$5 < T < 6$	9	4
$6 < T$		25

a. Use the Kaplan–Meier plotting method and determine whether a Weibull distribution is proper.
b. From the plot, estimate the parameters of the distribution.
c. From the plot, calculate the 90% reliable life.
d. From the plot, estimate the probability the component will fail in the interval $5 < T < 6$.

3.66 In a mission, 11 of 2012 switches in a plant are defective. Estimate the reliability of a single component and the associated 90% confidence limits of the true reliability. Use a uniform prior estimate of the probability of a single failure to be in the range of 0.001–0.05 and find the posterior estimate of the reliability including the 90% probability bounds. Compare the results with the maximum likelihood method.

3.67 Experts estimate the reliability of a new device at the end of its warranty period as more than 0.9. Assume that this estimate can be expressed as a noninformative prior (e.g., in form of a uniform distribution) for the reliability. One has performed tests of seven such components and has observed one failure over the same warranty period. What is the posterior estimate of the reliability?

BIBLIOGRAPHY

1. Barlow, R. E. and F. Proschan, *Statistical Theory of Reliability and Life Testing: Probability Models*, To Begin With, Silver Spring, MD, 1981.
2. Johnson, N. L. and S. Kotz, *Distribution in Statistics*, Wiley, New York, 1970.
3. Mann, N. R. E., R. E. Schafer, and N. D. Singpurwalla, *Methods for Statistical Analysis of Reliability and Life Data*, Wiley, New York, 1974.
4. Provan, J. W., Probabilistic approaches to the material-related reliability of fracture-sensitive structures, in *Probabilistic Fracture Mechanics and Reliability*, J. W. Provan, ed., Martinus Nijhoff Publishers, Dordrecht, 1987.
5. Castillo, E., *Extreme Value Theory in Engineering*, Academy Press, San Diego, CA, 1988.
6. Frechet, M., Sur la loi de probabilite de l'ecart maximum, *Ann. Soc. Polon. Math, Cracow*, 6, 93, 1927.
7. Fisher, R. A. and L. H. C. Tippet, Limiting forms of the frequency distributions of the largest or smallest member of a sample, *Proc. Cambridge Philos. Soc.*, 24, 180–190, 1928.
8. Gnedenko, B. V., Limit theorems for the maximal term of a variational series, *Comptes Rendus de l'Academie des Sciences de l'URSS*, 32, 7–9, 1941.
9. Nelson, W., *Applied Life Data Analysis*, Wiley, New York, 1982.
10. Gumble, E. J., *Statistics of Extremes*, Columbia University Press, New York, 1958.
11. Blom, G., *Statistical Estimates and Transformed Beta Variables*, Wiley, New York, 1958.
12. Kimbal, B. F., On the choice of plotting position on reliability paper, *J. Am. Stat. Assoc.*, 55, 546–560, 1960.
13. Kapur, K. C. and L. R. Lamberson, *Reliability in Engineering Design*, John Wiley and Sons, New York, NY, 1977.
14. Nelson, W., *How to Analyze Data with Simple Plots*, ASQC Basic Reference in Quality Control: Statistical Techniques, Am. Soc. Quality Control, Milwaukee, WI, 1979.
15. Martz, H. F. and R. A. Waller, *Bayesian Reliability Analysis*, Wiley, New York, 1982.
16. Kececioglu, D., *Reliability Engineering Handbook*, Prentice-Hall, NJ, 1991.

17. Barlow, R. E. and R. A. Campo, *Total Time on Test Processes and Applications to Failure Data Analysis, Reliability and Fault Tree Analysis*, F. Barlow and N. Singpurwalla, eds, SIAM, Philadelphia, pp. 451–481, 1975.
18. Barlow, R. E., Analysis of retrospective failure data using computer graphics, *Proceedings of the 1978 Annual Reliability and Maintainability Symposium*, pp. 113–116, 1978.
19. Davis, D. J., An analysis of some failure data, *J. Am. Stat. Assoc.*, 47, 113–150, 1952.
20. Lawless, J. F., *Statistical Models and Methods for Lifetime Data*, Wiley, New York, 1982.
21. Epstein, B., Estimation from life test data, *Technometrics*, 2, 447, 1960.
22. Welker, E. L. and M. Lipow, *Estimating the exponential failure rate dormant data with no failure events, Proc. Rel. Maint. Symp.*, 1 (2), 1194, 1974.
23. Bain, L. J., *Statistical Analysis of Reliability and Life-Testing Models: Theory and Methods*, Marcel Dekker, New York, 1978.
24. Leemis, L. R., *Reliability: Probabilistic Models, and Statistical Methods*, Prentice-Hall, Englewood Cliffs, NJ, 1995.
25. Soland, R., *Use of the Weibull Distribution in Bayesian Decision Theory*, Report No. RAC-TP-225, Research Analysis Corporation, McLean, VA, 1966.
26. Soland, R., Bayesian analysis of the Weibull process with unknown scale and shape parameters, *IEEE Trans. Reliab.*, R-18, 181–184, 1969.
27. Kaminskiy, M. and V. Krivtsov, A simple procedure for Bayesian estimation of Weibull distribution, *IEEE Trans. Reliab*, 54, 612–616, 2005.
28. Ayyub, B. M., *Elicitation of Expert Opinions for Uncertainty and Risks*, CRC Press, Boca Raton, FL, 2002.
29. Box, G. E. P and C. Tiao, *Bayesian Inference in Statistical Analysis*, Addison-Wesley, Reading, MA, 1973.
30. Kaminskiy, M. and V. Krivtsov, A simple procedure for Bayesian estimation of Weibull distribution, *IEEE Transactions on Reliability*, 54 (4), 612–616, 2005.
31. Gradshteyn, I. S. and I. M. Ryzhik, *Table of Integrals, Series, and Products*, A. Jeffrey, ed., Translated from the Russian by Scripta Technica, Inc., Academic Press, New York, 2000.
32. Kaminsky, M. and V. Krivtsov, Bayesian probability papers, *Reliability: Theory Appl*, 1 (2), 57–62, 2006.
33. Lawless, J. F., *Statistical Models and Methods for Lifetime Data*, Wiley, New York, 2003.
34. Zelliner, A., *An Introduction to Bayesian Inference in Econometrics*, Wiley, New York, 1996
35. Gelman, A., J. B. Carlin, H. S. Stern, and D. B. Rubin, *Bayesian Data Analysis*, Chapman & Hall, London, 1995.
36. Salemi, S., L. Yang, J. Dai, J. Qin, and J. Bernstein, *Physics-of-Failure Based Handbook of Microelectronic Systems*, Reliability Information Analysis Center, Utica, NY, 2008.
37. Pecht, M., *Product Reliability, Maintainability and Supportability Handbook*, CRC Press, Boca Raton, FL, 1995.
38. IEEE Std. 500, *Guide to the Collection and Presentation of Electrical, Electronic, Sensing Component and Mechanical Equipment Reliability Data for Nuclear Power Generating Stations*, IEEE Standards, New York, 1984.
39. Center for Chemical Process Safety of the American Institute of Chemical Engineer, *Guidelines for Process Equipment Data*, New York, 1989.
40. IEC/TR 62380, *Reliability Data Handbook*—Universal Model for Reliability Prediction of Electronics Components, PCBs and Equipment, IEC, 2004.
41. NUREG/CR-4450, *Analysis of Core Damage Frequency From Internal Events*, Vol. 1, U.S. Nuclear Regulatory Commission, Washington, DC, 1990.
42. Keller, A. and A. Amendola, *Reliability Data Bases*, A. Z. Keller, ed., D. Reidel, Dordrecht, Holland, 1987.
43. Hahn, G. J. and S. S. Shapiro, *Statistical Models in Engineering*, Wiley, New York, 1967.
44. MIL-HDBK-217F, Notice #2, *Military Handbook, Reliability Prediction of Electronic Equipment*, 1995.
45. O'Connor, P. D. T., *Practical Reliability Engineering*, 3rd edition, Wiley, New York, 1991.

4 System Reliability Analysis

Assessment of the reliability of a system based on its basic elements is one of the most important aspects of reliability analysis. A system is a collection of items (subsystems, components, software, human operators, etc.) whose proper, coordinated operation leads to the proper functioning of the system. In reliability analysis, it is therefore important to model the relationship between various items as well as the reliability of the individual items themselves to determine the reliability of the system as a whole. In Chapter 3, we elaborated on the reliability analysis at a basic item level (one for which enough information is available to predict its reliability). In this chapter, we discuss methods to model the relationship between system components, which allow us to determine overall system reliability.

The physical configuration of an item that belongs to a system is often used to model system reliability. In some cases, the manner in which an item fails is important to system failure and should be considered in the system reliability analysis. For example, in a system composed of two parallel electronic units, if a unit fails short (i.e., its failure mode), the system will fail, but for most other types of failures of the unit, the system will still be functional, since the other unit works properly.

There are several system modeling methods for reliability analysis. In this chapter, we describe the following modeling methods: the reliability block diagram method, which includes parallel, series, standby, shared load, and complex systems; Boolean logic-based methods, including (1) the fault-tree and success tree methods and the method of construction and evaluation of the tree and (2) the event tree method, which includes modeling of multisystem designs and complex systems whose individual units should work in a chronological or approximately chronological manner to achieve a mission; failure mode and effect analysis (FMEA); and master logic diagram (MLD) analysis. We assume here that items composing a system are statistically independent (according to the definition provided in Chapter 2). In Chapter 7, we will elaborate on system reliability considerations when components are statistically dependent.

4.1 RELIABILITY BLOCK DIAGRAM METHOD

Reliability block diagrams are frequently used to model the effect of item failures on system performance. Item failures often correspond to the physical arrangement of items in the system. However, in certain cases, they may not correspond. For instance, when two resistors are in parallel, the system fails if one fails short. Therefore, the reliability block diagram of this system for the "fail short" mode of failure would be composed of two series blocks. However, for other modes of failure of one unit, such as the "open" failure mode, the reliability block diagram is composed of two parallel blocks. In the remainder of this section, we discuss the reliability of the system for several types of the system's functional configurations.

4.1.1 SERIES SYSTEM

A reliability block diagram is a series configuration in which the failure of any one item (according to the failure mode of each item, based on which the reliability block diagram is developed) results in the failure of the system. Accordingly, for the functional success of a series system, all of its blocks (items) must successfully function during the intended mission time of the system. Figure 4.1 shows the reliability block diagram of a series system consisting of N blocks.

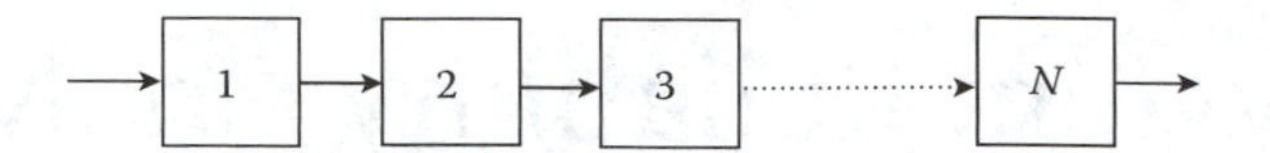

FIGURE 4.1 Series system reliability block diagram.

The reliability of the system in Figure 4.1 is the probability that all N units succeed during its intended mission time t. Thus, probabilistically, the system reliability $R_s(t)$ for independent units is obtained from

$$R_s(t) = R_1(t) \cdot R_2(t) \cdots R_N(t) = \prod_{i=1}^{N} R_i(t), \tag{4.1}$$

where $R_i(t)$ represents the reliability of the ith block. The hazard rate (instantaneous failure rate) for a series system is also a convenient expression. Since $\lambda(t) = -\mathrm{d}\ln R(t)/\mathrm{d}t$, according to Equation 4.1, the hazard rate of the system $\lambda_s(t)$ is

$$\lambda_s(t) = \frac{-\mathrm{d}\ln \prod_{i=1}^{N} R_i(t)}{\mathrm{d}t} = \sum_{i=1}^{N} \frac{-\mathrm{d}\ln R_i(t)}{\mathrm{d}t} = \sum_{i=1}^{N} \lambda_i(t). \tag{4.2}$$

Let us assume a constant failure rate model for each block (e.g., assume an exponential time to failure for each unit). Thus, $\lambda_i(t) = \lambda_i$. According to Equation 4.2, the system failure rate is

$$\lambda_s = \sum_{i=1}^{N} \lambda_i. \tag{4.3}$$

Expression 4.3 can also be easily obtained from Equation 4.1 by using the constant failure rate reliability model for each unit, $R_i(t) = \exp(-\lambda_i t)$.

$$R_s(t) = \prod_{i=1}^{N} \exp(-\lambda_i t) = \exp\left(-t \sum_{i=1}^{N} \lambda_i\right) = \exp(-\lambda_s t). \tag{4.4}$$

Using Equations 4.2 and 4.3, the MTTF of the system can be obtained as follows:

$$\mathrm{MTTF}_s = \frac{1}{\lambda_s} = \frac{1}{\sum_{i=1}^{N} \lambda_i}. \tag{4.5}$$

Example 4.1

A system consists of three units whose reliability block diagram is in a series. The failure rate for each unit is constant as follows: $\lambda_1 = 4.0 \times 10^{-6}\,\mathrm{h}^{-1}$, $\lambda_2 = 3.2 \times 10^{-6}\,\mathrm{h}^{-1}$, and $\lambda_3 = 9.8 \times 10^{-6}\,\mathrm{h}^{-1}$. Determine the following parameters of the system:

a. λ_s
b. R_s (1000 h)
c. MTTF_s

Solution:

a. According to Equation 4.3, $\lambda_s = 4.0 \times 10^{-6} + 3.2 \times 10^{-6} + 9.8 \times 10^{-6} = 1.7 \times 10^{-5}\,\mathrm{h}^{-1}$.
b. $R_s(t) = \exp(-\lambda_s t) = \exp[-1.7 \times 10^{-5}(1000)] = 0.983$, or unreliability of $\bar{R}(1000) = 0.017$.
c. According to Equation 4.5, $\mathrm{MTTF}_s = 1/\lambda_s = 1/(1.7 \times 10^{-5}) = 58{,}823.5\,\mathrm{h}$.

4.1.2 Parallel Systems

In a parallel configuration, the failure of all units results in a system failure. Accordingly, success of only one unit would be sufficient to guarantee the success of the system. Figure 4.2 shows a parallel system consisting of N units.

For a set of N independent units,

$$F_s(t) = F_1(t) \cdot F_2(t) \cdots F_N(t) = \prod_{i=1}^{N} F_i(t). \tag{4.6}$$

Since $R_i(t) = 1 - F_i(t)$, then

$$R_s(t) = 1 - F_s(t) = 1 - \prod_{i=1}^{N} [1 - R_i(t)]. \tag{4.7}$$

The system hazard rate can also be derived by using $h(t) = -\mathrm{d} \ln R(t)/\mathrm{d}t$.

For consideration of various characteristics of system reliability, let us analyze a special case in which the failure rate is constant for each unit (exponential time-to-failure model), and the system is composed of only two units. Since $R_i(t) = \exp(-\lambda_i t)$, then according to Equation 4.7,

$$\begin{aligned} R_s(t) &= 1 - \left[1 - \exp(-\lambda_1 t)\right]\left[1 - \exp(-\lambda_2 t)\right] \\ &= \exp(-\lambda_1 t) + \exp(-\lambda_2 t) - \exp[-(\lambda_1 + \lambda_2)t]. \end{aligned} \tag{4.8}$$

Since $h_s(t) = f_s(t)/R_s(t)$ and $f_s(t) = -\mathrm{d}[R_s(t)]/\mathrm{d}t$, then using Equation 4.8,

$$f_s(t) = \lambda_1 \exp(-\lambda_1 t) + \lambda_2 \exp(-\lambda_2 t) - (\lambda_1 + \lambda_2) \exp[-(\lambda_1 + \lambda_2)t]. \tag{4.9}$$

Thus,

$$h_s(t) = \frac{\lambda_1 \exp(-\lambda_1 t) + \lambda_2 \exp(-\lambda_2 t) - (\lambda_1 + \lambda_2) \exp[-(\lambda_1 + \lambda_2)t]}{\exp(-\lambda_1 t) + \exp(-\lambda_2 t) - \exp[-(\lambda_1 + \lambda_2)t]}. \tag{4.10}$$

The MTTF of the system can also be obtained as

$$\begin{aligned} \mathrm{MTTF}_s &= \int_0^{\infty} R_s(t)\,\mathrm{d}t = \int_0^{\infty} \exp(-\lambda_1 t) + \exp(-\lambda_2 t) - \exp[-(\lambda_1 + \lambda_2)t]\,\mathrm{d}t \\ &= \frac{1}{\lambda_1} + \frac{1}{\lambda_2} - \frac{1}{\lambda_1 + \lambda_2}. \end{aligned} \tag{4.11}$$

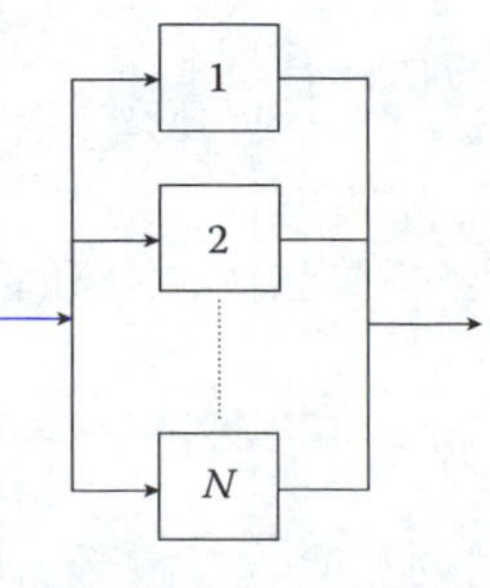

FIGURE 4.2 Parallel system block diagram.

Accordingly, one can use the binomial expansion to derive the MTTF for the system of N parallel units:

$$\mathrm{MTTF_s} = \left(\frac{1}{\lambda_1} + \frac{1}{\lambda_2} + \cdots + \frac{1}{\lambda_N}\right) - \left(\frac{1}{\lambda_1 + \lambda_2} + \frac{1}{\lambda_1 + \lambda_3} + \cdots + \frac{1}{\lambda_{N-1} + \lambda_N}\right)$$
$$- \left(\frac{1}{\lambda_1 + \lambda_2 + \lambda_3} + \cdots + \frac{1}{\lambda_{N-2} + \lambda_{N-1} + \lambda_N}\right) \cdots + (-1)^{N+1} \frac{1}{\lambda_1 + \lambda_2 + \cdots + \lambda_N}. \tag{4.12}$$

In the special case where all units are identical with a constant failure rate (e.g., in an active redundant system), Equation 4.7 simplifies to the following form:

$$R_s(t) = 1 - \left[1 - \exp(-\lambda t)\right]^N \tag{4.13}$$

and from Equation 4.12,

$$\mathrm{MTTF_s} = \mathrm{MTTF}\left(1 + \frac{1}{2} + \cdots + \frac{1}{N}\right). \tag{4.14}$$

It can be seen from Equation 4.14 that in the design of active redundant systems, the MTTF of the system exceeds the MTTF of an individual unit. However, the contribution to the MTTF of the system from the second unit, the third unit, and so on would have a diminishing return as N increases. That is, there would be an optimum number of parallel units by which a designer can maximize the reliability and at the same time minimize the life cycle cost by adding a redundant component.

Let us consider a more general structure of series and parallel systems: the K-out-of-N system. In this type of system, if any combination of K units out of N independent units works, it guarantees the success of the system. For simplicity, assume that all units are identical (which, by the way, is often the case). The binomial distribution can easily represent the probability that the system functions:

$$R_s(t) = \sum_{r=K}^{N} \binom{N}{r} [R(t)]^r [1 - R(t)]^{N-r} = 1 - \sum_{r=0}^{N-r} \binom{N}{r} [R(t)]^r [1 - R(t)]^{N-r}. \tag{4.15}$$

Example 4.2

A system is composed of the same units as in Example 4.1. However, these units are in parallel. Find the time-to-failure cdf (unreliability) and $\mathrm{MTTF_s}$.

Solution:

According to Equation 4.7,

$$F_s(t) = \left[1 - \exp(-\lambda_1 t)\right]\left[1 - \exp(-\lambda_2 t)\right]\left[1 - \exp(-\lambda_3 t)\right],$$

$$F_s(1000) = \left\{1 - \exp\left[-4 \times 10^{-6}(1000)\right]\right\}\left\{1 - \exp\left[-3.2 \times 10^{-6}(1000)\right]\right\}$$
$$\left\{1 - \exp\left[-9.8 \times 10^{-6}(1000)\right]\right\}$$
$$= 1.25 \times 10^{-7}.$$

$$\mathrm{MTTF_s} = \left(\frac{1}{\lambda_1} + \frac{1}{\lambda_2} + \frac{1}{\lambda_3}\right) - \left(\frac{1}{\lambda_1 + \lambda_2} + \frac{1}{\lambda_1 + \lambda_3} + \frac{1}{\lambda_2 + \lambda_3}\right) + \left(\frac{1}{\lambda_1 + \lambda_2 + \lambda_3}\right)$$
$$= 4.35 \times 10^5 \text{ h}.$$

EXAMPLE 4.3

How many components should be used in an active redundancy design to achieve a reliability of 0.999 such that, for successful system operation, a minimum of two components is required? Assume a mission of $t = 720$ h for a set of components that are identical and have a failure rate of 0.00015 per hour.

Solution:
For each component $R(t) = \exp(\lambda t) = \exp[(0.00015(720)] = 0.8976$. According to Equation 4.15,

$$0.999 = 1 - \sum_{r=0}^{1} \binom{N}{r} (0.8976)^r (0.1024)^{N-r} = 1 - (0.1024)^N - N(0.8976)(0.1024)^{N-1}.$$

From the above equation, $N = 5$, which means that at least five components should be used to achieve the desired reliability over the specified mission time.

4.1.3 STANDBY REDUNDANT SYSTEMS

A system is called a standby redundant system when some of its units remain idle until they are called for service by a sensing and switching device. For simplicity, let us consider a situation where only one unit operates actively, while the others are in standby, as shown in Figure 4.3.

In this configuration, Unit 1 operates constantly until it fails. The sensing and switching device recognizes a unit failure in the system and switches to another unit. This process continues until all standby units have failed, in which case the system is considered failed. Since Units 2–N do not operate constantly (as is the case in active parallel systems), we would expect them to fail at a much slower rate. This is because the failure rate for components is usually higher when the components are operating than when they are idle or dormant.

System reliability is totally dependent on the reliability of the sensing and switching device. The reliability of a redundant standby system is the reliability of Unit 1 over the mission time t (i.e., the probability that it succeeds the whole mission time) plus the probability that Unit 1 fails at time t_1 prior to t and the probability that the sensing and switching unit does not fail by t_1 and the probability that standby Unit 2 does not fail by t_1 (in the standby mode) and the probability that standby Unit 2 successfully functions for the remainder of the mission in an active operation mode, and so on.

Mathematically, the reliability function for a two-unit standby device according to this definition can be obtained as

$$R_s(t) = R_1(t) + \int_0^t f_1(t_1)\, \mathrm{d}t_1 \cdot R_{ss}(t_1) \cdot R_2'(t_1) \cdot R_2(t - t_1), \tag{4.16}$$

where $f_1(t)$ is the pdf for the time to failure of Unit 1, $R_{ss}(t_1)$ is the reliability of the sensing and switching device, $R_2'(t_1)$ is the reliability of Unit 2 in the standby mode of operation, and $R_2(t - t_1)$ is the

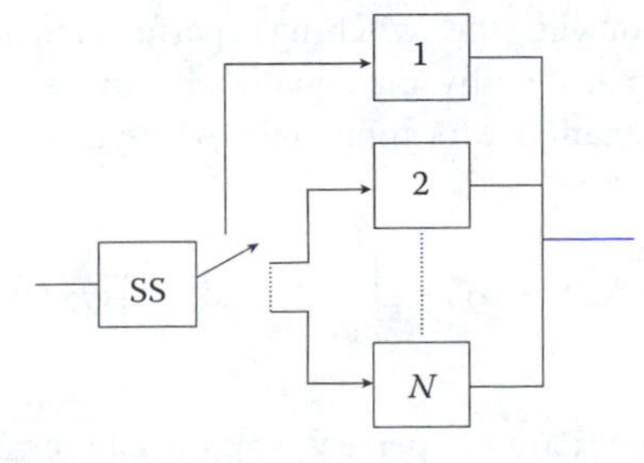

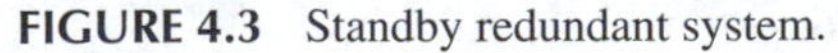
FIGURE 4.3 Standby redundant system.

reliability of Unit 2 after it started to operate at time t_1. Let us consider a case where time to failure of all units follows an exponential distribution.

$$R_s(t) = \exp(-\lambda_1 t) + \int_0^t \left[\lambda_1 \exp(-\lambda_1 t_1)\right]\left[\exp(-\lambda_{ss} t_1)\right]\left[\exp(-\lambda_2' t_1)\right]\{\exp[-\lambda_2 (t - t_1)]\}\, dt_1$$
$$= \exp(-\lambda_1 t) + \frac{\lambda_1 \exp(-\lambda_2 t)}{\lambda_1 + \lambda_{ss} + \lambda_2' - \lambda_2}\left\{1 - \exp\left[-\left(\lambda_1 + \lambda_{ss} + \lambda_2' - \lambda_2\right) t\right]\right\}. \quad (4.17)$$

For the special case of perfect sensing and switching and no standby failures, $\lambda_{ss} = \lambda_2' = 0$,

$$R_s(t) = \exp(-\lambda_1 t) + \int_0^t \left[\lambda_1 \exp(-\lambda_1 t_1)\right]\{\exp[-\lambda_2 (t - t_1)]\}\, dt_1$$
$$= \exp(-\lambda_1 t) + \frac{\lambda_1}{\lambda_1 - \lambda_2}[\exp(-\lambda_2 t) - \exp(-(\lambda_1 t))]. \quad (4.18)$$

If the two units are identical, that is, $\lambda_1 = \lambda_2 = \lambda$, then

$$R_s(t) = \exp(-\lambda t) + \lambda t \exp(-\lambda t) = (1 + \lambda t)\exp(-\lambda t). \quad (4.19)$$

In the case of perfect switching, a standby system possesses the same characteristic as the "shock model." That is, one can assume that the Nth shock (i.e., the Nth unit failure) causes the system to fail. Thus, a gamma distribution can represent the time to failure of the system such that

$$R_s(t) = 1 - \int_0^t \frac{\lambda^N}{\Gamma(N)} x^{N-1} \exp(-\lambda x)\, dx$$
$$= \exp(-\lambda t)\left[1 + \lambda t + \frac{(\lambda t)^2}{2!} + \cdots + \frac{(\lambda t)^{N-1}}{(N-1)!}\right]. \quad (4.20)$$

Accordingly, the MTTF of the above system is given by

$$\mathrm{MTTF}_s = \frac{N}{\lambda}, \quad (4.21)$$

which is N times the MTTF of a single unit. Expression 4.21 explains why high reliability can be achieved through a standby system when the switching is perfect and no failure occurs during standby.

When more than two units are in standby, the equation becomes somewhat more difficult, but the concept is almost the same. For example, for three units with perfect switching,

$$R_s(t) = R_1(t) + \int_0^t f_1(t_1)\, dt_1 \cdot R_2(t - t_1) + \int_0^t f_1(t_1)\, dt_1 \cdot \int_0^t f_2(t_2)\, dt_2 \cdot R_3(t - t_1 - t_2). \quad (4.22)$$

If the sensing and switching devices are not perfect, appropriate terms should be added to Equation 4.22 to account for their unreliability, similar to Equation 4.16.

Example 4.4

Consider two identical independent units with $\lambda = 0.01\ \text{h}^{-1}$. Mission time $t = 24$ h. Compare the reliability of a system made of these units if they are placed in

a. Parallel configuration.
b. Series configuration.
c. Standby configuration with perfect switching.
d. Standby configuration with imperfect switching and standby failure rates of $\lambda_{SS} = 1 \times 10^{-6}$ and $\lambda' = 1 \times 10^{-5}$ per hour, respectively.

Solution:
Let us assume an exponential time-to-failure model for each unit

$$R(t) = \exp(-\lambda t) = \exp[-0.01(24)] = 0.7866.$$

Then

a. For the parallel system, using Equation 4.13,

$$R_S(24) = 1 - (1 - 0.7866)^2 = 0.9544.$$

b. For the series system, using Equation 4.1,

$$R_S(24) = 0.7866(0.7866) = 0.6187.$$

c. For the standby system with perfect switches, using Equation 4.19,

$$R_S(24) = (1 + 0.24)\exp[-0.01(24)] = 0.9755.$$

d. For the standby system with imperfect switching and standby failure rate using Equation 4.18,

$$R_S(24) = 0.7866 + \frac{0.01(0.7866)}{1.1 \times 10^{-5}}\left\{1 - \exp\left[-1.1 \times 10^{-5}(24)\right]\right\} = 0.9754.$$

4.1.4 Load-Sharing Systems

A load-sharing system refers to a parallel system whose units equally share the system function. For example, if a set of two identical parallel pumps delivers x gpm of water to a reservoir, each pump delivers $x/2$ gpm. If a minimum of x gpm is required at all times, and one of the pumps fails at a given time, t_0, then the other pump's speed should be increased to provide x gpm alone. Other examples of load sharing are multiple load-bearing units (such as those in a bridge) and load-sharing multiunit electric power plants. In these cases, when one of the units fails, the others should carry its load. Since these other units would then be working under a more stressful condition, they would experience a higher rate of failure.

Load-sharing system reliability models can be divided into two groups: time-independent models and time-dependent models. Note that most of the reliability models discussed in this book are time-dependent. The time-independent reliability models are considered in the framework of stress–strength analysis, which is briefly discussed in Chapter 6. Historically, the first time-independent load-sharing system model was developed by Daniels [1] and is known as the Daniels model. This model was originally applied to textile strength problems; now, it is also applied to composite materials.

To illustrate the basic ideas associated with this kind of model, consider a simple parallel system composed of two identical components [2]. Let $F(s)$ be the time-independent failure probability for the component subjected to load (stress) s. Denoted by $F_2(s)$, the failure probability for a parallel system of two identical components can be found. The reliability function of the system, $R_2(s)$, is $1 - F_2(s)$. Initially, both components are subjected to an equal load s. When one component fails, the nonfailed component takes on the full load, $2s$.

The probability of the system failure, $F_2(s)$, can be modeled as follows. Let A be the event when the first component fails under load s and the second component fails under load $2s$; let B be the event in which the second component fails under load s. The first component fails under load $2s$. Finally, let $A \cap B$ be the event that both components fail under load s:

$$\Pr(A \cup B) = \Pr(A) + \Pr(B) - \Pr(A \cap B). \tag{4.23}$$

It is evident that

$$\Pr(A) = \Pr(B) = F(s)\,F(2s), \quad \Pr(A \cap B) = F^2(s), \tag{4.24}$$

hence

$$F_2(s) = 2F(s)\,F(2s) - F^2(s) \tag{4.25}$$

and

$$R_2(s) = 1 - 2F(s)\,F(2s) + F^2(s). \tag{4.26}$$

A similar equation for reliability of a three-component load-sharing system contains seven terms, and the problem becomes more difficult as the number of components increases. For such situations, different recursive procedures have been developed [2].

Now, consider a simple example of the time-dependent load-sharing system model. Let us assume again that two components share a load (i.e., each component carries half the load) and that the time-to-failure distribution for both components is $f_h(s, t)$. When one component fails (i.e., one component carries the full load), the time-to-failure distribution is $f_f(2s, t)$. Let us also assume that the corresponding reliability functions during full-load and half-load operation are $R_f(2s, t)$ and $R_h(s, t)$, respectively. The system will succeed if both components carry half the load, or if component 1 fails at time t_0 and component 2 carries a full load thereafter, or if component 2 fails at time to failure and component 1 carries the full load thereafter. Accordingly, the system reliability function $R_s(t)$ can be obtained from Kapur and Lamberson [3]:

$$R_s(t) = [R_h(s,t)]^2 + 2\int_0^t f_h\,(s,t_1)\cdot R_h\,(s,t_1)\cdot R_f\,(2s,\,t-t_1)\,\mathrm{d}t_1. \tag{4.27}$$

In Equation 4.27, the first term shows the contribution from both components working successfully, with each carrying a half load; the second term represents the two equal probabilities that component 1 fails first and component 2 takes the full load at time t_0 or vice versa.

If there are switching or control mechanisms involved to shift the total load to the nonfailed component when one component fails, then, similar to Equation 4.16, the reliability of the switching mechanism can be incorporated into Equation 4.27.

In the special situation where exponential time-to-failure models with failure rates λ_f and λ_h can be used for the two components under full and half loads, respectively, then Equation 4.27 can be simplified to

$$R_s(t) = \exp(-2\lambda_h t) + \frac{2\lambda_h \exp(-\lambda_f t)}{(2\lambda_h - \lambda_f)}\{1 - \exp[-(2\lambda_h - \lambda_f)\,t]\}. \tag{4.28}$$

The reader is referred to Crowder et al. [2] for a review of more sophisticated time-dependent load-sharing models.

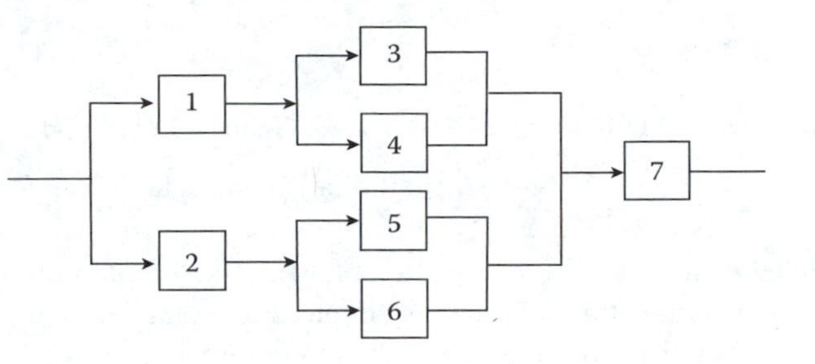

FIGURE 4.4 Complex parallel-series system.

4.1.5 Complex Systems

Most practical systems are neither parallel nor series, but exhibit some hybrid combination of the two. These systems are often referred to as parallel-series systems. Figure 4.4 shows an example of such a system.

Another type of complex system is one that is neither series nor parallel alone, nor parallel-series. Figure 4.5 shows an example of such a system.

A parallel-series system can be analyzed by dividing it into its basic parallel and series modules and then determining the reliability function for each module separately. The process can be continued until a reliability function for the whole system is determined. For an analysis of all types of complex systems, Shooman [4] describes several analytical methods for complex systems. These are the inspection method, event space method, path-tracing method, and decomposition method. These methods are good only when there are not many units in the system. For analysis of large number of units, fault trees would be more appropriate. In the following, we discuss the decomposition and path-tracing methods.

The decomposition method relies on the conditional probability concept to decompose the system. The reliability of a system is equal to the reliability of the system given a chosen unit's work multiplied by the reliability of the unit plus the reliability of the system given the unit's failure multiplied by the unreliability of the unit. For example, using Unit 3 in Figure 4.5,

$$R_s(t) = R_s\,(t \mid \text{unit 3 work}) \cdot R_3(t) + R_s\,(t \mid \text{unit 3 fail})\;[1 - R_3(t)]\,. \tag{4.29}$$

If Equation 4.29 is applied to all units that make the system a nonparallel series (such as Units 3 and 6 in Figure 4.5), the system would reduce to a simple parallel-series system. Thus, for Figure 4.5 and for the conditional reliability terms in Equation 4.29, it follows that

$$\begin{aligned} R_s\,(t \mid \text{unit 3 work}) = {} & R_s\,(t \mid \text{unit 6 work} \cap \text{unit 3 work})\, R_6(t) \\ & + R_s\,(t \mid \text{unit 6 fail} \cap \text{unit 3 work})\,[1 - R_6(t)] \end{aligned} \tag{4.30}$$

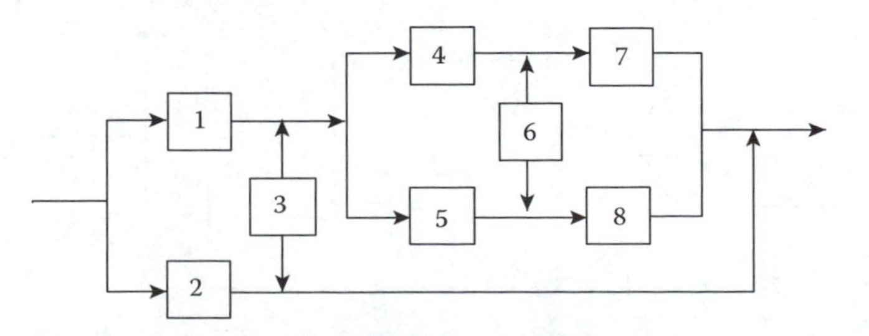

FIGURE 4.5 Complex nonparallel-series system.

or

$$R_s\,(t \mid \text{unit 3 fail}) = R_s\,(t \mid \text{unit 6 work} \cap \text{unit 3 fail})\,R_6(t) + R_s\,(t \mid \text{unit 6 fail} \cap \text{unit 3 fail})\,[1 - R_6(t)] \tag{4.31}$$

Each of the conditional reliability terms in Equations 4.30 and 4.31 represents a purely parallel-series system, the reliability determination of which is simple. For example, R_s (t | unit 6 work ∩ unit 3 fail) corresponds to a reliability block diagram shown in Figure 4.6.

The combination of Equations 4.29 through 4.31 results in an expression for $R(s)$.

A more computationally intensive method for determining the reliability of a complex system involves the use of path-set and cut-set methods (path-tracing methods). A path set (or tie set) is a set of units that form a connection between input and output when traversed in the direction of the reliability block diagram arrows. Thus, a path set merely represents a "path" through the graph. A minimal path set (or minimal tie set) is a path set containing the minimum number of units needed to guarantee a connection between the input and output points. For example, in Figure 4.5, path set $P_1 = (1, 3)$ is a minimal path set, but $P_2 = (1, 3, 6)$ is not, since Units 1 and 3 are sufficient to guarantee a path.

A cut set is a set of units that interrupt all possible connections between the input and output points. A minimal cut set is the smallest set of units needed to guarantee an interruption of flow. In practice, minimal cut sets show a combination of unit failures that cause a system to fail. For example, in Figure 4.5, the minimal path sets are $P_1 = (2)$, $P_2 = (1, 3)$, $P_3 = (1, 4, 7)$, $P_4 = (1, 5, 8)$, $P_5 = (1, 4, 6, 8)$, and $P_6 = (1, 5, 6, 7)$. The minimal cut sets are $C_1 = (1, 2)$, $C_2 = (4, 5, 3, 2)$, $C_3 = (7, 8, 3, 2)$, $C_4 = (4, 6, 8, 3, 2)$, and $C_5 = (5, 6, 7, 3, 2)$. If a system has m minimal path sets denoted by $P_1, P_2, \ldots, P_m$, then the system reliability is given by

$$R_s(t) = \Pr\left(P_1 \cup P_2 \cup \cdots \cup P_m\right), \tag{4.32}$$

where each path set P_i represents the event that units in the path set survive during the mission time t. This guarantees the success of the system. Since many path sets may exist, the union of all these sets gives all possible events for successful operation of the system. The probability of this union clearly represents the reliability of the system. It should be noted here that in practice the path sets P_is are not disjoint. This poses a problem for determining the left-hand side of Equation 4.32. In Section 4.2, we will explain formal methods to deal with this problem. However, an upper bound on the system reliability may be obtained by assuming that the P_is are highly disjoint. Thus,

$$R_s(t) = \Pr\left(P_1\right) + \Pr\left(P_2\right) + \cdots + \Pr\left(P_m\right). \tag{4.33}$$

Expression 4.33 yields better answers when we deal with small reliability values. Since this is not usually the case, Equation 4.33 is not a good bound for use in practical applications.

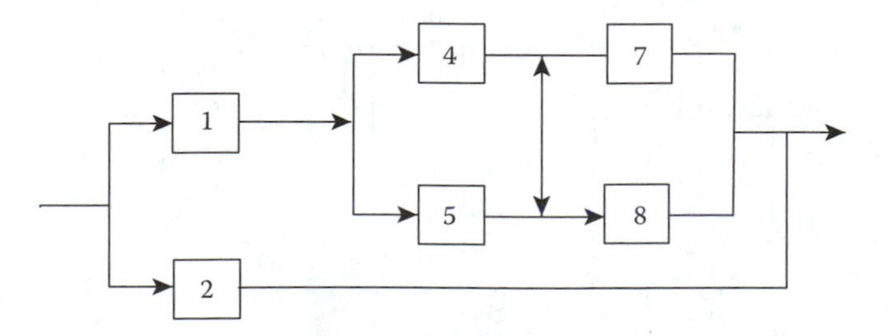

FIGURE 4.6 Representation of $R_s(t \mid \text{unit 6 work} \cap \text{unit 3 fail})$.

Similarly, system reliability can be determined through minimal cut sets. If the system has n minimal cut sets denoted by $C_1, C_2, \ldots, C_n$, then the system reliability is obtained from

$$R_s(t) = 1 - \Pr(C_1 \cup C_2 \cup \cdots \cup C_n), \tag{4.34}$$

where C_i represents the event that units in the cut set fail sometime before the mission time t. This guarantees system failure. The $\Pr(\cdot)$ term on the right-hand side of Equation 4.34 shows the probability that at least one of all possible minimal cut sets exists before time t. Thus $\Pr(\cdot)$ represents the probability that the system will fail sometime before t. By subtracting this probability from 1, the reliability of the system is obtained. Similar to the union of path sets, the union of cut sets is not a disjoint function. Again, Equation 4.34 can be written in the form of its lower bound, which is a much simpler expression given by

$$R_s(t) \geq 1 - [\Pr(C_1) + \Pr(C_2) + \cdots + \Pr(C_n)]. \tag{4.35}$$

Notice that each element of a path set represents the success of a unit operation, whereas each element of a cut set represents the failure of a unit. Thus, for probabilistic evaluations, the reliability function of each unit should be used in connection with path-set evaluations, that is, Equation 4.33, while the unreliability function should be used in connection with cut-set evaluations, that is, in Equation 4.35.

The bounding technique used in Equation 4.35, in practice, yields a much better representation of the reliability of the system than Equation 4.33 because most engineering units have reliability greater than 0.9 over their mission time, making the use of Equation 4.35 appropriate.

EXAMPLE 4.5

Consider the reliability block diagram in Figure 4.5. Determine the lower bound of the system reliability function if the hazard rates of each unit are constant and are $\lambda_1, \lambda_2, \ldots, \lambda_8$.

Solution:

Using the system cut sets discussed earlier and Equation 4.35,

$$R_s(t) \geq 1 - \left[\Pr(C_1) + \Pr(C_2) + \cdots + \Pr(C_5)\right]$$

assuming C_1 and C_2 are independent, and

$$\Pr(C_1) = \left[1 - \exp(-\lambda_1 t)\right]\left[1 - \exp(-\lambda_2 t)\right]$$

and so on. Therefore,

$$\begin{aligned} R_s(t) \geq 1 - \{&\left[1 - \exp(-\lambda_1 t)\right]\left[1 - \exp(-\lambda_2 t)\right] \\ &+ \left[1 - \exp(-\lambda_2 t)\right]\left[1 - \exp(-\lambda_3 t)\right]\left[1 - \exp(-\lambda_4 t)\right]\left[1 - \exp(-\lambda_5 t)\right] \\ &+ \left[1 - \exp(-\lambda_2 t)\right]\left[1 - \exp(-\lambda_3 t)\right]\left[1 - \exp(-\lambda_7 t)\right]\left[1 - \exp(-\lambda_8 t)\right] \\ &+ \left[1 - \exp(-\lambda_2 t)\right]\left[1 - \exp(-\lambda_3 t)\right]\left[1 - \exp(-\lambda_4 t)\right]\left[1 - \exp(-\lambda_6 t)\right]\left[1 - \exp(-\lambda_8 t)\right] \\ &+ \left[1 - \exp(-\lambda_2 t)\right]\left[1 - \exp(-\lambda_3 t)\right]\left[1 - \exp(-\lambda_5 t)\right]\left[1 - \exp(-\lambda_6 t)\right]\left[1 - \exp(-\lambda_7 t)\right]\}. \end{aligned}$$

continued

For some typical values of λ, the lower bound for $R_s(t)$ can be compared to the exact value of $R_s(t)$. Here, "exact" means that the cut sets are not assumed disjoint. For example, Figure 4.7 shows the exact and the lower probability bound of system reliability for $\lambda_1 = 1 \times 10^{-6}\,\mathrm{h}^{-1}$, $\lambda_2 = 1 \times 10^{-5}\,\mathrm{h}^{-1}$, $\lambda_3 = 2 \times 10^{-5}\,\mathrm{h}^{-1}$, and $\lambda_4 = \lambda_5 = \lambda_6 = \lambda_7 = \lambda_8 = 1 \times 10^{-4}\,\mathrm{h}^{-1}$.

Figure 4.7 illustrates that as time increases, the reliability of the system decreases (unit failure probability increases), causing Equation 4.35 to yield a poor approximation. At this point, it is more appropriate to use Equation 4.33. Again, notice that Equations 4.33 and 4.35 assume the path sets and cut sets are disjoint.

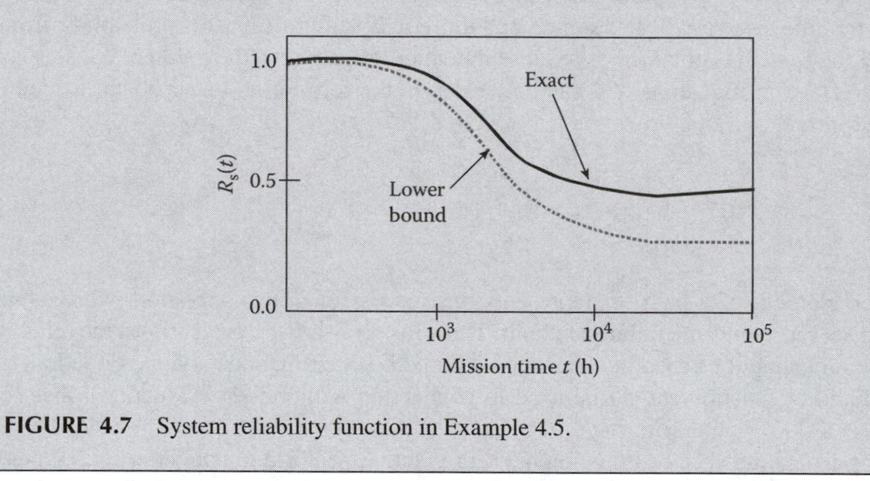

FIGURE 4.7 System reliability function in Example 4.5.

In cases of very complex systems that have multiple failure modes for each unit and complex physical and operational interactions, the use of reliability block diagrams becomes difficult. The logic-based methods such as the fault tree and success tree analyses are more appropriate in this context. We will elaborate on this topic in the next section.

4.2 FAULT TREE AND SUCCESS TREE METHODS

The operation of a system can be considered from two opposite viewpoints: the various ways that a system fails or the various ways that a system succeeds. Most of the construction and analysis methods used are, in principle, the same for both fault trees and success trees. First we will discuss the fault tree method and then describe the success tree method.

4.2.1 Fault Tree Method

The fault tree approach is a deductive process by means of which an undesirable event, called the top event, is postulated, and the possible ways for this event to occur are systematically deduced. For example, a typical top event might look like, "Failure of control circuit *A* to send a signal when it should." The deduction process is performed so that the fault tree embodies all component failures (i.e., failure modes) that contribute to the occurrence of the top event. It is also possible to include individual failure modes of each component as well as human and software errors (and the interaction between the two) during the system operation. The fault tree itself is a graphical representation of the various combinations of failures that lead to the occurrence of the top event.

A fault tree does not necessarily contain all possible failure modes of the components (or units) of the system. Only those failure modes that contribute to the existence or occurrence of the top event are modeled. For example, consider a fail-safe control circuit. If loss of the DC power to the circuit causes the circuit to open a contact, which in turn sends a signal to another system for operation, a top event of “control circuit fails to generate a safety signal” would not include “failure of DC power source” as one of its events, even though the DC power source (e.g., batteries) is part of the control circuit. This is because the top event would not occur due to the loss of the DC power source.

The postulated fault events that appear on the fault tree may not be exhaustive. Only those events considered important can be included. However, it should be noted that the decision for inclusion

Primary Event Symbols

BASIC EVENT—A basic event requiring no further development

CONDITIONING EVENT—Specific conditions or restrictions that apply to any logic gate (used primary with PRIORITY AND and INHIBIT gate)

UNDEVELOPED EVENT—An event which is not further developed either because it is of insufficient consequence or because information is unavailable

EXTERNAL EVENT—An event which is normally expected to occur

Intermediate Event Symbols

INTERMEDIATE EVENT—An event that occurs because of one or more antecedent causes acting through logic gates

Gate Symbols

AND—Output occurs if all of the input events occur.

OR—Output occurs if at least one of the input events occurs

EXCLUSIVE OR—Output occurs if exactly one of the input events occurs

PRIORITY AND—Output occurs if all of the input events occur in a specific sequence (the sequence is represented by a CONDITIONING EVENT drawn to the right of the gate)

INHIBIT—Output occurs if a single event input to produce output only if a CONDITIONING EVENT input is met

Not—OR—Output occurs if at least one of the input events does not occur

Not—AND—Output occurs if all of the input events do not occur

Transfer Symbols

TRANSFER IN—Indicates that the tree is developed further at the occurrence of the corresponding TRANSFER OUT (e.g., on another page)

TRANSFER OUT—Indicates that this portion of the tree must be attached at the corresponding TRANSFER IN

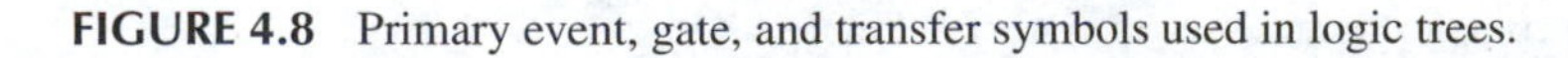

FIGURE 4.8 Primary event, gate, and transfer symbols used in logic trees.

of failure events is not arbitrary; it is influenced by the fault tree construction procedure, system design and operation, operating history, available failure data, and experience of the analyst. At each intermediate point, the postulated events represent the immediate, necessary, and sufficient causes for the occurrence of the intermediate (or top) events.

The fault tree itself is a logical model, and thus represents the qualitative characterization of the system logic. There are, however, many quantitative algorithms to evaluate fault trees. For example, the concept of cut sets discussed earlier can also be applied to fault trees by using the Boolean algebra method. By using Pr $(C_1 \cup C_2 \cup \cdots \cup C_m)$, the probability of occurrence of the top event can be determined using Equation 4.34.

To understand the symbology of logic trees, including fault trees, consider Figure 4.8. In essence, there are three types of symbols: events, gates, and transfers. Basic events, undeveloped events, condition events, and external events are sometimes referred to as primary events. When postulating events in the fault tree, it is important to include not only the undesired component states (e.g., applicable failure modes), but also the time when they occur.

To better understand the fault tree concept, let us consider the complex block diagram shown in Figure 4.4. Let us also assume that the block diagram models a circuit in which the arrows show the direction of electric current. A top event of "no current at point F" is selected, and all events that cause this top event are deductively postulated. Figure 4.9 shows the results.

As another example, consider the pumping system shown in Figure 4.10. Sufficient water is delivered from the water source T_1 when only one of the two pumps, P-1 or P-2, works. All the valves V-1 through V-5 are normally open. The sensing and control system S senses the demand for the pumping system and automatically starts both P-1 and P-2. (If one of the two pumps fails to start or fails during operation, the mission is still considered successful if the other pump functions properly.)

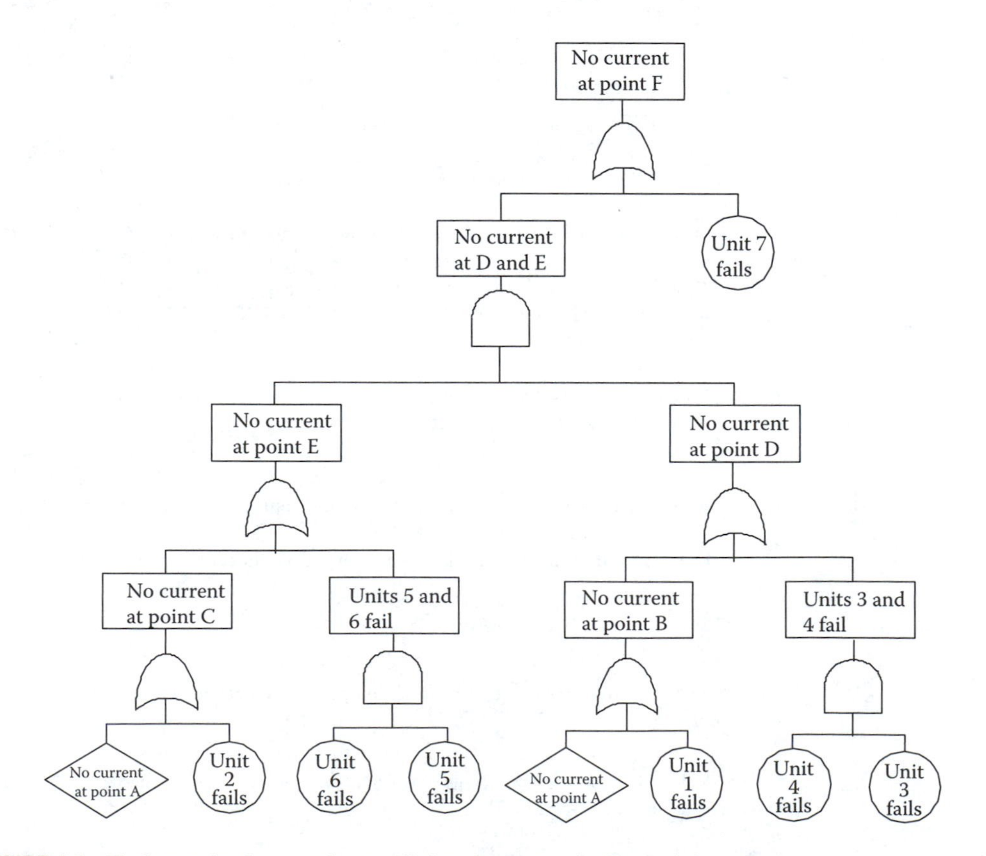

FIGURE 4.9 Fault tree for the complex parallel-series system in Figure 4.4.

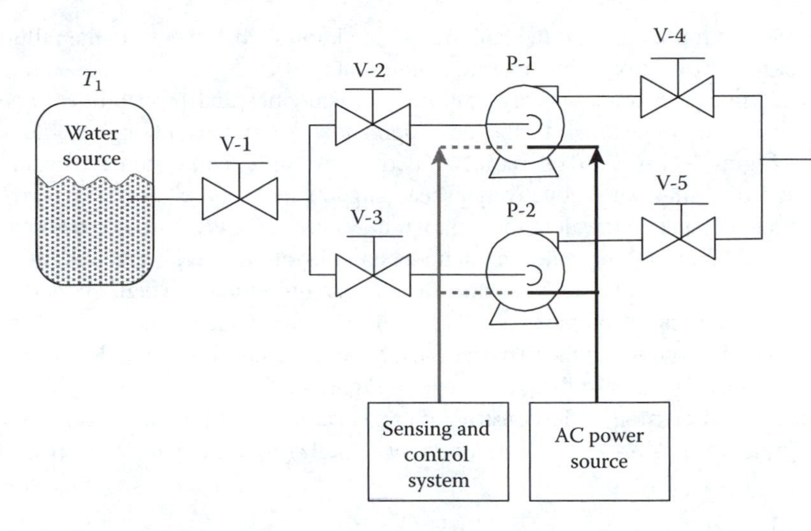

FIGURE 4.10 Example of a pumping system.

The two pumps and the sensing and control system use the same AC power source, AC. Assume the water content in T_1 is sufficient and available, there are no human errors, and no failure in the pipe connections is considered important.

The system's mission is to deliver sufficient water when needed. Therefore, the top event of the fault tree for this system should be "no water is delivered when needed." Figure 4.11 shows the fault tree for this example. In Figure 4.11, the failures of AC and S are shown with undeveloped events.

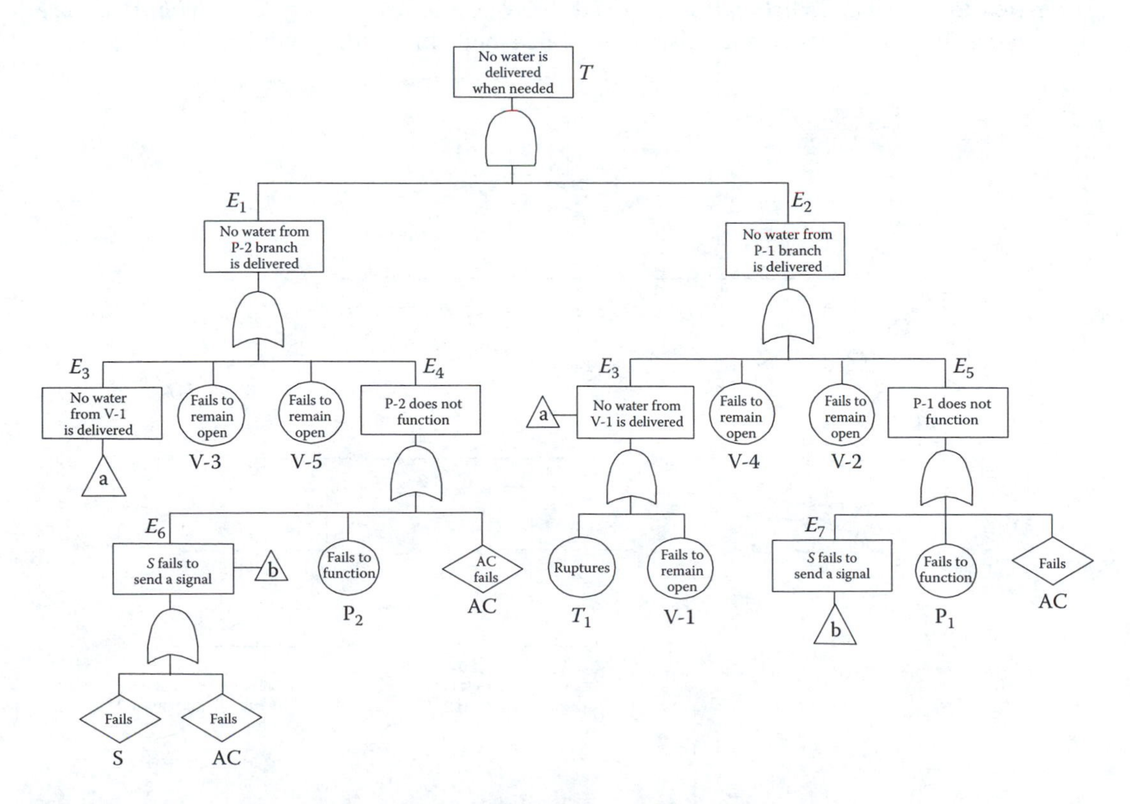

FIGURE 4.11 Fault tree for the pumping system in Figure 4.10.

This is because one can further expand the fault tree if one knows what makes up the failures of AC and *S*, in which case these events will be intermediate events.

However, since enough information (e.g., failure characteristics and probabilities) about these events is known, we have stopped their further development at this stage. Although the development of the fault tree in Figure 4.11 is based on a strict deductive procedure (i.e., systematic decomposition of failures starting from "sink" and deductively proceeding toward "source"), one can rearrange it to the more concise and compact equivalent form shown in Figure 4.12. While the development of the fault tree in Figure 4.11 requires only a minimum understanding of the overall functionality and logic of the system, direct development of more compact versions requires a much better understanding of the overall system logic. If more complex logical relationships are required, other logical representations can be described by combining the two basic AND and OR gates. For example, the *K*-out-of-*N* and exclusive OR logics can be described, as shown in Figure 4.13.

For a more detailed discussion of the construction and evaluation of fault trees, refer to the *Fault Tree Handbook* [5] and the *NASA Fault Tree Handbook with Aerospace Applications* [6].

4.2.2 Evaluation of Logic Trees

The evaluation of logic trees (e.g., fault trees, success trees, and MLDs) involves two distinct aspects: logical or qualitative evaluation and probabilistic or quantitative evaluation. Qualitative evaluation involves the determination of the logic tree cut sets, path sets, or logical evaluations to rearrange the tree logic for computational efficiency (similar to the rearrangement presented in Figure 4.12 for a fault tree). Determining the logic tree cut sets or path sets involves some straightforward Boolean manipulation of events that we describe here. However, there are many types of logical rearrangements and evaluations, such as fault tree modularization, that are beyond the scope of this book. The reader is referred to the *Fault Tree Handbook* [5] and the *NASA Fault Tree Handbook with Aerospace Applications* [6] for a more detailed discussion of this topic. In addition to the traditional Boolean

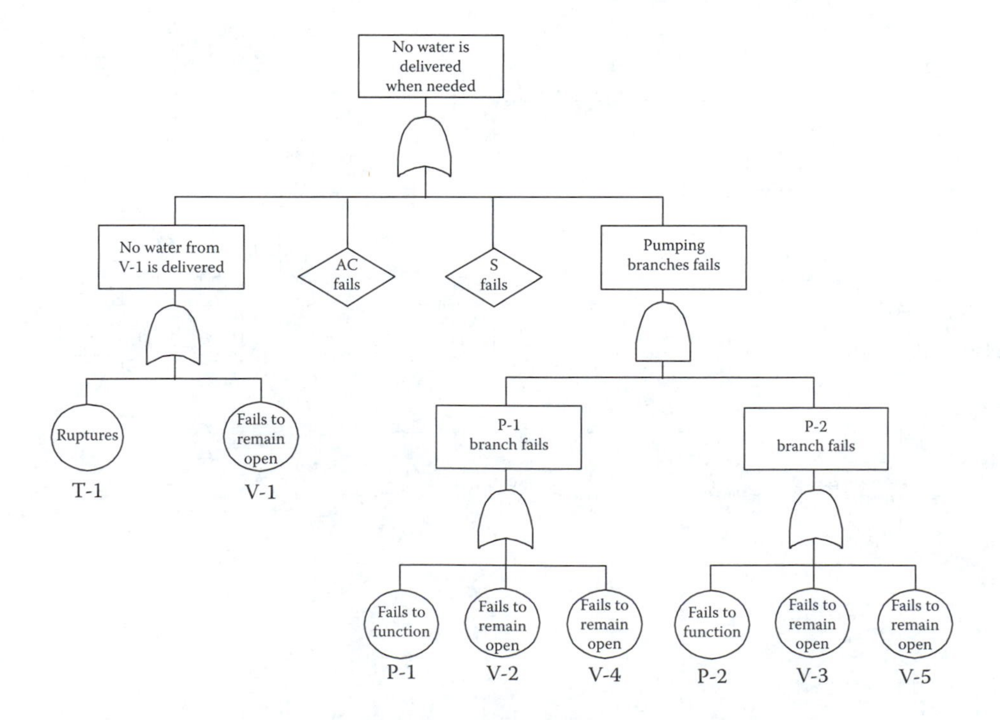

FIGURE 4.12 More compact form of the fault tree in Figure 4.11.

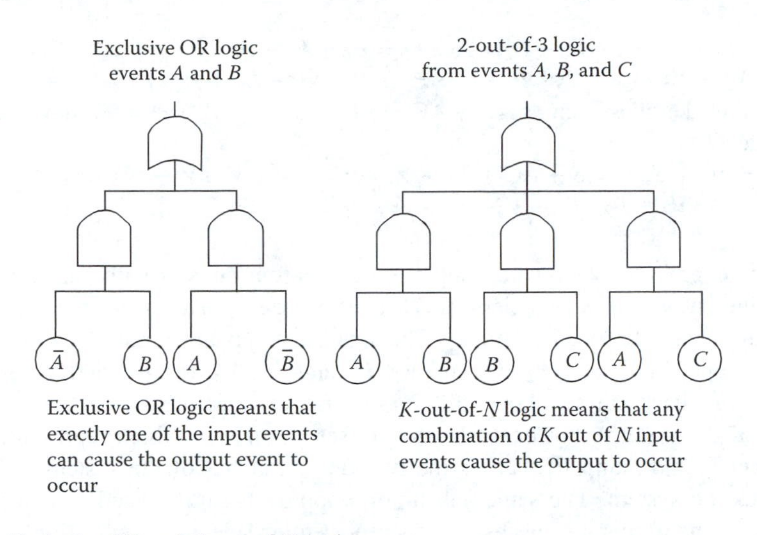
Exclusive OR logic
events A and B
2-out-of-3 logic
from events A, B, and C
Ā
B
A
B̄
A
B
B
C
A
C
Exclusive OR logic means that exactly one of the input events can cause the output event to occur
K-out-of-N logic means that any combination of K out of N input events cause the output to occur

FIGURE 4.13 Exclusive OR and K-out-of-N logics.

analysis of logic trees, a combinatorial approach will also be discussed. This technique generates mutually exclusive cut or path sets.

4.2.2.1 Boolean Algebra Analysis of Logic Trees

The quantitative evaluation of logic trees involves the determination of the probability of the occurrence of the top event. Accordingly, unreliability or reliability associated with the top event can also be determined. The qualitative evaluation of logic trees through the generation of cut or path sets is conceptually very simple. The tree OR-gate logic represents the union of the input events. That is, any of the input events, if they occur, will cause the output event to occur. For example, an OR gate with two inputs, events A and B, and the output event Q can be represented by its equivalent Boolean expression, $Q = A \cup B$. Either A or B or both must occur for the output event Q to occur. Instead of the union symbol $\cup$, the equivalent "+" symbol is often used. Thus, $Q = A + B$ (read A or B). Generally, for an OR gate with n inputs, $Q = A_1 + A_2 + \cdots + A_n$. The AND gate can be represented by the intersect logic. Therefore, the Boolean equivalent of an AND gate with two inputs A and B would be $Q = A \cap B$ [or $Q = A \cdot B$ (read A and B)].

Determination of cut sets using the above expressions is possible through several algorithms. These algorithms include the top-down or bottom-up successive substitution methods, the modularization approach, and Monte Carlo simulation. The *NASA Fault Tree Handbook with Aerospace Applications* [6] describes the underlying principles of these qualitative evaluation algorithms. The most widely used and straightforward algorithm is the successive substitution method. In this approach, the equivalent Boolean representation of each gate in the logic tree is determined such that only primary events remain. Various Boolean algebra rules are applied to reduce the Boolean expression to its most compact form, which represents the minimal path or cut sets of the logic tree. The substitution process can proceed from the top of the tree to the bottom or vice versa. Depending on the logic tree and its complexity, either the former or the latter approach, or a combination of the two, can be used.

As an example, let us consider the fault tree shown in Figure 4.11. Each node represents a failure. The step-by-step, top-down Boolean substitution of the top event is presented below.

Step 1. $T = E_1 \cdot E_2$,

Step 2. $E_1 = E_3 + V_3 + V_5 + E_4$, $E_2 = E_3 + V_4 + V_2 + E_5$, $T = E_3 + V_3 \cdot V_4 + V_3 \cdot V_2 + V_5 \cdot V_4 + V_5 \cdot V_2 + E_4 \cdot V_4 + E_4 \cdot V_2 + E_4 \cdot E_5 + V_3 \cdot E_5 + V_5 \cdot E_5$. ($T$ has been reduced by using the Boolean identities $E_3 \cdot E_3 = E_3$, $E_3 + E_3 \cdot X = E_3$, and $E_3 + E_3 = E_3$.)

Step 3. $E_3 = T_1 + V_1$, $E_4 = E_6 + P_2 + \mathrm{AC}$, $E_5 = E_6 + P_1 + \mathrm{AC}$, $T = T_1 + V_1 + \mathrm{AC} + V_3 \cdot V_4 + V_3 \cdot V_2 + V_5 \cdot V_4 + V_5 \cdot V_2 + V_4 \cdot P_2 + P_2 \cdot V_2 + E_6 + P_2 \cdot P_1 + V_3 \cdot P_1 + V_5 \cdot P_1$. (Again, identities such as $\mathrm{AC} + \mathrm{AC} = \mathrm{AC}$ and $E_6 + V_3 \cdot E_6 = E_6$ have been used to reduce T.)

Step 4. $E_6 = \mathrm{AC} + S$, $T = S + \mathrm{AC} + T_1 + V_1 + V_3 \cdot V_4 + V_3 \cdot V_2 + V_5 \cdot V_4 + V_5 \cdot V_2 + V_4 \cdot P_2 + P_2 \cdot V_2 + P_2 \cdot P_1 + V_3 \cdot P_1 + V_5 \cdot P_1$.

The Boolean expression obtained in Step 4 represents four cut sets with one element (cut set of size 1), and nine cut sets with two elements (cut set of size 2). The size-1 cut sets are the occurrences of failure events S, AC, T_1, and V_1. The size-2 cut sets are events V_3 and V_4, V_3 and V_2, V_5 and V_4, V_5 and V_2, V_4 and P_2, P_2 and V_2, P_2 and P_1, V_3 and P_1, and V_5 and P_1. A simple examination of each cut set shows that its occurrence guarantees the occurrence of the top event (failure of the system). For example, the cut set V_5 and P_1, which represents simultaneous failure of valve V_5 and pump P_1, causes the two flow branches of the system to be lost, which in turn disables the system. The same substitution approach can be used to determine the path sets. In this case, the events are success events representing adequate realization of the described functions.

This fault tree example shows that the evaluation of a large logic tree by hand can be a formidable job. A number of computer programs are available for the analysis of logic trees. Specter and Modarres [7] elaborate on the important characteristics of these software programs. Quantitative evaluation of the cut sets or path sets has already been discussed under the context of the reliability block diagram. For example, Expression 4.34 forms the basis for quantitative evaluation of the cut sets.

That is, the probability that the top event, T, occurs in a mission time t is

$$\Pr(T) = \Pr(C_1 \cup C_2 \cup \cdots \cup C_n). \tag{4.36}$$

The probability of the top event in a system reliability framework can be thought of as the unreliability of the system. To understand the complexities discussed earlier for the determination of $\Pr(T)$, let us consider the case where the following two cut sets are obtained:

$$C_1 = A \cdot B$$

and

$$C_2 = A \cdot C.$$

Then,

$$\Pr(T) = \Pr(A \cdot B + A \cdot C). \tag{4.37}$$

According to Equation 4.7,

$$\begin{aligned} \Pr(T) &= \Pr(A \cdot B) + \Pr(A \cdot C) - \Pr(A \cdot B \cdot A \cdot C) \\ &= \Pr(A \cdot B) + \Pr(A \cdot C) - \Pr(A \cdot B \cdot C). \end{aligned} \tag{4.38}$$

If A, B, and C are independent, then

$$\Pr(T) = \Pr(A) \cdot \Pr(B) + \Pr(\mathrm{A}) \cdot \Pr(C) - \Pr(A) \cdot \Pr(\mathrm{A}) \cdot \Pr(C). \tag{4.39}$$

The determination of the cross-product terms, such as Pr(A) Pr(B) Pr(C) in Equation 4.39, poses a dilemma in the quantitative evaluation of cut sets, especially when the number of cut sets is large. In general, there are 2^{n-1} such terms in cut sets.

For example, in the 13 cut sets generated for the pumping example, there are 8191 such terms. For large logic trees, this can be a formidable job even for powerful mainframe computers.

Fortunately, when dealing with cut sets, evaluation of these cross-product terms is often not necessary, and the boundary approach shown in Equation 4.35 is quite adequate. As discussed earlier, this is true whenever we are dealing with small probabilities, which is often the case for the probability of failure events. In these cases, for example, in Equation 4.39, $\Pr(A) \cdot \Pr(B) \cdot \Pr(C)$ is substantially smaller than $\Pr(A) \cdot \Pr(B)$ and $\Pr(A) \cdot \Pr(C)$. Thus the bounding result can also be used as an approximation of the true reliability or unreliability value of the system. This is often called the rare event approximation. Let us assume that

$$P(A) = \Pr(B) = \Pr(C) = 0.1. \tag{4.40}$$

Then,

$$\Pr(A) \cdot \Pr(B) = \Pr(A) \cdot \Pr(C) = 0.01 \tag{4.41}$$

and

$$\Pr(A) \cdot \Pr(B) \cdot \Pr(C) = 0.001. \tag{4.42}$$

The latter is smaller than the former by an order of magnitude. Although $\Pr(T) = 0.019$, the rare event approximation yields $\Pr(T) \approx 0.02$. Obviously, smaller probabilities of the events lead to better approximations.

As another example, consider the simple block diagram shown in Figure 4.14, which represents a system that has three paths from point X to point Y.

The equivalent fault tree is shown in Figure 4.15. The equivalent Boolean substitution equations are

$$\begin{aligned} T &= A \cdot B \cdot G_1, \\ G_1 &= C + D, \\ T &= A \cdot B \cdot (C + D), \\ T &= A \cdot B \cdot C + A \cdot B \cdot D. \end{aligned} \tag{4.43}$$

If the probability of events A, B, and C is 0.1, and the probability of event D is 0.2, the top event probability is evaluated as follows.

Using the rare event approximation discussed earlier,

$$\Pr(T) \approx \Pr(A) \cdot \Pr(B) \cdot \Pr(C) + \Pr(A) \cdot \Pr(B) \cdot \Pr(D). \tag{4.44}$$

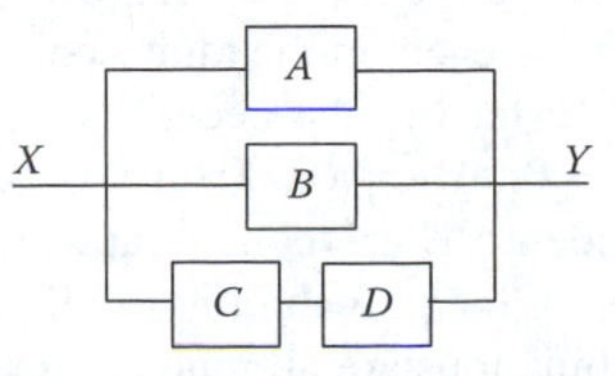

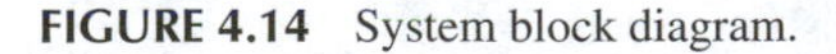
FIGURE 4.14 System block diagram.

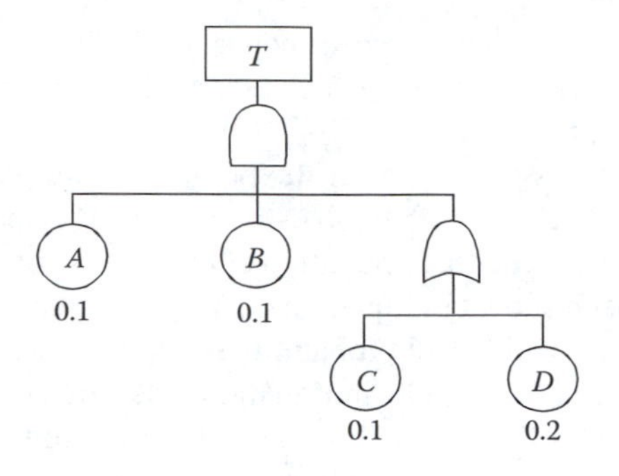

FIGURE 4.15 Fault tree representation of Figure 4.14.

Therefore,

$$\Pr(T) \approx 0.1 \times 0.1 \times 0.1 + 0.1 \times 0.1 \times 0.2 = 0.003. \tag{4.45}$$

Note that the terms $A \cdot B \cdot C$ and $A \cdot B \cdot D$ are not mutually exclusive and, therefore, the value of $\Pr(T)$ is approximate, since the rare event approximation has been used.

When all events are independent, in order to calculate the exact failure probability using minimal cut sets, their cross-product terms must also be included in the calculation of $\Pr(T)$,

$$\Pr(T) = \Pr(A) \cdot \Pr(B) \cdot \Pr(C) + \Pr(A) \cdot \Pr(B) \cdot \Pr(D) - \Pr(A) \cdot \Pr(B) \cdot \Pr(C) \cdot \Pr(D). \tag{4.46}$$

Accordingly,

$$\Pr(T) = 0.1 \times 0.1 \times 0.1 + 0.1 \times 0.1 \times 0.2 - 0.1 \times 0.1 \times 0.1 \times 0.2 = 0.0028. \tag{4.47}$$

4.2.2.2 Combinatorial (Truth Table) Technique for Evaluation of Logic Trees

Unlike the substitution technique, which is based on Boolean reduction, the combinatorial method does not convert the tree logic into Boolean equations to generate cut or path sets. Rather, this method relies on a combinatorial algorithm to exhaustively generate all probabilistically significant combinations of both "failure" and "success" events and subsequently to propagate the effect of each combination on the logic tree to determine the state of the top event. Because successes and failures are combined, all combinations are mutually exclusive. The quantification of logic trees based on the combinatorial method yields a more exact result and these are associated with a specific physical state of the system.

To illustrate the combinatorial approach, consider the fault tree in Figure 4.15. All possible combinations of success or failure events should be generated. Because there are four events and two states (success or failure) for each event, there are $2^4 = 16$ possible system states (i.e., actual physical states). Some of these states constitute system operation (when top event T does not happen), and some states constitute failure (when top event T does happen). These 16 states are illustrated in Table 4.1.

In Table 4.1, the subscript S refers to the nonoccurrence of an event (success), and subscript F refers to the failure or occurrence of the event in the fault tree.

Only combinations 14, 15, and 16 lead to the occurrence of the top event T, which results in system failure probability of $\Pr(T) = 0.0018 + 0.0008 + 0.0002 = 0.00028$. This is the exact value (provided that the events are independent). This is consistent with the exact calculation by the Boolean reduction method. Note that the sum of the probabilities of all possible combinations (16 of them in this case) is unity because the combinations are all mutually exclusive and cover all event space (a universal set). Combinations 14, 15, and 16 are mutually exclusive cut sets.

TABLE 4.1
Combinatorial Method (Truth Table) of Evaluating Event Tree

Combination Number	Combination Definition (System States)	Probability of C_i	System Operation T
1	$A_S B_S C_S D_S$	0.5832	S
2	$A_S B_S C_S D_F$	0.1458	S
3	$A_S B_S C_F D_S$	0.0648	S
4	$A_S B_S C_F D_F$	0.0162	S
5	$A_S B_F C_S D_S$	0.0648	S
6	$A_S B_F C_S D_F$	0.0162	S
7	$A_S B_F C_F D_S$	0.0072	S
8	$A_S B_F C_F D_F$	0.0018	S
9	$A_F B_S C_S D_S$	0.0648	S
10	$A_F B_S C_S D_F$	0.0162	S
11	$A_F B_S C_F D_S$	0.0072	S
12	$A_F B_S C_F D_F$	0.0018	S
13	$A_F B_F C_S D_S$	0.0072	S
14	$A_F B_F C_S D_F$	0.0018	F
15	$A_F B_F C_F D_S$	0.0008	F
16	$A_F B_F C_F D_F$	0.0002	F
		$\sum C_i = 1.0000$	

In order to visualize the difference between the results generated from the Boolean reduction and the combinatorial approach, the Venn diagram technique is helpful. Again consider the simple system in Figure 4.14 consisting of four events A, B, C, and D. The Boolean reduction process results in the minimal cut sets corresponding to system failure. These are $A \cdot B \cdot C$ and $A \cdot B \cdot D$.

The left side of Figure 4.16 represents a Venn diagram for the two cut sets above. Each cut set is represented by one shaded area. The two shaded areas are overlapping, indicating that the cut sets are not mutually exclusive. Now consider how combinations 14, 15, and 16 are represented in the Venn diagram (the right side of Figure 4.16). Again, each shaded area corresponds to a combination. In this case, there is no overlapping of the shaded areas. That is, the combinatorial approach generates mutually exclusive sets, and those sets that lead to system failure are called exclusive cut

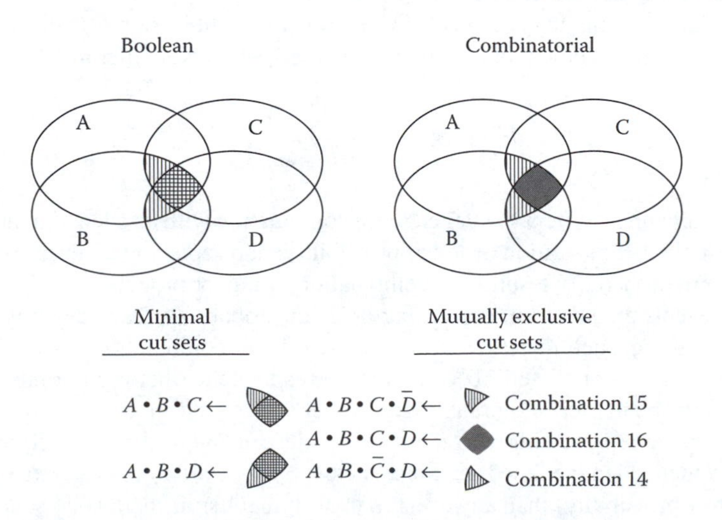

FIGURE 4.16 Boolean and combinatorial diagrams of events.

sets. Therefore, when the rare event approximation is not used, the contributions generated by the combinatorial approach have no overlapping area and produce the exact probability. Since for sizable problems, usually the rare event approximation is the only practical choice, if the exact probabilities are desired, or failure probabilities are greater than 0.1, then the combinatorial approach is preferred.

A typical logic model may contain hundreds of events. For n events, there are 2^n combinations. Obviously, for a large n (e.g., $n > 20$), the generation of this large number of combinations is impractical; a more efficient method would be needed. An algorithm to generate combinations whose probabilities exceed some cutoff limit (e.g., 10^{-7}) has been proposed by Dezfuli et al [8]. The algorithm generates combinations that are referred to as probabilistically significant combinations.

In this combinatorial algorithm, the total number of events is first determined. Each event has an associated probability of failure occurrence. A combination represents the status (i.e., failed or not failed) of every event in the entire logic diagram. The collection of all failed blocks within a combination is referred to as a failure set (FS). A FS may have zero elements, meaning that there are no failure events in the combination. This set is called the nil combination. The objective is to generate other probabilistically significant combinations. The following assumptions are made:

1. The failure events are independent.
2. The nil combination is a significant combination.

Given a combination C, the assumption of independence implies that the probability of the combination is

$$P_C = \prod_{i \in FS} P_i \prod_{i \notin FS} (1 - P_i), \tag{4.48}$$

where P_i is the probability of an individual failure event.

Consider the combination C', which differs from the combination C in that an event j is added to its FS (i.e., transition of a success event to a failure event). From the above results, it can be concluded that

$$P'_C = P_C \times \frac{P_j}{1 - P_j}. \tag{4.49}$$

Note that adding a block j to the failed set increases the probability of a combination if $P_j > 0.5$, and decreases the probability of a combination if $P_j < 0.5$.

Consider also the combination C'', which differs from the combination C in that block j is replaced with block k (i.e., the replacement of a block in the failed set with another block). Therefore,

$$P''_C = P_C \times \frac{P_j}{1 - P_j} \times \frac{1 - P_k}{P_k}. \tag{4.50}$$

This shows that replacing an event of a failed set in combination with an event that has a lower failure probability results in a combination of lower probability; and replacing an event with an event that has a higher failure probability results in a combination of higher probability.

As such, the events are sorted in decreasing order of probability. Each event is identified by its position in this ranking, such that $P_i > P_j$ when $i < j$. Each combination is identified by a list of events it contains in the failed set. To make the correspondence between combinations and lists unique, the list must be in ascending rank order, which corresponds to decreasing probability order.

Now consider a list representing a combination. A descendant of the list is defined as a list with one extra event appended to the failed set. Since the list must be ordered, this extra event must have a higher rank (lower probability) than any event in the original list. If there is no such event, there are no descendants. One should generate all descendants of the input list, and recursively generate all subsequent descendants. Since the algorithm begins with an empty list, the algorithm will generate all possible lists. Figure 4.17 illustrates this scheme for the simple case of four events.

To generate only significant lists, we first need to prove that if a list is not significant, its descendants are not significant. According to Equation 4.49, at least one item of the list must have a probability <0.5. Any failure event added to form the descendant would also have a probability <0.5. Therefore, the probability of the descendant would be lower than that of the original set, and, therefore, cannot be significant. The algorithm takes advantage of this property. The descendants are generated in an increasing rank (decreasing probability) order of the added events. Equation 4.50 shows that the probability of the generated combinations is also decreasing. Each list is checked to see whether it is significant. If it is not significant, the routine exits without any recursive operation and without generating any further descendants of the original input list. Figure 4.17 shows the effect if the state consisting of events a, c, and d is found to be insignificant; all of the indicated combinations are immediately excluded from further consideration.

4.2.2.3 Binary decision diagrams

This technique, as described in detail in the *NASA Fault Tree Handbook with Aerospace Applications* [6] is one of the most recently developed techniques for evaluation of logic trees. Advances in binary logic have recently led to the development of a new logic manipulation procedure for fault trees called binary decision diagrams (BDDs), which work directly with the logical expressions. A BDD in reality is a graphical representation of the tree logic. The BDD used for logic tree analysis is referred to as the reduced ordered BDD (i.e., the minimal form in which the events appear in the same order at each path). Sinnamon and Andrews [9] provide more information on the BDD approach to fault tree analysis.

The BDD is assembled using the logic tree recursively in a bottom-up fashion. Each basic event in the fault tree has an associated single-node BDD. For example, the BDD for a basic event B is shown in Figure 4.18. Starting at the bottom of the tree, a BDD is constructed for each basic event, and then combined according to the logic of the corresponding gate. The BDD for the OR gate "B or C" is constructed by applying the OR Boolean function to the BDD for event B and the BDD for event C. Since B is first in the relation, it is considered as the "root" node. The union of C BDD is with each "child" node of B.

First consider the terminal node 0 of event B in Figure 4.18. Accordingly, since $0 + X = X$ and $1 + X = 1$, the left child reduces to C, and the right child reduces to 1, as shown in Figure 4.19.

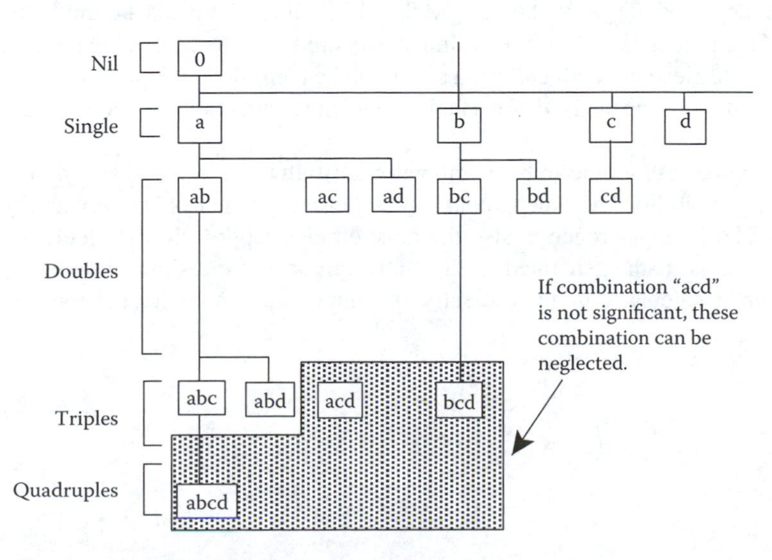

FIGURE 4.17 Computer algorithm for combinatorial approach.

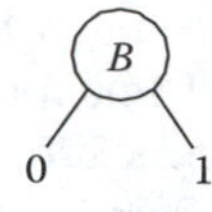

FIGURE 4.18 BDD for basic event B.

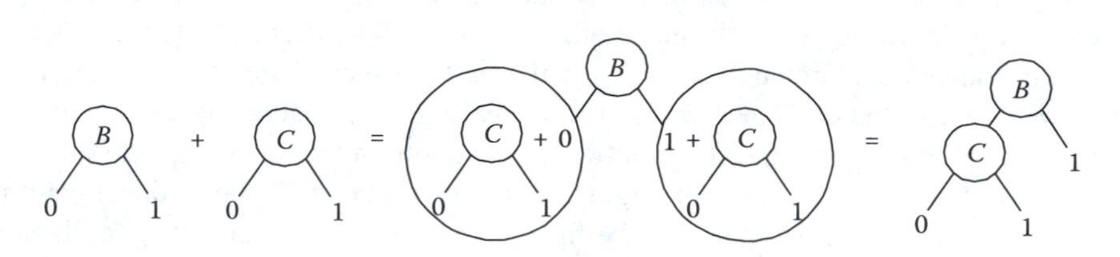

FIGURE 4.19 BDDs for the OR gate involving events B and C (step 1).

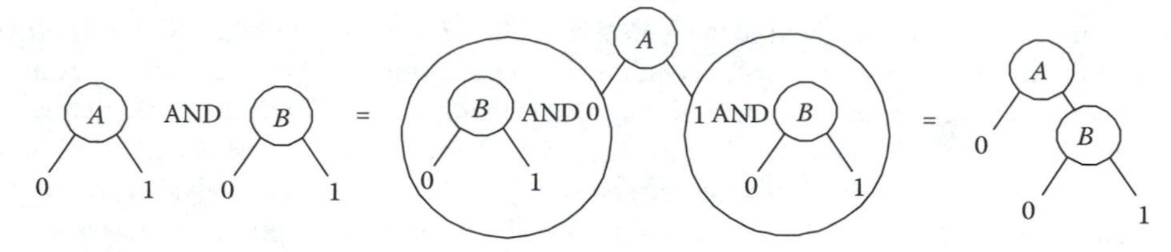

FIGURE 4.20 BDDs for the AND gate involving A and B (step 2).

Now consider the intersection operation applied to events A and B, shown in Figure 4.20. Note that $0 \cdot X = 0$ and that $1 \cdot X = X$. Thus, the reduced BDD for event $A \cdot B$ is shown in Figure 4.20.

Consider the Boolean logic $A \cdot B + C$, which is the union of the AND gate operation, in Figure 4.20, with event C. The BDD construction and reduction is shown in Figure 4.21. Since A comes before C, A is considered the root node, and the union operation is applied to A's children.

The reduced BDD for $A \cdot B + C$ can further be reduced because node C appears two times. Each path from the root node A to the lowest terminal node 1 represents a disjoint combination of events that constitute failure (occurrence) of the root node. Thus, in Figure 4.21, the failure paths (those leading to a bottom node 1) would be $\bar{A}C + AB + A\bar{B}C$. Since the paths are mutually exclusive, the calculation of the probability of failure is simple and similar to the truth table approach.

The failure paths leading to all end values of 1 can be identified and quantified by summing their probabilities. In developing the BDD for a fault tree, various reduction techniques are used to simplify the BDD.

The BDD approach is similar to the combinatorial (truth table) method and yields an exact value of the top event probability. The exact probability is useful when many high-probability events appear in the model. The BDD approach is also the most efficient approach for calculating probabilities. Because the minimal paths generated in the BDD approach are disjoint, important measures and sensitivities can be calculated more efficiently. It is important to note that generation of the minimal

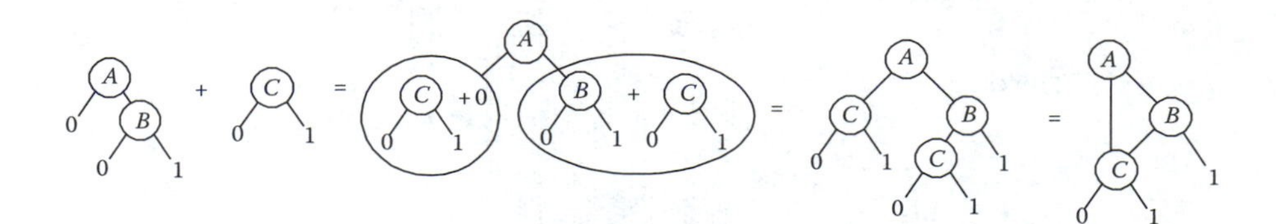

FIGURE 4.21 BDD construction and reduction for $A \cdot B + C$.

cut sets, obtained, for example, through the substitution method, provides important qualitative information as well as quantitative information. For example, minimal cut sets can be used to highlight the most significant failure combinations (cut sets of the systems or scenarios) and to show where design changes can remove or reduce certain failure combinations. Minimal cut sets also help to validate fault tree modeling of the system or subsystem by determining whether such combinations are physically meaningful, and by examining the tree logic to see whether they would actually cause the top event to occur. Minimal cut sets are also useful in investigating the effect or dependencies among the basic events modeled in the fault tree.

Example 4.6

Find the cut sets for the following fault tree through the BDD approach, and compare the results with the combinatorial (truth table) method.

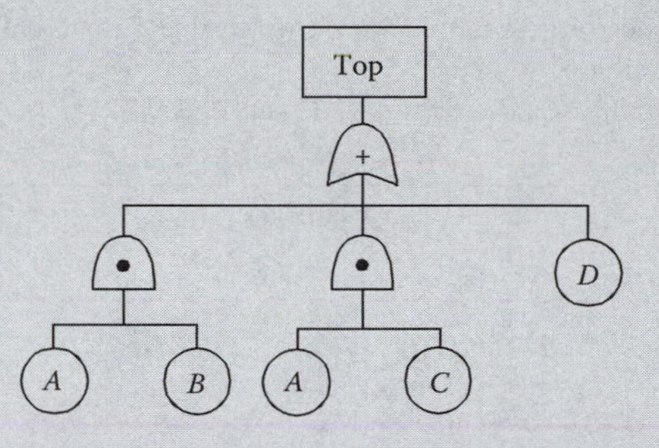

Solution: First, consider the basic events "*A* and *B*"; and "*A* and *C*":

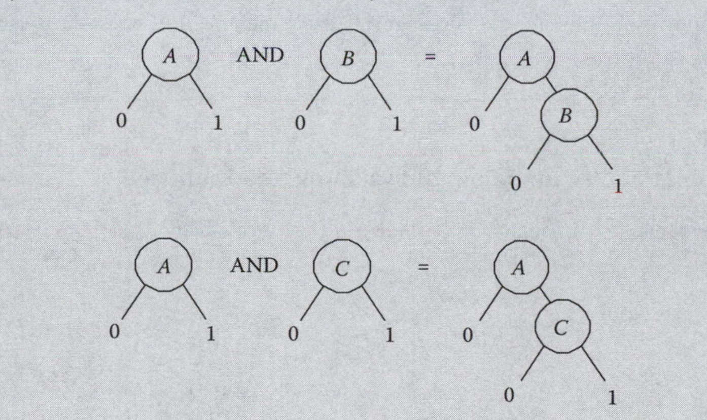

Now consider the union operation:

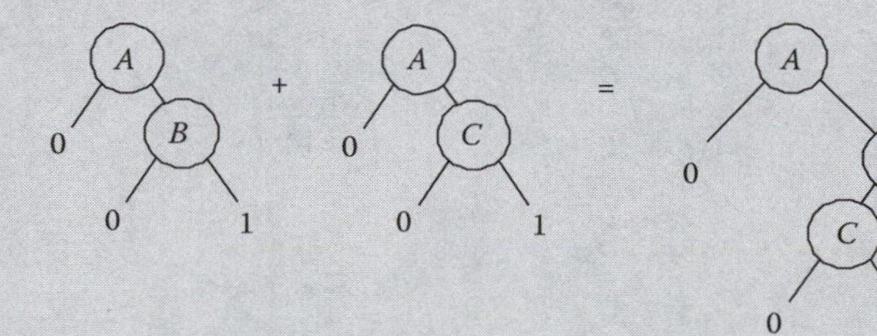

continued

Finally, considering event D, the BDD structure for the fault tree is

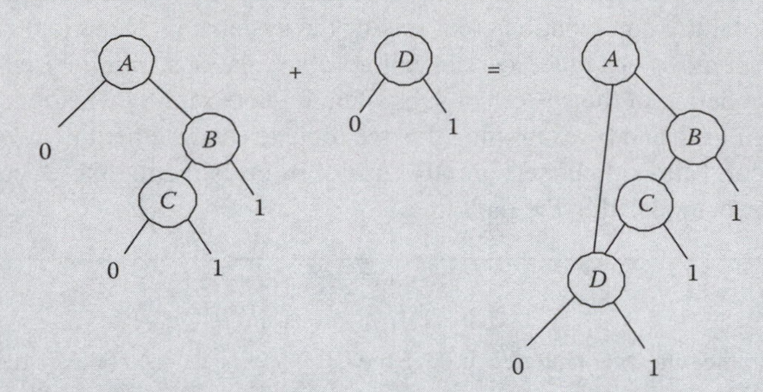

Each path from the root node, A, to a terminal node with value 1 represents a disjoint combination of events that causes system failure.

Assuming that the probability of failure for A, B, C, and D is 0.2,

Disjoint Path	Probability of Failure	
AB	$\Pr(A)\Pr(B)$	0.0400
$A\overline{B}C$	$\Pr(A)(1-\Pr(B))\Pr(C)$	0.0320
$A\overline{B}\,\overline{C}D$	$\Pr(A)(1-\Pr(B))(1-\Pr(C))\Pr(D)$	0.0256
$\overline{A}D$	$(1-\Pr(A))\Pr(D)$	0.1600
		0.2576

The system failure probability for the top event is 0.2576. Table 4.2 illustrates the results for the truth table method.

TABLE 4.2
Combinatorial (Truth Table) of Evaluating the Fault Tree

Combination Number	Combination Definition (System states)	Probability of C_i	System Operation T
1	$A_S B_S C_S D_S$	0.4096	S
2	$A_S B_S C_S D_F$	0.1024	F
3	$A_S B_S C_F D_S$	0.1024	S
4	$A_S B_S C_F D_F$	0.0256	F
5	$A_S B_F C_S D_S$	0.1024	S
6	$A_S B_F C_S D_F$	0.0256	F
7	$A_S B_F C_F D_S$	0.0256	S
8	$A_S B_F C_F D_F$	0.0064	F
9	$A_F B_S C_S D_S$	0.1024	S
10	$A_F B_S C_S D_F$	0.0256	F
11	$A_F B_S C_F D_S$	0.0256	F
12	$A_F B_S C_F D_F$	0.0064	F
13	$A_F B_F C_S D_S$	0.0256	F
14	$A_F B_F C_S D_F$	0.0064	F
15	$A_F B_F C_F D_S$	0.0064	F
16	$A_F B_F C_F D_F$	0.0016	F
		$\sum C_i = 1.0000$	

Note: S = success; F = failure.

Combinations 2, 4, 6, 8, 10, 11, 12, 13, 14, 15, and 16 lead to the occurrence of the top event T, which results in system failure probability of $\Pr(T) = 0.2576$. This result is similar to the BDD approach.

4.2.3 Success Tree Method

The success tree method is conceptually the same as the fault tree method. By defining the desirable top event, all intermediate and primary events that guarantee the occurrence of this desirable event are deductively postulated. Therefore, if the logical complement of the top event of a fault tree is used as the top event of a success tree, the Boolean structure represented by the fault tree is the Boolean complement of the success tree. Thus, the success tree, which shows the various combinations of success events that guarantee the occurrence of the top event, can be logically represented by path sets instead of cut sets.

To better understand this problem, consider the simple block diagram shown in Figure 4.22a. The fault tree for this system is shown in Figure 4.22b and the success tree in Figure 4.22c. Figure 4.23 shows an equivalent representation of Figure 4.22c.

From Figures 4.22b and 4.22c, it can be seen that changing the logic of one tree (changing AND gates to OR gates and vice versa) and changing all primary and intermediate events to their logical complements yields the other tree. This is also true for cut sets and path sets. That is, the logical complement of the cut sets of the fault tree yields the path sets of the equivalent success tree. This can be seen in Figure 4.22. The complement of the cut sets is

$$
\begin{aligned}
\overline{A + B \times C} &= \text{(apply de Morgan's Theorem)},\\
\overline{A} \times \overline{B \times C} &= \text{(apply de Morgan's Theorem)},\\
\overline{A} \times (\overline{B} + \overline{C}) &= \overline{A} \times \overline{B} + \overline{A} \times \overline{C},
\end{aligned}
\tag{4.51}
$$

which are the path sets.

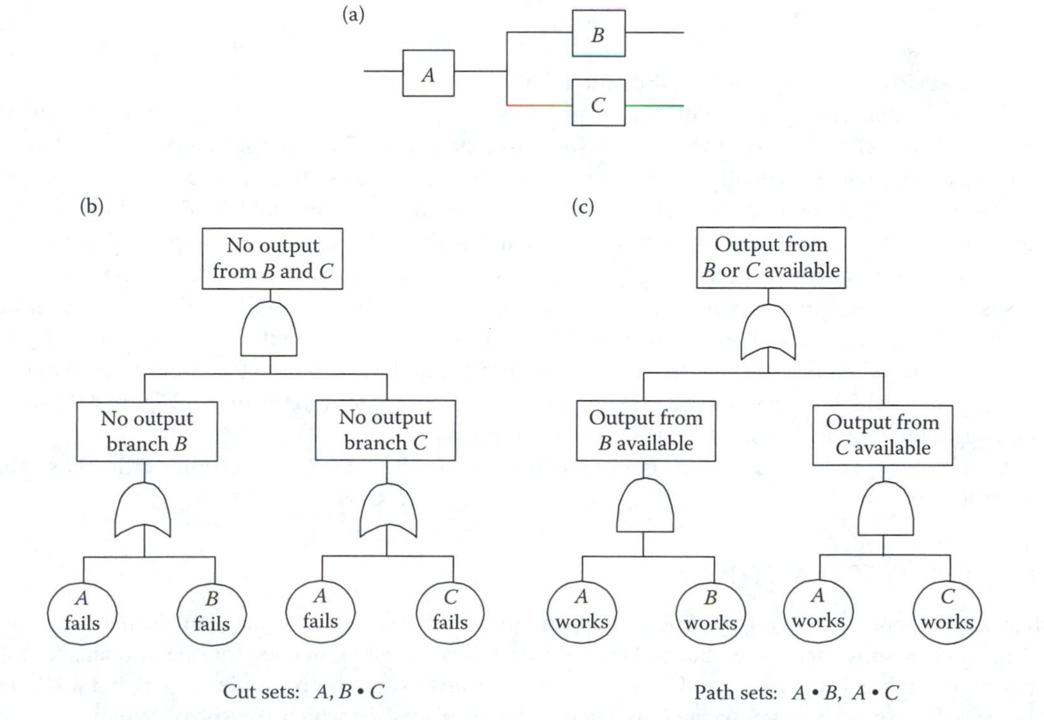

FIGURE 4.22 Correspondence between a fault and success tree. (a) Block diagram; (b) fault tree; and (c) success tree.

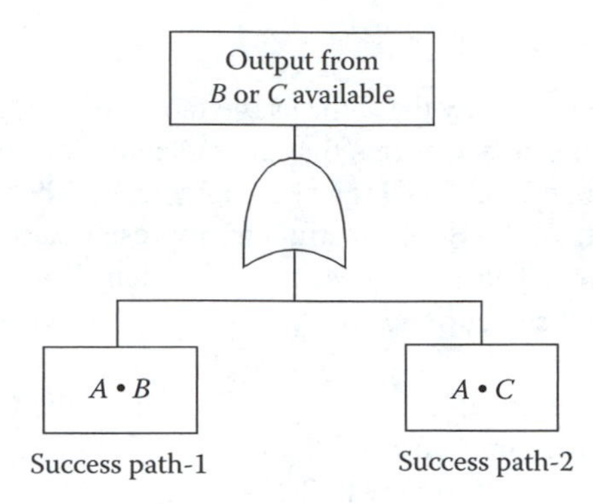

FIGURE 4.23 Equivalent representation of a success tree in Figure 4.22c.

Qualitative and quantitative evaluations of success paths are mechanistically the same as those of fault trees. For example, the top-down successive substitution of the gates and reduction of the resulting Boolean expression yields the minimal path sets. Accordingly, the use of Equation 4.32, or its lower bound Equation 4.33, allows us to determine the top-event probability (in this case, reliability). As noted earlier, Equation 4.32 poses a computational problem. In the context of using path sets, Wang and Modarres [10] have described several options for efficiently dealing with this problem.

A convenient way to reduce complex Boolean equations, especially the paths sets, is to use the following expressions:

$$\begin{aligned}\Pr(T) &= \Pr\left(P_1 \cup P_2 \cup \cdots \cup P_n\right)\\ &= \Pr\left(P_1\right) + \Pr\left(\overline{P}_1 \cap P_2\right) + \Pr\left(\overline{P}_1 \cap \overline{P}_2 \cap P_3\right) + \cdots + \Pr\left(\overline{P}_1 \cap \overline{P}_2 \cap \cdots \cap \overline{P}_{n-1} \cap P_n\right).\end{aligned} \tag{4.52}$$

For further discussions on applying Equation 4.52, see [11].

The combinatorial approach discussed in Section 4.2.2 is far superior for generating mutually exclusive path sets that assure the system's successful operation. For example, combinations 1–13 in Table 4.1 represent all mutually exclusive path sets for the system shown in Figure 4.14.

Success trees, as opposed to fault trees, provide a better understanding and display of how a system functions successfully. While this is important for designers and operators of complex systems, fault trees are more powerful for analyzing failures associated with systems and determining the causes of system failures. The minimal path sets of a system show the system user how the system operates successfully. A collection of events in a minimal path set is sometimes referred to as a success path. A logical equivalent of a success tree can also be represented by using the top event as an output to an OR gate in which input to the gate would show the success paths. For example, Figure 4.23 shows the equivalent representation of the success tree in Figure 4.22c.

In complex systems, the type of representation given in Figure 4.23 is useful for efficient system operation.

4.3 EVENT TREE METHOD

If successful operation of a system depends on an approximately chronological, but discrete, operation of its units or subsystems (i.e., the units should work in a defined sequence for operational success), then an event tree is appropriate. This may not always be the case for a simple system, but it is often the case for complex systems, such as nuclear power plants, in which the subsystems must work according to a given sequence of events to achieve a desirable outcome. Event trees are particularly useful in these situations.

4.3.1 Construction of Event Trees

Let us consider the event tree built for a nuclear power plant and shown in Figure 4.24. Event trees are horizontally built structures that start on the left, where the initiating event is modeled. This event describes a situation when a legitimate demand for the operation of a system(s) occurs. Development of the tree proceeds chronologically, with the demand on each unit (or subsystem) being postulated. The first unit demanded appears first, as shown on the top of the structure. In Figure 4.24, the events (referred to as event tree headings) are as follows:

RP = Operation of the reactor-protection system to shut down the reactor;
ECA = Injection of emergency coolant water by pump A;
ECB = Injection of emergency coolant water by pump B;
LHR = Long-term heat removal.

At each branch point, the upper branch shows the success of the event heading, and the lower branch shows its failure. In Figure 4.24, following the occurrence of the initiating event A, RP needs to work (event B). If RP does not work, the overall system will fail (as shown by the lower branch of event B). If RP works, then it is important to know whether ECB functions or not. If ECB does not function, even though RP has worked, the overall system would still fail. However, if ECB functions properly, it is important for LHR to function. Successful operation of LHR leads the system to a successful operating state, and failure of LHR (event E) leads the overall system to a failed state. Likewise, if ECA functions, it is important that it be followed by proper operation of LHR. If LHR fails, the overall system will be in a failed state. If LHR operates successfully, the overall system will be in a success state. However, the operation of certain subsystems may not be necessarily dependent on the occurrence of some preceding events. For example, if ECA operates successfully it does not matter for the overall system success whether or not ECB operates.

The outcome of each of the sequences of events is shown at the end of each sequence. This outcome, in essence, describes the final outcome of each sequence, whether the overall system succeeds, fails, initially succeeds but fails at a later time, or vice versa. The logical representation of each sequence can also be shown in the form of a Boolean expression. For example, for sequence 5 in Figure 4.24, events A, C, and D have occurred, but event B has not occurred (shown by $\bar{B}$). Clearly, these sequences are mutually exclusive.

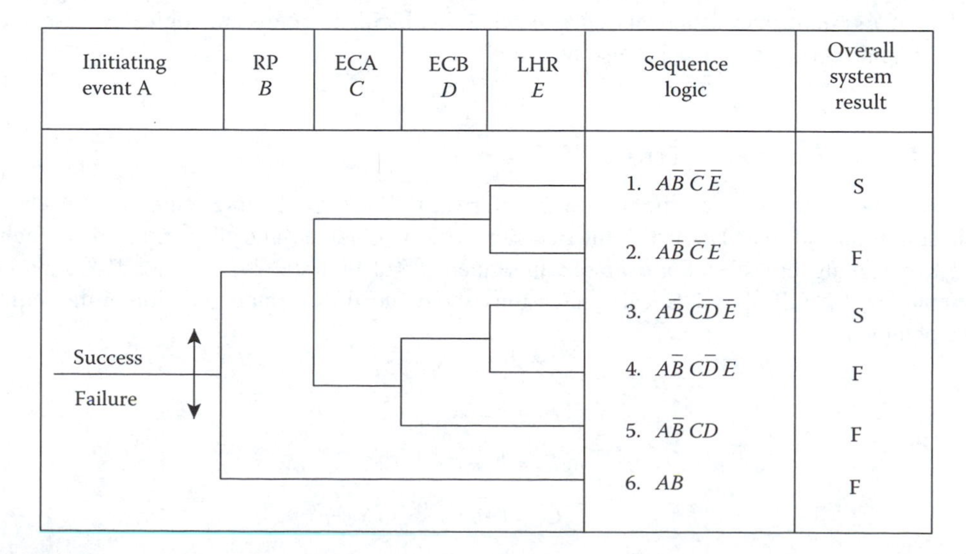

FIGURE 4.24 Example of an event tree.

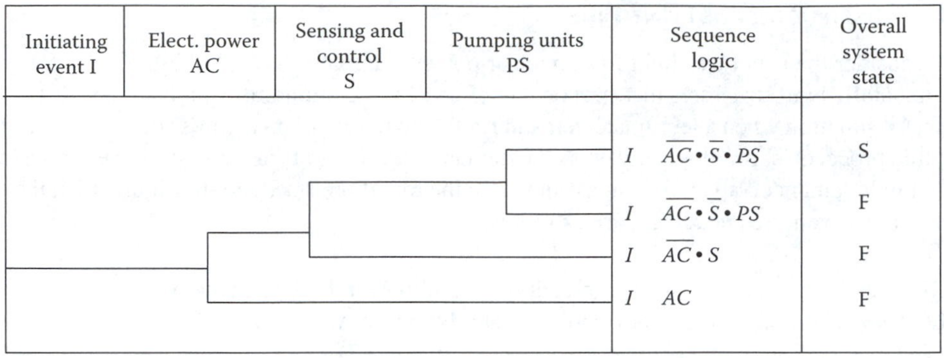

FIGURE 4.25 Event tree for the pumping system.

The event trees are usually developed in a binary format; that is, the heading events are assumed to either occur or not occur. In cases where a spectrum of outcomes is possible, the branching process can proceed with more than two outcomes. In these cases, the qualitative representation of the event tree branches in a Boolean sense would not be possible.

The development of an event tree, although somewhat deductive in principle, requires a good deal of inductive thinking by the analyst. To demonstrate this issue and further understand the concept of event tree development, let us consider the system shown in Figure 4.10. One can think of a situation where the sensing and control system device S initiates one of the two pumps. At the same time, the AC power source AC should always exist to allow S and pumps P-1 and P-2 to operate. Thus, if we define three distinct events S, AC, and pumping system PS for a sequence of events starting with the initiating event, an event tree that includes these three events can be constructed. Clearly, if AC fails, both PS and S fail and if S fails, only PS fails. This would lead to placing AC as the first event tree heading followed by S and PS. This event tree is illustrated in Figure 4.25.

Events represent discrete states of the systems. The logic of these states can be modeled by fault trees. This way, the event tree sequences and the logical combinations of events can be considered. This is a powerful aspect of the event tree technique. If the event tree headings represent complex subsystems or units, using a fault tree for each event tree heading can conveniently model the logic. Clearly, other system analysis models, such as reliability block diagrams and logical representations in terms of cut sets or path sets, can also be used.

4.3.2 Evaluation of Event Trees

Qualitative evaluation of event trees is straightforward. The logical representation of each event tree heading, and ultimately each event tree sequence, is obtained and then reduced through the use of Boolean algebra rules. For example, in sequence 5 of Figure 4.24, if events B, C, and D are represented by the following Boolean expressions, the reduced Boolean expression of the sequence can be obtained:

$$\begin{aligned} A &= a, \\ B &= b + c \cdot d, \\ C &= e + d, \\ D &= c + e \cdot h. \end{aligned} \tag{4.53}$$

The simultaneous Boolean expression and reduction proceeds as follows:

$$\begin{aligned} A \cdot \overline{B} \cdot C \cdot D &= a \cdot \left(\overline{b + c \cdot d}\right) \cdot (e + d) \cdot (c + e \cdot h) \\ &= a \cdot \left(\overline{b} \cdot \overline{c} + \overline{b} \cdot \overline{d}\right) \cdot (e \cdot c + e \cdot h + d \cdot c) \\ &= a \cdot \overline{b} \cdot \overline{c} \cdot e \cdot h + a \cdot \overline{b} \cdot c \cdot \overline{d} \cdot e + a \cdot \overline{b} \cdot \overline{d} \cdot e \cdot h. \end{aligned} \tag{4.54}$$

If an expression explaining all failed states is desired, the union of the reduced Boolean equations for each sequence that leads to failure should be obtained and reduced.

Quantitative evaluation of event trees is similar to the quantitative evaluation of fault trees. For example, to determine the probability associated with the sequence $A \cdot \overline{B} \cdot C \cdot D$, one would consider

$$\begin{aligned} \Pr\left(A \cdot \overline{B} \cdot C \cdot D\right) &= \Pr\left(a \cdot \overline{b} \cdot \overline{c} \cdot e \cdot h + a \cdot \overline{b} \cdot c \cdot \overline{d} \cdot e + a \cdot \overline{b} \cdot \overline{d} \cdot e \cdot h\right) \\ &= \Pr\left(a \cdot \overline{b} \cdot \overline{c} \cdot e \cdot h\right) + \Pr\left(a \cdot \overline{b} \cdot c \cdot \overline{d} \cdot \mathrm{e}\right) + \Pr\left(a \cdot \overline{b} \cdot \overline{d} \cdot e \cdot h\right) \\ &= \Pr(a) \cdot [1 - \Pr(b)]\,[1 - \Pr(c)]\,\Pr(e) \cdot \Pr(h) \\ &\quad + \Pr(a) \cdot [1 - \Pr(b)]\,\Pr(c) \cdot [1 - \Pr(d)]\,\Pr(e) \\ &\quad + \Pr(a) \cdot [1 - \Pr(b)]\,[1 - \Pr(d)]\,\Pr(e) \cdot \Pr(h). \end{aligned} \tag{4.55}$$

Since the two terms are disjoint, the above probability is exact. However, if the terms are not disjoint, the rare event approximation can be used here.

For more information on construction and evaluation of event trees, see NUREG/CR-2300 [12].

4.4 MASTER LOGIC DIAGRAM

For complex systems such as a nuclear power plant, modeling for reliability analysis or risk assessment may become very difficult. In complex systems, there are always several functionally separate subsystems that interact with each other, each of which can be modeled independently. However, it is necessary to find a logical representation of the overall system interactions with respect to the individual subsystems. The MLD is such a model [13].

Consider a functional block diagram of a complex system in which all of the functions modeled are necessary in one way or another to achieve a desired objective. For example, in the context of a nuclear power plant, the independent functions of heat generation, normal heat transport, emergency heat transport, reactor shutdown, heat-to-mechanical conversion, and mechanical-to-electrical conversion collectively achieve the goal of safely generating electric power. Each of these functions, in turn, is achieved through the design and operating function from others. For example, emergency heat transport may require internal cooling, which is obtained from other support functions.

The MLD clearly shows the interrelationships among the independent functions (or systems) and the independent support functions. The MLD (in the success space) can show the ways various functions, subfunctions, and hardware interact to achieve the overall system objective. On the other hand, an MLD in a failure space can show the logical representation of the causes for failure of functions (or systems). The MLD (in the success or failure space) can easily map the propagation of the effect of failures, that is, establish the trajectories of event failure propagation.

In essence, the hierarchy of the MLD is displayed by the dependency matrix. For each function, subfunction, subsystem, and hardware item shown on the MLD, the effect of the failure or success of all combinations of items is established and explicitly shown by "·". Consider the MLD shown in a success space in Figure 4.26. In this diagram, there are two major functions (or systems), F_1 and F_2.

Together, they achieve the system objective. Each of these functions, because of reliability concerns, is further divided into two identical subfunctions, each of which can achieve the respective parent functions. This means that both subfunctions must be lost for F_1 or F_2 to be lost. Suppose

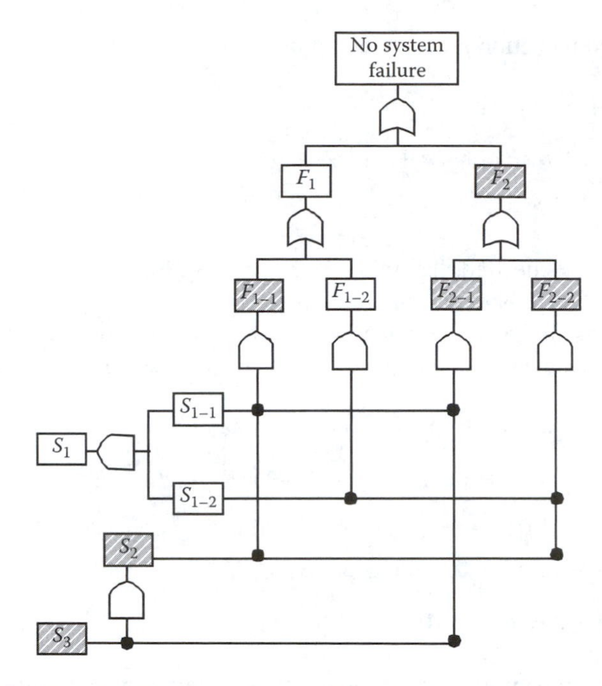

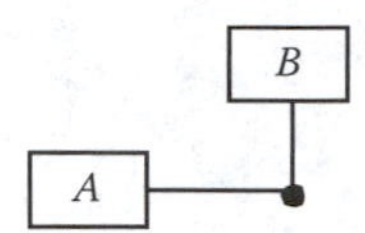

FIGURE 4.26 MLD showing the effect of failure of S_3.

the development of the subfunctions (or systems) can be represented by their respective hardware, which interfaces with other support functions (or support systems) S_1, S_2, and S_3. Support functions are those that allow the main functions to be realized. For example, if a pump function is to "provide pressure," then functions "provide a.c. power," "cooling and lubrication," and "activation and control" are called support functions. However, function (or system) S_1 can be divided into two independent subfunctions (or subsystems) (S_{1-1} and S_{1-2}), so that each can interact independently with the subfunctions (or subsystems) of F_1 and F_2. The dependency matrix is established by reviewing the design specifications or operating manuals that describe the relationship between the items shown in the MLD, in which the dependencies are explicitly shown by "·". For instance, the dependency matrix shows that failure of S_3 leads directly to failure of S_2, which in turn results in failures of F_{1-1}, F_{2-2}, and F_{2-1}. This failure is highlighted on the MLD in Figure 4.26.

A key element in the development of an MLD is the assurance that the items for which the dependency matrix is developed (e.g., S_{1-1}, S_{1-2}, S_3, F_{2-1}, F_{1-2}, and F_{2-2}) are all physically independent. Physically independent means they do not share any other system parts. Each element may have other dependencies, such as common-cause failure (see Chapter 6). Sometimes it is difficult to distinguish *a priori* between main functions and supporting functions. In these situations, the dependency matrix can be developed irrespective of the main and supporting functions. Figure 4.27 shows an example of such a development.

However, the main functions can be identified easily by examining the resulting MLD; they are those functions that appear, hierarchically, at the top of the MLD model and do not support other items.

The analysis of an MLD is straightforward. One must determine all possible 2^n combinations of failures of independent items (elements), map them onto the MLD and propagate their effects, using the MLD logic. The combinatorial approach discussed in Section 4.2.2 is the most appropriate method for that purpose, although the Boolean reduction method can also be applied. Table 4.3 shows the combinations for the example in Figure 4.26.

For reliability calculations, one can combine those end-state effects that lead to the system success. Suppose independent items (here, systems or subsystems) S_{1-1}, S_{1-2}, S_2, and S_3 have a failure probability of 0.01 for a given mission, and the probability of independent failure of F_{1-1}, F_{1-2}, F_{2-1}, and F_{2-2} is also 0.01. Table 4.4 shows the resulting probability of the end-state effects.

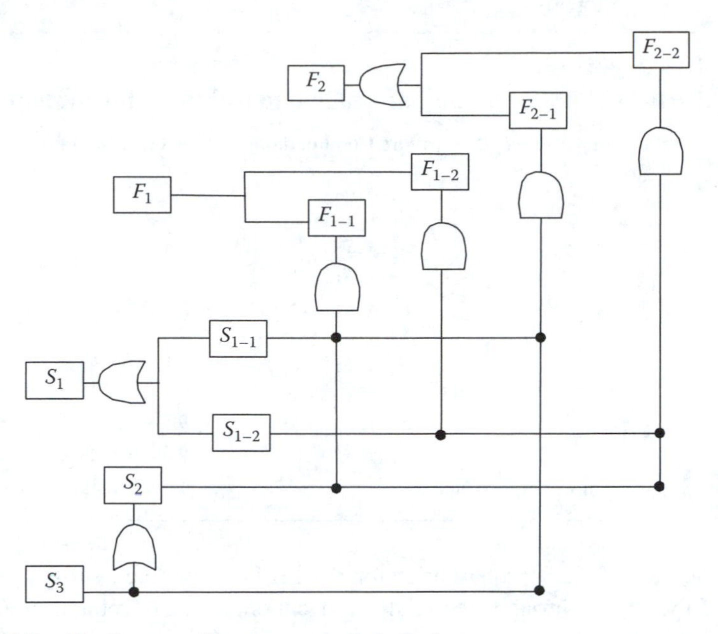

FIGURE 4.27 MLD with all system functions treated similarly.

TABLE 4.3
Dominant Combinations of Failure and Their Respective Probabilities

Combination Number (*i*)	Failed Items	Probability of Failed Items (and Success of Other Items)	End State (Actually Failed and Casually Failed Elements)
1	None	9.3×10^{-1}	Success
2	S_3	9.3×10^{-3}	$F_{1-1}, F_2, F_{2-1}, F_{2-2}, S_2, S_3$
3	F_{2-2}, S_3	9.4×10^{-5}	$F_{1-1}, F_2, F_{2-1}, F_{2-2}, S_2, S_3$
4	S_2, S_3	9.4×10^{-5}	$F_{1-1}, F_2, F_{2-1}, F_{2-2}, S_2, S_3$
5	F_{1-1}, S_3	9.4×10^{-5}	$F_{1-1}, F_2, F_{2-1}, F_{2-2}, S_2, S_3$
6	F_{2-1}, S_3	9.4×10^{-5}	$F_{1-1}, F_2, F_{2-1}, F_{2-2}, S_2, S_3$
7	S_{1-2}	9.3×10^{-3}	$F_{2-1}, F_{2-2}, S_{1-2}, S_1$
8	F_{1-2}, S_{1-2}	9.4×10^{-5}	$F_{1-2}, F_{2-2}, S_{1-2}, S_1$
9	F_{2-2}, S_{1-2}	9.4×10^{-5}	$F_{1-2}, F_{2-2}, S_{1-2}, S_1$
10	S_{1-1}	9.3×10^{-3}	$F_{1-2}, F_{2-2}, S_{1-2}, S_1$
11	F_{2-1}, S_{1-1}	9.4×10^{-5}	$F_{1-1}, F_{2-1}, S_{1-1}, S_1$
12	F_{1-1}, S_{1-1}	9.4×10^{-5}	$F_{1-1}, F_{2-1}, S_{1-1}, S_1$
13	S_2	9.3×10^{-3}	F_{1-1}, F_{2-2}, S_2
14	F_{1-1}, S_2	9.4×10^{-5}	F_{1-1}, F_{2-2}, S_2
15	F_{2-2}, S_2	9.4×10^{-5}	F_{1-1}, F_{2-2}, S_2
16	F_{2-2}	9.3×10^{-3}	F_{2-2}
17	F_{2-1}	9.3×10^{-3}	F_{2-1}
18	F_{1-2}	9.3×10^{-3}	F_{1-2}
19	F_{1-1}	9.3×10^{-3}	F_{1-1}
20	S_{1-2}, S_3	9.4×10^{-5}	$F_1, F_{1-1}, F_{1-2}, F_2, F_{2-1}, F_{2-2}, S_{1-2}, S_2, S_3, S_1$

TABLE 4.4
Partial List of Combinations Leading to Failure of the System

Combination Number (*i*)	Failure Combination	Probability of State
1	F_{1-2}, S_3	9.4×10^{-5}
2	S_{1-1}, S_{1-2}	9.4×10^{-5}
3	S_{1-2}, S_3	9.4×10^{-5}
4	F_{2-1}, S_{1-2}, S_3	9.5×10^{-7}
5	$F_{2-2}, S_{1-1}, S_{1-2}$	9.5×10^{-7}
6	F_{2-1}, S_{1-2}, S_2	9.5×10^{-7}
7	$F_{2-1}, S_{1-1}, S_{1-2}$	9.5×10^{-7}
8	$F_{1-1}, F_{2-1}, S_{1-2}$	9.5×10^{-7}
9	S_{1-1}, S_{1-2}, S_3	9.5×10^{-7}
10	S_{1-2}, S_2, S_3	9.5×10^{-7}
Total probability of system failure		2.9×10^{-4}

If needed, calculation of failure probabilities for the MLD items (e.g., subsystems) can proceed independently of the MLD, through one of the conventional system reliability analysis methods (e.g., fault tree analysis).

Table 4.3 shows all possible mutually exclusive combinations of items modeled in the MLD (i.e., S_{1-1}, S_{1-2}, S_2, S_3, F_{1-1}, S_{1-2}, and F_{2-2}). Those combinations that lead to a failure of the system are mutually exclusive cut sets.

One may select only those combinations that lead to a complete failure of the system. The sum of the probabilities of occurrence of these combinations determines the failure probability of the system. If one selects the combinations that lead to the system's success, then the sum of the probabilities of occurrence of these combinations determines the reliability of the system. Table 4.4, for example, shows dominant combinations that lead to the system's failure.

Another useful analysis that can be performed via MLD is the calculation of the conditional system probability of failure. In this case, a particular element of the system is set to failure, and all other combinations that lead to the system's failure can be identified. Table 4.5 shows all combinations within the MLD that lead to the system's failure when element S_{1-2} is set to failure.

TABLE 4.5
Partial List of Combinations Leading to the System Failure When S_{1-2} is known to Have Failed

Combination Number (*i*)	Failure Combination	Probability of State
1	S_3	9.6×10^{-3}
2	S_{1-1}	9.6×10^{-3}
3	S_2, S_3	9.7×10^{-5}
4	F_{2-1}, S_3	9.7×10^{-5}
5	F_{1-1}, F_{2-1}	9.7×10^{-5}
6	F_{2-1}, S_2	9.7×10^{-5}
7	F_{1-2}, S_{1-1}	9.7×10^{-5}
8	S_{1-1}, S_2	9.7×10^{-5}
9	S_{1-1}, S_3	9.7×10^{-5}
10	F_{1-2}, S_3	9.7×10^{-5}
11	F_{2-1}, S_{1-1}	9.7×10^{-5}
Total probability of system failure		2.01×10^{-2}

EXAMPLE 4.7

Consider the H-Coal process shown in Figure 4.28. In case of an emergency, a shutdown device (SDD) is used to shut down the hydrogen flow. If the reactor temperature is too high, an emergency cooling system (ECS) is also needed to reduce the reactor temperature. To protect the process plant when the reactor temperature becomes too high, both ECS and SDD must succeed. The SDD and ECS are actuated by a

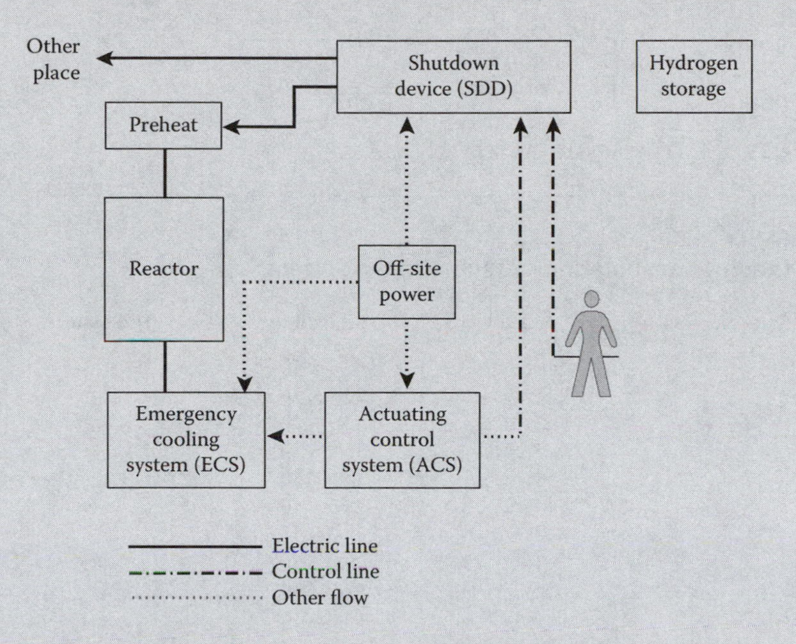

FIGURE 4.28 Simplified diagram of the safety systems.

control device. If the control device fails, the ECS will not be able to work. However, an operator (OA) can manually operate the SDD and terminate the hydrogen flow. The power for the SDD, ECS, and control device comes from an outside electric company (off-site power—OSP). The failure data for these systems are listed in Table 4.6. Draw an MLD and use it to find the probability of losing both the SDD and ECS.

Solution: The MLD is shown in Figure 4.29. Important combinations of independent failures and their impacts on other components are listed in Table 4.7. The probability of losing both the ECS and SDD for

TABLE 4.6
Failure Probability of Each System

System Failure	Failure Probability
OSP	2.0×10^{-2}
OA	1.0×10^{-2}
ACS	1.0×10^{-3}
SDD	1.0×10^{-3}
ECS	1.0×10^{-3}

each end state is calculated and listed in the third column of Table 4.8. Combinations that exceed 1×10^{-6} are included in Table 4.8. The combinations that could lead to failure of both SDD and ECS are shown in Table 4.8. Using Equation 4.49, the probability of losing both systems is calculated as 4.99×10^{-3}.

continued

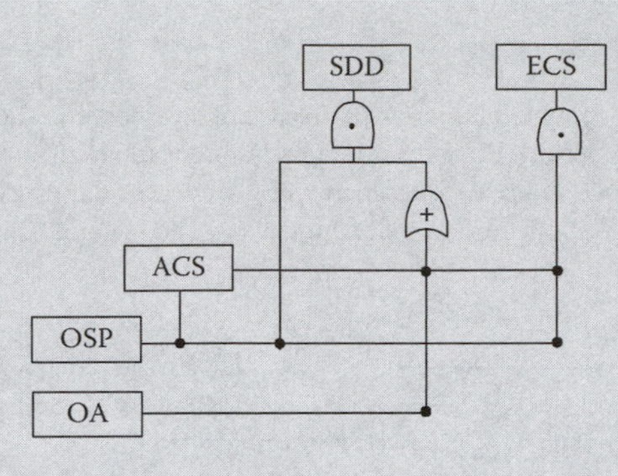

FIGURE 4.29 MLD for the safety system in Figure 4.28.

TABLE 4.7
Leading Combination of Failure in the System

State Number (i)	Failed Units	Probability[a]	End State
	None	9.94×10^{-1}	Success
1	OSP	1.99×10^{-3}	
	OSP, ECS	1.99×10^{-6}	
	OSP, SDD	1.99×10^{-6}	
2	OSP, ACS	1.99×10^{-6}	SDD, ECS, ACS, OSP
	OSP, ACS, SDD	2.00×10^{-9}	
	OSP, ECS, SDD	2.00×10^{-9}	
	OSP, ECS, ACS	2.00×10^{-9}	
3	ACS	9.95×10^{-4}	ECS, ACS
	ECS, ACS	9.96×10^{-7}	
4	SDD	9.95×10^{-4}	SDD
5	OA	9.95×10^{-4}	OA
6	ECS	9.95×10^{-4}	ECS
	OSP, OA	1.99×10^{-6}	
7	OSP, ACS, OA	2.00×10^{-9}	SDD, ECS, ACS
	OSP, ECS, OA	2.00×10^{-9}	OSP, OA
8	OSP, SDD, OA	2.00×10^{-9}	ECS, OA
	ECS, OA	9.96×10^{-7}	ECS, OA
9	ACS, OA	9.96×10^{-7}	SDD, ECS, ACS, OA
10	ACS	9.96×10^{-7}	SDD

[a] Includes probability of success of elements not affected.

TABLE 4.8
Probability of Losing Two Systems

Combination Number	Units Failed	Probability[a]	Contribution to Total Probability (%)
1	OSP	1.99×10^{-3}	39.9
2	ACS	9.95×10^{-4}	19.9
3	ECS	9.95×10^{-4}	19.9
4	SDD	9.95×10^{-4}	19.9
5	OSP and ECS	1.99×10^{-6}	Negligible
6	OSP and SDD	1.99×10^{-6}	Negligible
7	OSP and ACS	1.99×10^{-6}	Negligible
8	OSP and OA	1.99×10^{-6}	Negligible

[a] Includes probability of success of elements not affected.

EXAMPLE 4.8

The simple event tree shown in Figure 4.30 has five events (A, B, C, D, and E) that make up the headings of the event tree. The initiating event is labeled I.

Consider sequence no. 5, which is highlighted with a bold line. The logical equivalent of the sequence is

$$S_5 = I \cdot \overline{A} \cdot B \cdot \overline{C} \cdot D,$$

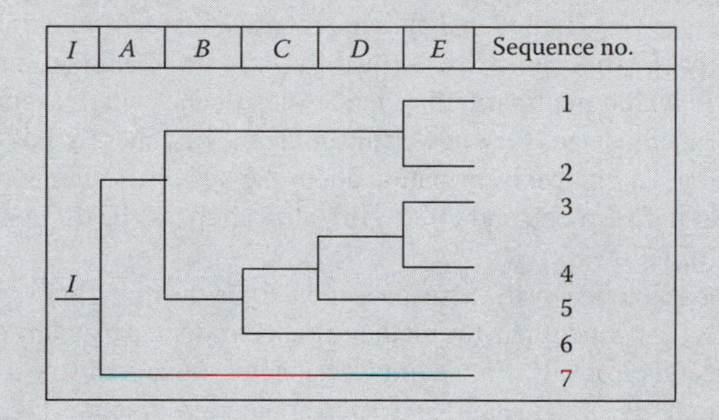

FIGURE 4.30 Simple event tree.

where S_5 is the fifth sequence and I is the initiating event. Develop an equivalent MLD representation of this event tree.

Solution: Sequence 5 occurs when the expression $\overline{A} \cdot B \cdot \overline{C} \cdot D$ is true. Note that the above Boolean expression involves two failed elements (i.e., B and D). We can express these terms, in the success space, through the complement of $\overline{A} \cdot B \cdot \overline{C} \cdot D$, which is

$$\overline{\overline{A} \cdot B \cdot \overline{C} \cdot D} = A + \overline{B} + C + \overline{D}.$$

The last expression represents every event in a success space (e.g., $\overline{A} \cdot \bar{B} \cdot \overline{C}$) and its equivalent MLD logic is shown in Figure 4.31.

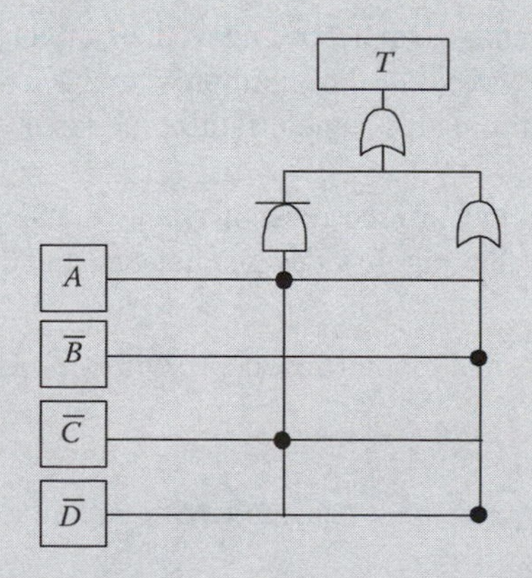

FIGURE 4.31 MLD equivalent of sequence 5 shown in Figure 4.30.

4.5 FAILURE MODE AND EFFECT ANALYSIS

FMEA is a powerful technique for reliability analysis. This method is inductive in nature. In practice, it is used at all stages of system failure analysis, from concept to implementation. The FMEA analysis describes inherent causes of events that lead to a system failure, determines their consequences, and devises methods to minimize their occurrence or recurrence.

The FMEA proceeds from one level or a combination of levels of abstraction, such as system functions, subsystems, or components. The analysis assumes that a failure has occurred. The potential effect of the failure is then postulated, and its potential causes are identified. A criticality rating or the risk priority number (RPN) rating may also be determined for each failure mode and its resulting effect. The rating is normally based on the probability of the failure occurrence, the severity of its effect(*s*), and its detectability. Failures that score high in this rating represent areas of greatest risk, and their causes should be mitigated.

Although the FMEA is an essential reliability task for many types of system design and development processes, it provides very limited insight into probabilistic representation of system reliability. Another limitation is that FMEA can be performed for only one failure at a time. This may not be adequate for systems in which multiple failure modes can occur, with reasonable likelihood, at the same time. (Deductive methods are very powerful for identifying these kinds of failures.) However, FMEA provides valuable qualitative information about the system design and operation. An extension of FMEA is called failure mode and effect criticality analysis (FMECA), which provides more quantitative treatment of failures.

The FMEA was first developed by the aerospace industry in the mid-1960s. The standard reference is US MIL-STD-1629A [14]. Since then, the method has been adopted by many other industries, which have modified it to meet their needs. For example, the automotive industry uses the FMEA refined by the Society of Automotive Engineers (SAE) recommended Practice J1739 [15]. The methods of FMEA and FMECA are briefly discussed in this section. For more information, the readers are referred to the above-mentioned publications.

4.5.1 Types of FMEA

Depending on the stage in product development, one may perform two types of FMEA (SAE Recommended Practice J1739 [15]): design FMEA and process FMEA.

Design FMEA is used to evaluate the failure modes and their effects for a product before it is released to production and is normally applied at the subsystem and the component abstraction levels. The major objectives of a design FMEA are to

1. Identify failure modes and rank them according to their effect on the product performance, thus establishing a priority system for design improvements.
2. Identify design actions to eliminate potential failure modes or reduce the occurrence of the respective failures.
3. Document the rationale behind product design changes and provide future reference for analyzing field concerns, evaluating new design changes, and developing advanced designs.

Process FMEA is used to analyze manufacturing and assembly processes. The major objectives of a process FMEA are to

1. Identify failure modes that can be associated with manufacturing or assembly process deficiencies.
2. Identify highly critical process characteristics that may cause particular failure modes.
3. Identify the sources of manufacturing/assembly process variations (equipment performance, material, operator, and environment) and establish a strategy to reduce it.

4.5.2 FMEA/FMECA Procedure

Outlined below is a logical sequence of steps by which FMEA/FMECA is usually performed.

- Define the system to be analyzed. Identify the system decomposition (indenture) level that will be subject to analysis. Identify internal and interface system functions, restraints, and develop failure definitions.
- Construct a block diagram of the system. Depending on system complexity and the objectives of the analysis, consider at least one of these diagrams: structural (hardware), functional, combined, or MLD. (The latter method is discussed in greater detail in Section 4.4.)
- Identify all potential item failure modes and define their effects on the immediate function or item, on the system, and on the mission to be performed.
- Evaluate each failure mode in terms of the worst potential consequence, and assign a severity classification category.
- Identify failure detection methods and compensating provision(s) for each failure mode.
- Identify corrective designs or other actions required to eliminate the failure or control the risk.
- Document the analysis and identify the problems that could not be corrected by design.

4.5.3 FMEA Implementation

4.5.3.1 FMEA for Aerospace Applications

The FMEA is usually performed using a tabular format. A worksheet implementation of a typical MIL-STD-1629A FMEA procedure is shown in Table 4.9. The major steps of the analysis are described below.

4.5.3.1.1 System Description and Block Diagrams

It is important to first describe the system in a manner that allows the FMEA to be performed efficiently and to be understood by others. This description can be done at different levels of abstraction. For example, at the highest level (i.e., the functional level), the system can be represented by a functional block diagram. The functional block diagram is different from the reliability block diagram discussed earlier in this chapter. Functional block diagrams illustrate the operation, interrelationship, and interdependence of the functional entities of a system. For example, the pumping system of Figure 4.10 can be represented by its functional block diagram, as shown in Figure 4.32. In this figure, the components that support each system function are also described.

4.5.3.1.2 Item/Functional Identification

Provide the descriptive name and the nomenclature of the item under analysis. If the failures are postulated at a lower abstraction level, such levels should be shown. A fundamental item of the current FMEA may be subject to a separate FMEA that further decomposes this item into more basic parts. The lower the abstraction level, the greater the level of detail required for the analysis. This step provides necessary information for the identification number, functional identification (nomenclature), and function columns in the FMEA.

4.5.3.1.3 Failure Modes and Causes and Mission Phase/Operational Mode

The manner of failure of the function, subsystem, component, or part identified in the second column of the table is called the failure mode and is listed in the failure mode and causes column of the FMEA table. The causes (a failure mode can have more than one cause) of each failure mode should also be identified and listed in this column. The failure modes applicable to components and parts are often known *a priori*. Typical failure modes for electronic components are open, short, corroded, drifting, misaligned, and so on. Some representative failure modes for mechanical components include deformed, cracked, fractured, sticking, leaking, and loosened. However, depending on the specific system under analysis, the environmental design, and other factors, only certain failure modes may apply. These should be known and specified by the analyst.

TABLE 4.9
U.S. MIL-STD-1629A FMEA Worksheet Format

FAILURE MODE AND EFFECTS ANALYSIS

System ______________ Date ______________
Indenture level ________ Sheet ____ of __________
Reference drawing ______ Compiled by ___________
Mission _____________ Approved by ___________

IDENTIFICATION NUMBER	ITEM/FUNCTIONAL IDENTIFICATION (NOMENCLATURE)	FUNCTION	FAILURE MODES AND CAUSES	MISSION PHASE/ OPERATIONAL MODE	FAILURE EFFECTS			FAILURE DETECTION METHOD	COMPENSATING PROVISIONS	SEVERITY CLASS	REMARKS
					LOCAL EFFECTS	NEXT HIGHER LEVEL	END EFFECT				

TABLE 4.10
MIL-STD-1629A Security Levels

Effect	Rating	Criteria
Catastrophic	1	A failure mode that may cause death or complete mission loss
Critical	2	A failure mode that may cause severe injury or major system degradation, damage, or reduction in mission performance
Marginal	3	A failure that may cause minor injury or degradation in system or mission performance
Minor	4	A failure that does not cause injury or system degradation but may result in system failure and unscheduled maintenance or repair

4.5.3.1.4 Failure Effects

The consequences of each failure mode on the item's operation should be carefully examined and recorded in the column labeled "Failure Effects." The effects can be distinguished at three levels: local, next higher abstraction level, and end effect. Local effects specifically show the impact of the postulated failure mode on the operation and function of the item under consideration. The consequence of each failure mode on the operation and functionality of an item under consideration is described as its local effect. It should be noted that sometimes no local effects can be described beyond the failure mode itself. However, the consequences of each postulated failure on the output of the item should be described along with second-order effects. End-effect analysis describes the effect of postulated failure on the operation, function, and status of the next higher abstraction level and ultimately on the system itself. The end effects shown in this column may be the result of multiple failures. For example, the failure of a supporting subsystem in a system can be catastrophic if it occurs along with another local failure. These cases should be clearly recognized and discussed in the end-effect column.

4.5.3.1.5 Failure Detection Method

Failure detection features for each failure mode should be described. For example, previously known symptoms, based on the item's behavior pattern(s), can indicate that a failure has occurred. The described symptom can cover the operation of a component under consideration (logical symptom) or can cover both the component and the overall system, or equipment evidence of failure.

4.5.3.1.6 Compensating Provision

A detected failure should be corrected so as to eliminate its propagation to the whole system so as to maximize reliability. Therefore, at each abstraction level, provisions that will alleviate the effect of a malfunction or failure should be identified. These provisions include items such as (a) redundant

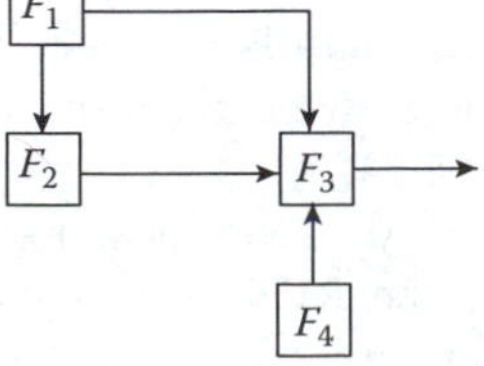

Function	Functional description	Components involved
F_1	Provide AC power	AC
F_2	Sensing and control	S
F_3	Provide pumping	V-2, V-3, V-4 V-5, P-1, P-2
F_4	Maintain source	T-1, V-1

FIGURE 4.32 Functional block diagram for the pumping system.

elements for continued and safe operation, (b) safety devices, and (c) alternative modes of operation, such as backup and standby units. Any action that may require operator action should be clearly described.

4.5.3.1.7 Severity

Severity classification is used to provide a qualitative indicator of the worst potential effect resulting from the failure mode. For FMEA purposes, MIL-STD-1629A classifies severity levels as given in Table 4.10.

4.5.3.1.8 Remarks

Any pertinent information, clarifying items, or notes should be entered in the column labeled "Remarks."

4.5.3.2 FMEA for Transportation Applications

The SAE J1739 FMEA procedure is, in principle, similar to the MIL-STD-1629A FMEA above. However, some definitions and ratings differ from those discussed so far. The key criterion for identifying and prioritizing potential design deficiencies here is the RPN, defined as the product of the severity, occurrence, and detection ratings. An example of an SAE J1739 FMEA format is shown in Table 4.11.

The contents of the item/function, potential failure mode, potential effect(s) of failure, potential cause(s)/failure mechanism(s), and the recommended actions steps of this FMEA procedure are similar to the respective parts of the MIL-STD-1629A FMEA discussed above.

Severity is evaluated on a 10-grade scale as shown in Table 4.12. Note that in contrast to the MIL-STD-1629A FMEA, a higher rating here corresponds to a higher severity (and, consequently, a higher RPN).

Occurrence is defined as the likelihood that a specific failure cause/mechanism will occur. The rating is based on the estimated or expected failure frequency as shown in Table 4.13.

4.5.3.2.1 Current Design Controls

Before the design is finalized and released to production, the engineer has complete control over it in terms of possible design changes. Three types of design control are usually considered: those that (1) prevent the failure cause/mechanism or mode from occurring or reduce their rate of occurrence, (2) detect the cause/mechanism and lead to corrective actions, or (3) detect the failure mode.

The preferred approach is to first use type 1 controls, if possible; second, to use the type 2 controls; and third, to use type 3 controls. The initial occurrence ranking is affected by the type 1 controls, provided they are integrated as a part of the design intent. The initial detection ranking is based on the type 2 or 3 controls, provided the prototypes and models being used are representative of design intent.

Detection is defined as the ability of the proposed type 2 design controls to detect a potential cause/mechanism, or the ability of the proposed type 3 design controls to detect the respective failure mode before the system/component is released to production (Table 4.14).

In the FMEA worksheet format, RPN is the product of the severity, occurrence, and detection ratings and is used to rank the potential design concerns. While the RPN is a major measure of design risk, special attention should be given to the high-severity failure modes irrespective of the resultant RPN number. Action Results columns describe the implemented corrective actions along with the estimated reduction in severity, occurrence, and detection ratings and the resultant RPN.

TABLE 4.11
SAE J1739 FMEA Worksheet Format

Potential Failure Mode and Effects Analysis (Design FMEA)

System: ____________

Subsystem: ____________

Component: ____________

Model Year/Vehicle(s): ____________

Core Team:

Design Responsibility: ____________

Key Date: ____________

FMEA Number: ____________

Page ___ of ___

Prepared by: ____________

FMEA Date (Orig): ____________ (Rev.) ________

Item/Function	Potential Failure Mode	Potential Effect(s) of Failure	Sever	Potential Cause(s)/ Failure Mechanism(s)	Occur	Current Design Controls	Detec	RPN	Recommended Actions	Responsibility and Target Completion Date	Action Results				
											Actions Taken	Sever	Occur	Detec	RPN

Example 4.9

Based on the functional block diagram of the vehicle generic front lighting system (see Figure 4.33), develop a design FMEA on the system abstraction level.

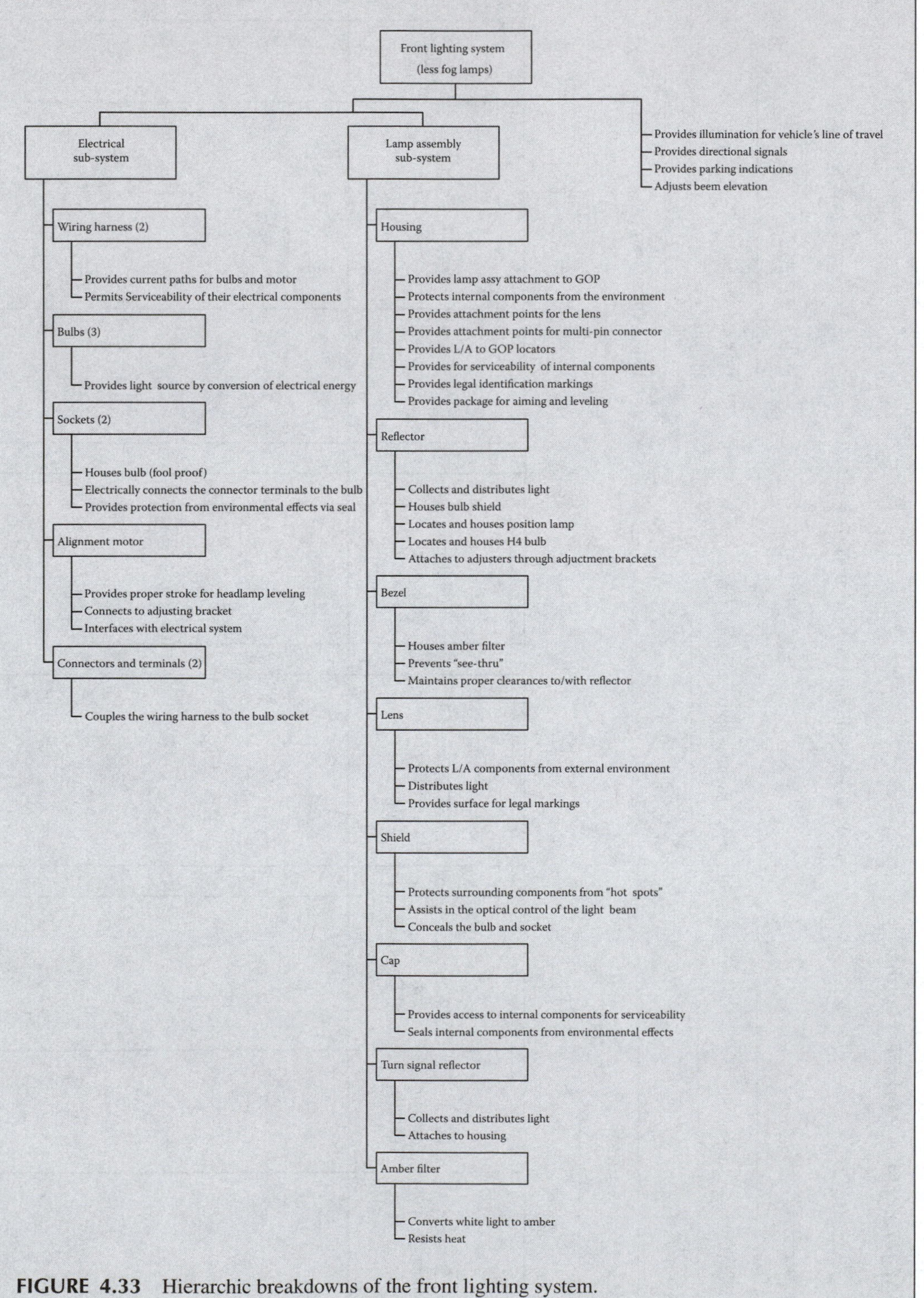

FIGURE 4.33 Hierarchic breakdowns of the front lighting system.

Solution: The FMEA of the vehicle generic front lighting system is shown in Table 4.15. As seen from the table, the highest RPN corresponds to the failure mode potentially caused by a defective light bulb. The corrective action of pursuing the cost–benefit analysis (CBA) on a more reliable bulb reduces the occurrence rating of this failure mode from 5 to 1, which, in turn, decreases the RPN to 27.

4.5.4 FMECA Procedure: Criticality Analysis

Criticality analysis is the combination of a probabilistic determination of a failure mode occurrence combined with the impact it has on the system mission success. Table 4.16 shows an example of a criticality analysis worksheet format. The criticality analysis part of this worksheet is explained below.

- Failure Effect Probability β. The β value represents the conditional probability that the failure effect with the specified criticality classification will occur given that the failure mode occurs.

TABLE 4.12
Ten-Grade Scale of Severity

Effect	Rating	Criteria
Hazardous	10	Safety-related failure modes causing noncompliance with government regulations without warning
Serious	9	Safety-related failure modes causing noncompliance with government regulations with warning
Very high	8	Failure modes resulting in loss of primary vehicle/system/component function
High	7	Failure modes resulting in a reduced level of vehicle/system/component performance and customer dissatisfaction
Moderate	6	Failure modes resulting in loss of function by comfort/convenience systems/components
Low	5	Failure modes resulting in a reduced level of performance of comfort/convenience systems/components
Very low	4	Failure modes resulting in loss of fit and finish, squeak, and rattle functions
Minor	3	Failure modes resulting in partial loss of fit and finish, squeak, and rattle functions
Very minor	2	Failure modes resulting in minor loss of fit and finish, squeak, and rattle functions
None	1	No effect

TABLE 4.13
Rate of Occurrence of Failure

Likelihood of Failure	Estimated or Expected Failure Frequency	Rating
Very high (failure is almost inevitable)	>1 in 2	10
	1 in 3	9
High (frequently repeated failures)	1 in 8	8
	1 in 20	7
Moderate (occasional failures)	1 in 80	6
	1 in 400	5
	1 in 2000	4
Low (rare failures)	1 in 15,000	3
	1 in 150,000	2
Remote (failures are unlikely)	<1 in 150,000	1

TABLE 4.14
Detection Ratings Based on Design Control Criteria

Detection	Rating	Criteria
Uncertain	10	Design control will not and/or cannot detect a potential cause/mechanism and subsequent failure mode
Very remote	9	Very remote chance the design control will detect a potential cause/mechanism and subsequent failure mode
Remote	8	Remote chance the design control will detect a potential cause/mechanism and subsequent failure mode
Very low	7	Very low chance the design control will detect a potential cause/mechanism and subsequent failure mode
Low	6	Low chance the design control will detect a potential cause/mechanism and subsequent failure mode
Moderate	5	Moderate chance the design control will detect a potential cause/mechanism and subsequent failure mode
Moderately high	4	Moderately high chance the design control will detect a potential cause/mechanism and subsequent failure mode
High	3	High chance the design control will detect a potential cause/mechanism and subsequent failure mode.
Very high	2	Very high chance the design control will detect a potential cause/mechanism and subsequent failure mode.
Almost certain	1	The design control will almost certainly detect a potential cause/mechanism and subsequent failure mode

For complex systems, it is difficult to determine β unless a comprehensive logic model of the system (e.g., a fault tree or an MLD) exists. Therefore, in many cases, estimation of β becomes primarily a matter of judgment greatly driven by the analyst's prior experience. The general guidelines shown in Table 4.17 can be used for determining β.

- Failure Mode Ratio α. The fraction of the item's (component's, part's, etc.) failure rate, λ, related to the particular failure mode under consideration is evaluated and recorded in the failure mode ratio (α) column. The failure mode ratio is the probability that the item will fail in the identified mode of failure. If all potential failure modes of an item are listed, the sum of their corresponding α values should be equal to 1. The values of α should normally be available from a data source (e.g., [16]). However, if not available, the values can be assessed based on the analyst's judgment.
- Failure Rate λ. The generic or specific failure rate for each failure mode of the item should be obtained and recorded in the failure rate (λ) column. The estimates of λ can be obtained from the test or field data, or from generic sources of failure rates discussed in Section 3.7.
- Operating Time T. The operating time, in hours, or the number of operating cycles of the item should be listed in the corresponding column.

Failure Mode Criticality Number C_r is used to rank each potential failure mode based on its occurrence and the consequence of its effect. For a particular severity classification, the C_r of an item is the sum of the failure mode criticality numbers C_m that have the same severity classification.

Thus,

$$C_r = \sum_{i=1}^{n} (C_m)_i \tag{4.56}$$

and

$$C_m = \beta\alpha\lambda T, \tag{4.57}$$

TABLE 4.15
FMEA in Example 4.9

Potential Failure Mode and Effects Analysis (Design FMEA)

__x__ System
____ Subsystem
____ Component: Generic Front Lighting System
Model Year/Vehicle(s): 2000/LITTLE TRUCKS
Core Team:

Design Responsibility: Electrical Engineering Dept.
Key Date:

FMEA Number:
Page __1__ of __5__
Prepared by:
FMEA Date (Orig): __97.02____________ (Rev.) __________

Item/Function	Potential Failure Mode	Potential Effect(s) of Failure	Sever	Potential Cause(s)/Failure Mechanism(s)	Occur	Current Design Controls	Detec	RPN	Recommended Actions	Responsibility and Target Completion Date	Action Results: Actions Taken	Sever	Occur	Detec	RPN
1	2	3	4	5	6	7	8	9	10	11	12	13	14	15	16
Provide Illumination for vehicle's line of travel, as defined by a. beam width b. intensity c. vertical aim d. horizontal aim	System does not provide adequate illumination including high beam and low beam.	Customer dissatisfaction and/or non-compliance with government regulation(s).	9	Inadequate reflector size	1	System Analysis Modeling Vehicle Integration Testing	2	18							
				Defective bulb	5	Supplier Bulb Durability Testing Lighting System Durability Testing Vehicle Durability Testing	3	135	Pursue CBA on High Rel Bulb	John Doe 11/98	Engineering change to 104317 at var. cost penalty of $.45	9	1	3	27
				Defective Wiring Harness - Bulb circuit [includes multi project chip (MPC) and bulb connector]	2	Supplier Bulb Durability Testing Lighting System Durability Testing Vehicle Durability Testing	3	54							

continued

TABLE 4.15 (continued)

Item/Function	Potential Failure Mode	Potential Effect(s) of Failure	Sever	Potential Cause(s)/Failure Mechanism(s)	Occur	Current Design Controls	Detec	RPN	Recommended Actions	Responsibility and Target Completion Date	Action Results				
											Actions Taken	Sever	Occur	Detec	RPN
1	2	3	4	5	6	7	8	9	10	11	12	13	14	15	16
				Inadequate vertical alignment setting specified	1	Specification Review Assembly Drawing Review	1	9							
				Inadequate horizontal alignment setting specified (includes tolerances)	2	Specification Review Assembly Drawing Review	1	18							
Provide directional (turn) signals	System does not provide adequate turn signal indication	Non-compliance with government regulation(s)	9	Incorrect reflector size	2	SAM - Sys. Anal. Model Vehicle Integration Testing	2	36							
				Defective Bulb	1	Supplier Bulb Durability Testing Lighting System Durability Testing Vehicle Durability Testing	3	27							
				Defective Socket	2	Supplier Bulb Durability Testing Lighting System Durability Testing Vehicle Durability Testing	3	54							

Item/ Function	Potential Failure Mode	Potential Effect(s) of Failure	Sever	Potential Cause(s)/ Failure Mechanism(s)	Occur	Current Design Controls	Detec	RPN	Recommended Actions	Responsibility and Target Completion Date	Action Results				
											Actions Taken	Sever	Occur	Detec	RPN
1	2	3	4	5	6	7	8	9	10	11	12	13	14	15	16
Provide a lighted indication of vehicle's position while parked	System does not provide adequate parking indication.	Non-compliance with government regulation(s).	9	Incorrect reflector geometry	3	System Analysis Modeling Vehicle Integration Testing	2	54							
				Defective Position Bulb	3	Supplier Bulb Durability Testing Lighting System Durability Testing Vehicle Durability Testing	3	81							
				Defective Position Socket	1	Supplier Bulb Durability Testing Lighting System Durability Testing Vehicle Durability Testing	3	27							
				Defective Wiring Harness - Position circuit	2	Supplier Bulb Durability Testing Lighting Syste m Durability Testing Vehicle Durability Testing	4	72							
Adjusts beam elevation when commanded by driver to compensate for load effects on vehicle attitude.	Driver is unable to adjust beam for load conditions, or control is inadequate.	Driver's ability to see the road may not be optimal and non-compliance with govt. regulations.	5	Defective alignment motor/shaft	1	Motor and Shaft Durability Testing Lighting System Durability Testing Vehicle Durability Testing	5	25							
				Broken (molded) attachment points	1										

TABLE 4.16
FMECA Worksheet Format

CRITICALITY ANALYSIS

System ____________

Indenture level ____________

Reference drawing ____________

Mission ____________

Date ____________
Sheet ____ of ____________
Compiled by ____________
Approved by ____________

Identification Number	Item/Functional Identification (Nomenclature)	Function	Failure Modes And Causes	Mission Phase/ Operational Mode	Severity Class	Failure Probability Failure Rate Data Source	Failure Effect Probability (β)	Failure Mode Ratio (α)	Failure Rate (λ–ρ)	Operating Time (T)	Failure Mode Crit # $C_m = \beta\alpha\lambda_\rho T$	Item Crit # $C_r = \Sigma(C_m)$	Remark

*NOTE: Both criticality number (C_r) and probability of occurrence level are shown for convenience

TABLE 4.17
Failure Effect Probabilities for Various Failure Effects

Failure Effect	Value
Actual loss	1.00
Probable loss	0.1 < 1.0
Possible loss	0.0 < 0.1
No effect	0

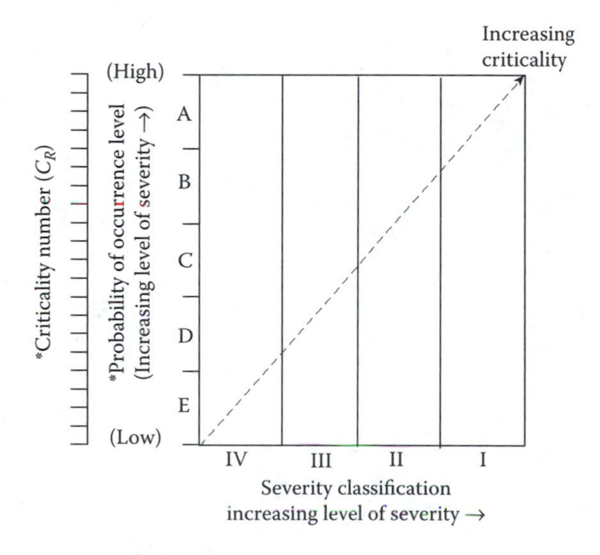

FIGURE 4.34 Example of criticality matrix.

where C_m is the criticality of an individual failure mode, and n is the number of failure modes of an item with the same severity classification.

Based on the criticality number, a so-called criticality matrix is usually developed to provide a visual way of identifying and comparing each failure mode to all other failures with respect to severity. Figure 4.34 shows an example of such a matrix. This matrix can also be used for a qualitative criticality analysis in an FMEA-type study. Along the vertical dimension of the matrix, the probability of occurrence level (subjectively estimated by the analyst in an FMEA study) or the criticality number C_r (calculated in an FMECA study) is entered. Along the horizontal dimension of the matrix, the severity classification of an effect is entered. The severity increases from left to right. Each item on the FMEA or FMECA could be represented by one or more points on this matrix. If the item's failure modes correspond to more than one severity effect, each failure mode will correspond to a different point in the matrix. Clearly, those severities that fall in the upper-right quadrant of the matrix require immediate attention for reliability or design improvements.

Example 4.10

Develop an FMECA for a system of two amplifiers A and B in parallel configuration. In a given mission, this system should function for a period of 72 h.

continued

TABLE 4.18
FMECA for the Amplifier System in Example 4.10

Indent Number	CKT Name	Failure Mode	Effects: Local	Effects: System	Severity Class	β	α	λ	Mission Time (Hours)	Criticality Number C_m	Remarks
1.	A	a. Open	Circuit A failure	Degraded	III	0.0699[1]	0.90	1×10^{-3}	72	4.47E−3	May cause
		b. Short	Both A and B	Failure	II	1.00	0.05			3.60E−3	secondary
		c. Other	Circuit failure A lost	Degraded	IV	0.0093[2]	0.05			3.35E−5	failure
2.	B	a. Open	Circuit B failure	Degraded	III	0.069	0.90	1×10^{-3}	72	4.47E−3	May cause
		b. Short	Both A and B	Failure	II	1.00	0.05			3.60E−3	secondary
		c. Other	Circuit failure B lost	Degraded	IV	0.0093	0.05			3.35E−5	failure
$\Sigma\lambda_{II} = 1.8 \times 10^{-4}$, $\Sigma\lambda_{III} = 1 \times 10^{-4}$, $\Sigma\lambda_{IV} = 1 \times 10^{-4}$								$\Sigma\lambda = 2 \times 10^{-5}$	$C_r = \Sigma C_m = 16.21 \times 10^{-3}$		

Solution: The summary of this analysis is displayed in Table 4.18. One can draw the following conclusions for this mission of the system:

1. The probability of a critical failure of the system is $0.0036 + 0.0036 = 0.0072$.
2. The probability of a failure resulting in system degradation is $3.35 \times 10^{-5}(2) = 6.7 \times 10^{-5}$.
3. The probability of a critical failure of the system due to "open" circuit failure mode is $4.47 \times 10^{-3}(2) = 8.94 \times 10^{-3}$.

The above approximate probabilities can only hold true if the product of α, β, λ, and T is small (e.g., <0.1). Normally, criticality numbers are used as a measure of severity and not as a prediction of system reliability. Therefore, the most effective design would allocate more engineering resources to the areas with high criticality numbers, and would minimize the Class I and Class II severity failure modes.

NOTES

1. Pr(System failure | A open) = for "A" open mode of failure $= 1 - R_B(72) = 1 - \exp[-1.0 \times 10^{-3}(72) = 1 - 0.931 = 0.069$.
2. Assume failure rate doubles due to degradation: $R_A = \exp[(-2 \times 10^{-3}(72)] = 0.866$, then Pr(System failure | A degraded)$= 1 - [0.931 + 0.866 - (0.931)(0.866)] = 1 - 0.99075 = 0.00925$.

EXERCISES

4.1 For the circuit below, assume the reliability of each circuit $R(x_i) = \exp(-\lambda_i t)$, and $\lambda_i = 2.0 \times 10^{-4}$ per hour for all i. Find the following:

a. Minimal path sets.
b. Minimal cut sets.
c. MTTF.
d. Reliability of the system at 1000 h.

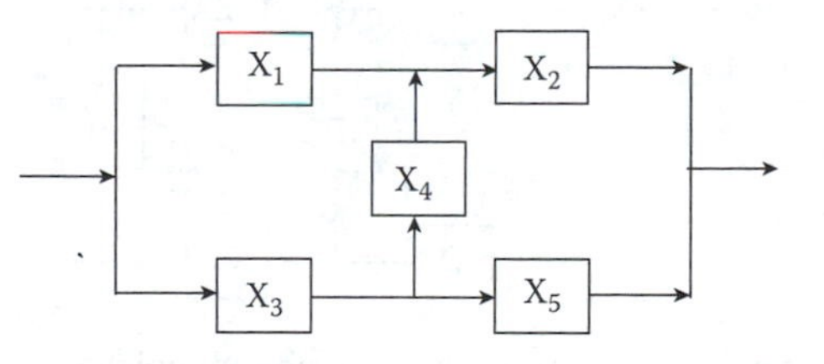

4.2 For the circuit below, find the following with $\lambda = 1.0 \times 10^{-4}$:

a. Minimal path sets.
b. Minimal cut sets.
c. Reliability of the system at 1000 h.
d. Probability of failure at 1000 h, using cut sets, to verify results from (c).
e. Accuracy of the results of (d) and/or (c), using an approximate method.
f. MTTF of the system.

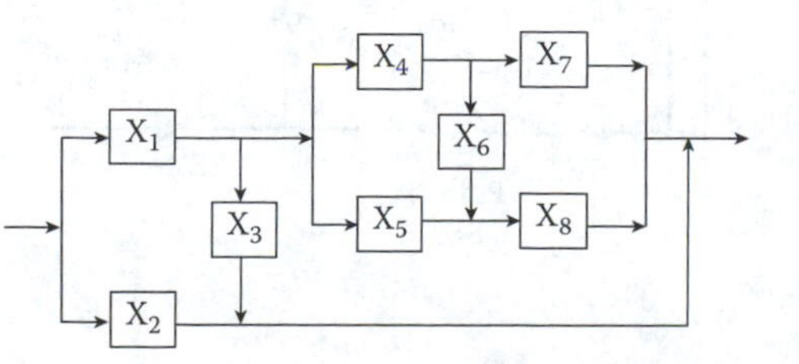

4.3 Calculate the reliability of the system shown in the figure below for a 1000-h mission. What is the MTTF for this system?

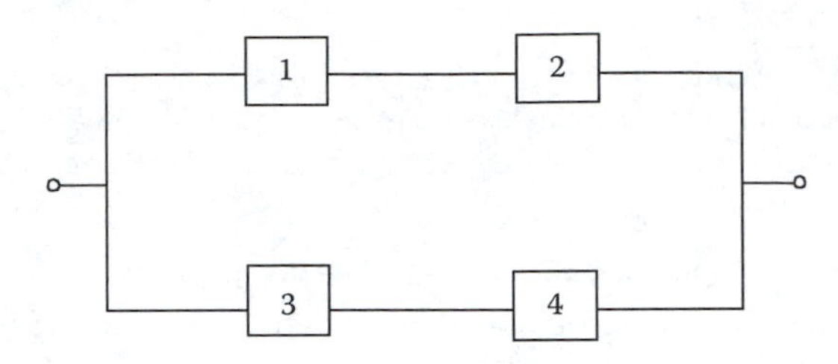

4.4 The purpose of the pumping system shown below is to pump water from point A to point B. The time to failure of all the valves and the pump can be represented by the exponential distributions with failure rates λ_v and λ_p, respectively.

a. Calculate the reliability function of the system.
b. If $\lambda_v = 1.0 \times 10^{-3}$ and $\lambda_p = 2.0 \times 10^{-3}$ per hour, and the system has survived for 10 h, what is the probability that it will survive another 10 h?

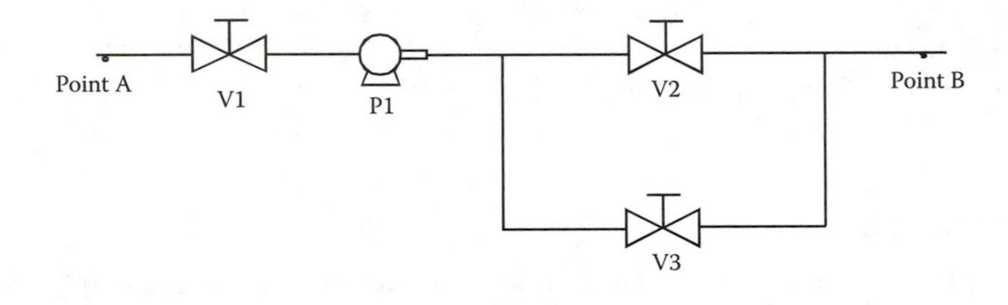

4.5 Estimate reliability of the system represented by the following reliability block diagram for a 2500-h mission. Assume a failure rate of 1.0×10^{-6} per hour for each unit.

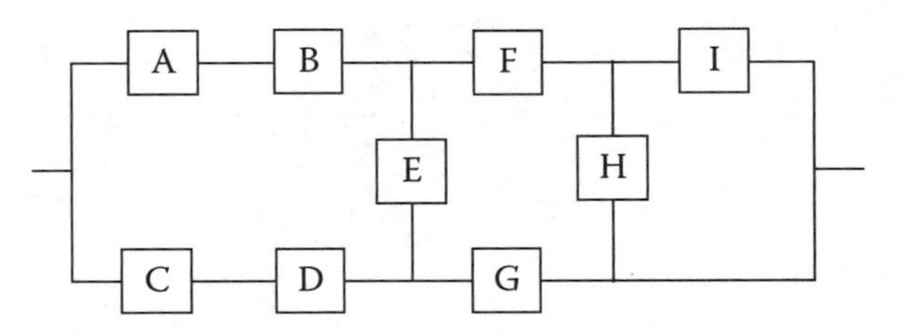

4.6 A containment spray system is used to scrub and cool the atmosphere around a nuclear reactor during an accident. Develop a fault tree using "No H_2O spray" as the top event.

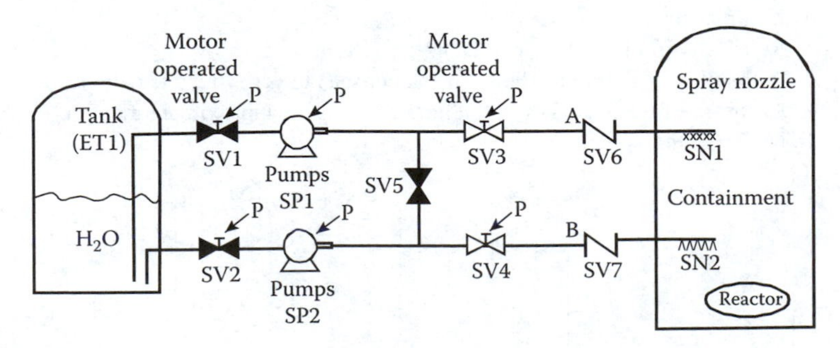

Assume the following conditions:

- There are no secondary failures.
- There is no testing and maintenance.
- There are no passive failures.
- There are independent failures.
- One of the two pumps and one of the two spray heads are sufficient to provide spray. (Only one train is enough.)
- One of the valves SV_1 or $SV_2$is opened after demand. However, SV_3 and SV_4 are always normally open.
- Valve sv_5 is always in the closed position.
- There is no human error.
- SP_1, SP_2, SV_1, SV_2, SV_3, and SV_4 use the same power source P to operate.

4.7 For the fault tree below, find the following.

a. Minimal cut sets.
b. Minimal path set.
c. Probability of the top event if the following probabilities apply:

$$\Pr(A) = \Pr(C) = \Pr(E) = 0.01.$$

$$\Pr(B) = \Pr(D) = 0.0092.$$

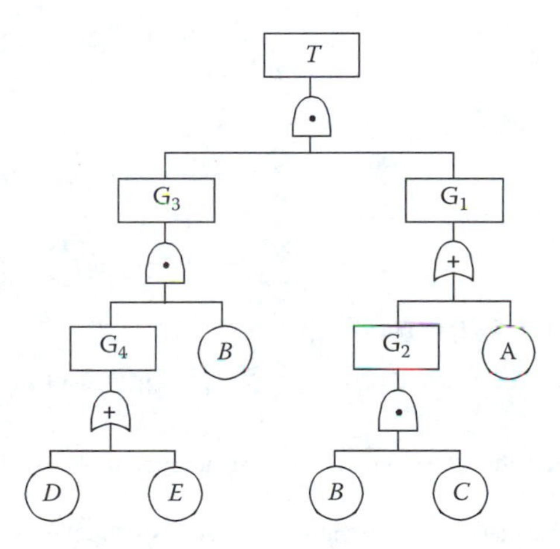

4.8 In the pumping system below, the system cycles every hour. Ten minutes are required to fill the tank. A timer is set to open the contact 10 min after the switch is closed. The operator opens switch or the tank emergency valve if he/she notices an overpressure alarm. Develop a fault tree for this system with the top event "Tank ruptures."

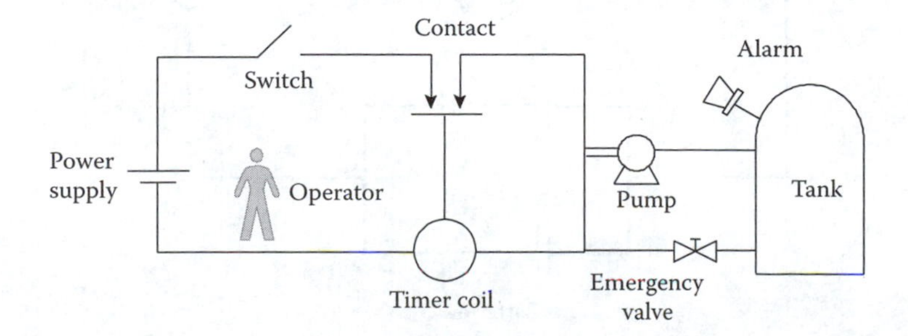

4.9 Find the cut sets and path sets of the fault tree shown below using the top-down substitution method.

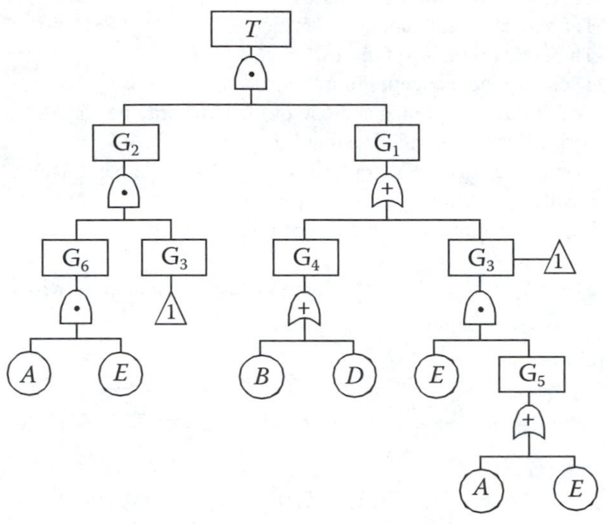

4.10 Compare Design 1 and Design 2 below.

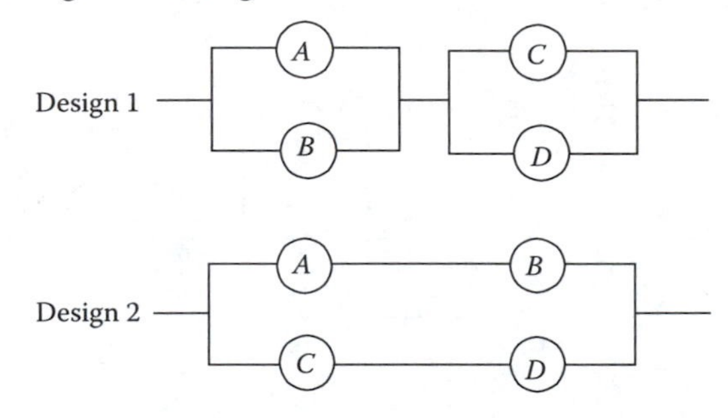

Failure rates (per hour)

$$\lambda_A = 10^{-6}, \lambda_C = 10^{-3}, \lambda_B = 10^{-6}, \text{ and } \lambda_D = 10^{-3}$$

a. Assume that the components are nonrepairable. Which is the better design?
b. Assume that the system failure probability cannot exceed 10^{-2}. What is the operational life for Design 1 and for Design 2?

4.11 The following electric circuit provides emergency light during a blackout.

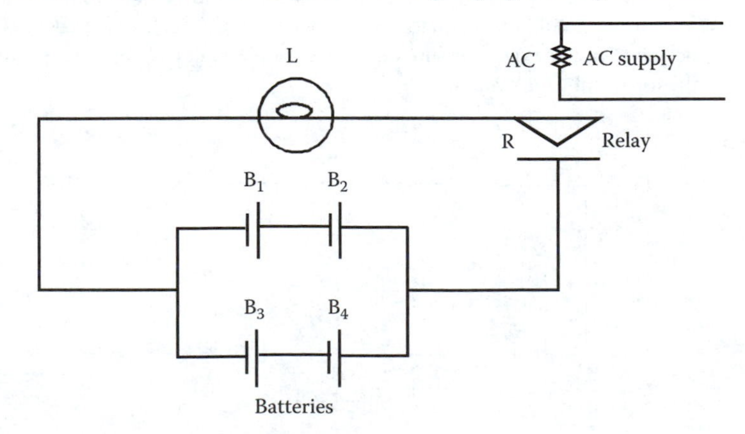

In this circuit, the relay is held open as long as AC power is available, and any of the four batteries is capable of supplying light power. Start with the top event "No Light When Needed."

a. Draw a fault tree for this system.
b. Find the minimal cut sets of the system.
c. Find the minimal path sets of the system.

4.12 The following pumping system consists of three identical parallel pumps and a valve in series.

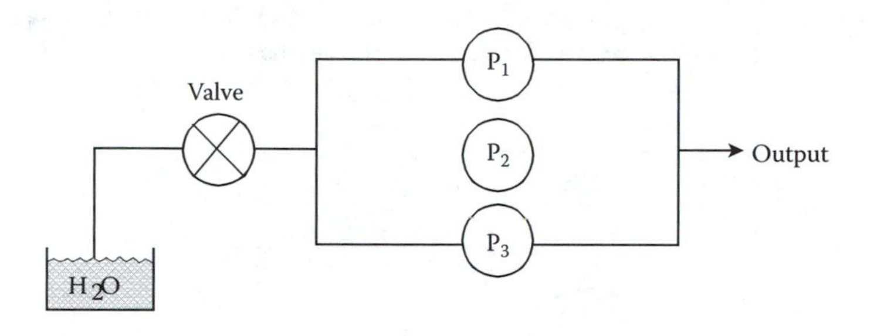

The pumps have a constant failure rate of λ_p (per hour) and the valve has a constant failure rate of λ_v (per hour) (accidental closure).

a. Develop an expression for the reliability of the system using the success tree. (Assume nonrepairable components.)
b. Find the average reliability of the system over time period T.
c. Repeat questions (a) and (b) for the case that $\lambda t < 0.1$, and then approximate the reliability functions. Find the average reliability of the system when $\lambda = 0.001$ per hour and $T = 10\,\text{h}$.

4.13 In the following system, which uses active redundancy, what is the probability that there will be no failures in the first year of operation? Assume constant failure rates given below.

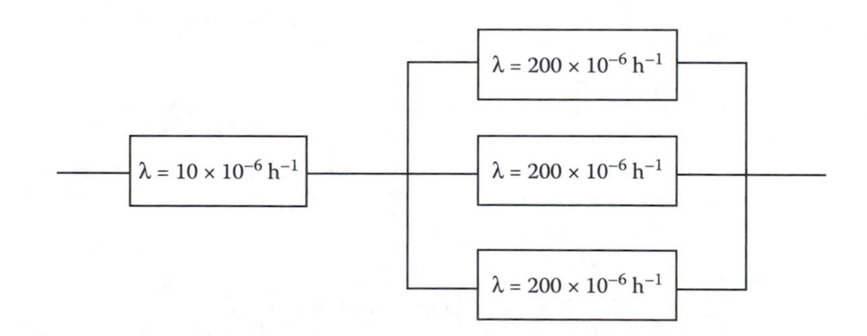

4.14 A filter system is composed of 30 elements, each with a failure rate of 2×10^{-4} per hour. The system will operate satisfactorily with two elements failed. What is the probability that the system will operate satisfactorily for 1000 h?

4.15 For the reliability diagram below,

a. Find all minimal path sets.
b. Find all minimal cut sets.
c. Assuming each component has a reliability of 0.90 for a given mission time, compute the system reliability over mission time.

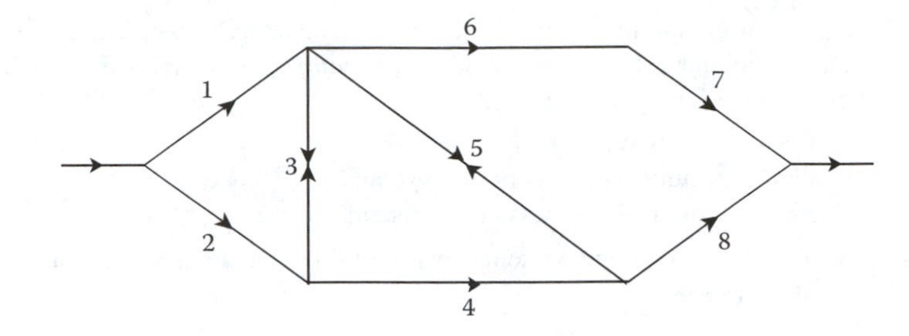

4.16 In the following fault tree, find all minimal cut sets and path sets. Assuming all component failure probabilities are 0.01, find the top event probability.

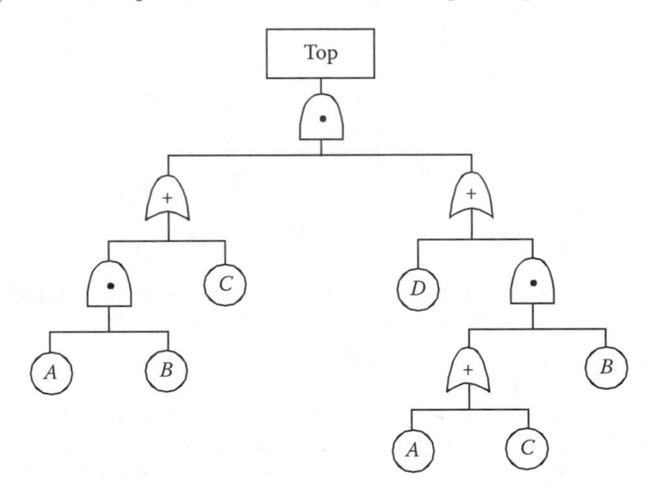

4.17 An event tree is used in reactor accident estimation as shown below.

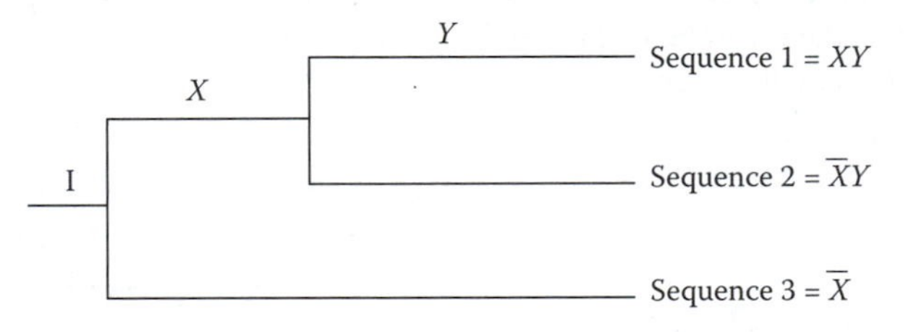

where sequence 1 is a success and sequences 2 and 3 are failures. The cut sets of system X and Y are

$$\bar{X} = A \cdot B + A \cdot C + D$$

and

$$\bar{Y} = B \cdot D + E + A$$

Find cut sets of sequence 2 and sequence 3.

4.18 In a cement production factory, a system such as the one shown below is used to provide cooling water to the outside of the furnace. Develop an MLD for this system.

System bounds:	$S_{12}, S_{11}, S_{10}, S_9, S_8, S_7, S_6, S_5, S_4, S_3, S_2, S_1$
Top event:	Cooling from legs 1 and 2
Not-allowed events:	Passive failures and external failures
Assumptions:	Only one of the pumps or legs is sufficient to provide the necessary cooling. Only one of the tanks is sufficient as a source.

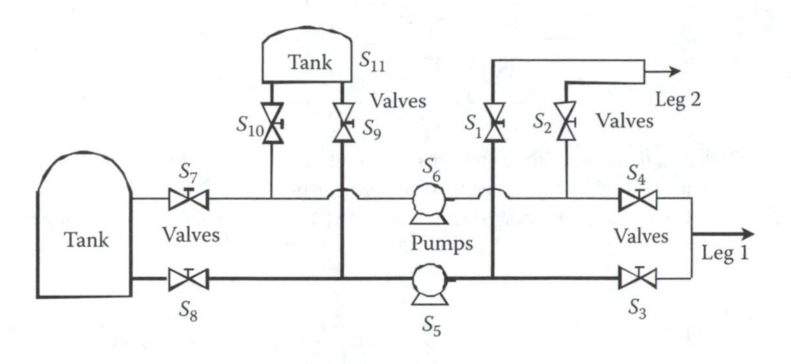

4.19 Develop an MLD model of the following system.

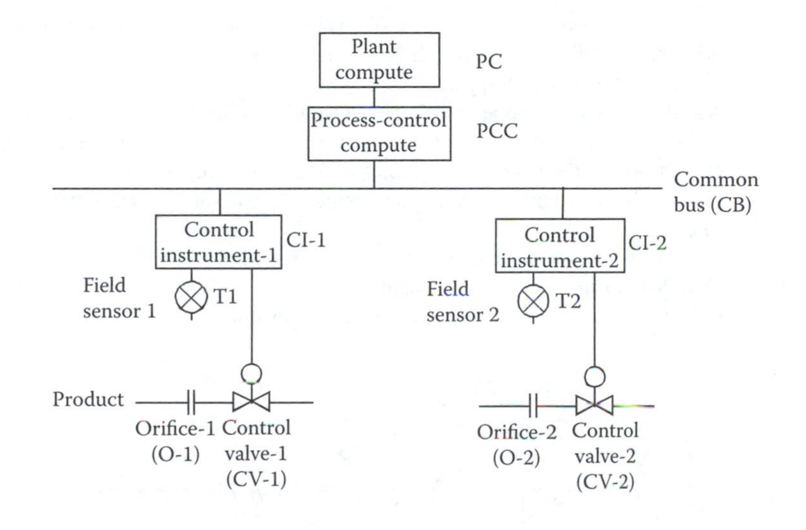

Assume the following:

- One of the two product lines is sufficient for success.
- Control instruments feed the sensor values to the process-control computer, which calculates the position of the control valves.
- The plant computer controls the process-control computer.

a. Develop a fault tree for the top event "inadequate product feed."
b. Find all the cut sets of the top event.
c. Find the probability of the top event.
d. Determine which components are critical to the design.

4.20 Perform an FMECA analysis for the system described in Exercise 4.19. Compare the results with part (d) of Exercise 4.19.

4.21 For the standby system shown below, assume that components a, b, and c are identical with a constant failure rate of 1×10^{-3} (per hour) and a constant standby failure rate of 1×10^{-5} (per hour). The probability that the switch fails to operate if component "a" fails is 5×10^{-2}. Calculate the reliability of this system at $t = 200$ h. (Note that either "a" or "b and c" is required for system operation.)

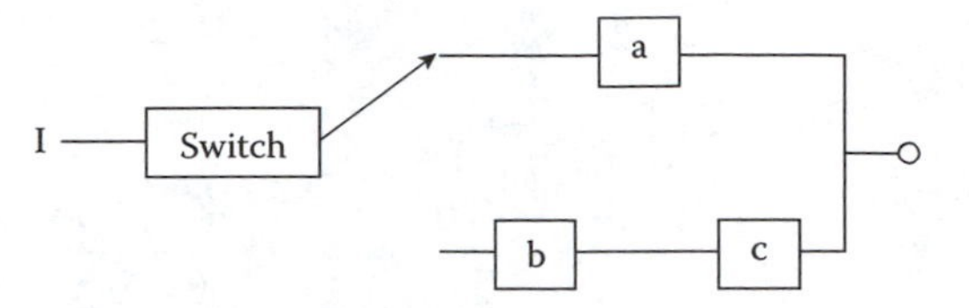

4.22 A supercomputer requires subsystem "A" and either subsystem "B" or subsystem "C" to function so as to save some critical data following a sudden loss of power to the computer. Subsystems A, B, and C are configured as shown below.

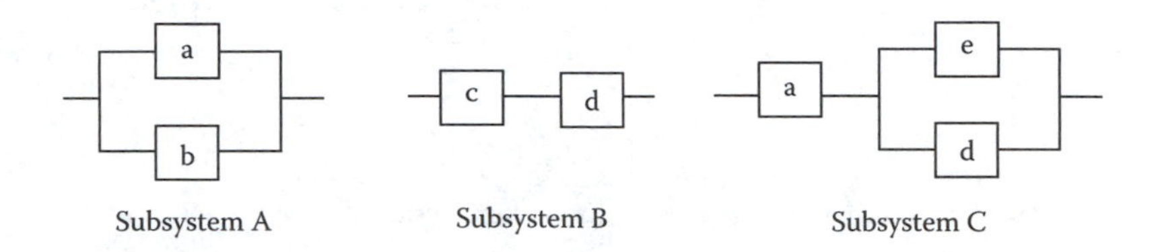

Use the following component reliability values to determine the probability that critical data will not be saved following a loss of power: $R_a = 0.99$, $R_b = 0.98$, $R_c = 0.999$, $R_d = 0.998$, and $R_e = 0.99$.

4.23 A system of two components arranged in parallel redundancy has been observed to fail every 1000 h, on average. Data for the individual components, which come from the field experience, indicate a failure rate of about 0.01 per hour. Is there an inconsistency between the component-level and system-level failure rates?

4.24 For the system shown below, develop a fault tree with the top event of "No Output from the System." Calculate the reliability of this system for a 100-h mission. Assume MTTF of 600 h for all components.

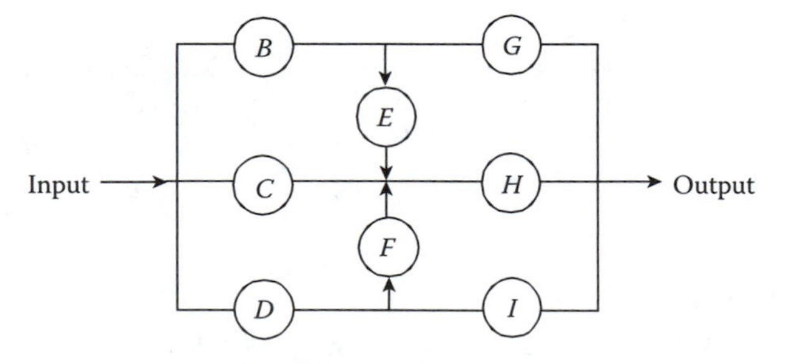

E and F are unidirectional

4.25 Consider the system below.

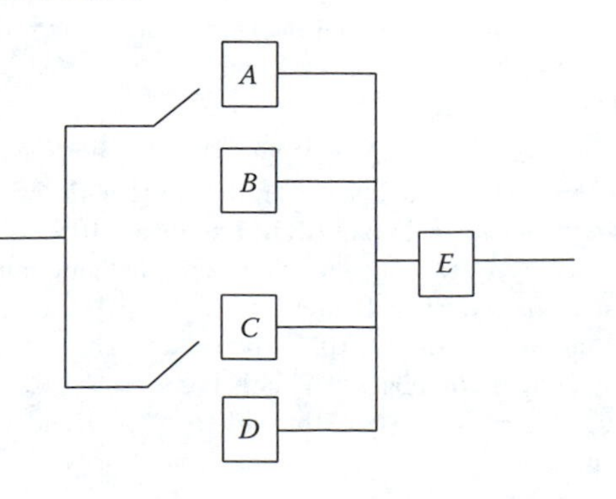

Two standby subsystems (A and B) and (C and D) are placed in parallel. Unit E should be in series with the two standby subsystems. Assume time to failure of A, B, C, D, and E are exponentially distributed and switching is perfect for the standby subsystems.

If $\lambda A = \lambda B = \lambda C = \lambda D =$ 10–4/h determine the reliability of each standby subsystem i.e., reliability of (A and B) and reliability of (C and D).

As a designer, if you are asked to select Unit E such that its contribution to the failure of the whole system is less than 10% of the total system failure probability for a mission of $t =$ 1000 h, what should be the minimum acceptable constant failure rate for Unit E?

4.26 For the system below, draw a success tree and derive the path sets of the top event (system "output" occurs) from the success tree logic using the Boolean substitution method and the truth table approach. Compare the results.

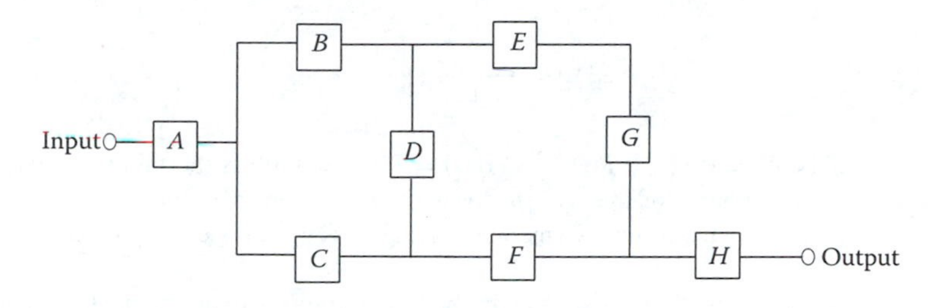

4.27 Consider the system below.

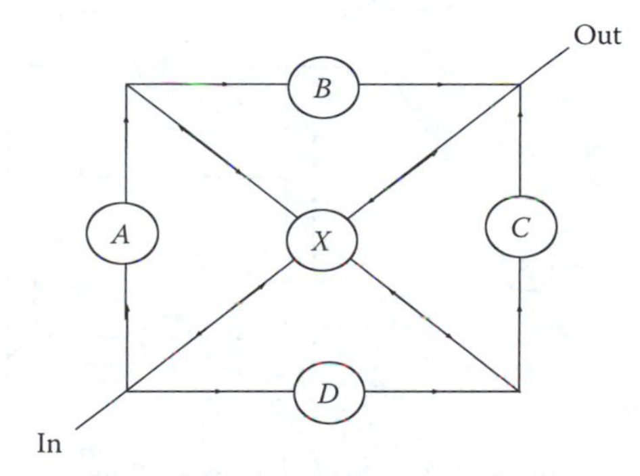

Components A, B, C, D, and X are identical and have a Weibull time-to-failure distribution model with shape parameter $\beta = 1.6$.

a. Determine the reliability of the system as a function of time.
b. Using the results of part (a), determine the scale parameter if we desire that the total probability of system failure not exceed 10^{-2} over a 10-year period.
c. Determine the minimal cut sets of this system by inspection.
d. Use the minimal cut sets to verify that the probability of failure of this system meets the 10^{-2} over a 10-year period criterion. Explain the reason for any discrepancies.

4.28 Consider the following system.

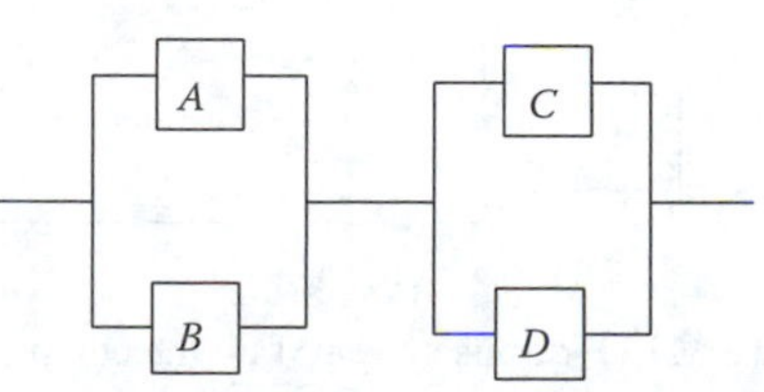

Components A and B are share load, but C and D are parallel. Components A and B are identical and their time to failure follows exponential distribution with the full-load failure rate being twice that of half-load failure rate. If $\lambda C = 0.001$/h and $\lambda D = 0.002$/h, determine the MTTF of this system if the total system reliability should be at least 0.99 for a mission of 10 h length.

4.29 Consider the system below

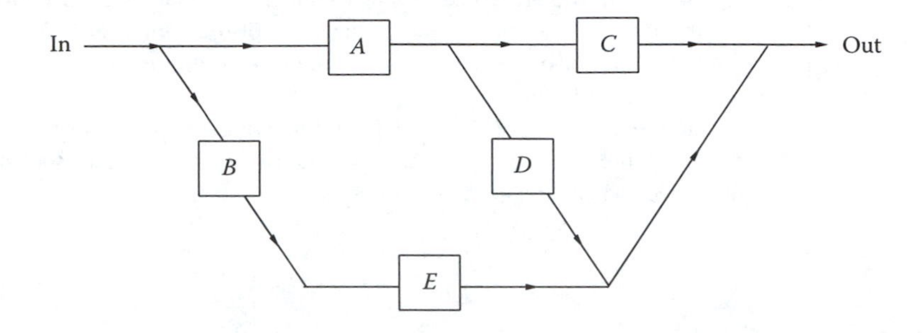

If $\Pr(A) = \Pr(B) = \Pr(C) = \Pr(D) = \Pr(E) = 0.05$ is the failure probability of components of this system for a given mission, determine system failure probability using any two methods and discuss which one is more accurate and why?

4.30 For the following standby share-load system block diagram, write and solve the equation for determining the reliability of the system.

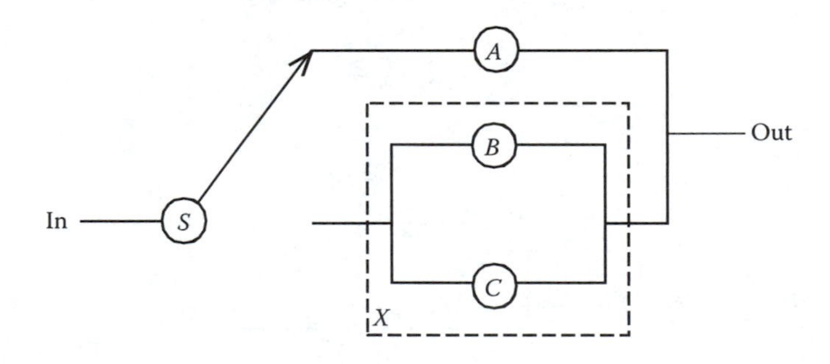

Subsystem X is made of share-load components B and C (with identical constant failure rates λh and λf for half- and full-load operation, respectively). Upon failure of component A, subsystem X is used by switching via a perfect switch S. Assume no standby failure rate. Assume a constant failure rate for component A (i.e., λA).

4.31 Consider the engineering system as shown below:

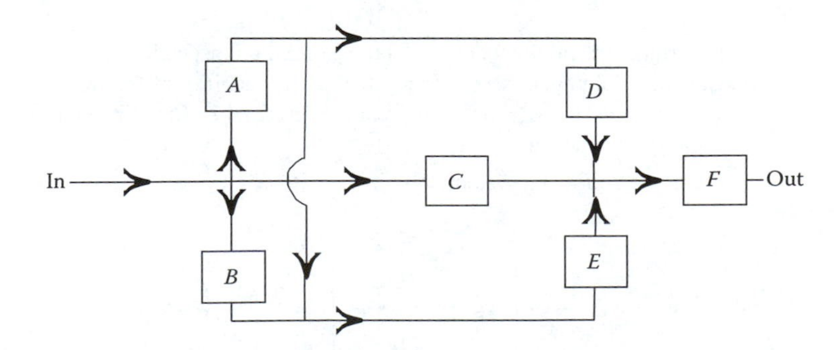

Determine the MTTF of this system if all the components are identical with a failure rate of $\lambda = 0.005$/month.

4.32 Consider the following two fault trees:

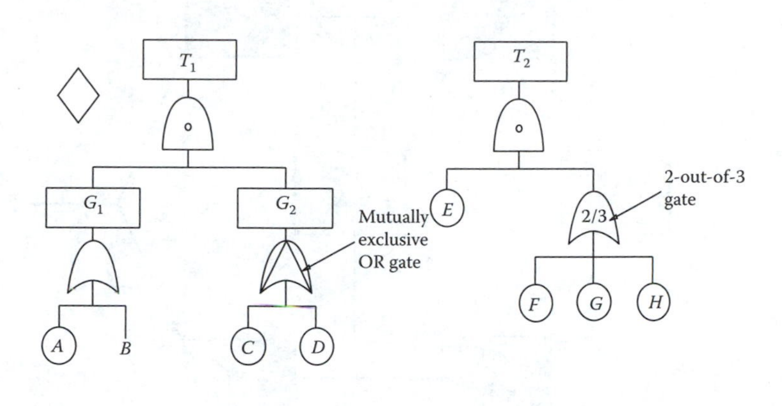

For both trees T_1 and T_2:

a. Find the minimal cut sets.
b. Find the minimal path sets
c. Draw the equivalent success tree.
d. Calculate the unreliability function of the two systems based on the constant failure rates below:

Component	*A*	*B*	*C*	*D*	*E*	*F*	*G*	*H*
Failure rate	0.08	0.02	0.07	0.05	0.01	0.05	0.05	0.05

4.33 Consider the standby system composed of three identical units. It is known that failure probability of each block follows a normal distribution. Note that the following censored failure data has been obtained in a life test involving 10 of the same blocks. Assume perfect switching and no failure in standby mode.

Use Kaplan–Meier plotting method to address the following questions.

Test no.	1	2	3	4	5	6	7	8	9	10
Time to failure	298	492	692	850	980	1200+	1200+	1200+	1200+	1200+

a. What is the mean time-to-failure distribution of the system as a whole?
b. Estimate the standard deviation of the unit's time-to-failure.

4.34 The reliability block diagram of a system is shown in the following figure with component reliability noted in each block. What is the reliability of the system?

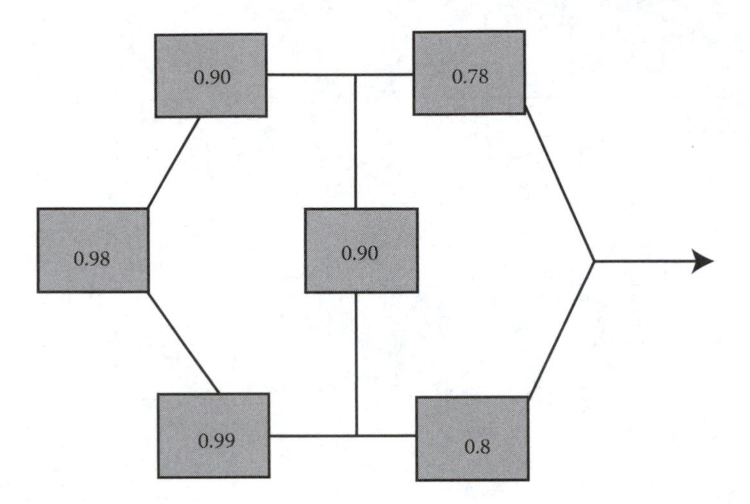

4.35 Consider the system below.

a. Write the reliability function, $R(t)$, of the following shared load system. Assume that each unit is identical and can provide 1/3 load, 2/3 load, or full load. With the following constant failure rates

$$\lambda_{1/3\text{-load}} = 10^{-4}\text{h}^{-1},\ \lambda_{2/3\text{-load}} = 10^{-3}\text{h}^{-1},\ \text{and } \lambda_{\text{full-load}} = 10^{-2}\text{h}^{-1}$$

b. estimate the reliability of the system at 10,000 h

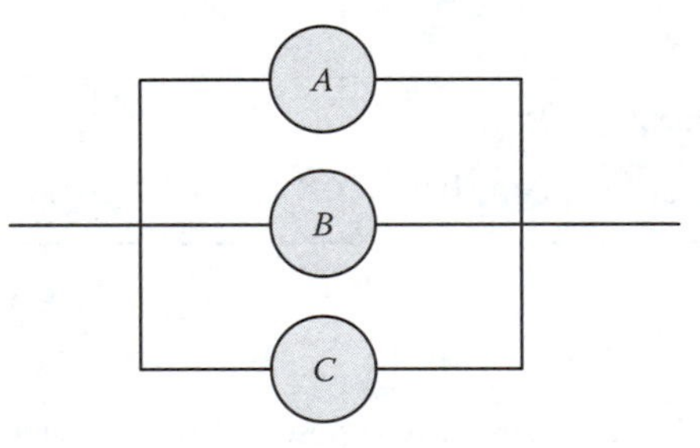

4.36 Write the reliability function of the standby system below as a function of time. Functions $f(t)$ and $g(t)$ are pdf of time to failure and failure probability while in standby is constant and expressed by F' and G'. Assume a perfect switch.

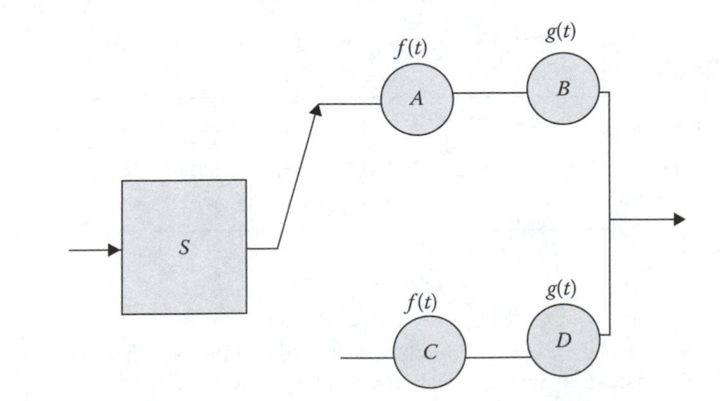

4.37 Draw a success tree for the system shown below (with "system output available" as the top event).

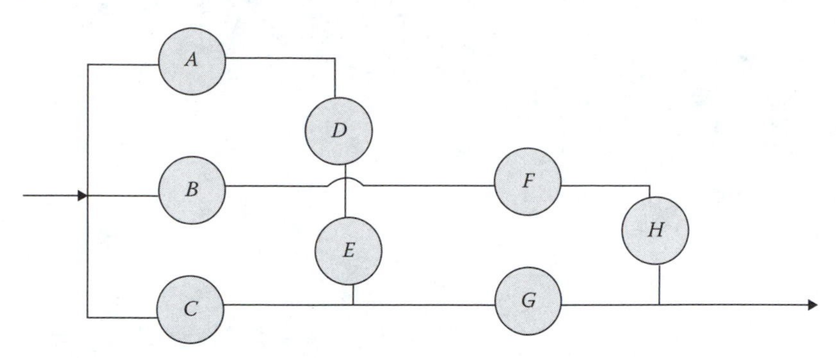

REFERENCES

1. Daniels, H. E., The statistical theory of the strength of bundles of threads, I. *Proc. R. Soc. London*, A183, 404–435, 1945.
2. Crowder, M. J., A. C. Kimber, R. L. Smith, and T. J. Sweeting, *Statistical Analysis of Reliability Data*, Chapman & Hall, London, New York, 1991.
3. Kapur, K. and L. Lamberson, *Reliability in Engineering Design*, Wiley & Sons, New York, NY, 1977.
4. Shooman, M. L., *Probabilistic Reliability: An Engineering Approach*, 2nd edition, Krieger, Melbourne, FL, 1990.
5. Vesely, W. E., F. Goldberg, N. Roberts, and D. Haasl, *Fault Tree Handbook*, NUREG-0492, U.S. Nuclear Regulatory Commission, Washington, DC, 1981.
6. Stamatelatos, M., W. Vesely, J. Dugan, J. Fragola, J. Minarick, and J. Railsback, *NASA Fault Tree Handbook with Aerospace Applications*, National Aeronautics and Space Administration, Washington, DC, 2002.
7. Specter, H. and M. Modarres, *Functional Specifications for a PRA Based Design Making Tool*, Empire State Electric Energy Research Corporation, EP 95-14, New York, 1996.
8. Dezfuli, H., *Application of REVEAL_W to Risk-based Configuration Control*, 1994.
9. Sinnamon, R. M. and J. D. Andrews, Fault Tree Analysis and Binary Decision Diagrams, in Annual Reliability and Maintainability Symposium Proceedings. The International Symposium on Product Quality and Integrity (Cat. No. 96CH35885). New York, 1996, p. 215.
10. Wang, J. and M. Modarres, REX: An intelligent decision and analysis aid for reliability and risk studies, *Reliab. Eng. Syst. Saf. J.*, 30, 185–239, 1990.
11. Fong, C. C. and J. A. Buzacoot, An algorithm for symbolic reliability combination with path-sets or cut-sets, *IEEE Trans. Reliab.*, 36 (1), 34–37, 1987.
12. PRA Procedures Guide, U.S. Nuclear Regulatory Commission, NUREG/CR-2300, Washington, DC, 1982.
13. Modarres, M. *Application of the Master Plant Logic Diagram in Risk-Based Evaluations*, American Nuclear Society Topical Meeting on Risk Management, Boston, MA, 1992.
14. MIL-STD-1629A, *Procedure for Performing a Failure Mode, Effects, and Criticality Analysis*, Department of Defense, NTIS, Springfield, Virginia, 1980.
15. SAE Reference Manual J1739, *Potential Failure Mode and Effect Analysis in Design and Manufacturing*, 1994.
16. MIL-STD-338, *Electronic Reliability Design Handbook*, NTIS, Springfield, Virginia, 1980.

5 Reliability and Availability of Repairable Components and Systems

When we perform reliability studies, it is important to distinguish between repairable and nonrepairable items. The reliability analysis methods discussed in Chapters 3 and 4 are largely applicable to nonrepairable items. In this chapter, we examine the peculiar aspects of repairable systems and discuss methods used to determine the failure characteristics of these systems, as well as the methods for predicting their reliability and availability.

Nonrepairable components and systems are discarded or replaced (i.e., taken out of their sockets and replaced) with new ones when they fail. For example, light bulbs, transistors, contacts, unmanned satellites, and small appliances are nonrepairable items. The reliability of a nonrepairable item is expressed in terms of its time-to-failure distribution, which can be represented by its respective cdf, pdf, or hazard (failure) rate function, as discussed in Chapter 3.

On the other hand, repairable components and systems, generally speaking, are not replaced following the occurrence of a failure; rather, they are repaired and put into operation again. However, if a nonrepairable item is a component of a repairable system, the distribution of the number of component replacements over a given time interval is estimated within the framework of repairable systems reliability. In contrast to nonrepairable items, reliability problems associated with repairable items use, basically, different *random* (*stochastic*) *process* models that will be discussed later.

Repair should not be confused with maintenance. *Maintenance* is a broader term used to describe all sorts of renewal processes (RPs), and can be performed on an item that has not necessarily failed. It is carried out to prevent, protect, or mitigate progression of the degradation process that ultimately leads to failure (e.g., preventive maintenance, predictive maintenance associated with condition monitoring, and corrective maintenance). During the maintenance process, repair may be necessary to correct incipient failures. A special kind of maintenance, corrective maintenance, reacts to failures or symptoms that point to the existence of a failure. Thus, repair is inevitable in corrective maintenance.

Both maintenance and repair lead to downtime of the components and systems, which affects their availability and unavailability. Items can be repaired, and repair activities take time. The probability that an item (system) is up (functioning) can be measured by a probability value called *availability*. Conversely, the probability that the system is down is called *unavailability*.

We will begin this chapter with probabilistic models and statistical methods that are used to determine the failure characteristics of repairable items and their reliability. We will then define the concept of availability and explain availability evaluation methods for repairable items. Although the presentation of the material in this chapter focuses on system reliability and availability, the methods are equally applicable to components.

5.1 REPAIRABLE SYSTEM RELIABILITY

5.1.1 Basics of Point Processes

In this book, the term *repairable item* is used as a synonym for any repairable or renewable object, for example, a system, subsystem, or component (part). The term *repair* applies to replacing an old system with a new one, which is equivalent to a perfect repair (as-good-as-new restoration condition), and also applies to other types of restoration or fixes of incipient failures and to design/manufacturing changes (as in reliability growth modeling), which will be discussed in Chapter 6.

Time to repair (or maintenance) is assumed to be negligible compared with the mean time between (successive) failures (MTBF). This assumption makes it possible to apply different point processes as candidate models for real-life failure processes. For example, a minor automobile repair might take only a few hours to perform. In this case, the assumption is rather realistic. On the other hand, an opposite example could be the time it takes to repair an aviation system, which might be comparable with its predicted MTBF for the so-called "soft" (noncritical) failures.

It is worth mentioning that the applications of the point processes are not limited to repairable systems. Typical applications of these models can be divided into two groups. The first group includes cases in which undesirable events (incidents) are associated with a failure process: for example, a failure process of different engineering systems (ship or nuclear power plant pump failures, or automobile or electrical appliance failures) due to inherited defects, such as microcracks in solids, or inherited diseases (in humans), or to aging processes. The second group includes cases in which undesirable events are associated with external phenomena, such as strong winds, floods, or earthquakes.

Note that in contrast to repairable systems, any typical reliability model of *nonrepairable* identical units is based on a positively defined *r.v.*, such as time to the first (and the only) failure or the number of cycles to the first failure. In this case, the lifetime of the unit is treated as the realization of an identical and independently distributed (IID) r.v. In a sense, the r.v. can be considered as a special case of the failure process, in which the process is terminated just after the first failure.

Observations of repairable systems are typically of the times between successive failures: time to the first failure (TTFF), time between the first and second failures, time between the second and third failures, and so on. Observations of a point process as a series of successive events observed during some time interval are referred to as *realizations* of a given point process. Most often the assumption that these successive failure times are IID r.v.s turns out to be not applicable, and different point process models are considered an appropriate alternative approach to modeling the repairable system failure processes.

It should be noted that the assumption about negligible time to repair might not be critical for some problems related to repairable systems. Let us consider a series of failure-repair events in which each failure is *instantaneously* followed by a repair action. The respective data are the times between failures and repair action durations, which can be represented as

$$t_{0\,1}, \tau_1, t_{1\,2}, \tau_2, \ldots, t_{k-1\,k}, \tau_k, \ldots, t_{n-1\,n}, \tau_n, \tag{5.1}$$

where $t_{0\,1}$ is the TTFF; $t_{k-1\,k}$ is the time between $(k-1)$th and kth failures; and τ_k is the repair time just after kth failure $(k = 1, 2, \ldots, n)$.

In the framework of repairable system analysis, we can analyze the whole data set (Equation 5.1) using the *alternating RP*, which is closely related to the notion of system availability, discussed later in Section 5.2. On the other hand, deleting the repair times from the data set (Equation 5.1), one obtains the following series of times between successive failures, including the TTFF:

$$t_{1\,0}, t_{1\,2}, \ldots, t_{k-1\,k}, \ldots, t_{n-1\,n}. \tag{5.2}$$

This data set is typically needed for solving different reliability and risk analysis problems related to repairable systems, such as estimating whether the system is improving, deteriorating, or revealing a constant failure rate, predicting the number of failures observed in a given time interval, developing the respective logistics, or analyzing reliability growth.

Similarly, picking the repair times out of the data set (Equation 5.1), one obtains the following series of successive repair times:

$$\tau_1, \tau_2, \ldots, \tau_k, \ldots, \tau_n. \tag{5.3}$$

Based on this type of data, it is possible to determine whether the duration of repair actions of interest are decreasing or not. In other words, one can perform a trend analysis of the repair time similar to the reliability trend analysis.

A *point process* can be informally defined as a mathematical model for highly localized events distributed randomly in time. The major r.v. of interest related to such processes is the number of failures (or, generally speaking, *events*) $N(t)$ observed in the time interval $[0, t]$, which is why such processes are also referred to as *counting processes*. Using the nondecreasing integer-valued function $N(t)$, the point process $\{N(t), t \geq 0\}$ is defined as the one satisfying the following conditions:

1. $N(t) \geq 0$;
2. $N(0) = 0$;
3. If $t_2 > t_1$, then $N(t_2) \geq N(t_1)$;
4. If $t_2 > t_1$, then $[N(t_2) - N(t_1)]$ is the number of events (e.g., failures) that occurred in the interval $(t_1, t_2]$.

It is important to note that for most of the point processes considered below, the origin of time, counting $t = 0$, is the moment when the system starts functioning and its age is equal to zero.

A *trajectory* (*sample path*) or *realization* of a point process is the successive failure times of an item: $T_1, T_2, \ldots, T_k \ldots$. It is expressed in terms of the integer-valued function $N(t)$; that is, the number of events observed in the time interval $[0, t]$, as illustrated in Figure 5.1

$$N(t) = \max(k | T_k \leq t). \tag{5.4}$$

$N(t)$ is a random function. The mean value $E(N(t))$ of the number of failures $N(t)$ observed in the time interval $(0, t]$ is called the *cumulative intensity function* (CIF), the *mean cumulative function* (MCF), or the *renewal function*. In the following, the term CIF is used. The CIF is usually denoted by W; that is,

$$W(t) = E(N(t)). \tag{5.5}$$

Similar to the cdf for a r.v., from now on we will assume that CIF $W(t)$ is related to one system. The system is supposed to be a member of a population (finite or infinite) consisting of identical (from reliability standpoint) systems.

Another important characteristic of point processes is the *rate of occurrence of failures* (ROCOF), which is defined as the derivative of CIF with respect to time. That is,

$$w(t) = \frac{\mathrm{d}W(t)}{\mathrm{d}t}. \tag{5.6}$$

Based on the above definition of ROCOF, the CIF is sometimes called the *cumulative* ROCOF.

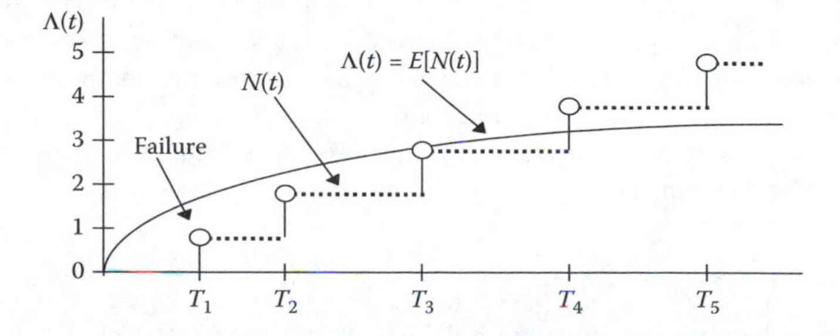

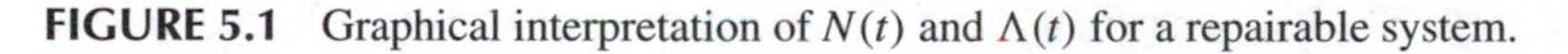
FIGURE 5.1 Graphical interpretation of $N(t)$ and $\Lambda(t)$ for a repairable system.

Most of the processes discussed below have monotone ROCOF. The system modeled by a point process with an increasing ROCOF is called an *aging* (or *degrading*) system. Analogously, the system modeled by a point process with a decreasing ROCOF is called a *rejuvenating* (or *improving*) system. The distribution of TTFF of a point process is called *the underlying distribution*. For some point processes, this distribution coincides with the distribution of time between successive failures (TBF) (which is also called *interarrival time*); for others it does not. The *underlying distribution* is included in the definition of any particular point process used as a model for the failure/repair process of reparable systems. The underlying distribution is, in a sense, a "nonrepairable" component of the respective point process definition.

5.1.2 Homogeneous Poisson Process

At this point, we need to introduce some other terms related to the point processes. A point process is said to have *independent increments* if the numbers of events (failures) in mutually exclusive intervals are independent r.v.s. A point process is called *stationary* if the distribution of the number of events in any time interval depends on the length of the time interval only.

The HPP is probably the oldest as well as the simplest failure process model. A point process having independent increments is called HPP with parameter $\lambda > 0$ if the number of events (failures) N in any interval of length $t = (t_2 - t_1)$ has the Poisson distribution with mean $E(N) = \lambda t$. That is,

$$\Pr[n(t_2) - n(t_1) = n] = \frac{[\lambda(t_2 - t_1)]^n}{n!} \exp[-\lambda(t_2 - t_1)], \tag{5.7}$$

where $t_2 > t_1 \geq 0$. Note that the location of the time interval on the time axis does not matter.

It is not difficult to show that for the HPP, the times between successive failures are independent, identically distributed r.v.s having the exponential distribution with parameter λ, which is the failure rate for this exponential distribution.

In terms of failure processes, parameter λ is the ROCOF. The mean number of failures observed in interval $(0, t]$ $W(t)$, that is, the CIF of this point process, is

$$W(t) = \int_0^t \lambda(\tau)\, d\tau. \tag{5.8}$$

The ROCOF of the HPP is obviously constant, so that

$$W(t) = \lambda t. \tag{5.9}$$

Due to the memoryless property of the exponential distribution, a repairable system's failure behavior modeled by HPP cannot be expressed in terms of system age. Consequently, any preventive action does not make any sense in the framework of the HPP model.

In many situations, it is necessary to model the failure behavior of different items *simultaneously* (the items must be put into service or on a test at the same moment). Such situations can be modeled by the *superposition* of the respective point processes. The superposition of several point processes is the ordered sequence of all failures that occur in any of the individual point processes.

The superposition of several HPP processes with parameters $\lambda_1, \lambda_2, \ldots, \lambda_k$ is the HPP with $\lambda = \lambda_1 + \lambda_2 + \cdots + \lambda_k$. A well-known example of this is a series system, introduced in Chapter 4, with exponentially distributed elements.

5.1.3 Renewal Process

Another important failure process model for reliability applications is the RP. The RP can be considered as a generalization of the HPP for the case in which the TBF distribution—that is, the *underlying distribution*—is an arbitrary continuous distribution of a positively defined r.v. In this case, strictly speaking, the number of failures in any interval of length t no longer follows the Poisson distribution.

The time origin is used for the RP and all other processes considered below. It is assumed (unless another assumption is made) that the time origin coincides with the time at which a new (perfectly renewed or repaired) unit is put in a functioning state. In other words, it means that at the time origin the age of the unit is equal to zero.

For any underlying distribution, the RP models *as-good-as-new* restorations (an equivalent term is *same-as-new*) or operates under the so-called *perfect repair assumption*. It should be noted that the term *as-good-as-new* (as well as *perfect*) might be misleading. For example, for the DFR distribution, a restoration to the *as-good-as-new* condition cannot be called a perfect repair because the unit having the DFR lifetime distribution has its highest failure rate when it is new (at zero age). In this case, the repair returning our unit to the state corresponding to its zero age is difficult to call *perfect*. As previously discussed, the notion of "age" is not applicable for exponential distribution, so that any product having the exponential TBF is always as-good-as-new.

The perfect repair assumption is definitely not appropriate for a multicomponent system if only a few of the system components are repaired or replaced upon failure. On the other hand, if a failed system (or a separate component) is replaced by an identical new one, the same-as-new repair assumption is quite appropriate for this system (or component).

In the following, it is assumed that the cdf of underlying distribution, $F(t)$, is continuous together with its pdf $f(t)$. The respective CIF, $W(t)$, for the considered RP can be found as a solution of the so-called *renewal equation*,

$$W(t) = F(t) + \int_0^t F(t-\tau)\,\mathrm{d}W(\tau). \tag{5.10}$$

The respective ROCOF $w(t)$ can be found as a solution of the equation obtained by taking the derivatives of both sides of Equation 5.10 with respect to t. That is,

$$w(t) = f(t) + \int_0^t f(t-\tau)w(\tau)\,\mathrm{d}\tau. \tag{5.11}$$

Closed-form solutions of the above integral equations are known as the *exponential underlying distribution* (the case of the HPP) and the *gamma underlying distribution*, respectively.

A number of numeric solutions to Equation 5.10 have been developed. Smith and Leadbetter [1] found an iterative solution for the case when the underlying distribution is the Weibull one. Another iterative solution with the Weibull-distributed time between failures was reported by White [2]. Baxter et al. [3] offered a numerical integration approach that covers the cases of the following underlying distributions: Weibull, gamma, lognormal, truncated normal, and inverse normal. Garg and Kalagnanam [4] proposed the Pade approximation approach to solve the renewal equation for the case of inverse normal underlying distribution. Blischke and Murthy [5] used a Monte Carlo simulation, which provides a universal numerical approach to the problem.

The CIFs for the RP with the Weibull underlying distributions having the scale parameter equal to 1 and the different shape parameters are given in Figure 5.2.* Note that the case $\beta = 1$ is the HPP. For the case $\beta = 1.5$ (IFR underlying distribution), the CIF is concave upward, whereas for the case $\beta = 0.5$ (DFR underlying distribution), the CIF is concave downward, and for the case $\beta = 1$, it is a straight line (HPP). This property corresponds to the increasing (for the case $\beta = 1.5$) and decreasing (for the case $\beta = 0.5$) ROCOFs of the corresponding RPs. The respective ROCOF plots are given in Figure 5.3. It is important to note that the upper time axis limit in Figures 5.2 and 5.3 is the 63rd percentile for all the underlying Weibull distributions.

*Figure 5.2, as well as many of the following figures, were obtained using the Monte Carlo simulation.

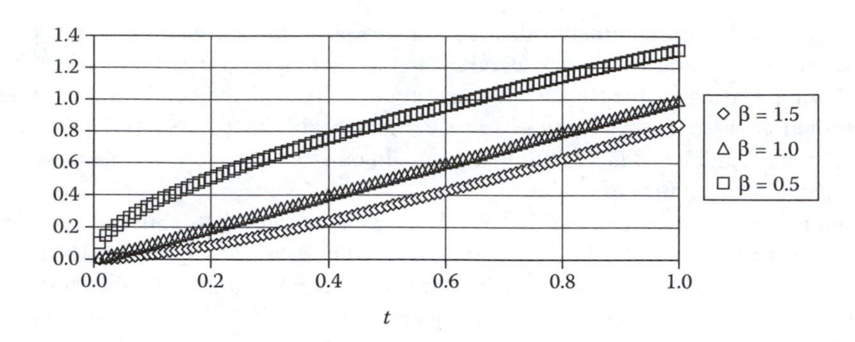

FIGURE 5.2 CIFs of RPs with Weibull underlying distributions having scale parameter equal to 1 and different shape parameters.

At this point, it is rational to introduce the notions of *short-term* and *long-term behavior* for point processes [6]. *Short-term* behavior implies that a process is observed during an interval limited by a time close to the mean (or the median) of the respective underlying distribution—that is, for the time intervals for which the probability of the second and successive failures is much lower compared to the probability of the first failure. As opposed to short-term behavior, *long-term behavior* is the behavior of a failure process observed during a time much longer than the mean or median of its underlying distribution. Another way to define long-term behavior is to define the observation time interval long enough to successfully apply asymptotic theorems, which are discussed later.

Table 5.1 contains the medians and means of the underlying distributions for the RPs illustrated by Figures 5.2 and 5.3, which illustrate the short-term behavior of the respective RPs. In contrast to these figures, Figure 5.4 illustrates the long-term behavior of the CIFs of the same RPs. The CIFs in Figure 5.4 look like straight lines, implying that the corresponding ROCOF might be constant. The respective ROCOF is shown in Figure 5.5. The figure shows that the ROCOF gets closer to a constant limit as time increases. This limit is given by Blackwell's theorem. The theorem states that for an RP with an underlying continuous distribution $F(t)$ with mean $E(t)$, the following *limiting* relationship holds

$$\lim_{t\to\infty}(W(t+a)-W(t))=\frac{a}{E(t)}, \tag{5.12}$$

where $a > 0$.

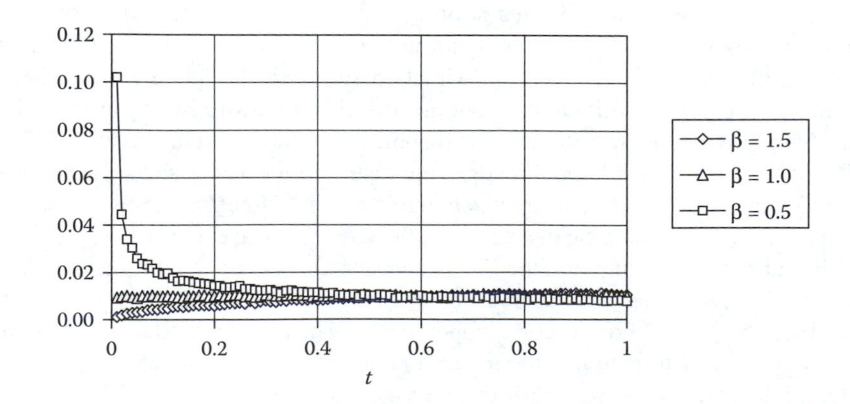

FIGURE 5.3 ROCOF functions of RPs with Weibull underlying distributions having scale parameter equal to 1 and different shape parameters.

TABLE 5.1
Medians and Means of Underlying Distributions of RPs Illustrated by Figures 5.2 and 5.3

Distribution	Parameters	Median	Mean
Weibull	$\alpha = 1.0,\ \beta = 1.5$	0.78	0.90
Weibull	$\alpha = 1.0,\ \beta = 0.5$	0.48	2.00
Weibull (exponential)	$\alpha = 1.0,\ \beta = 1.0$	0.69	1.00

The approximate values of the ROCOF used in Figure 5.5 may be expressed as

$$w(\hat{t}) = \frac{W(t+a) - W(t)}{a}, \tag{5.13}$$

with $a = 1.0$.

The average value of the ROCOF for the RP illustrated by Figures 5.4 and 5.5 and the values of the ROCOF estimated using Equation 5.13 turn out to be very close.

Example 5.1

For the RP with the underlying distribution having the shape parameter $\beta = 1.5$, the average value of the ROCOF is estimated as 1.11, which practically coincides with the value $1/E(t) = 1/0.90 = 1.11$, based on Equation 5.12 with $a = 1$. In other words, this example shows that the long-term behavior of the RP can be accurately described by Blackwell's theorem.

5.1.4 Nonhomogeneous Poisson Process

A point process having independent increments is called the nonhomogeneous Poisson process (NHPP) with time-dependent ROCOF $\lambda(t) > 0$, if the probability that exactly n events (failures) occur in any interval (a, b) and have the Poisson distribution with the mean equal to $\int_a^b \lambda(t)\,\mathrm{d}t$. That is,

$$P[N(b) - N(a) = n] = \frac{\left[\int_a^b \lambda(t)\,\mathrm{d}t\right]^n \mathrm{e}^{-\int_a^b \lambda(t)\,\mathrm{d}t}}{n!}, \tag{5.14}$$

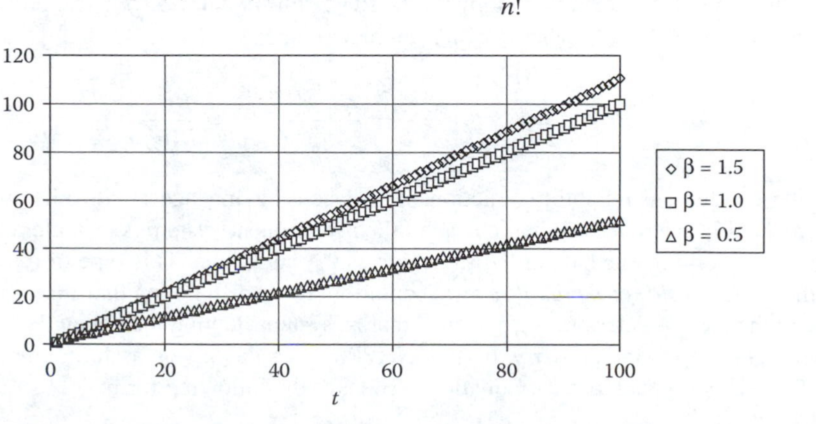

FIGURE 5.4 Long-term behavior of RP. CIFs of RPs having the Weibull underlying distributions with scale parameter $\alpha = 1$.

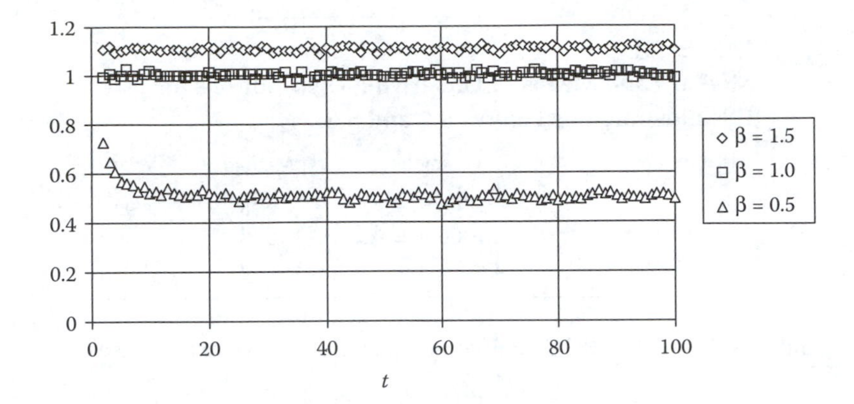

FIGURE 5.5 Long-term behavior of RP. ROCOF functions of RPs having the Weibull underlying distributions with scale parameter $\alpha = 1$.

for $n = 0, 1, 2, \ldots, \infty$, and $N(0) = 0$. Opposite to the RP, the times between successive events (e.g., failures) in the framework of the NHPP model are neither independent nor identically distributed.

Similar to the RP discussed in Section 5.1.3, it is further assumed that for the NHPP applications, the time origin (or zero age) is the time when a new (renewed or repaired) unit has been put into operation.

Based on the above definition, the CIF and ROCOF of NHPP can be written as follows:

$$W(t) = \int_0^t \lambda(\tau)\,\mathrm{d}\tau \tag{5.15}$$

and

$$w(t) = \lambda(t). \tag{5.16}$$

The cdf of TTFF (i.e., the cdf of the underlying distribution) for the NHPP can be found as

$$F(t) = 1 - \Pr[N(t) - N(0) = 0] = 1 - \exp(-W(t)), \tag{5.17}$$

where $W(t)$ is given by Equation 5.15.

Let us consider a series of failures occurring according to the NHPP with ROCOF $\lambda(t)$. Let t_k be the time to the kth failure; so, at this moment, the ROCOF is equal to $\lambda(t_k)$. The probability that no failure occurs in interval $(t_k, t]$, where $t > t_k$, can be written as

$$R(t_k, t) = \mathrm{e}^{-\int_{t_k}^{t} \lambda(\tau)\,\mathrm{d}\tau} = \frac{\mathrm{e}^{-\int_0^t \lambda(\tau)\,\mathrm{d}\tau}}{\mathrm{e}^{-\int_0^{t_k} \lambda(\tau)\,\mathrm{d}\tau}} = \frac{R(t)}{R(t_k)}, \tag{5.18}$$

which is the conditional reliability function of a system having age t_k. In other words, we can consider the NHPP as a process in which each failed component/system is instantaneously replaced by an identical, working one having the same age as the failed one. This type of restoration model is called the *same-as-old* (or *minimal repair*) condition. It is worth noting that *same-as-old* does not mean *bad* in the case of deteriorating systems, that is, systems having increasing ROCOF.

Another very important property of the NHPP under the given assumptions follows from Equation 5.18. If t_k is equal to zero, Equation 5.18 takes the following form:

$$R(t) = \exp\left(-\int_0^t \lambda(\tau)\,\mathrm{d}\tau\right), \tag{5.19}$$

which means that the ROCOF of the NHPP coincides with the failure (hazard) rate function of the underlying TTFF distribution. In other words, all future behavior of a repairable system is completely defined by this distribution. It also means that just after any repair/maintenance action carried out at time t, the ROCOF is equal to the failure rate of the TTFF distribution $\lambda(t)$. So, we can also consider the NHPP as a process in which each failed system is instantaneously replaced by an identical one having the same failure rate as the failed one.

Depending on the application, assumption of "as old restoration" may and may not be realistic. Applied to a single-component system, it is definitely not a realistic assumption. For a complex system, composed of many components having close reliability functions, this assumption is much more realistic, because only a small fraction of the system's components is repaired, which results in a small change of the system failure rate [7].

An important particular case of the NHPP is the case when the ROCOF is a power function of time. That is,

$$w(t) = \frac{\beta}{\alpha}\left(\frac{t}{\alpha}\right)^{\beta-1} \quad t \geq 0, \alpha, \beta > 0, \tag{5.20}$$

which obviously results in the Weibull TTFF (underlying) distribution. Accordingly, the NHPP process with the ROCOF given by Equation 5.20 is sometimes referred to as the Weibull, the power law NHPP process, or the "Crow-AMSAA model."

Statistical procedures for this model were developed by Crow [8,9], based on suggestions of Duane [10]. These procedures can also be found in MIL-HDBK-781 and IEC Standard 1164 (1995). The main applications of the power law model are associated with reliability monitoring (which is optimistically called "reliability growth") problems for repairable products as well as for nonreparable ones.

Another important particular case of NHPP is the case when the ROCOF is a simple log-linear function. That is,

$$w(t) = \exp(\beta_0 + \beta_1 t) \quad t \geq 0. \tag{5.21}$$

This model was proposed and statistically developed by Cox and Lewis [11], which is why the model is often referred to as the *Cox–Lewis* model.

Similar to the power-law model, the log-linear model has a monotonic ROCOF, which can be increasing, decreasing, or constant (if $\beta_1 = 0$). The respective underlying (TTFF) distribution is the so-called *truncated Gumbel* (smallest extreme value) distribution. This distribution has the following cdf:

$$F(t) = 1 - \exp\left(-\alpha(e^{\beta_1 t} - 1)\right), \tag{5.22}$$

where $\alpha = \exp(\beta_0)/\beta_1$ and $t \geq 0$.

The NHPP with the simple linear ROCOF, that is,

$$w(t) = \beta_0 + \beta_1 t \quad t \geq 0, \tag{5.23}$$

is not as popular as the previous ones. This might be explained by the fact that for any time interval of interest (0, t] the parameters of Equation 5.23 must obviously satisfy the following inequality:

$$w(t) = \beta_0 + \beta_1 t \geq 0. \tag{5.24}$$

If both parameters are positive, the respective underlying distribution is the TTF distribution of the series system (competing risk model) composed of a component with the exponential TTF distribution and a component with the Rayleigh distribution (i.e., the Weibull distribution with shape parameter equalling 2). Risk analysis applications of the linear model are discussed by Vesely [12] and Atwood [13].

Statistical procedures for NHPP data analysis are well developed and can be found in many books on reliability data analysis, for example, see [7,11,14,15].

5.1.5 General Renewal Process

The validity of the minimal repair condition and the perfect repair condition as realistic assumptions has been criticized by many specialists: for example, see [16–18]. Thompson [18] notes that the NHPP "is a nonintuitive fact that is casting doubt on the NHPP as a realistic model for repairable systems. Use of an NHPP model implies that if we are able to estimate the failure rate of the time to the first failure, such as for a specific types of automobiles, we at the same time have an estimate of the ROCOF of the entire life of the automobile." On the other hand, Ascher and Feingold [16], discussing the RP, point out, "If an automobile were modeled by a renewal process, its age at any instant in time would be the backward recurrence time (mileage) to the most recent failure/repair!" Finally, Lindquist [17] notes, "For many applications it is more reasonable to model the repair action by something in between the two given extremes (RP & NHPP)."

For many years, these processes were the most commonly used models for the failure process. As mentioned above, the RP can be used to model the situations with restoration to "good-as-new" condition (perfect repair assumption); meanwhile, the NHPP is applied to the situations with the "same as old" restoration (minimal repair assumption). In a sense, these two assumptions can be considered extreme, from both theoretical and practical standpoints. In order to avoid this extremism, several generalizing models have been introduced in recent years [17,19–21].

Among these models, the General Renewal Process (GRP) introduced by Kijima and Sumita [20] is the most attractive one, since it covers not only the RP and the NHPP, but also the intermediate "younger-than-old-but-older-than-new" repair assumption. The introduced GRP results in the so-called G-renewal equation, which is a generalization of the ordinary renewal Equation 5.10.

It should be noted that being a rather new model, the GRP model has been used in many applications, including the automobile industry [22], the oil industry [23], and reliability studies of hydro-electric power plants [24].

The GRP operates on the notion of *virtual age*. Let A_n be the virtual age of a system immediately after the nth repair. If $A_n = y$, then the system has time to the $(n+1)$th failure X_{n+1}, which is distributed according to the following cfd:

$$F(X|A_n = y) = \frac{F(X+y) - F(y)}{1 - F(y)}, \tag{5.25}$$

where $F(X)$ is the cdf of the TTFF distribution of the system when it was new (the underlying distribution). The cdf (Equation 5.25) is the conditional cdf of the system at age y.

The following sum

$$S_n = \sum_{i=1}^{n} X_i, \tag{5.26}$$

with $S_0 = 0$, is called the *real age* of the system.

In the framework of the considered GRP model it is assumed that the nth repair can remove the damage incurred only during the time between the $(n-1)$th and the nth failures, so that the respective virtual age after the nth repair is

$$A_n = A_{n-1} + qX_n = qS_n, \quad n = 1, 2, \ldots, \tag{5.27}$$

where q is the so-called *parameter of rejuvenation* (or *repair effectiveness parameter*) and the virtual age of a new system $A_0 = 0$, so that the TTFF is distributed according to $F(t|0) \equiv F(t)$, which is the underlying distribution.

The time between the first and second failures is distributed according to Equation 5.24, where $A_1 = qX_1$. The time between the second and third failure is distributed according to Equation 5.27 with $A_2 = q(X_1 + X_2)$, and so on.

It is clear that when $q = 0$, this process coincides with an ordinary RP, thus modeling the same-as-new repair assumption. When $q = 1$, the system is restored to the same-as-old condition, which is the case of NHPP. The case of $0 < q < 1$ falls between the same-as-new and same-as-old repair assumptions. Finally, when $q > 1$, the virtual age $A_n > S_n$. In this case, the repair either damages the system (if the failure process has an increasing ROCOF) to a higher degree than it was just before the respective failure or improves the system (if failure the process has a decreasing ROCOF), which corresponds to the older-than-it was repair assumption. It is important to note that the above interpretation is good for the system with monotone (increasing or decreasing) ROCOF.

For the GRP, the expected number of failures in (0, t), that is, CIF $W(t)$, is given by a solution of the so-called G-renewal equation [20]:

$$W(t) = \int_0^t \left(g(\tau|0) + \int_0^\tau h(x)g(\tau - x|x)\,\mathrm{d}x \right) \mathrm{d}\tau, \tag{5.28}$$

where

$$g(t|x) = \frac{f(t + qx)}{1 - F(qx)}, \quad t, x \geq 0$$

is the conditional function such that $g(t|0) = f(t)$, $F(t)$ and $f(t)$ are the cdf and pdf of the TTFF (underlying) distribution. By definition $w(t) = \mathrm{d}(W(t))/\mathrm{d}t$, therefore one obtains the following equation for the ROCOF of the GRP:

$$w(t) = g(t|0) + \int_0^t w(x)g(t - x|x)\,\mathrm{d}x. \tag{5.29}$$

Kijima and Sumita [20] showed that the Volterra integral of Equation 5.29 has a unique solution. It should be noted that the closed-form solutions of Equations 5.28 and 5.29, and even numerical solutions, are difficult to obtain, since each equation contains a recurrent infinite system [25]. A Monte Carlo-based solution is, however, possible and was discussed by Kaminskiy and Krivtsov [26]. The respective Monte Carlo approach is discussed later in this chapter.

Kijima et al. [27] point out that numerical solution of the G-renewal equation is very difficult in the case of the Weibull underlying distribution. This statement is not appropriate in the situations where the Monte Carlo approach is applied, which is illustrated by the examples below.

Compared to the ordinary renewal Equation 5.10, Equation 5.28 has one additional parameter, which is the repair effectiveness parameter q. Similar to Section 5.1.3 (*Renewal Process*), we will discuss below the short-term and the long-term behavior of the GRP.

Figure 5.6 illustrates the short-term behavior of GRP CIFs having the aging (IFR) Weibull underlying distribution (with scale parameter $\alpha = 1$ and shape parameter $\beta = 1.5$) and different rejuvenation parameters q including $q = 0$, which corresponds to the RP, and $q = 1$, which corresponds to the NHPP. For this case, the NHPP CIF is the upper bound and the RP CIF is the lower bound for any GRP CIF with $0 < q < 1$.

Figure 5.7 shows the *short-term behavior* of the respective GRP ROCOFs. Similar to Figure 5.6, the NHPP ($q = 1$) exhibits the highest ROCOF, and the RP ($q = 0$) the lowest.

Figures 5.8 and 5.9 are similar to Figures 5.6 and 5.7, except for the underlying distributions, which are the DFRs for the processes depicted in Figures 5.8 and 5.9. For the GRPs with DFR underlying

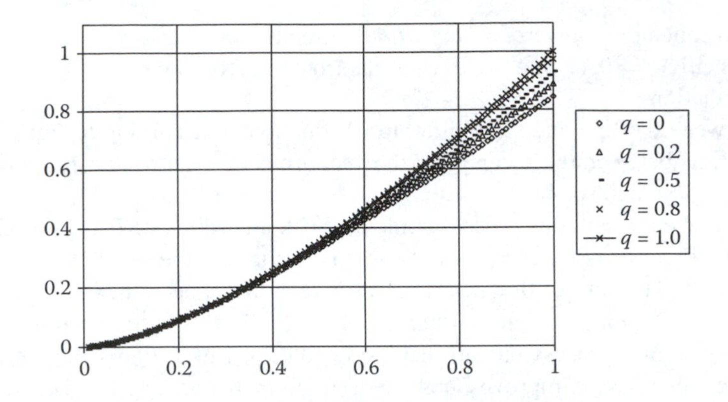

FIGURE 5.6 Short-term behavior of GRP. CIFs of the processes having the aging (IFR) Weibull underlying distribution (with scale parameter $\alpha = 1$ and shape parameter $\beta = 1.5$) and different rejuvenation parameters q.

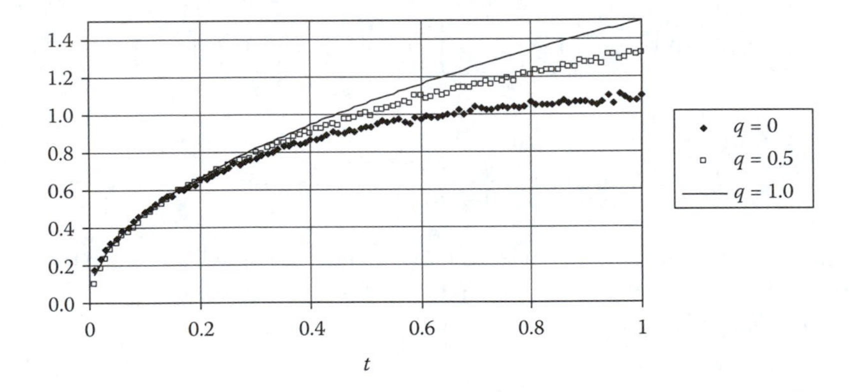

FIGURE 5.7 Short-term behavior of GRP. The ROCOFs of GRPs having the same aging (IFR) Weibull underlying distribution as in Figure 5.6 and different rejuvenation parameters q. The ROCOFF of NHPP ($q = 1$) is based on its explicit expression.

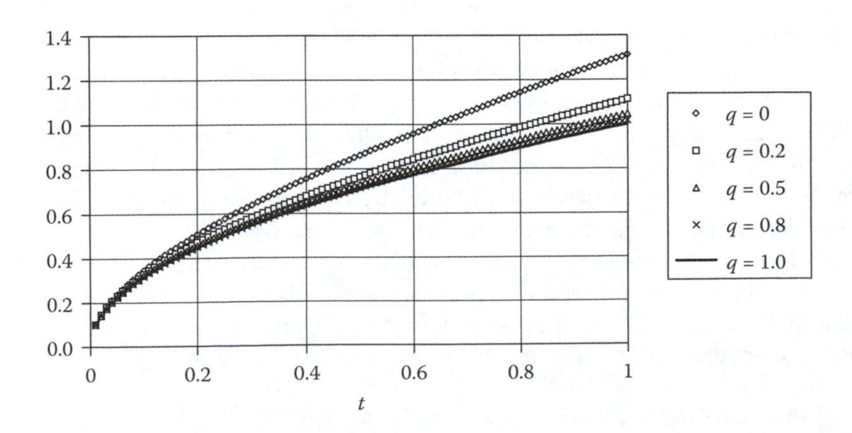

FIGURE 5.8 Short-term behavior of GRP. CIFs of the processes having the DFR Weibull underlying distribution (with scale parameter $\alpha = 1$ and shape parameter $\beta = 0.5$) and different rejuvenation parameters q.

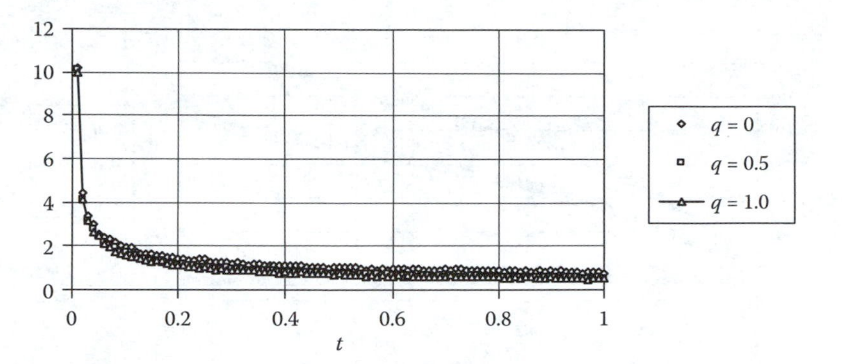

FIGURE 5.9 Short-term behavior of GRP. The ROCOFs of GRPs having the same DFR Weibull underlying distribution as in Figure 5.8 and different rejuvenation parameters q.

distributions (in contrast to the GRP with IFR underlying distributions), the NHPP CIF is the lower bound and the RP CIF is the upper bound for any GRP CIF with rejuvenation parameter $0 < q < 1$.

Figure 5.10 illustrates the *long-term behavior* of GRP CIFs having the aging (IFR) Weibull underlying distribution with scale parameter $\alpha = 1$ and shape parameter $\beta = 1.5$ (IFR distribution) and different rejuvenation parameters q, including $q = 0$, which corresponds to the RP, and $q = 1$, which corresponds to the NHPP.

As in the case of the short-term behavior, the NHPP CIF is an upper bound and the RP CIF is the lower bound for any GRP CIF with $0 < q < 1$. Figure 5.11 shows the long-term behavior of the respective GRP ROCOFs. Similar to Figure 5.7, the NHPP ($q = 1$) exhibits the highest ROCOF, and the RP ($q = 0$) the lowest one. Note that unlike the RP, the ROCOF of the NHPP and GRP do not reveal the limiting behavior similar to the one given by Blackwell's theorem (Equation 5.12).

Figures 5.12 and 5.13 are similar to Figures 5.10 and 5.11, except for the underlying distributions, which are the DFR for the processes depicted in Figures 5.12 and 5.13. Note that for the GRPs with DFR underlying distributions (in contrast to the GRP with IFR underlying distributions), the NHPP CIF is the lower bound and the RP CIF is the upper bound for any GRP CIF with rejuvenation parameter $0 < q < 1$.

5.1.6 Probabilistic Bounds

From the above discussion, it is clear that finding the CIF of the RP and GRP is far from trivial, which is why applying a simple bound on the respective CIFs can be considered as a useful short cut.

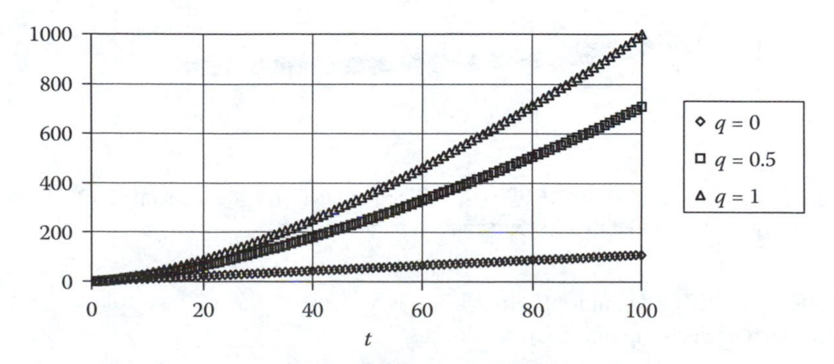

FIGURE 5.10 Long-term behavior of GRP. CIFs of the processes having the aging (IFR) Weibull underlying distribution (with scale parameter $\alpha = 1$ and shape parameter $\beta = 1.5$) and different rejuvenation parameters q.

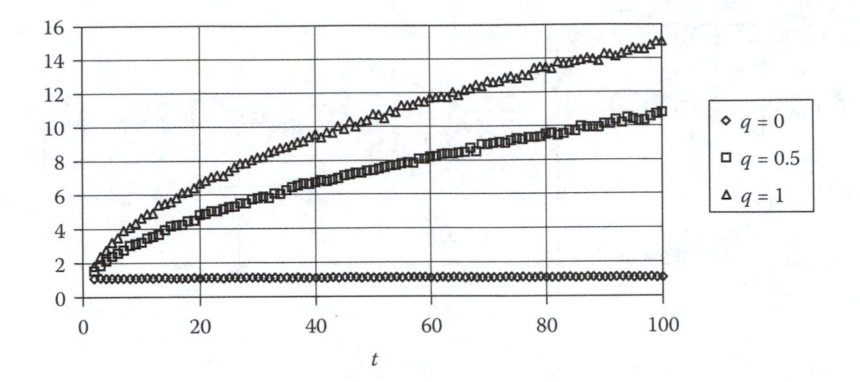

FIGURE 5.11 Long-term behavior of GRP. The ROCOFs of GRPs having the same aging (IFR) Weibull underlying distribution as in Figure 5.10 and different rejuvenation parameters q.

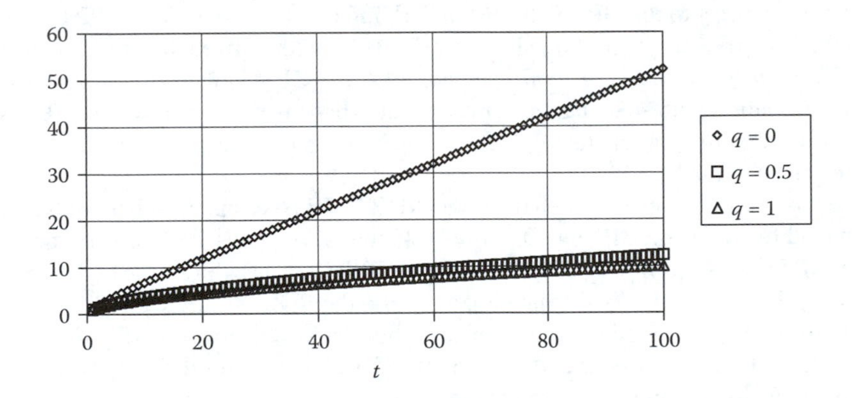

FIGURE 5.12 Long-term behavior of GRP. CIFs of the processes having the DFR Weibull underlying distribution (with scale parameter $\alpha = 1$ and shape parameter $\beta = 0.5$) and different rejuvenation parameters q.

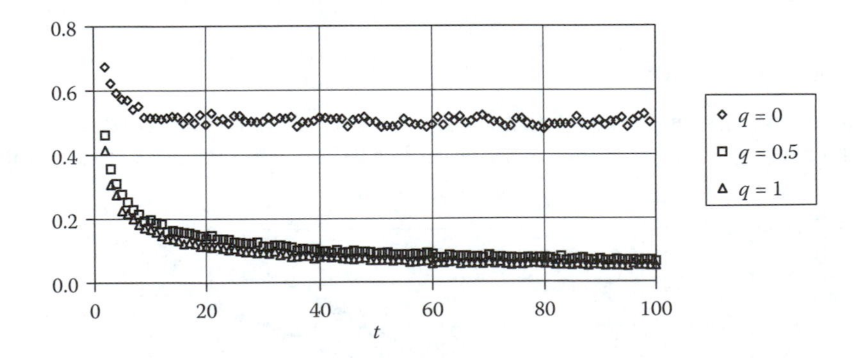

FIGURE 5.13 Long-term behavior of GRP. The ROCOFs of GRPs having the same DFR Weibull underlying distribution as in Figure 5.12 and different rejuvenation parameters q.

Hoyland and Rausand [7] give, among other bounds suggested earlier, a simple upper bound on the CIF of the RP with underlying distribution $F(t)$ as

$$UB(t) = \frac{F(t)}{1 - F(t)}. \tag{5.30}$$

Let us consider an NHPP having the ROCOF equal to the failure rate $r(t)$ of the underlying distribution $F(t)$, which is assumed to be an IFR distribution. That is,

$$r(t) = \frac{f(t)}{1 - F(t)}. \tag{5.31}$$

Let $M_{\text{NHPP}}(t|F(t))$ be the CIF of the NHPP with ROCOF given by Equation 5.31. That is,

$$M_{\text{NHPP}}(t|F(t)) = \int_0^t r(\tau)\,\mathrm{d}\tau = -\ln(1 - F(t)) \tag{5.32}$$

and let $M_{\text{RP}}(t|F(t))$ be the CIF of the RP having the same IFR underlying distribution $F(t)$. In this case, the CIF of the NHPP is always larger than the CIF of the RP having the same underlying IFR distribution, $F(t)$. That is,

$$M_{\text{RP}}(t|F(t)) < -\ln(1 - F(t)) \quad 0 < t < \infty \tag{5.33}$$

Based on inequality (Equation 5.33), one can use the CIF of the introduced NHPP as an upper bound on the CIF of the RP having any underlying IFR distribution $F(t)$. That is,

$$UB(t) = -\ln(1 - F(t)). \tag{5.34}$$

It is easy to show that this upper bound is always more strict than the upper bound given by Equation 5.30. That is,

$$-\ln(1 - F(t)) \le \frac{F(t)}{1 - F(t)} \quad 0 \le t < \infty, \tag{5.35}$$

and should therefore be used instead of Equation 5.30, when the underlying distribution is IFR. The same upper bound (Equation 5.35) can be applied to the CIF of GRP with the IFR underlying distribution and repair effectiveness parameter $q < 1$.

The above-discussed bounds might be practically applicable only to the short-term observation intervals, introduced in Section 5.1.3. In this section, the long-term behavior was defined as the behavior of a failure process observed during a time much longer than the mean or median of its underlying distribution. Another possible way to define the long-term behavior is to define the observation time interval long enough to successfully apply asymptotic theorems.

For the nonformal definition of the long-term observation interval, recall the upper bound for the RP introduced by Barlow and Proschan [28]. Their nonparametric upper bound on the CIF of the RP with a NBUE* underlying distribution is given by

$$UB(t) = \frac{t}{E(t)}, \tag{5.36}$$

where $E(t)$ is the mean of the underlying distribution, that is, the mean TTFF. This bound is associated with the Elementary Renewal Theorem expressing the asymptotic behavior of any RP. The theorem states

$$\lim_{t\to\infty} \frac{M_{\text{RP}}(t)}{t} = \frac{1}{E(t)}. \tag{5.37}$$

*NBUE is *new better than used in expectation.* This class includes IFR distributions.

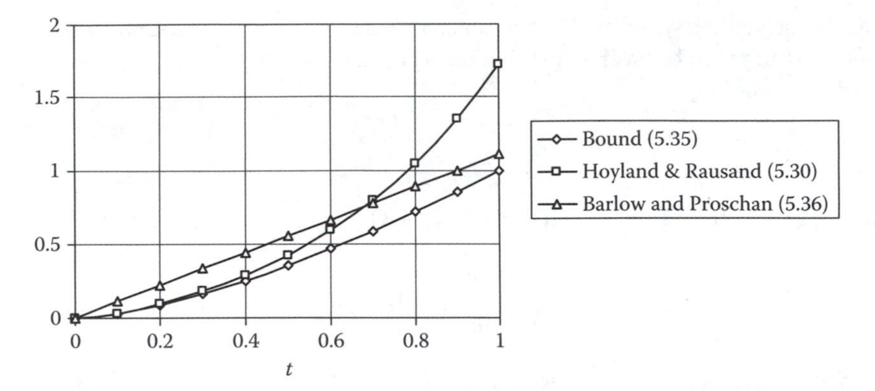

FIGURE 5.14 Upper bounds for RP with Weibull underlying distribution with scale parameter equal to 1 and shape parameter equal to 1.5.

Figure 5.14 shows the upper bounds given by Equations 5.30, 5.33, and 5.36 for the RP having the same Weibull underlying distribution with the scale parameter equal to 1 and the shape parameter equal to 1.5. It is easy to show that the upper bound given by Equation 5.35 is also sharper than the upper bound on the CIF of RP (Equation 5.36) in the interval $(0, t^*)$, where t^* is given by a solution of the following equation:

$$\frac{-\ln(1 - F(t))}{t} = \frac{1}{E(t)}, \tag{5.38}$$

for the bound depicted in Figure 5.14, $t^* \approx 1.227$. Thus, the simple upper bound given by Equation 5.35 can be effectively used for short-term applications.

For the application considered, the point t^* defined as a solution of Equation 5.38 can also be used as a "boundary point" between the short- and long-term random process behavior. Figure 5.15 depicts the long-term behavior of some of the failure process models, including those shown in Figure 5.14, but with a time interval 10 times larger than that in Figure 5.14. Note again that the CIF of the NHPP with the same IFR underlying distribution $F(t)$ is the upper bound on the CIF, not only for the RP, but also [in contrast to the Barlow and Proschan upper bound (Equation 5.36)] for the GRP with repair effectiveness parameter $q < 1$ as well.

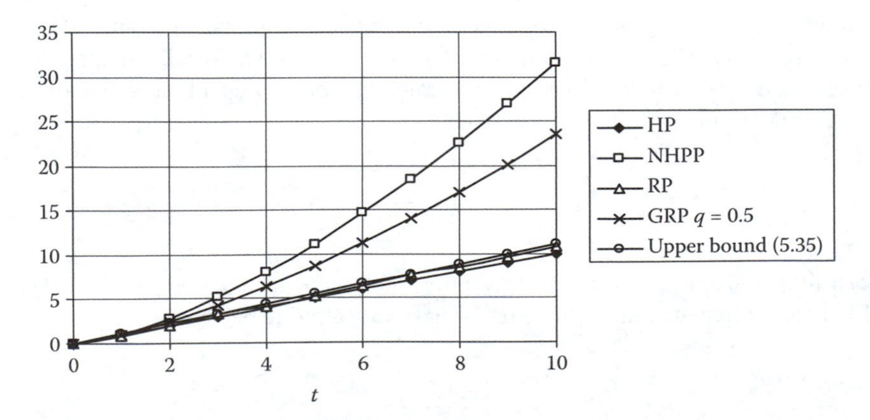

FIGURE 5.15 The CIFs for RP, HPP, NHPP [the upper bound (5.35)], GRP with rejuvenation parameter $q = 0.5$, and Barlow–Proschan upper bound (5). The underlying distribution is Weibull one with the scale parameter equal to 1 and the shape parameter equal to 1.5 (except for the HPP).

As illustrated by Figure 5.15, closely associated with Elementary Renewal Theorem, the Barlow–Proschan upper bound on the CIF of the RP Equation 5.36 is practically applicable for the RP observed in a long-term interval (i.e., for $t > t^*$).

It should be noted that, from the standpoint of the applications considered, neither Blackwell's theorem (see Section 5.1.3) nor the Elementary Renewal theorem addresses the question of how large the value of t must be in order to apply the respective limit. In a sense, for practical use, the solution of (38), t^*, can be used as a lower time limit, beyond which the "long-term" Barlow and Proschan bound Equation 5.36 can be effectively used.

Now, let us consider an analogy between a positively defined r.v. used as a model for TTF of nonrepairable objects, and random events considered in the framework of a counting process used as a model for the failure process of repairable objects. In nonparametric TTF distribution estimation (applied to nonrepairable objects), several special classes of distribution can be used, such as the IFR/DFR class, the IFRA/DFRA class, the new-better/worse-than-used (NBU/NWU) class, and the new-better/worse-than-used-in-expectation (NBUE/NWUE) class [28] (see Chapter 3).

Likewise, in Section 5.1, the CIF and its derivative and the ROCOF were introduced as counting processes used for modeling failure processes of repairable objects. A counting process is known as an *increasing/decreasing ROCOF process*, that is, an IRP/DRP if its ROCOF exists and it is an increasing/decreasing function. The quantile bounds based on a known IFRA/DFRA cdf were introduced in Chapter 3. Note that using these bounds, Barlow and Gupta [29] developed single-stage sampling plans for quantiles of IFRA distributions. Below, the similar bounds for the IRP/DRR CIF are introduced [30].

It is worth mentioning that all known bounds (including those discussed above) for the RP and GRP are based on some assumptions about the respective underlying distribution. The bounds, which are introduced below, are *nonparametric* and *are not* expressed in terms of underlying distributions. They are applicable to any counting process with an increasing/decreasing ROCOF.

Consider a counting process with strictly increasing CIF $W(t) = E(N(t))$, which is the mean number of events (failures) observed in the interval $(0, t]$. Let us define pth *quantile of the counting process* as the time τ_p at which $W(\tau_p) = p$, where $p > 0$. Note that, in contrast to the quantile of a r.v., the quantile of a counting process can be greater than 1. Below, we will consider a counting process for which ROCOF exists and is an increasing/decreasing function. In other words, the process is IRP/DRP.

Let us consider the IRP (DRP) counting process with pth quantile τ_p [i.e., $W(\tau_p) = p$]. Then,

$$\begin{aligned} W(t) &\le (\ge) p\frac{t}{\tau_p} \quad \text{for } 0 \le t < \tau_p \\ W(t) &\ge (\le) p\frac{t}{\tau_p} \quad \text{for } t \ge \tau_p. \end{aligned} \tag{5.39}$$

Inequalities Equation 5.39 are not difficult to prove. Consider the HPP with the same pth quantile as the IRP of interest (Figure 5.16). The CIF of the HPP is linear. According to the IRP definition, its CIF is concave upward. Both functions pass through the origin, and the CIF of the IRP crosses the CIF of the respective HPP at τ_p. Thus, the segment of the CIF of the HPP in the time interval $[0, \tau_p]$ is the chord joining the point (0, 0) to (τ_p, p) and lying above the graph of the CIF of IRP, and so Equation 5.39 holds. For DRP, the proof is similar. The example considered below is, to an extent, similar to the example given by Barlow and Proschan [28] for nonrepairable objects having IFR life distribution.

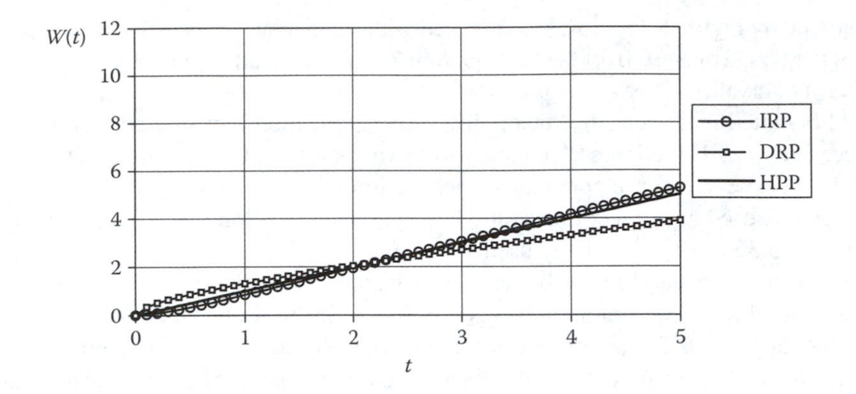

FIGURE 5.16 The CIFs for IRP (DRP), which is the RP with underlying Weibull distribution with the scale parameter equal to 1 and the shape parameter equal to 1.5 (0.5). The HPP line is the CIF for the HPP (the exponential underlying distribution with scale parameter equal to 1).

EXAMPLE 5.2

A contractor is required to produce a repairable system admitting, on average, five failures during a mission length of 10,000 h. Time to repair is considered negligible. Assuming the contractor just meets the requirement, what is a conservative prediction of the expected number of failures during the first 3000 h of the mission?

Solution: Using Equation 5.34, $t = 3000$ h, $p = 5$, and $\tau_p = 10{,}000$. Noting that $t < \tau_p$, and evaluating pt/τ_p, we find that $W(3000) \leq 1.5$. Thus, given the above-described assumptions, the contractor may conservatively claim that the system will require on average not more than two repairs during the first 3000 h of its mission.

5.1.7 Nonparametric Data Analysis

5.1.7.1 Estimation of the Cumulative Intensity Function

In this section, estimation of the CIF for a point process is discussed. Observations of a point process during some time interval are referred to as *realizations* of a given point process $\{N(t), t \geq 0\}$ [31]. Each realization can be related to its duration or observation time, which is the time during which the process was observed and a given realization was recorded.

5.1.7.2 Estimation Based on One Realization (Ungrouped Data)

Let us consider a realization of a point process observed during time interval $[0, T_{\text{obs}}]$. Suppose that a repairable system was put in operation at time $t = 0$. The observed successive failure times $t_1, t_2, \ldots, t_n$ constitute a realization of the corresponding failure process. It is obvious that $t_1 < t_2 < \cdots < t_n \leq T_{\text{obs}}$. Based on this realization, the so-called *natural estimate* of the respective CIF $\hat{W}(t)$ is given by

$$\hat{W}(t) = \text{Number of failures in } (0, t], \tag{5.40}$$

where $t \leq T_{\text{obs}}$. Similar to the empirical cumulative distribution function (ECDF) of a r.v., estimate Equation 5.40 is called the *empirical cumulative intensity function* (ECIF). Figure 5.17 illustrates a realization of a point process, the respective CIF $W(t)$ and ECIF $\hat{W}(t)$, based on the above formulae and the data given in Table 5.2.

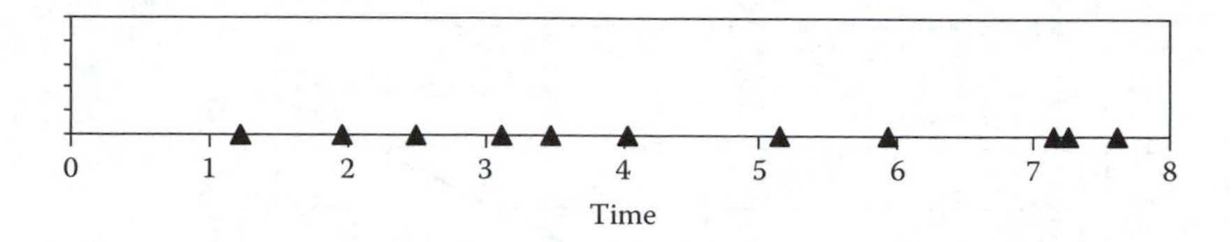

FIGURE 5.17 A realization of a point process consisting of 11 events (failures). Failure times are given in Table 5.2.

TABLE 5.2
Failure Times of a Realization of a Point Process and Respective ECIF

Order Number	Failure Times (in Arbitrary Units)	ECIF
1	0.51	1
2	0.60	2
3	1.33	3
4	2.19	4
5	2.98	5
6	3.06	6
7	3.61	7
8	4.12	8
9	4.77	9
10	5.71	10
11	7.43	11

The estimate of CIF Equation 5.40 is, to an extent, similar to the empirical (sample) cumulative distribution function (ECF), which is the commonly used estimate of the cdf of r.v.s, based on uncensored samples. Both of these are upward stepwise functions. This rather limited similarity is discussed later. At this point, we will note only that, as distinct from the CIF, the values of the ECF are always limited from above by unity.

A realization of a point process in which 11 failures were observed is illustrated by Figure 5.17. The respective failure times are displayed in Table 5.2. Figure 5.18 shows the corresponding ECIF $\hat{W}(t)$ based on this realization.

5.1.7.3 Estimation Based on One Realization (Grouped Data)

This case is considered using the example based on the same data given in Table 5.2 and illustrated in Figure 5.19. The figure shows the CIF and the ECIF for the grouped data* obtained using the intervals of the same length of 2. The figure illustrates the fact that grouping results in a less accurate estimation.

5.1.7.4 Estimation Based on Several Realizations

In the case of ungrouped data including several realizations, CIF is estimated using the so-called Nelson–Aalen procedure (also called Nelson's procedure), which provides an unbiased, nonparametric estimate of the CIF. The meaning of the term *nonparametric* is twofold in the given context. It is applicable to any point process, and in the framework of a given point process it is applicable to any underlying distribution.

---

*In many real problems, failure time data are often initially available in the grouped form.

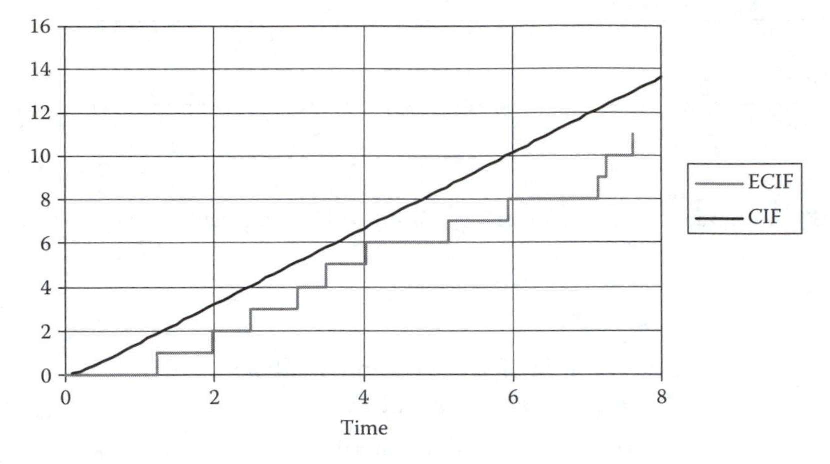

FIGURE 5.18 CIF (smooth) of a point process and ECIF (stepwise) based on one realization consisting of 11 events of the same process.

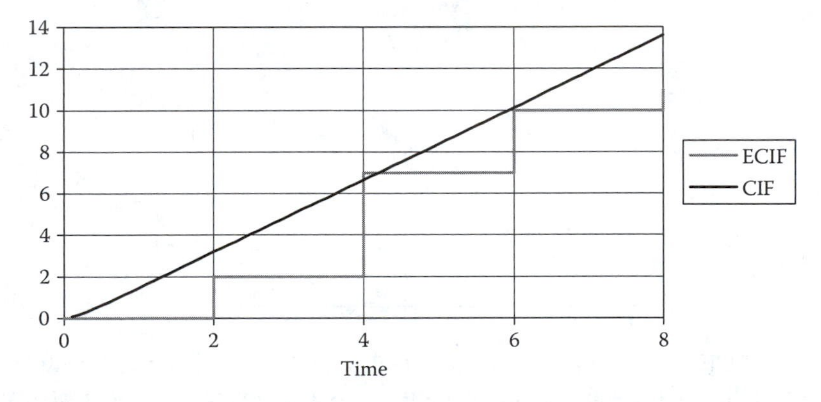

FIGURE 5.19 CIF (smooth) and ECIF (stepwise) based on the same as in Figure 5.18 realization, but the data are grouped.

Different formulae expressions exist for the estimate: for example, see [7,32–36]. Below, the simplest expression of the estimate [7] is given and is followed by a numerical example based on the algorithm given in [34].

The data are the mth realization of a point process. The ith realization was observed in the time interval $(a_i, b_i]$, where $i = 1, 2, \ldots, m$, and n_i is the number of failures that have occurred by time t_{ij}, where $j = 1, 2, \ldots, n_i$. The estimate $\hat{W}(t)$ is given by

$$\hat{W}(t) = \sum_{t_{ij} \leq t} \frac{1}{Y(t_{ij})}, \tag{5.41}$$

where $Y(t_{ij})$ denotes the number of trajectories that were observed immediately before time t_{ij}.

Example 5.3

Three realizations of a point process are given in Table 5.3. The realizations have different durations, so the most general case is considered. These realizations, pooled in one sample and sorted in ascending

TABLE 5.3
Failure Times of Three Realizations of a Point Process

Order Number	Failure Times		
	Realization 1	Realization 2	Realization 3
1	0.27	0.68	1.23
2	0.75	0.94	1.97
3	0.78	2.15	2.49
4	1.37	3.26	3.12
5	2.03	4.27	3.49
6		4.62	4.03
7		5.12	5.14
8		5.45	5.93
9			7.16
10			7.27
11			7.63

TABLE 5.4
Calculation of ECIF Based on Data Consisting of Three Realizations

Order Number (i)	Failure Times (t_i)	Number of Realizations (r)	Increment = $1/r$	ECIF = Sum of Increments	Comments
1	0.27	3	0.333	0.333	
2	0.68	3	0.333	0.667	
3	0.75	3	0.333	1.000	
4	0.78	3	0.333	1.333	
5	0.94	3	0.333	1.667	
6	1.23	3	0.333	2.000	
7	1.37	3	0.333	2.333	
8	1.97	3	0.333	2.667	
9	2.03	3	0.333	3.000	The last failure in Realization 1
10	2.15	2	0.500	3.500	
11	2.49	2	0.500	4.000	
12	3.12	2	0.500	4.500	
13	3.26	2	0.500	5.000	
14	3.49	2	0.500	5.500	
15	4.03	2	0.500	6.000	
16	4.27	2	0.500	6.500	
17	4.62	2	0.500	7.000	
18	5.12	2	0.500	7.500	
19	5.14	2	0.500	8.000	
20	5.45	2	0.500	8.500	The last failure in Realization 2
21	5.93	1	1.000	9.500	
22	7.16	1	1.000	10.500	
23	7.27	1	1.000	11.500	
24	7.63	1	1.000	12.500	The last failure in Realization 3

continued

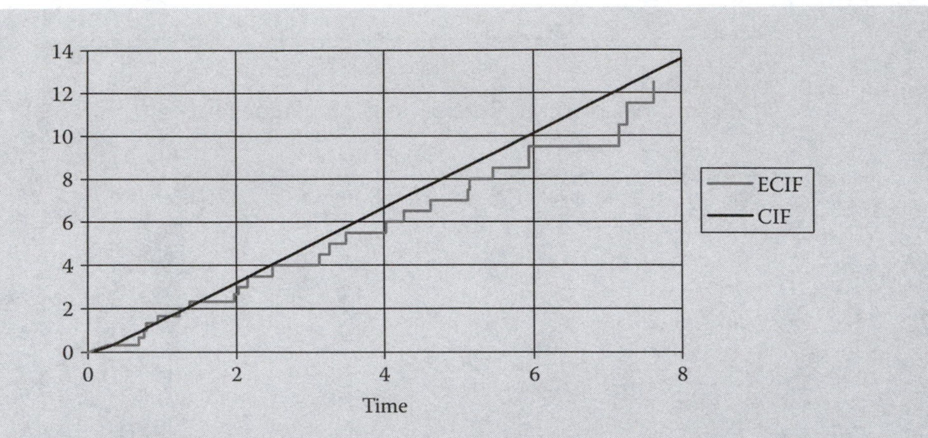

FIGURE 5.20 CIF (smooth) and ECIF (stepwise) based on three realizations given in Table 5.3.

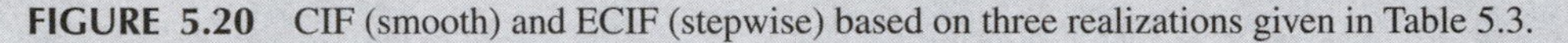

order, are given in Table 5.4, as well as some calculation details of the respective ECIF, which is also shown in Figure 5.20. Pay particular attention to the decreasing accuracy of the estimate along with the decreasing number of realizations participating in the calculation of $\hat{W}(t)$.

5.1.7.5 Confidence Limits for Cumulative Intensity Function

In this section, only the *point limits* on CIF are discussed. Our discussion is limited to the Poisson limits suggested by Nelson [34], which are simple and more robust for practical applications compared to the Nelson and naive approximate limits [34]. It should be noted that, unlike the Nelson and naive approximate limits, the lower Poisson limits are always positive.

The Poisson two-sided $(1-\alpha)100\%$ confidence limits for CIF $W(t_i)$ are given by the following equations:

$$W_u(t_i) = \frac{\chi^2_{1-\alpha/2}(2i+2)}{2r_i} \tag{5.42a}$$

for the upper limit and

$$W_l(t_i) = \frac{\chi^2_{\alpha/2}(2i)}{2r_i} \tag{5.42b}$$

for the lower limit, where t_i is the failure (event) time having order number i, and r_i is the number of trajectories observed at t_i. The number r_i is also called *number at risk* [34], and $\chi^2_\beta(n)$ is the quantile of level β of the χ^2 distribution with n degrees of freedom.

Consider a simple case of one realization $(r_i = 1)$ observed during time T_{obs} and consisting of the failure times $t_1 < t_2 < \cdots < t_{i-1} < t_i < t_{i+1} < \cdots < t_n \leq T_{\text{obs}}$. In this case, Equations 5.42a and 5.42b are reduced to the following:

$$W_u(t_i) = \frac{\chi^2_{1-\alpha/2}(2i+2)}{2} \tag{5.43a}$$

for the upper limit and

$$W_l(t_i) = \frac{\chi^2_{\alpha/2}(2i)}{2} \tag{5.43b}$$

for the lower limit. These equations are the upper and lower limits for unknown mean number of events of the Poisson distribution, based on i events observed in the interval $(0, t_i]$.

Example 5.4

Use the realization displayed in Table 5.2 to construct the 90% upper and lower confidence limits for the CIF. To illustrate the calculations, the 90% confidence limits are constructed for $t_5 = 2.98$. The upper limit W_u is

$$W_u(t_5) = \frac{\chi^2_{0.95}(10+2)}{2} = \frac{21.103}{2} = 10.55$$

and the lower limit W_l is

$$W_l(t_5) = \frac{\chi^2_{0.05}(10)}{2} = \frac{3.94}{2} = 1.97.$$

The respective confidence limits for the whole realization (displayed in Table 5.2) are depicted in Figure 5.21. According to Nelson [34], the Poisson limits Equation 5.42 are applicable for restrictions $i \leq 10$ and $r_i \geq 20$ (Equation 5.36).

The inequalities (Equation 5.43) impose a serious restriction on the use of the Poisson limits. The next example illustrates the situation in which these inequalities are violated.

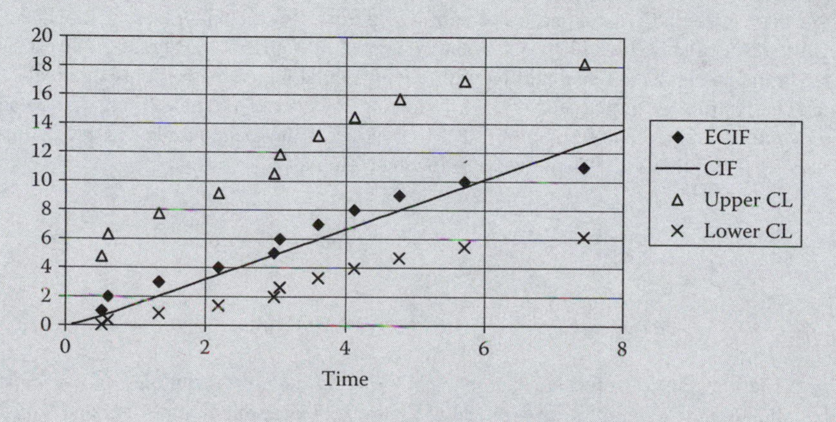

FIGURE 5.21 CIF of a point process and ECIF and 90% confidence limits based on one realization consisting of 11 events (failures) of the same process.

Example 5.5

Now, consider a case in which more than one realization is available during time T_{obs}. We will use the data from Table 5.3, which include 3 realizations ($r_{max} = 3$). To illustrate the calculations, the 90% confidence limits are constructed for the failure time $t_6 = 4.62$ from realization 2, which is also the failure time $t_{17} = 4.62$ in the pooled sample, as given in Table 5.4. So, according to Table 5.4, for this failure time, one obtains $i = 17$ and $r_{17} = 2$; that is, both inequalities (Equation 5.43) are violated. Applying Equation 5.40, one obtains

$$W_u(t_{17}) = \frac{\chi^2_{0.95}(34+2)}{4} = \frac{51.00}{4} = 12.75,$$

$$W_l(t_{17}) = \frac{\chi^2_{0.05}(34)}{4} = \frac{21.66}{4} = 5.42.$$

continued

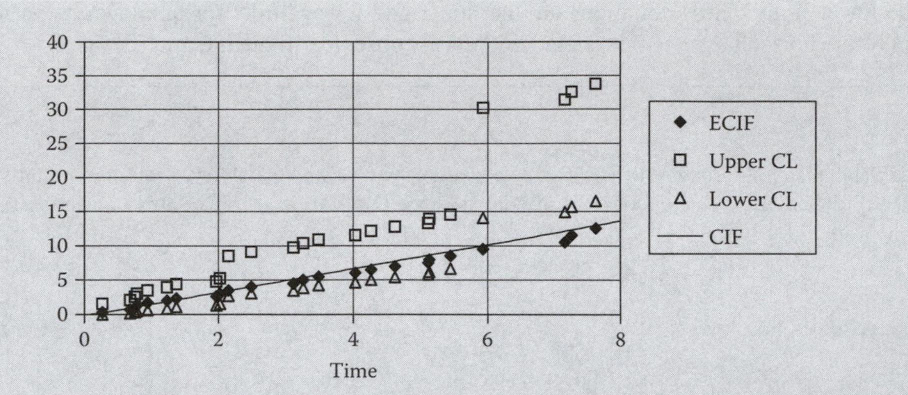

FIGURE 5.22 CIF, ECIF, and 90% upper and lower confidence limits (Equation 5.32) for data given in Tables 5.2 and 5.3.

Figure 5.22 depicts the 90% upper and lower confidence limits for the data given in Tables 5.3 and 5.4, as well as the respective ECIF, which was already depicted in Figure 5.20. Figure 5.22 reveals that the confidence limits do not make sense starting with $t_{21} = 5.93$, at which there is only one realization left.

To avoid the restrictions Equation 5.43, the following Poisson confidence limits can be suggested. These limits are based on the definition of CIF $W(t)$, which is the mean number of events observed in the interval $(0, t]$. The limits are given by the following formulas:

$$W_u(t_i) = \frac{\chi^2_{1-\alpha/2}(2W(t_i) + 2)}{2} \quad \text{for the upper limit,} \tag{5.44a}$$

$$W_l(t_i) = \frac{\chi^2_{\alpha/2}(2W(t_i))}{2} \quad \text{for the lower limit.} \tag{5.44b}$$

The only restriction related to the lower limit is $W(t_i) > 1/2$. This restriction is not very strong for practical applications. These limits are depicted in Figure 5.23 as applied to the data given in Tables 5.2 and 5.3. Note the different axis scales of Figures 5.22 and 5.23.

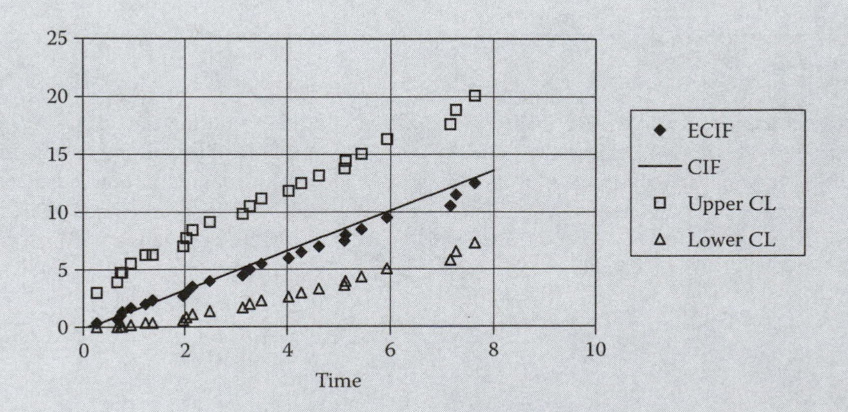

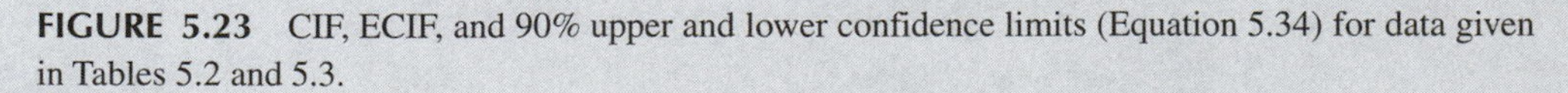

FIGURE 5.23 CIF, ECIF, and 90% upper and lower confidence limits (Equation 5.34) for data given in Tables 5.2 and 5.3.

5.1.8 Data Analysis for the HPP

5.1.8.1 Procedures Based on the Poisson Distribution

Suppose that a failure process is observed for a predetermined time t_0 during which n failures have been recorded at times $t_1 < t_2 \cdots < t_n$, where $t_n < t_0$. The process is assumed to follow an HPP. The corresponding likelihood function can be written as

$$L = \lambda^n e^{-\lambda t_0}. \tag{5.45}$$

It is clear that, with t_0 fixed, the number of events, n, is a *sufficient statistic* (note that one does not need to know $t_1, t_2, \ldots, t_n$ to construct our likelihood function). Thus, the statistical inference can be based on the Poisson distribution of the number of events. As a point estimate of λ, one usually takes n/t_0, which is the unique unbiased estimate based on the sufficient statistic.

A typical problem associated with repairable systems, in which the failure behavior follows the HPP, is to test for the null hypothesis $\lambda = \lambda_0$ (or the mean number of events, $\mu = \mu_0 = \lambda_0 t_0$) against the alternative $\lambda > \lambda_0 (\mu > \mu_0)$. The alternative hypothesis has the exact level of significance P_+ corresponding to the observed number of failures n, given by Cox and Lewis [11].

$$P_+(n, \mu_0) = \Pr(N \geq n | \mu = \mu_0) = \sum_{r=n}^{\infty} \frac{\mu_0^r e^{-\mu_0}}{r!}. \tag{5.46}$$

For the alternatives $\lambda < \lambda_0 (\mu < \mu_0)$, the exact level of significance corresponding to an observed value n is given by

$$P_-(n, \mu_0) = \Pr(N \leq n | \mu = \mu_0). \tag{5.47}$$

If the two-sided alternatives are considered, the level of significance is defined as

$$P(n, \mu_0) = 2 \min[P_+(n, \mu_0),\ P_-(n, \mu_0)]. \tag{5.48}$$

If the normal approximation to the Poisson distribution is used (see Section 2.3.2), the corresponding statistic having the standard normal distribution is

$$\frac{|n - \mu_0| - 0.5}{\sqrt{\mu_0}}, \tag{5.49}$$

where 0.5 is a correction term.

Example 5.6

Twelve failures of a new repairable unit were observed during a 3-year period. From the past experience, it is known that for similar units, the ROCOF, λ_0, is 3.33 per year. Check the hypothesis that the ROCOF of the new unit λ is equal to λ_0.

Solution: Choose 5% significance level. Using Table A.1, find the respective acceptance region for statistic Equation 5.49 as interval $(-1.96, 1.96)$. Keeping in mind that $\mu_0 = \lambda_0 t = 3.33(3) = 10$, calculate statistic Equation 5.49

$$\frac{|12 - 10| - 0.5}{\sqrt{10}} \approx 0.47,$$

which is inside the acceptance region. Thus, the hypothesis that the ROCOF of the new unit is equal to the rate of similar units, λ_0, is not rejected.

Another typical problem associated with repairable systems whose failure behavior can be modeled by the HPP is the comparison of two HPPs. Such problems can appear, for example, when two identical units are operated in different plants or by different personnel, and one is interested in the corresponding ROCOF comparison.

For instance, assume that our data are the observations of two independent HPPs and the goal is to compare the corresponding rates of occurrence of failure, λ_1 and λ_2. Let the data collected be the numbers of failures n_1 and n_2, observed in nonrandom time intervals T_1 and T_2, respectively. The random numbers of events n_1 and n_2 are observed values of independent r.v.s with Poisson distributions having the means $\mu_1 = \lambda_1 T_1$ and $\mu_2 = \lambda_2 T_2$, so that we can write

$$\Pr(N_1 = n_1, N_2 = n_2) = \frac{\exp(-\mu_1)\mu_1^{n_1}}{n_1!}\frac{\exp(-\mu_2)\mu_2^{n_2}}{n_2!}. \tag{5.50}$$

To compare the ROCOFs for the processes considered, one may use the following statistic [11]:

$$\rho = \frac{\mu_2}{\mu_1} = \frac{T_2\lambda_2}{T_1\lambda_1}. \tag{5.51}$$

Since the nonrandom time intervals T_1 and T_2 are known, the inference about ρ is identical to the inference about the ratio λ_2/λ_1. The inference about ρ can be calculated based on the conditional distribution of N_2 (or N_1), given $N_2 + N_1 = n_2 + n_1$. This probability can be written as

$$\begin{aligned}\Pr(N_2 = n_2 | N_1 + N_2 = n_1 + n_2) &= \frac{\Pr(N_1 = n_1, N_2 = n_2)}{\Pr(N_1 + N_2 = n_1 + n_2)} \\ &= \frac{\mu_1^{n_1}\mu_2^{n_2}/n_1!n_2!\exp[-(\mu_1+\mu_2)]}{(\mu_1+\mu_2)^{n_1+n_2}/(n_1+n_2)!\exp[-(\mu_1+\mu_2)]} \\ &= \binom{n_1+n_2}{n_1}\theta^{n_2}(1-\theta)^{n_1},\end{aligned} \tag{5.52}$$

where $\theta = \rho/(1+\rho)$.

In the case when $\lambda_1 = \lambda_2$, the probability Equation 5.52 is binomial with parameter $T_1/(T_1 + T_2)$, and this parameter is 0.5 in an important particular case of equal time intervals. Thus, the same procedures for the binomial distribution or its normal approximations can be used for making inferences about ρ.

EXAMPLE 5.7

In nuclear power plants, *Accident Sequence Precursors* are defined as "those operational events which constitute important elements of accident sequences leading to severe core damage." The occurrence of precursors is assumed to follow an HPP. There were 32 events observed in 1984 and 39 in 1993. Test the hypothesis that the rate of occurrence of events (per year) is the same for the years given.

Solution: For the data given, $n_1 = 32$ and $n_2 = 39$. Because $T_1 = T_2 = 1$ year, our null hypothesis is $H_0 : \rho_0 = 1$, so that $\theta_0 = 0.5$. Using the normal approximation (similar to Equation 5.49), calculate the following statistic:

$$\frac{|n_2 - n\theta_0| - 0.5}{\sqrt{n\theta_0(1-\theta_0)}},$$

where $n = n_1 + n_2$. Thus, one obtains

$$\frac{|39 - 71/2| - 0.5}{\sqrt{71 \times 0.5 \times 0.5}} \approx 0.71,$$

which is inside the acceptance region for any reasonable significance level, α. In other words, the data do not show any significant change in the rate of precursor occurrence (i.e., H_0 is not rejected).

5.1.8.2 Procedures Based on the Exponential Distribution of Time Intervals

In Section 3.2.1, it was shown that under the HPP model, the intervals between successive failures have the exponential distribution. Therefore, data analysis procedures for the exponential distribution considered in Chapter 3 (classical as well as Bayesian) can be used. Some special techniques applicable to the HPP are considered in the next section, where the data analysis for the HPP is treated as a particular case of the data analysis for the NHPP.

Assume again that the failure data are the observations from two independent HPPs and our goal is to compare the corresponding ROCOFs, λ_1 and λ_2.

Let t_1 and t_2 be the times at which predetermined (nonrandom) numbers of failures, n_1 and n_2, occur for the corresponding processes. It is clear that t_1 and t_2 can be considered as realizations (observed values) of independent r.v.s, T_1 and T_2, for which the quantity $2\lambda T$ has the χ^2 distribution with $2n$ degrees of freedom (see Section 3.4.2). We can introduce statistic

$$R = \frac{(2\lambda_2 T_2/2n_2)}{(2\lambda_1 T_1/2n_1)}, \tag{5.53}$$

which follows the F distribution with $(2n_2, 2n_1)$ degrees of freedom [11]. Based on this statistic, the confidence intervals for the ratio λ_2/λ_1 can be written as

$$\begin{gathered} \Pr(F_{1-\alpha/2} < R < F_{\alpha/2}) = 1 - \alpha, \\ \Pr\left(F_{1-\alpha/2}\frac{t_1 n_2}{t_2 n_1} < \frac{\lambda_2}{\lambda_1} < F_{\alpha/2}\frac{t_1 n_2}{t_2 n_1}\right) = 1 - \alpha, \end{gathered} \tag{5.54}$$

where F_α is the upper α quantile of the F distribution with $(2n_2, 2n_1)$ degrees of freedom. Substituting the observed values, t_1 and t_2, one obtains the confidence interval corresponding to the confidence probability $1 - \alpha$ as

$$F_{1-\alpha}\frac{t_1 n_2}{t_2 n_1} < \frac{\lambda_2}{\lambda_1} < F_\alpha \frac{t_1 n_2}{t_2 n_1}. \tag{5.55}$$

The corresponding null hypothesis that $\lambda_2/\lambda_1 = r_0$ can be tested using the two-tailed test for the statistic

$$\frac{r_0}{(n_2/t_2)/(n_1/t_1)}, \tag{5.56}$$

having under H_0 the F distribution with $(2n_2, 2n_1)$ degrees of freedom (see Table A.5).

EXAMPLE 5.8

The failure data on two identical items used at two different sites were collected. At the first site, observations continued until the eighth failure, which was observed at 1880 h. At the second site, observations continued until the 12th failure, which was observed at 1654 h. Assuming that the time-to-failure distributions of both items are exponential, check whether the items are identical from a reliability standpoint. That is, test the null hypothesis, $H_0 : \lambda_1 = \lambda_2$.

Solution: Calculate statistic (Equation 5.56) for $r_0 = 1$:

$$\frac{1}{(12/1654)/(8/1880)} = 0.586.$$

Using the 10% confidence level and Table A.5, find the acceptance region as (0.48, 2.24). So, our null hypothesis is not rejected.

5.1.9 DATA ANALYSIS FOR THE NHPP

The NHPP can be used to model improving and deteriorating systems: if the intensity function (ROCOF) is decreasing, the system is improving, and if the intensity function is increasing, the system is deteriorating. ROCOF trend analysis is of great importance simply because the HPP, with the memoryless property of its exponential time-to-failure distribution, is useless for making decisions about preventive actions.

Formally, we can test for trend by taking the null hypothesis of no trend, that is, that the events form the HPP, and applying a goodness-of-fit test for the exponential distribution of the intervals between successive failures and the Poisson distribution of the number of failures in the time intervals of constant (nonrandom) length. A simple graphical procedure based on this property is to plot the cumulative number of failures versus the cumulative time. Deviation from linearity indicates the presence of a trend.

However, this test is less sensitive than the NHPP alternatives, so it is better to apply the following methods [11].

5.1.9.1 Regression Analysis of Time Intervals

Suppose one has a reasonably long series of failures and the problem is to examine any gradual trend in the rate of failure occurrence. Choose an integer, l, which is recommended to be no less than 4, but such that no appreciable change in ROCOF arises during the interval of occurrence of l failures. Let t_1 be the observed time from the start to the lth failure, t_2 be the time from the lth failure to the $2l$th failure, and so on. This results in a series of intervals $t_1, t_2, \ldots, t_r$. Using the HPP, one can write

$$\begin{aligned} E(\ln t_i) &= -\ln \lambda_i + c_l, \\ \mathrm{var}(\ln t_l) &= \nu_l, \end{aligned} \tag{5.57}$$

where c_l and ν_l are known constants independent of λ. For example,

$$\nu_l = \frac{1}{l - 0.5}, \tag{5.58}$$

and $t_i(i = 1, 2, \Lambda$ are independently distributed). In comparison, assume that the observations are generated by a process satisfying all the conditions for an HPP, except that the ROCOF λ is *slowly* varying with time. Consider the approximation that λ is a constant, λ_i, within the period covered by t_i, and that an independent variable z_i can be attached to each t_i such that in the case of simplest model

$$\ln \lambda_i = \alpha + \beta z_i. \tag{5.59}$$

For example, z_i might be

- the midpoint of the interval t_i , if λ is being considered as a function of time, t;
- the value of any constant independent variable, or any constant independent variable averaged over the interval t_i, which could be responsible for ROCOF variation.

Under the above assumptions, we obtain the following linear regression model:

$$\begin{aligned} E(\ln t_i) &= -(\alpha' + \beta z_i), \\ \text{var}(\ln t_i) &= \nu_l, \end{aligned} \tag{5.60}$$

where $\alpha' = \alpha - c_l$ and β are unknown parameters and ν_l is a *known* constant. Using the standard regression procedures (see Section 2.8), one can

- obtain the standard least-squares estimates of parameters α' and β;
- test approximately the null hypothesis $\beta = 0$ and obtain approximate confidence limits for β;
- compare the residual variance with the respective theoretical value, ν_l, to check the adequacy of the model.

One can include in the model considered above additional independent variables. For example, we can generalize model 5.59 to a log-linear polynomial model,

$$\log \lambda_i = \alpha + \beta z_i + \gamma z_i^2 + \cdots$$

Another regression approach, performed in terms of counts of failures observed in successive equal time intervals, is considered in Cox and Lewis [11]. The regression procedures considered can also be performed in the framework of the Bayesian approach to regression, given, for example, in [37]. The maximum likelihood estimation for model 5.59 is considered by Lawless [38], who also applied this model to failure data on a set of similar air conditioning units.

EXAMPLE 5.9

Consider the following data in the form of successive times between failures of a repairable item. Let t_1 be the observed time from the start to the fourth failure, t_2 be the time from the fourth failure to the eighth failure, and so on, and let z_i be the time at the center of the interval t_i. Using the data listed below, fit the simple linear regression model 5.59 and determine whether there is any trend in the ROCOF.

Interval Number (i)	$\ln t_i$	z_i (in Relative Units)
1	0.151	0.581
2	0.157	1.748
3	0.275	2.991
4	−0.445	3.970
5	−0.983	4.478
6	−0.703	4.913

continued

Solution: Rewrite Equations 5.60 in the form

$$E(\ln t_i) = -(\alpha' + \beta z_i) = \gamma_0 + \gamma z,$$
$$\text{var}(\ln t_i) = \nu_4$$

where $\alpha' = \alpha - c_4$, and c_4 and ν_4 are given by Equation 5.3. That is,

$$c_4 \approx \ln 4 - \frac{1}{2 \cdot 4 - 1/3 + 1/16 \cdot 4} = 1.256,$$

$$\nu_4 \approx \frac{1}{4 - 0.5 + 1/10 \cdot 4} = 0.284.$$

Meanwhile, α and β are unknown parameters to be estimated. Using the standard least-squares estimates (Equation 2.139) for γ_0 and γ based on the data, one obtains

$$\hat{\gamma}_0 = 0.540, \quad \hat{\gamma} = -0.256$$

Therefore,

$$\alpha = \alpha' + c_4 = -0.540 + 1.256 = 0.716.$$
$$\beta = 0.256.$$

Finally,

$$\lambda(t) = \text{Exp}(0.761 + 0.256t).$$

To check the adequacy of the ROCOF model obtained, we need to check the hypothesis that the theoretical variance $\nu_4 = 0.284$ (having a number of degrees-of-freedom) is not less than the residual variance, which can be calculated using Equation 2.141. The value of the residual variance is 0.114 and has $6 - 2 = 4$ degrees of freedom. Using the significance level of 5% and the respective critical value from Table A.5, conclude that our hypothesis is not rejected, so the model obtained is adequate.

5.1.9.2 Maximum Likelihood Procedures

Under the NHPP model, the intervals between successive events are independently distributed and the probability that, starting from time t_i, the next failure will occur at $(t_{i+1}, t_{i+1} + \Delta t)$ can be approximated by Cox and Lewis [11]:

$$\lambda(t_{i+1})\Delta t \exp\left(-\int_{t_i}^{t_{i+1}} \lambda(x)\,\mathrm{d}x\right), \tag{5.61}$$

where the first multiplier is the probability of failure in $(t_{i+1}, t_{i+1} + \Delta t)$ and the second is the probability of a failure-free operation in the interval (t_i, t_{i+1}).

If the data are the successive failure times $t_1, t_2, \ldots, t_n$, $(t_1 < t_2 < \cdots < t_n)$ observed in the interval $(0, t_0)$, $t_0 > t_n$ (the data are Type I censored), the likelihood function for any $\lambda(t)$ dependence may be written as

$$\prod_{i=1}^{n} \lambda(t_i) \mathrm{e}^{-\int_0^{t_1} \lambda(x)\,\mathrm{d}x} \mathrm{e}^{-\int_{t_1}^{t_2} \lambda(x)\,\mathrm{d}x} \cdots \mathrm{e}^{-\int_{t_{n-1}}^{t_n} \lambda(x)\,\mathrm{d}x} \mathrm{e}^{-\int_{t_n}^{t_0} \lambda(x)\,\mathrm{d}x} = \prod_{i=1}^{n} \lambda(t_i) \mathrm{e}^{-\int_0^{t_n} \lambda(x)\,\mathrm{d}x} \mathrm{e}^{-\int_{t_n}^{t_0} \lambda(x)\mathrm{d}x}. \tag{5.62}$$

The corresponding log-likelihood function is given by

$$l = \sum_{i=1} n \ln \lambda(t_i) - \int_0^{t_0} \lambda(x)\,\mathrm{d}x. \tag{5.63}$$

To avoid complicated notation, consider the case when the ROCOF takes the simple form similar to Equation 5.59. That is,

$$\lambda(t) = \mathrm{e}^{\alpha+\beta t}. \tag{5.64}$$

Note that the model above is, in some sense, more general than the linear one, $\lambda(t) = \alpha + \beta t$, which can be considered as a particular case of Equation 5.64, when $\beta t \ll 1$.

Plugging Equation 5.64 into Equations 5.62 and 5.63, one obtains

$$L_1(\alpha, \beta) = \exp\left[n\alpha + \beta \sum_{i=1}^{n} t_i - \frac{\mathrm{e}^{\alpha}(\mathrm{e}^{\beta t_0} - 1)}{\beta} \right], \tag{5.65}$$

$$l = \ln[L_1(\alpha, \beta)] = n\alpha + \beta \sum_{i=1}^{n} t_i - \frac{\mathrm{e}^{\alpha}(\mathrm{e}^{\beta t_0} - 1)}{\beta}. \tag{5.66}$$

The conditional likelihood function can be found by dividing Equation 5.65 by the marginal probability of observing n failures, which is given by the respective term of the Poisson distribution with mean

$$\int_0^{t_0} \lambda(x)\,\mathrm{d}x = \frac{\mathrm{e}^{\alpha}(\mathrm{e}^{\beta t_0} - 1)}{\beta}. \tag{5.67}$$

The conditional likelihood function is given by Cox and Lewis [11]:

$$L_c = \frac{\exp\left(n\alpha + \beta \sum_{i=1}^{n} t_i\right) \exp(-(\mathrm{e}^{\alpha}\mathrm{e}^{\beta t_0} - 1)/\beta)}{(\mathrm{e}^{\alpha}\mathrm{e}^{\beta t_0} - 1)^n/\beta^n n! \exp(-(\mathrm{e}^{\alpha}\mathrm{e}^{\beta t_0} - 1)/\beta)} = \frac{\beta^n n!}{(\mathrm{e}^{\beta t_0} - 1/\mathrm{e}^{a})^n} \exp\left(\beta \sum_{i=1}^{n} t_i\right). \tag{5.68}$$

Because $0 < t_1 < t_2 < \cdots < t_n < t_0$, the conditional likelihood function Equation 5.68 is the pdf of an ordered sample of size n from the truncated exponential distribution having the pdf

$$f(t) = \frac{\beta}{\mathrm{e}^{\beta t_0} - 1}\mathrm{e}^{\beta t}, \quad 0 \le t \le t_0, \beta \neq 0. \tag{5.69}$$

Thus, for any β the conditional pdf of Σt_i is the same as for the sum of n independent r.v.s having the pdf (Equation 5.69).

EXAMPLE 5.10

In a repairable system, eight failures have been observed at 595, 905, 1100, 1250, 1405, 1595, 1850, and 1995 h. Assume the observation ends at the time when the last failure is observed, and that the time to repair is negligible. Test whether these data exhibit a trend in a form of Equation 5.64.

continued

Solution: Taking the derivative of Equation 5.69 with respect to β and the derivative of Equation 5.69 with respect to α and equating them to zero results in the following system of equations for the MLEs of these parameters

$$\sum_{i=1}^{n} t_i + \frac{n}{\beta} - \frac{nt_n}{1 - e^{-\beta t_n}} = 0,$$

$$e^{\alpha} = \frac{n\beta}{e^{-\beta t_n} - 1}.$$

For the data given, $n = 8$, $t_n = 1995$ h, and $\Sigma t_i = 10,695$ h. Solving these equations numerically, one obtains the following trend model:

$$\lambda(t) = \exp(-6.8134 + 0.0011t).$$

5.1.9.3 Laplace's Test

Next we will use the conditional pdf (Equation 5.58) to test the null hypothesis, H_0: $\beta = 0$, against the alternative hypothesis H_1: $\beta \neq 0$. This test is known as the Laplace test (sometimes called the Centroid test). As was mentioned above, under the condition of $\beta = 0$, pdf (Equation 5.58) is reduced to the uniform distribution over $(0, t_0)$, and $S = \Sigma t_i$ has the distribution of the sum of n independent, uniformly distributed r.v.s. Thus, one can use the distribution of the following statistic:

$$U = \frac{S - (nt_0/2)}{t_0\sqrt{(n/12)}} = \frac{\left(\sum_{i=1}^{n} t_i/n\right) - (t_0/2)}{t_0\sqrt{(1/12n)}}, \tag{5.70}$$

which has approximately the standard normal distribution [11].

If the alternative hypothesis is H_1: $\beta \neq 0$, then the large values of $|U|$ signify evidence against the null hypothesis. If the alternative hypothesis is $H_1 : \beta > (<) 0$, then the large values of $U(-U)$ also provide evidence against the null hypothesis. In other words, if U is close to 0, there is no evidence of trend in the data, and the process is assumed to be stationary (i.e., an HPP). If $U < 0$, the trend is decreasing; that is, the intervals between successive failures (interarrival values) are becoming larger. If $U > 0$, the trend is increasing. For the latter two situations, the process is not stationary (i.e., it is an NHPP).

If the data are failure terminated (Type II censored), statistic (Equation 5.70) is replaced by

$$U = \frac{\left(\sum_{i=1}^{n-1} t_i/n - 1\right) - (t_n/2)}{t_n\sqrt{(1/12(n-1))}}. \tag{5.71}$$

Example 5.11

Consider the failure arrival data for a motor-operated rotovalve in a process system. This valve is normally in standby mode and is demanded when overheating occurs in the process. The only major failure mode is "failure to start upon demand." The interarrivals of this failure mode are shown in the table below (in months). Determine whether an IFR is justified. Assume that a total of 5256 demands occurred during this time. Further assume a constant rate of occurrence for demands.

Failure Order Number	Time (days)	Failure Order Number	Time (days)
1	104	16	140
2	131	17	1
3	1597	18	102
4	59	19	3
5	4	20	393
6	503	21	96
7	157	22	232
8	6	23	89
9	118	24	61
10	173	25	37
11	114	26	293
12	62	27	7
13	101	28	165
14	216	29	87
15	106	30	99

Solution: Distribute the total number of demands (5256) over the period of observation. Also, calculate the interarrival of demands (the number of demands between two successive failures), and the demand arrival. These values are shown in Table 5.5. Since the observation ends at the last failure, the following results are obtained using Equation 5.71:

$$\sum t_i = 95898, \quad n-1=29, \quad \sum_{i=1}^{n-1} t_i/(n-1) = 3307, \quad \frac{t_n}{2} = 2628, \quad U = \frac{3307-2628}{5256\sqrt{1/12(29)}} = 2.41.$$

To test the null hypothesis that there is no trend in the data, and the ROCOF, λ, of rotovalves is constant, we would use Table A.1 with $U = 2.41$. Therefore, we can reject the null hypothesis at the 5% significance level; the respective acceptance region is $(-1.96, +1.96)$.

TABLE 5.5
Arrival and Interarrival for the Rotovalve Example

Interarrival Demand (Days)	Arrival Demand Demand (Days)
104	104
131	235
1597	1832
59	1891
4	1895
503	2398
157	2555
6	2561
118	2679
173	2852
114	2966
62	3028
101	3129
216	3345

continued

TABLE 5.5 (continued)

Interarrival Demand (Days)	Arrival Demand Demand (Days)
106	3451
140	3591
1	3592
102	3694
3	3697
393	4090
96	4186
232	4418
89	4507
61	4568
37	4605
293	4898
7	4905
165	5070
87	5157
99	5256

The existence of a trend in the data in Example 5.11 indicates that the interarrivals of rotovalve failures are not IID r.v.s, and thus the stationary process for evaluating reliability of rotovalves is incorrect. Rather, these interarrival times can be described in terms of the NHPP.

Another form of $\lambda(t)$ considered by Bassin [39,40] and Crow [8] is

$$\lambda(t) = \lambda\beta t^{\beta-1}. \tag{5.72}$$

Expression 5.61 has the same form as the failure (hazard) rate of nonrepairable items (3.18) for the Weibull distribution. The reliability function of a repairable system having ROCOF (Equation 5.30) for an interval $(t, t+t_1)$ can be obtained as follows:

$$R(t+t_1|t)\mathrm{e}^{-[\lambda(t+t_1)^{\beta}-\lambda t^{\beta}]}. \tag{5.73}$$

Crow [8] has shown that under the condition of a single system observed to its nth failure, the maximum likelihood estimator of β and λ can be obtained as

$$\hat{\beta} = \frac{n}{\sum_{i=1}^{n-1}\ln(t_n/t_i)}, \tag{5.74}$$

$$\hat{\lambda} = \frac{n}{t_n^{\hat{\beta}}}. \tag{5.75}$$

The $1-\alpha$ confidence limits for inferences on β and λ have been developed and discussed by Bain [41].

Example 5.12

Using the information in Example 5.11, calculate the maximum likelihood estimator of β and λ. Also, plot the demand failure rate.

Solution: Using Equations 5.74 and 5.75, we can calculate $\hat{\beta}$ and $\hat{\lambda}$ as 1.59 and 3.71×10^{-5}, respectively. Using $\hat{\beta}$ and $\hat{\lambda}$, the functional form of the demand failure rate can be obtained by using Equation 5.72 as

$$\lambda(d) = 3.71 \times 10^{-5} \times 1.59 d^{0.59},$$

where d represents the demand number (time in days).

The plot of the demand failure rate (ROCOF of NHPP) as a function of calendar time for the rotovalve is shown in Figure 5.24. For comparison purposes, the constant demand failure rate function (HPP case) is also shown. For the HPP, the point estimate of λ was obtained by dividing the number of failures by the number of demands. The upper and lower confidence intervals were obtained using the HPP assumption.

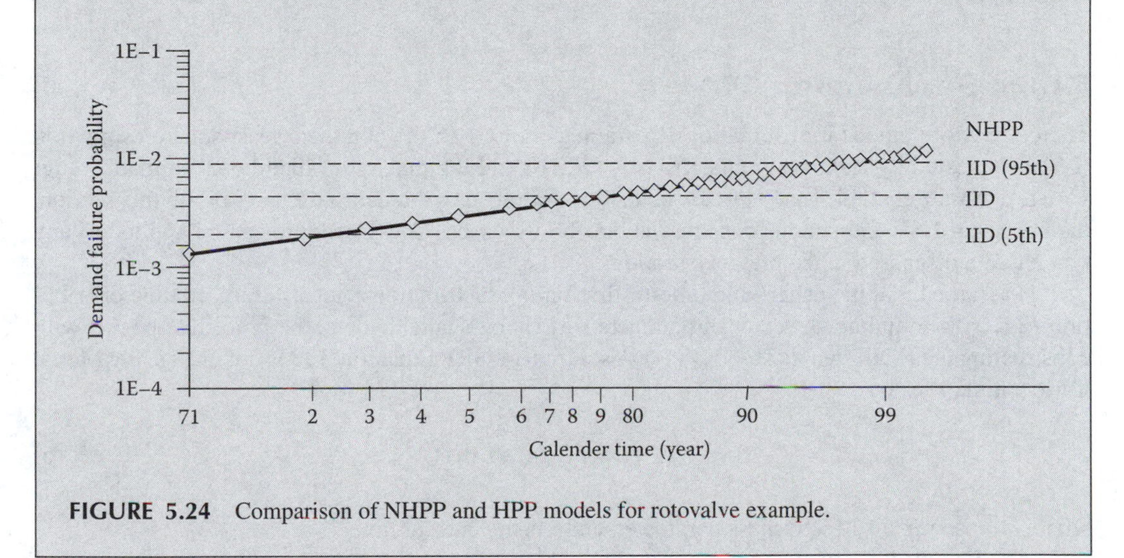

FIGURE 5.24 Comparison of NHPP and HPP models for rotovalve example.

Example 5.13

In a repairable system, the following six interarrival times between failures were observed: 16, 32, 49, 60, 78, and 182 (in hours). Assume the observation ends at the time when the last failure is observed.

a. Test whether these data exhibit a trend. If so, estimate the trend model parameters as given in Equation 5.72.
b. Find the probability that the interarrival time for the seventh failure will be greater than 200 h.

Solution: Use Laplace's test to test the null hypothesis that there is no trend in the data at the 10% significance level. The respective acceptance region is (−1.645, +1.645). From Equation 5.71, find

$$U = \frac{[16 + (16 + 32) + \cdots]/5 - (417/2)}{417\sqrt{1/12(5)}} = -1.82.$$

Notice that $t_n = 417$. The value of U obtained indicates that the NHPP can be applicable (H_0 is rejected), and the sign of U shows that the trend is decreasing.

continued

Using Equations 5.74 and 5.75, we can find

$$\hat{\beta} = \frac{6}{\ln\left(\frac{417}{16}\right) + \ln\left(\frac{417}{16+32}\right) + \cdots} = 0.712.$$

$$\hat{\lambda} = \frac{6}{(417)^{0.712}} = 0.0817\text{h}^{-1}.$$

Thus, $\hat{\lambda} = 0.058t^{-0.288}$. From Equation 5.73, with $t_1 = 200$,
$\Pr(\text{seventh failure occurs within } 200\,\text{h}) = 1 - \exp[-\lambda(t_0 + t_1)^{\beta} - \lambda(t_0)^{\beta}] = 0.85$.
The probability that the interarrival time is greater than 200 h is $1 - 0.85 = 0.15$.

Crow [42] has expanded estimates 5.74 and 5.75 to include situations where data originate from multiunit repairable systems.

5.1.10 Data Analysis for GRP

There are two methods for estimation of parameters of GRP: the nonlinear least-square estimation (LSE) suggested by Kaminskiy and Krivtsov [22,26] and the maximum-likelihood estimation suggested by Yanez et al. [43] and Hurtado et al. [44], followed by Mettas and Zhao [45]. In this section, the nonlinear LSE estimation procedure is briefly discussed. The procedure can be applied to any type of data discussed in the previous section.

It is assumed that the underlying time-to-first-failure distribution is an arbitrary lifetime distribution $F(\tau; \alpha)$ with unknown vector of parameters α. Using available data, the procedure begins with constructing the ECIF, denoted by $W_{\text{emp}}(t)$. A solution of GRP Equation 5.72 is obtained using Monte Carlo simulations by

$$W_{\text{mc}}(t) = f(F(\tau; \alpha), q, t), \tag{5.76}$$

where q is the repair-effectiveness parameter, and t is the current time.

The LSE of the GRP parameters α and q can be obtained as a solution of the following optimization problem:

$$\min_{\alpha, q}\left(\sum_{i=1}^{n}(W_{\text{emp}}(t_i) - W_{\text{mc}}(F(\tau; \alpha), q, t_i))^2\right), \tag{5.77}$$

where n is the number of points at which $W_{\text{emp}}(t)$ makes a step up.

5.1.10.1 Example Based on Simulated Data

The ECIF, $W_{\text{emp}}(t)$, was obtained by simulating a GRP with a Weibull distributed TTFF (shape parameter $\beta = 1.5$ and scale parameter $\alpha = 1$) and the GRP rejuvenation parameter $q = 0.5$ using $N_0 = 100$ realizations over the observation interval, $T = 50$.

Estimates of α, β, and q were obtained based on $n_0 = 1000$ realizations of GRP as follows: $\hat{\beta} = 1.48, \hat{\theta} = 1.00, \hat{q} = 0.49$. Tables 5.6 and 5.7, respectively, show the sample correlation and covariance matrices for the obtained estimates of GRP parameters for 30 simulated ECIFs, $W_{\text{emp}}(t)$.

With the ECIF simulated for $N_0 = 100{,}000$ realizations (which is a more typical case for a company concerned with mass production) and using $n_0 = 1000{,}000$, the estimation procedure returns the original GRP parameters with close to zero variance.

TABLE 5.6
Sample Correlation Matrix

	β	α	q
β	1.000		
α	0.702	1.000	
q	0.079	−0.523	1.000

TABLE 5.7
Sample Covariance Matrix

	β	α	q
β	2.6×10^{-3}		
α	1.9×10^{-3}	2.8×10^{-3}	
q	2.3×10^{-4}	-1.6×10^{-3}	3.2×10^{-3}

TABLE 5.8
Example of Warranty Data for a Repairable System

MIC (t)	3	6	9	12	15	18	21	24	27
Cumulative number of failures per system ($H_{emp}(t)$)	0.03	0.09	0.14	0.24	0.38	0.54	0.70	0.90	1.17

Note: Population Size, $N_0 = 100{,}000$.

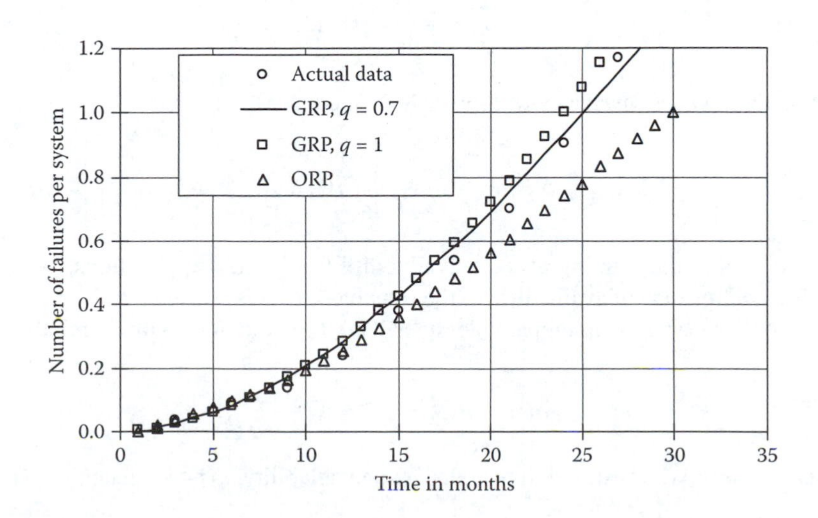

FIGURE 5.25 A least square fit of G-renewal function.

5.1.10.2 Example Based on Real Data

The warranty data collected on a system during its first 18 months (Table 5.8) were used to estimate GRP parameters. The Weibull distribution with the shape parameter β and the scale parameter α was assumed as the underlying TTFF distribution. The solid line in Figure 5.25 represents the least-squares fit from a family of GRP parameters simulated in the parameter domain $\{1 < \beta < 2, 10 < \alpha < 50, 0 < q < 1\}$. The GRP estimates are $\hat{\beta} = 1.8, \hat{\alpha} = 24, \hat{q} = 0.70$. The estimated GRP shows a good fit to the data not only in the interval (0, 18] months (used for estimation) but also in the remaining interval (18, 30] months (obtained by prediction), as illustrated in Figure 5.25. The figure also shows the extreme repair conditions modeled by the RP ($q = 0$) and the GRP ($q = 1$).

It is reasonable to conclude that the approach considered above is not only practically applicable for estimation of the GRP parameters, but also for prediction of the GRP, which is often essential in many situations. It should be noted that if the estimate of the rejuvenation parameter q turns out to be close enough to 0 or 1, the RP or the NHPP, respectively, must be considered as potential competing models.

5.2 AVAILABILITY OF REPAIRABLE SYSTEMS

We defined reliability as the probability that a component or system will perform its required function over a given time. The notion of availability is related to repairable items only. We define availability as the probability that a repairable system (or component) will function at time t as it is supposed to when called upon to do so. Conversely, the unavailability of a repairable item, $q(t)$, is defined as the probability that the item is in a failed state (down) at time t. There are several definitions of availability; the most common are as follows.

1. *Instantaneous (point) availability* of a repairable item at time t, $a(t)$, is the probability that the system (or component) is up at time t.
2. *Limiting availability*, a, is defined as the following limit of instantaneous availability, $a(t)$:

$$a = \lim_{t \to \infty} a(t). \tag{5.78}$$

3. *Average availability*, $\bar{a}$, is defined for a fixed time interval, T, as

$$\bar{a} = \frac{1}{T} \int_0^T a(t)\,\mathrm{d}t. \tag{5.79}$$

4. The respective *limiting average availability* is defined as

$$\bar{a}_l = \lim_{T \to \infty} \frac{1}{T} \int_0^T a(t)\,\mathrm{d}t. \tag{5.80}$$

It should be noted that the limiting average availability has limited applications. We elaborate on each of the three definitions of availability in the remainder of this section.

If a component or system is nonrepairable, its availability coincides with its reliability function, $R(t)$. That is,

$$a(t) = R(t) = \mathrm{e}^{-\int_0^t \lambda(\tau)\,\mathrm{d}\tau} \tag{5.81}$$

where $\lambda(t)$ is the failure (hazard) rate function. The unavailability, $q(t)$, is related to $a(t)$ as

$$q(t) = 1 - a(t). \tag{5.82}$$

From the modeling point of view, repairable systems can be divided into the following two groups:

1. Repairable systems for which failure is immediately detected (revealed faults).
2. Repairable systems for which failure is detected upon inspection, referred to as *periodically inspected (tested)* systems.

5.2.1 Instantaneous (Point) Availability

For the first group systems, it can be shown (see Section 5.3) that $a(t)$ and $q(t)$ are obtained from the following ordinary differential equations:

$$\begin{aligned}\frac{da(t)}{dt} &= -\lambda(t)a(t) + \mu(t)q(t),\\ \frac{dq(t)}{dt} &= \lambda(t)a(t) - \mu(t)q(t),\end{aligned} \tag{5.83}$$

where $\lambda(t)$ is the failure rate and $\mu(t)$ is the repair rate.

The most widely used models for availability are based on the exponential time-to-failure and repair time distribution. Based on Equation 5.72 it can be shown (see Section 5.3) that, in this case (no trend exists in the failure rate and repair rate), the point availability and unavailability of the system (or component) are given by

$$\begin{aligned}a(t) &= \frac{\mu}{\lambda+\mu} + \frac{\lambda}{\lambda+\mu}\exp[-(\lambda+\mu)t],\\ q(t) &= \frac{\mu}{\lambda+\mu} - \frac{\lambda}{\lambda+\mu}\exp[-(\lambda+\mu)t].\end{aligned} \tag{5.84}$$

Note that in Equation 5.84, $\mu = 1/\tau$, where τ is the average time interval per repair, sometimes referred to as *mean time to repair* (MTTR). Clearly, MTBF $= 1/\lambda$ in this case.

For the second type of repairable systems mentioned above, determining availability is a difficult problem. Caldorola [46] presents a form of $a(t)$ for cases where no trend in the failure rate exists, and the inspection interval (η), duration of inspection (θ), and duration of repair (τ) are fixed. In these cases,

$$a(t) = \exp[-(t-m\eta)\lambda_{\text{eff}}]\cdot\left\{1-\exp\left[-\left(\frac{t-m\eta}{\theta}\right)^{q}\right]\right\}, \tag{5.85}$$

where

$$\begin{aligned}\lambda_{\text{eff}} &= \frac{\eta-\theta}{\eta}\left(\frac{\eta-\theta}{\eta}+\frac{2\tau}{\eta}\right)+2\left[1-\frac{\Gamma(1/q)}{q}\right]\frac{\theta}{\eta^2},\\ q &= \ln[3-\ln(\theta\lambda)],\end{aligned}$$

and m is the inspection interval number $(1, 2, \ldots, n)$. When $t > m\eta + \theta$, it is easy to show that

$$a(t) \approx \exp[-(t-m\eta)\lambda_{\text{eff}}]. \tag{5.86}$$

In periodically tested components, if test and repair durations are negligible compared to operation time, and test and repair are assumed perfect without any trend, then average unavailability and availability can be estimated from Equation 5.87.

Example 5.14

Find the unavailability, as a function of time, for a system that is inspected once a month. Duration of the inspection is 1.5 h. Any required repair takes an average of 19 h. Assume the failure rate of the system is $3 \times 10^{-6}\,\text{h}^{-1}$.

Solution: Using Equation 5.86, for $\theta = 1.5$, $\tau = 19$, $\eta = 720$, $\lambda = 3 \times 10^{-6}$, we can obtain the plot of $q(t)$ as shown in Figure 5.26.

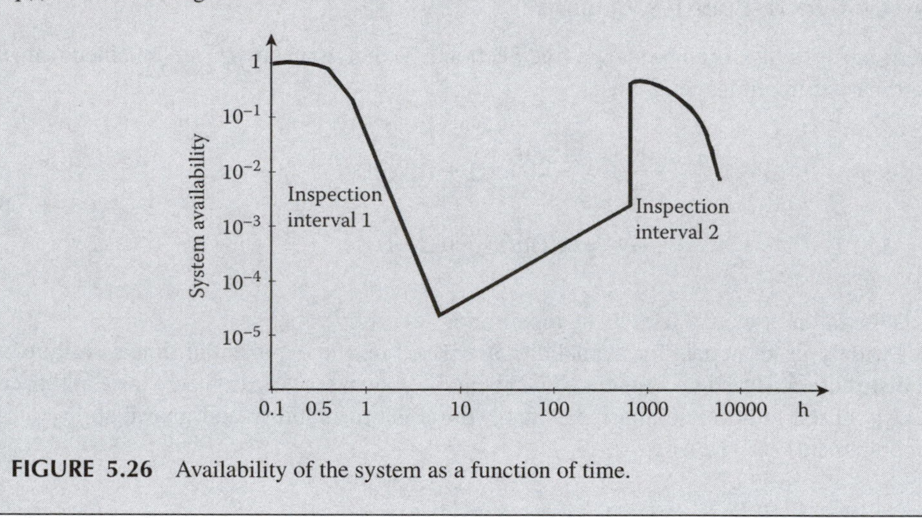

FIGURE 5.26 Availability of the system as a function of time.

For simplicity, the pointwise availability function can be represented in an approximate form. This simplifies availability calculations significantly. For example, for a periodically tested component, if the repair and test durations are very short compared with the operation time, and the test and repair are assumed perfect, one can neglect their contributions to unavailability of the system. This can be shown using Taylor's expansion of the unavailability equation [47]. In this case, for each test interval T, the availability and unavailability functions are

$$
\begin{aligned}
a(t) &\approx 1 - \lambda t, \\
q(t) &\approx \lambda t.
\end{aligned}
\tag{5.87}
$$

The plot of the unavailability as a function of time, using Equation 5.87, will take a shape similar to that in Figure 5.27.

Clearly, if the test and repair durations are long, one must include their effect. Vesely and Goldberg [48] have used the approximate pointwise unavailability functions for this case. The functions and their plot are shown in Figure 5.28. The average values of the approximate unavailability functions shown in Figures 5.27 and 5.28 are discussed in Section 5.2.3 and are presented in Table 5.9. It should be noted that, due to random imperfection in test and repair activities, it is possible that a residual

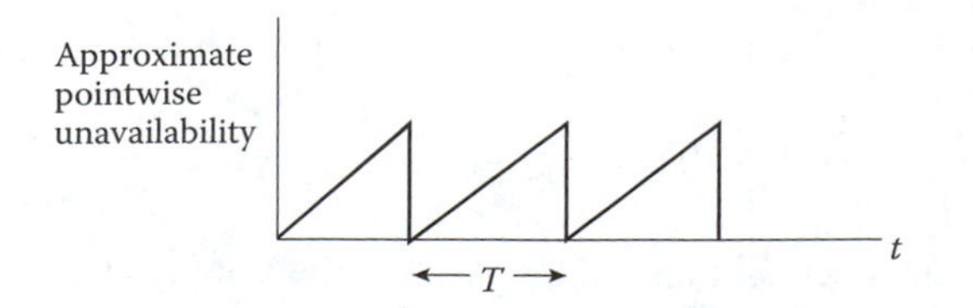

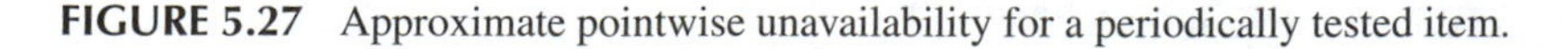

FIGURE 5.27 Approximate pointwise unavailability for a periodically tested item.

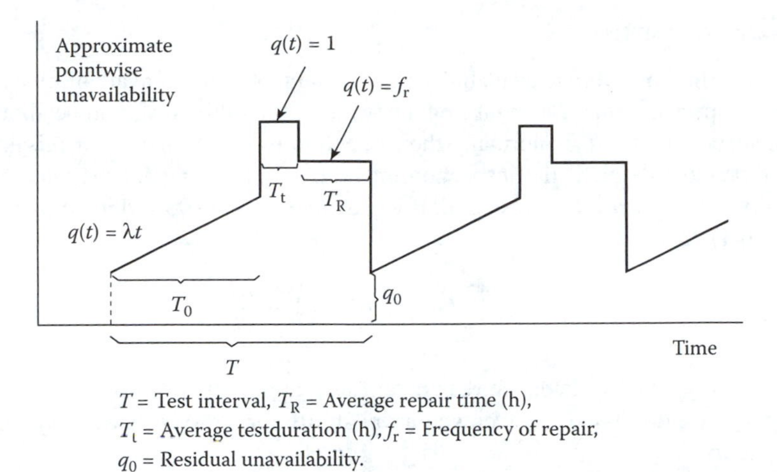

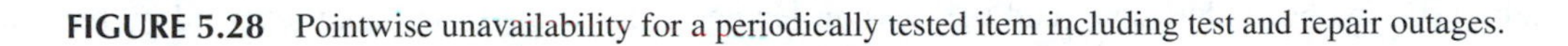

FIGURE 5.28 Pointwise unavailability for a periodically tested item including test and repair outages.

unavailability q would remain following a test and/or repair. Thus, unlike the unavailability function shown in Figure 5.27, the unavailability function in Figure 5.28 exhibits a residual unavailability q_0 due to these random imperfections.

5.2.2 Limiting Point Availability

Some of the pointwise availability equations discussed in Section 5.2.1 have limiting values. For example, Equation 5.73 has the following limiting value:

$$a = \lim_{t \to \infty} a(t) = \frac{\mu}{\lambda + \mu}, \tag{5.88}$$

or its equivalent

$$a = \frac{\text{MTBF}}{\text{MTTR} + MTBF}. \tag{5.89}$$

Equation 5.89 is also referred to as the asymptotic availability of a repairable system with a constant failure rate.

TABLE 5.9
Average Availability Functions

Type of Item	Average Unavailability	Average Availability
Nonrepairable	$\frac{1}{2}\lambda T_m$	$1 - \frac{1}{2}\lambda T_m$
Repairable revealed fault	$\frac{\lambda\tau}{1 + \lambda\tau}$	$\frac{1}{1 + \lambda\tau}$
Reparable periodically tested	$\frac{1}{2}\lambda T_0 + f_r\frac{T_R}{T} + \frac{T_t}{T}$	$1 - \frac{1}{2}\lambda T_0 + f_r\frac{T_R}{T} + \frac{T_t}{T}$

Note: λ = constant failure rate (h^{-1}), T_m = mission length (h), τ = average downtime or MTTR (h), T = test interval (h), T_R = average repair time (h), T_t = average test duration (h), f_r = frequency of repair per test intervals, T_0 = operating time (up time) = $T - f_r T_R - T_t$.

5.2.3 Average Availability

According to its definition, average availability is a constant measure of availability over a period of time T. For noninspected items, T can take on any value (preferably, it should be about the mission length). For inspected items, T is normally the inspection (or test) interval or mission length T_{m}. Thus, for nonrepairable items, if the inspection interval is T, then the approximate expression for point availability with constant λ can be used. If we assume $\bar{a} = 1 - \lambda t$ (which might be applicable, if at least $\lambda t < 0.1$), then

$$a = \frac{1}{T}\int_0^T (1 - \lambda t)\,\mathrm{d}t = 1 - \frac{\lambda T}{2}. \tag{5.90}$$

Accordingly, for all types of systems, one can obtain such approximations for average availabilities. Vesely et al. [49] have discussed the average unavailability for various types of systems. Table 5.9 shows these functions.

The equations in Table 5.9 can also be applied to standby equipment, with λ representing the standby (or demand) failure rate, and the mission length or operating time being replaced by the time between two tests. Barroeta and Modarres [50] have discussed equations and examples for availability cases with NHPP, including situations in which there are cost considerations associated with periodically tested and maintained units.

5.3 USE OF MARKOV PROCESSES FOR DETERMINING SYSTEM AVAILABILITY

Markov processes are useful tools for evaluating availability of systems that have multiple states (e.g., up, down, and degraded). For example, consider a system with the states shown in Figure 5.29.

In the framework of Markovian models, the transitions between various states are characterized by constant *transition rates*. These rates, generally speaking, may not necessarily be constant in practice. Consider a system with a given number of discrete states, n. Introduce the following characteristics of the system:

$$\Pr_i(t) = \Pr\,(\text{the system is in state } i \text{ at time } t),\ \sum_{i=1}^{n} \Pr_i(t) = 1$$

$$\rho_{ij} = \text{transition rate from state } i \text{ to state } j,\ (i,j = 1,2,\ldots,n)$$

Because ρ_{ij} is constant, the random time the system is at state i until the transition to state j follows the exponential distribution with rate ρ_{ij}. Assuming that $\Pr_i(t)$ is differentiable, it is possible to show [7] that

$$\frac{d\Pr_i(t)}{\mathrm{d}t} = -\Pr_i(t)\left(\sum_{j(j\neq i)} \rho_{ij}\right) + \left(\sum_{j(j\neq i)} \rho_{ji}\Pr_j(t)\right). \tag{5.91}$$

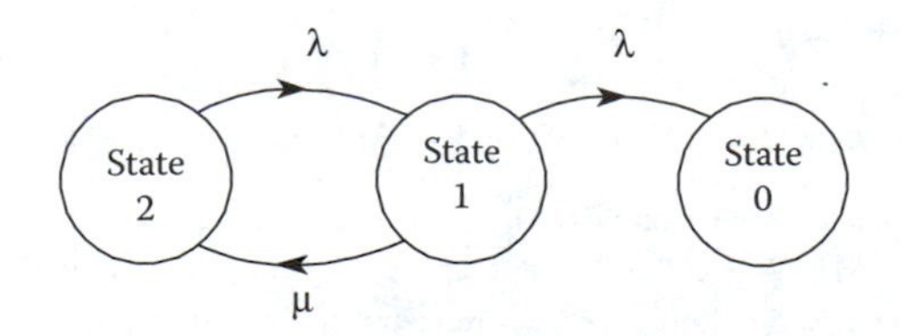

FIGURE 5.29 Markovian model for a system with the discrete states.

If a differential equation similar to Equation 5.91 is written for each state, and the resulting set of differential equations is solved, one can obtain the time-dependent probability of each state. This is demonstrated in the following example.

Example 5.15

Consider a system with constant failure rate λ and constant repair rate μ in a standby redundant configuration. When the system fails, its repair starts immediately, which puts it back into operation. The system has two states: state 0, when the system is down; and state 1, when the system is operating (Figure 5.30).

a. Find the probabilities of these states.
b. Determine the availability of this system.

Solution: Assuming that the system is functioning at time $t = 0$—that is, $\Pr_1(0) = 1$ and $\Pr_0(0) = 0$—and using the governing differential Equation 5.78, find

$$\frac{d\Pr_0(t)}{dt} = \lambda \Pr_1(t) - \mu \Pr_0(t),$$

$$\frac{d\Pr_1(t)}{dt} = -\lambda \Pr_1(t) + \mu \Pr_0(t).$$

For the above set of equations, matrix $A = \begin{pmatrix} \lambda & -\mu \\ -\lambda & \mu \end{pmatrix}$ is referred to as the *transition matrix*.

The above equations can be solved using the Laplace transformation. Below, we take the Laplace transform of both sides of the equations:

$$s\bar{P}_1(s) - 1 = -\lambda \bar{P}_1(s) + \mu \bar{P}_0(s),$$

$$s\bar{P}_0(s) = \lambda \bar{P}_1(s) - \mu \bar{P}_0(s).$$

The solution of the above system is given by

$$\bar{P}_0(s) = \frac{\lambda}{s(s+\lambda+\mu)},$$

$$\bar{P}_1(s) = \frac{s+\mu}{s(s+\lambda+\mu)}.$$

Finding the respective inverse Laplace transform, it follows that availability $a(t)$ is obtained from

$$a(t) = \Pr_1(t) = L^{-1}\left\{\frac{s+\mu}{s(s+\lambda+\mu)}\right\} = \frac{\mu}{\lambda+\mu} + \frac{\lambda}{\mu+\lambda}\exp[-(\lambda+\mu)t],$$

which coincides with Equation 5.84, discussed in Section 5.2. Accordingly, unavailability is

$$q(t) = \Pr_0(t) = 1 - a(t) = \frac{\lambda}{\lambda+\mu} - \frac{\lambda}{\mu+\lambda}\exp[-(\lambda+\mu)t].$$

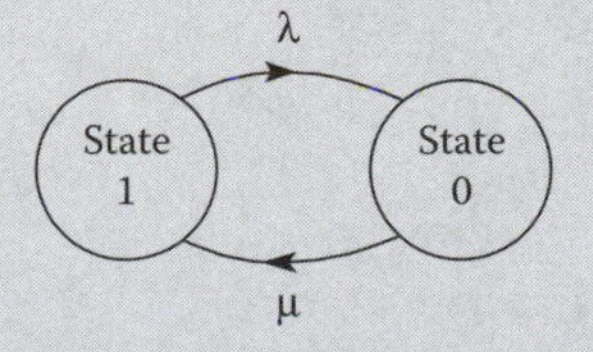

FIGURE 5.30 Markovian model for Example 5.15.

Example 5.16

A system that consists of two cooling units has the three states shown in the Markovian model in Figure 5.31. When one unit system fails, the other system takes over, and repair on the first starts immediately. When both systems are down, there are two repair crews to simultaneously repair the two systems. The three states are as follows:

1. State 0, when both systems are down.
2. State 1, when one of the systems is operating and the other is down.
3. State 2, when the first system is operating and the second is in standby (in an operating-ready condition).

   a. Determine the probability of each state.
   b. Determine the availability of the entire system.

Solution:

a. The governing differential equations are

$$\frac{d\mathrm{Pr}_2(t)}{\mathrm{d}t} = -\lambda \mathrm{Pr}_2(t) + \mu \mathrm{Pr}_1(t),$$

$$\frac{d\mathrm{Pr}_1(t)}{\mathrm{d}t} = \lambda \mathrm{Pr}_2(t) - (\mu + \lambda)\mathrm{Pr}_1(t) + 2\mu \mathrm{Pr}_0(t),$$

$$\frac{d\mathrm{Pr}_0(t)}{\mathrm{d}t} = \lambda \mathrm{Pr}_1(t) - 2\mu \mathrm{Pr}_0(t).$$

The Laplace transform of both sides of the equations yields

$$sP_2(s) - P_2(0) = -\lambda P_2(s) + \mu P_1(s),$$

$$sP_1(s) - P_1(0) = \lambda P_2(s) - (\mu + \lambda)P_1(s) + 2\mu P_0(s),$$

$$sP_0(s) - \mathrm{Pr}_0(0) = \lambda P_1(s) - 2\mu P_0(s)$$

$\mathrm{Pr}_2(0) = 1$ and $\mathrm{Pr}_1(0) = \mathrm{Pr}_0(0) = 0$. Solving the above set of equations, $\mathrm{Pr}_i(s)$ can be calculated as

$$\mathrm{Pr}_2(s) = \frac{1}{\lambda - s} + \frac{\mu\lambda(2\mu + s)}{s(s+\lambda)(s-k_1)(s-k_2)},$$

$$\mathrm{Pr}_1(s) = \frac{\lambda(2\mu + s)}{s(s-k_1)(s-k_2)},$$

$$\mathrm{Pr}_0(s) = \frac{\lambda^2}{s(s-k_1)(s-k_2)},$$

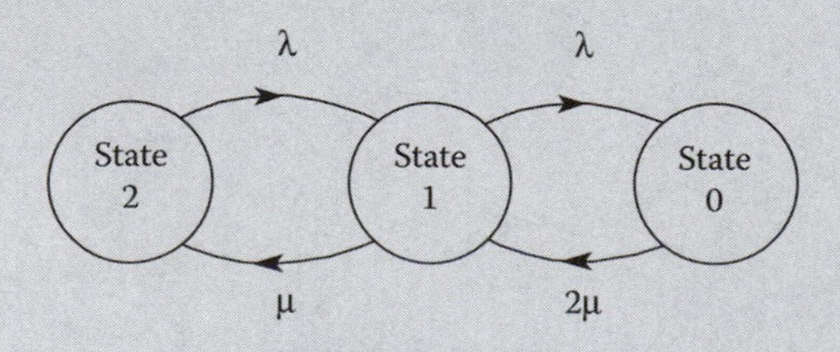

FIGURE 5.31 Markovian model for Example 5.16.

where

$$k_1 = \frac{-2\lambda - 3\mu - \sqrt{4\lambda\mu + \mu^2}}{2},$$

$$k_2 = \frac{2\mu\lambda + \lambda^2 + 2\mu^2}{k_1}.$$

If the inverses of the above Laplace transforms are taken, the probability of each state can be determined as

$$\Pr_2(t) = \exp(-\lambda t) + G_1 \exp(-\lambda t) + G_2 \exp(k_1 t) + G_3 \exp(k_2 t) + G_4,$$

where

$$G_1 = \frac{\mu(\lambda + 2\mu)}{(\lambda + k_1)(\lambda + k_2)}, \quad G_2 = \frac{\mu\lambda(2\mu + k_1)}{(k_1 + \lambda)(k_1 - k_2)(k_1)}$$

$$G_3 = \frac{\mu\lambda(2\mu + k_1)}{k_2(k_2 - k_1)(k_2 + \lambda)}, \quad G_4 = \frac{2\mu^2}{2\mu\lambda + \lambda^2 + 2\mu^2}$$

and

$$\Pr_1(t) = A_1 \exp(k_1 t) + A_2 \exp(k_2 t) + A_3,$$

where

$$A_1 = \frac{2\mu\lambda}{(k_1 - k_2)(k_1)} + \frac{\lambda}{k_1 - k_2},$$

$$A_2 = \frac{\lambda(2\mu + k_2)}{(k_1 - k_2)(k_2)},$$

$$A_3 = \frac{2\mu\lambda}{2\mu\lambda + \lambda^2 + 2\mu^2},$$

and

$$\Pr_0(t) = B_1 \exp(k_1 t) + B_2 \exp(k_2 t) + B_3,$$

where

$$B_1 = \frac{\lambda^2}{(k_1 - k_2)(k_1)},$$

$$B_2 = \frac{\lambda^2}{(k_2 - k_1)(k_1)},$$

$$B_3 = \frac{\lambda^2}{2\mu\lambda + \lambda^2 + 2\mu^2}.$$

b. The availability of the two units in the system is $a(t) = \Pr_2(t) + \Pr_1(t)$, and the unavailability of the entire system is $q(t) = \Pr_0(t)$.

It is possible simply to find the limiting pointwise availability from the governing equations of the system. For this purpose, consider the Markovian transition diagram shown in Figure 5.32.

It may be shown that

$$\Pr_{i+1}(\infty)\lambda_{i+1} = \mu_i \Pr_i(\infty), \tag{5.92}$$

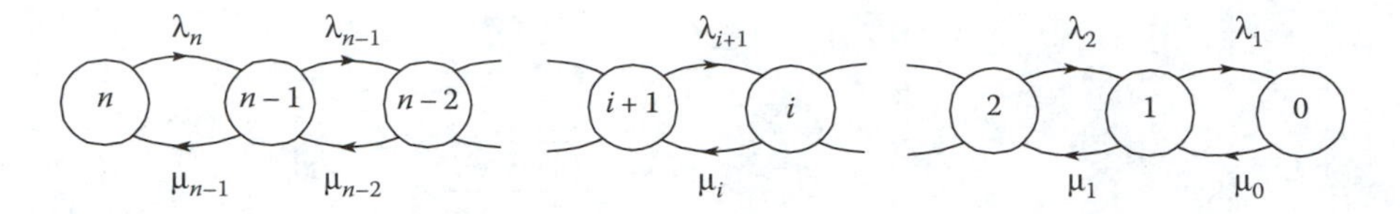

FIGURE 5.32 Markovian transition diagram with n states.

or

$$\Pr_i(\infty) = \frac{\lambda_{i+1} \times \lambda_{i+2} \times \cdots \times \lambda_n}{\mu_i \times \mu_{i+1} \times \cdots \times \mu_{n-1}} \Pr_n(\infty). \tag{5.93}$$

Since $\sum_{i=0}^{n} \Pr_i(\infty) = 1$, solving Equation 5.80 for $\Pr_n(\infty)$ yields

$$\Pr_i(\infty) = \frac{1}{1 + \sum_{i=0}^{n-1} \frac{(\lambda_{i+1} \times \lambda_{i+2} \times \cdots \times \lambda_n)}{(\mu_i \times \mu_{i+1} \times \cdots \times \mu_{n-1})}}. \tag{5.94}$$

Accordingly, the system's limiting pointwise unavailability (and similarly its availability) can be obtained as

$$q = \Pr_i(\infty) = \frac{\sum_{i=1}^{n} P_i \lambda_i}{\sum_{i=1}^{n-1} P_i \mu_i} \Pr_n(\infty). \tag{5.95}$$

If the system is unavailable when it is in any of the states $(0, 1, \ldots, r-1)$, then

$$q = \sum_{i=0}^{r-1} \Pr_i(\infty) = \Pr_n(\infty) \sum_{i=1}^{r-1} \frac{\lambda_{i+1} \times \lambda_{i+2} \times \cdots \times \lambda_n}{\mu_i \times \mu_{i+1} \times \cdots \times \mu_{n-1}}. \tag{5.96}$$

Example 5.17

For Example 5.16, determine the limiting pointwise unavailability from Equation 5.57 and confirm it with the results obtained in that example.

Solution: Since $\lambda_2 = \lambda_1 = \lambda$, $\mu_1 = \mu$, and $\mu_0 = 2\mu$ from Equation 5.93,

$$\Pr_1(\infty) = \frac{2\mu}{\lambda} \Pr_0(\infty)$$

and

$$\Pr_2(\infty) = \frac{\mu}{\lambda} \Pr_1(\infty) = \frac{2\mu^2}{\lambda^2} \Pr_0(\infty).$$

Since

$$\Pr_0(\infty) + \Pr_1(\infty) + \Pr_2(\infty) = 1$$

from Equation 5.96,

$$q = \Pr_0(\infty) = \frac{\lambda^2}{2\mu^2 + 2\mu\lambda + \lambda^2}.$$

Accordingly,

$$a = \Pr_1(\infty) + \Pr_2(\infty) = \frac{2\mu^2 + 2\mu\lambda}{2\mu^2 + 2\mu\lambda + \lambda^2}.$$

This can be verified by the solution for $\Pr_0(t)$. Since k_1 and k_2 are negative, and the exponential terms approach zero, then

$$\Pr_0(\infty) = B_3 = \frac{\lambda^2}{2\mu^2 + 2\mu\lambda + \lambda^2}.$$

Similarly,

$$\Pr_1(\infty) = A_3$$

and

$$\Pr_2(\infty) = G_4.$$

Thus

$$a = \frac{2\mu^2}{2\mu^2 + 2\mu\lambda + \lambda^2} + \frac{2\mu\lambda}{2\mu^2 + 2\mu\lambda + \lambda^2}.$$

Therefore, the results obtained in Examples 5.16 and 5.17 are consistent.

If a trend exists in the parameters that characterize system availability (e.g., failure rate and repair rate), one cannot use the Markovian method; only solutions of Equation 5.91 with time-dependent ρ can be used. Solving such equations may pose difficulty in systems with many states. However, with the emergence of efficient numerical algorithms and powerful computers, solutions to these equations are indeed possible.

5.4 USE OF SYSTEM ANALYSIS TECHNIQUES IN THE AVAILABILITY CALCULATIONS OF COMPLEX SYSTEMS

In Chapter 4, we discussed a number of methods for estimating the reliability of a system from the reliability of its individual components or units. The same concept applies here. That is, one can use the availability (or unavailability) functions for each component of a complex system and use, for example, system cut sets to obtain system availability (or unavailability). The method of determining system availability in these cases is the same as the system reliability estimation methods.

Example 5.18

Assume all components of the system shown in Figure 4.4 are repairable (revealed fault) with a failure rate of 10^{-3} (per hour) and a mean downtime of 15 h. Component 7 has a failure rate of 10^{-5} (per hour), with a mean downtime of 10 h. Calculate the average system unavailability.

Solution: The cut sets are (7), (1, 2), (1, 5, 6), (2, 3, 4), and (3, 4, 5, 6). The unavailability of components 1 through 6, according to Table 5.9, is

$$q_{1-6} = \frac{\lambda\tau}{1 + \lambda\tau} = \frac{10^{-3} \times 15}{1 + 10^{-3} \times 15} = 9.85 \times 10^{-3}.$$

Similarly,

$$q_7 = \frac{10^{-5} \times 10}{1 + 10^{-5} \times 10} = 9.99 \times 10^{-5}.$$

continued

Using the rare event approximation,

$$q_{sys} = q(\text{cut sets}) = q_7 + q_1 \cdot q_2 + q_1 \cdot q_5 \cdot q_6 + q_2 \cdot q_3 \cdot q_4 + q_3 \cdot q_4 \cdot q_5 \cdot q_6.$$

Thus,

$$q_{sys} = 9.99 \times 10^{-5} + 9.70 \times 10^{-5} + 9.56 \times 10^{-7} + 9.56 \times 10^{-7} + 9.41 \times 10^{-9} = 1.99 \times 10^{-4}.$$

Example 5.19

The auxiliary feedwater system in a pressurized water reactor (PWR) plant is used for emergency cooling of steam generators. The simplified piping and instrument diagram (P&ID) of a typical system like this is shown in Figure 5.33a.

Calculate the system unavailability. Assume all of the components are in standby mode and are periodically tested with the following characteristics. (Characteristics are shown collectively for each block.)

Block Name	Failure Rate (h^{-1})	Frequency of Repair	Average Test Duration (h)	Average Repair Time (h)	Test Interval (h)
A	1×10^{-7}	9.2×10^{-3}	0	5	720
B	1×10^{-7}	9.2×10^{-3}	0	5	720
C	1×10^{-6}	2.5×10^{-2}	0	10	720
D	1×10^{-6}	2.5×10^{-2}	0	10	720
E	1×10^{-6}	2.5×10^{-2}	0	10	720
F	1×10^{-6}	2.5×10^{-2}	0	10	720
G (G_1 and G_2)	1×10^{-7}	7.7×10^{-4}	0	15	720
H	1×10^{-7}	1.8×10^{-4}	0	24	720
I	1×10^{-4}	6.8×10^{-1}	2	36	720
J	1×10^{-4}	6.8×10^{-1}	2	36	720
K	1×10^{-5}	5.5×10^{-1}	2	24	720
L	1×10^{-7}	4.3×10^{-3}	0	10	720
M	1×10^{-4}	1.5×10^{-1}	0	10	720
N	1×10^{-7}	5.8×10^{-4}	0	5	720

Solution: According to Table 5.9, we can calculate the unavailability of each block.

Block Name	Unavailability	Block Name	Unavailability
A	1.0×10^{-4}	H	4.2×10^{-4}
B	1.0×10^{-4}	I	7.3×10^{-4}
C	7.0×10^{-4}	J	7.3×10^{-4}
D	7.0×10^{-4}	K	1.4×10^{-4}
E	7.0×10^{-4}	L	2.4×10^{-4}
F	7.0×10^{-4}	M	1.1×10^{-1}
G (G_1 and G_2)	5.2×10^{-4}	N	4.0×10^{-5}

The cut sets of the block diagram in Figure 5.33b are as follows

1. N
2. L M
3. H L

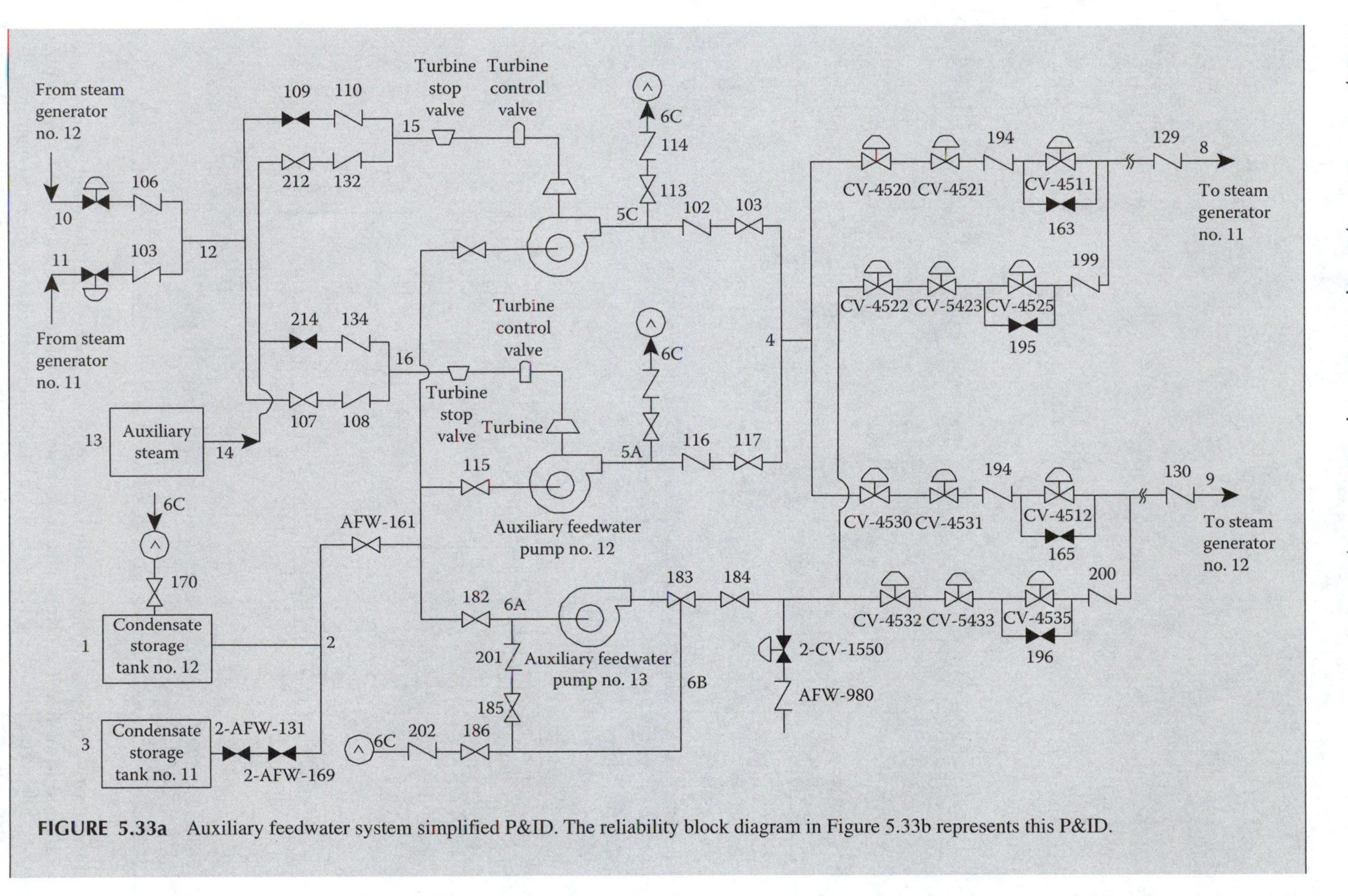

FIGURE 5.33a Auxiliary feedwater system simplified P&ID. The reliability block diagram in Figure 5.33b represents this P&ID.

4. G H
5. A B
6. H J I
7. G K M
8. D F L
9. D G F
10. C E H
11. B D L
12. B D G
13. B C H
14. B C D
15. A F L
16. A E F
17. A E H
18. A G F
19. J I K M
20. D F J I
21. C E K M
22. C D E H
23. B D J I
24. B C K M
25. A F J I
26. A E K M

Using the same procedure as the one used in Example 5.14 and rare event approximation, we can compute the average system unavailability as $q_{sys} = 7.49 \times 10^{-5}$.

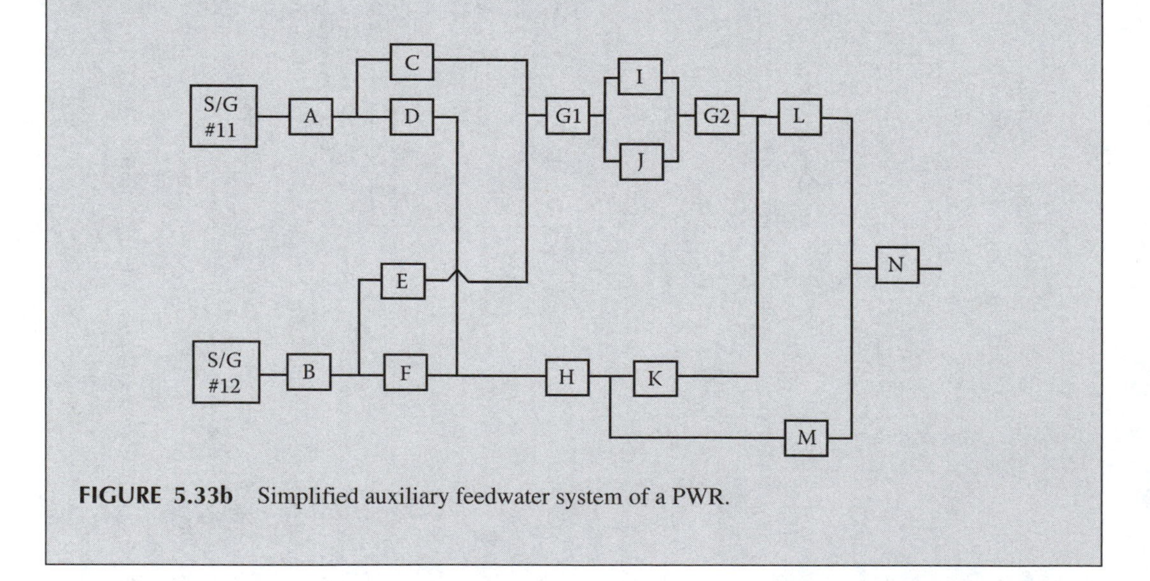

FIGURE 5.33b Simplified auxiliary feedwater system of a PWR.

In estimating the availability of redundant systems with periodically tested components, it is important to recognize that components whose simultaneous failures cause the system to fail (i.e., sets of components in each cut set of the system) should be tested in a *staggered* manner. This way the system will not become totally unavailable during the testing and repair of its components. For example, consider a system of two parallel units, each of which is periodically tested and has a pointwise unavailability behavior that can be approximated by the model shown in Figure 5.28. If

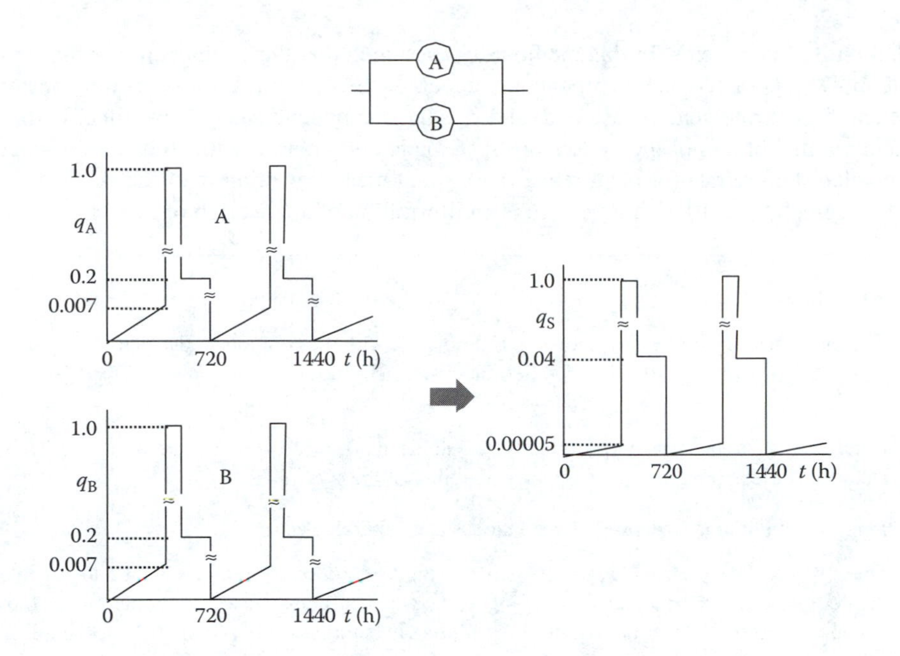

FIGURE 5.34 Unavailability of a parallel system using nonstaggered testing.

the components are not tested in a staggered manner, the system's pointwise unavailability exhibits the shape shown in Figure 5.34.

On the other hand, if the components are tested in a staggered manner, the system unavailability would exhibit the shape illustrated in Figure 5.35.

Clearly, the average unavailability in the case of staggered testing is lower. Vesely et al. [49] and Ginzburg and Vesely [51] discuss this subject in more detail. Also, to minimize unavailability, one can find an optimum value for test intervals as well as the optimum degree of staggering.

Modarres [52] has suggested a simple method for estimating approximate average system unavailability of a series–parallel system having a single input node and single output node, and repairable

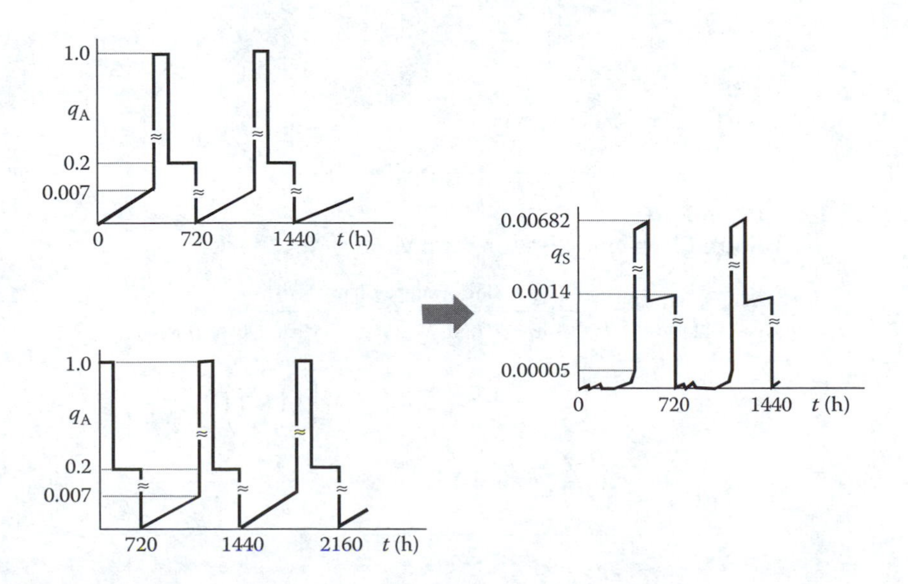

FIGURE 5.35 Unavailability of a parallel system using staggered testing.

(revealed fault) components. In this method, it is assumed that the components or blocks are independent and $\lambda_i \tau_i \ll 1$ for each component or block of the system, where λ_i is the constant failure rate (i.e., no failure rate trend is assumed) and τ_i is the component's mean downtime. In this method, series and parallel blocks of the system are systematically replaced with equivalent "superblocks." The equivalent failure rate (or occurrence rate) λ and mean downtime τ of the superblocks can be calculated from Table 5.10. Example 5.16 is an illustration of application of this method.

Example 5.20

Consider the series-parallel system shown in Figure 5.36, with the component data shown in Table 5.11. This system is composed of two parallel blocks. Each block is composed of subblock(s) and component(s).

a. Determine the approximate occurrence rate and mean downtime of this system.
b. Determine the approximate average unavailability of the system.

Solution: Assuming independence between blocks and superblocks,

a. The superblocks are enclosed by dotted lines in Figure 5.36. First, all of the blocks are resolved, and their equivalent λ and τ are obtained. Finally, the whole system is resolved. Equations in Table 5.10 are applied to the system along with the failure data summarized in Table 5.11 to obtain λ and τ values. The steps are illustrated in Figure 5.37.

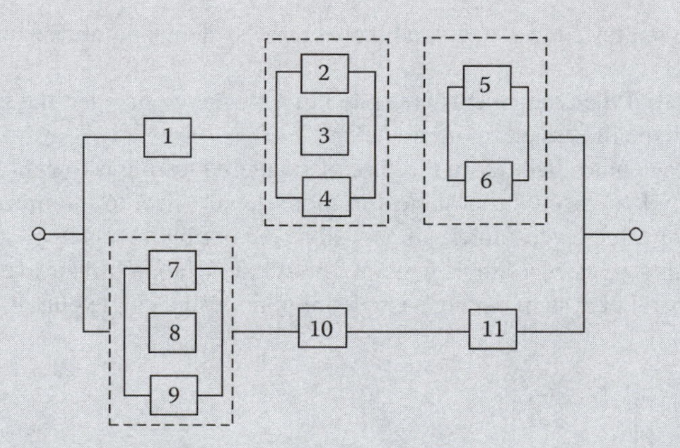

FIGURE 5.36 Sample series–parallel system.

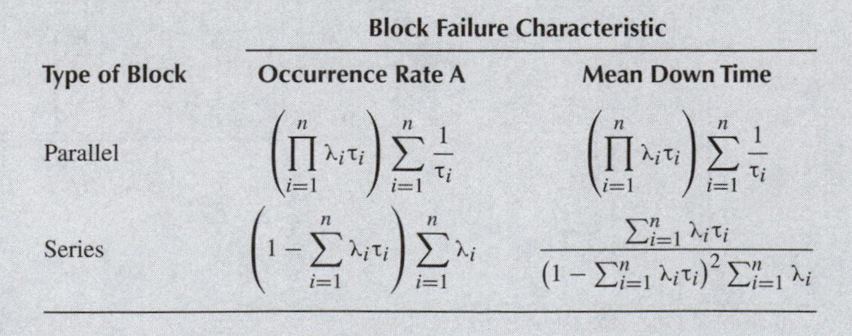

TABLE 5.10
Failure Characteristics for Parallel or Series Blocks

	Block Failure Characteristic	
Type of Block	**Occurrence Rate A**	**Mean Down Time**
Parallel	$\left(\prod_{i=1}^{n} \lambda_i \tau_i\right) \sum_{i=1}^{n} \frac{1}{\tau_i}$	$\left(\prod_{i=1}^{n} \lambda_i \tau_i\right) \sum_{i=1}^{n} \frac{1}{\tau_i}$
Series	$\left(1 - \sum_{i=1}^{n} \lambda_i \tau_i\right) \sum_{i=1}^{n} \lambda_i$	$\frac{\sum_{i=1}^{n} \lambda_i \tau_i}{\left(1 - \sum_{i=1}^{n} \lambda_i \tau_i\right)^2 \sum_{i=1}^{n} \lambda_i}$

b. The approximate unavailability of the system can be calculated using $q = \lambda\tau/(1+\lambda\tau)$ from Table 5.9. Thus, $q = 2.9\times10^{-5}(2.3)/[1+2.9\times10^{-5}(2.3)] = 6.67\times10^{-5}$. This can be compared with the direct calculation method using the cut set concept (similar to Examples 5.18 and 5.19), which yields the average system unavailability of 6.57×10^{-5}. The difference is due to the approximate nature of this approach and the assumption that the whole system's time to failure approximately follows an exponential distribution.

TABLE 5.11
Summary of Failure Data for the Components Shown in Figure 5.36

Component Serial Number	Failure Rate λ_i (per 1000 h)	Mean Downtime τ_i (h)
1	1	5.0
2	10	7.5
3	10	7.5
4	10	7.5
5	5	6.0
6	5	6.0
7	10	7.5
8	10	7.5
9	10	7.5
10	10	5.0
11	10	5.0

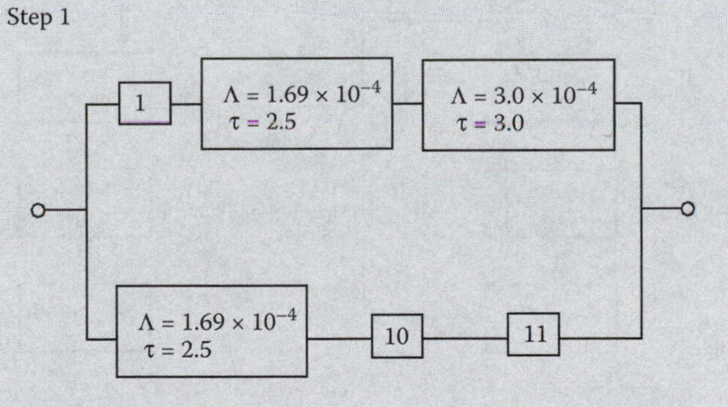

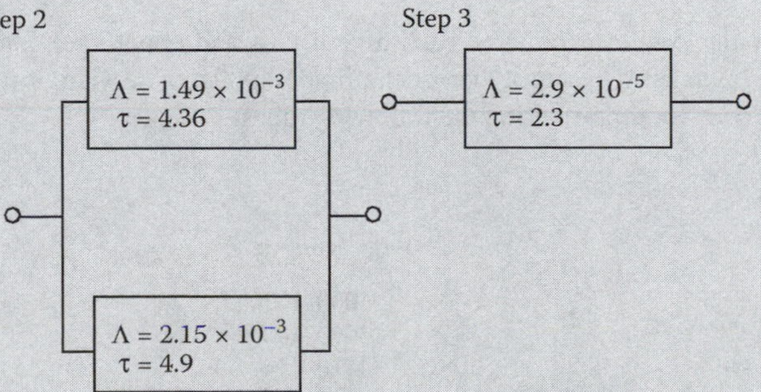

FIGURE 5.37 Step-by-step resolution of the system in Figure 5.36.

EXERCISES

5.1 The following table shows fire incidents during six equal time intervals of 22 chemical plants.

Time interval	1	2	3	4	5	6
Number of fires	6	8	16	6	11	11

Are the fire incidents time dependent? Prove your answer.

5.2 A simplified schematic of the electric power system at a nuclear power plant is shown in the figure below.

a. Draw a fault tree with the top event "Loss of Electric Power from Both Safety Load Buses."
b. Determine the unavailability of each event in the fault tree for 24 h of operation.
c. Determine the top event probability.

Assume the following:

- Either the main generator or one of the two diesel generators is sufficient.
- One battery is required to start the corresponding diesel generator.
- Normally, the main generator is used. If that is lost, one of the diesel generators provides the electric power on demand.

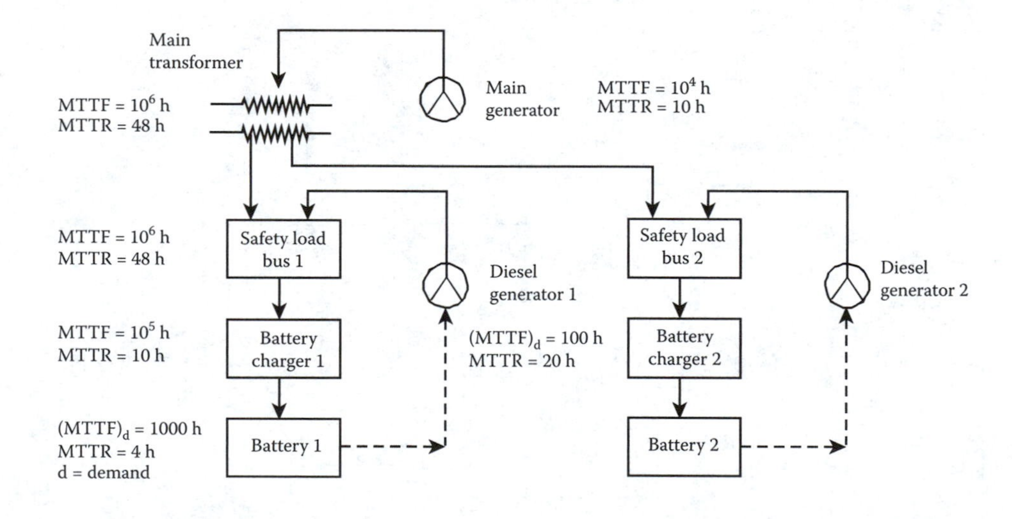

5.3 An operating system is repaired each time it fails and is put back into service as soon as possible (monitored system). During the first 10,000 h of service, it fails five times and is out of service for repair during the following times:

Hours
1000–1050
3960–4000
4510–4540
6130–6170
8520–8560

a. Is there a trend in the data?
b. What is the reliability of the system 100 h after the system is put into operation? What is the asymptotic availability assuming no trends in λ and μ?
c. If the system has been operating for 10 h without a failure, what is the probability that it will continue to operate for the next 10 h without a failure?
d. What is the 80% confidence interval for the MTTRs ($\tau = 1/\mu$)?

5.4 The following cycle-to-failure data have been obtained from a repairable component. The test stopped when the fifth failure occurred.

Repair number	1	2	3	4	5
Cycle to failure (interarrival of cycles)	5010	6730	4031	3972	4197

a. Is there any significant trend in these data?
b. Determine the ROCOF.
c. What is the reliability of the component 1000 cycles after the fifth failure is repaired?

5.5 Determine the limiting pointwise unavailability of the system shown below.

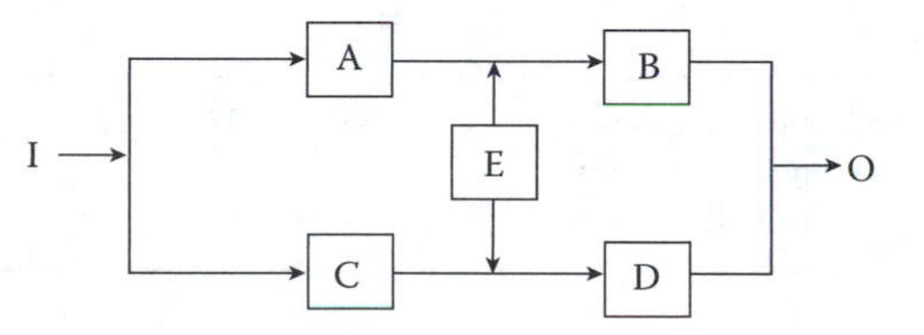

Assume that all components are identical and are repaired immediately after each of them experiences a failure. The rate of occurrence of the failure for each component is $\lambda = 0.001$ (per hour), and mean time to repair is 15 h.

5.6 For the system shown below, the following information are available:

- A and E are identical components, with $\lambda_A = \lambda_E = 1 \times 10^{-5}\,\text{h}^{-1}$, $\mu_A = \mu_E = 0.1\,\text{h}^{-1}$.
- B, C, and D are identical, periodically tested components with $\lambda_B = \lambda_C = \lambda_D = 1 \times 10^{-5}\,\text{h}^{-1}$. All test durations are equal ($\lambda t = 1$ h), all frequencies of repairs-per-cycle are equal ($f = 0.25$), and all durations of repair are equal ($\tau_r = 15$ h), $T_0 = 720$ h.

Given the above information, calculate the unavailability of the system assuming that all components are independent.

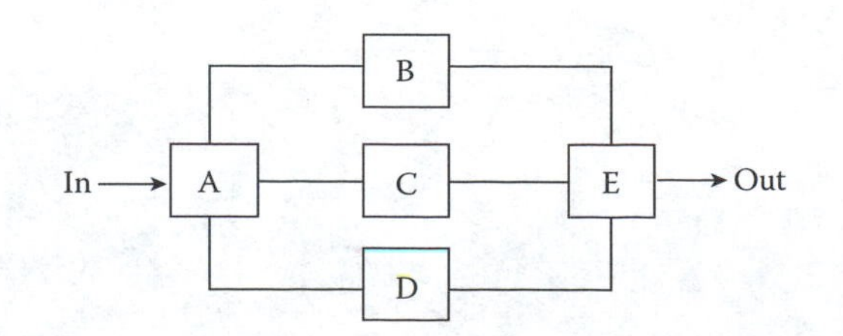

5.7 Consider the following system.

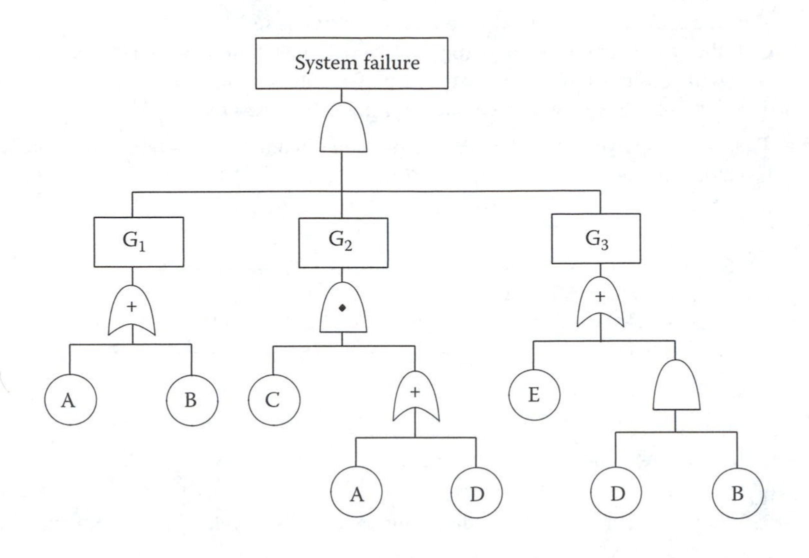

Suppose all components have the same availability value. Further, consider data obtained for one component in this set as follows:

Failure times from when component was new (h)	2823	3984	7232	9812	12,314
Repair duration (h)	17	11	28	14	32

Current time is 13,500 h

a. Is there a trend in the ROCOF and repair rate?
b. Determine ROCOF and repair rate as a function of time (if no trend in part (a) only determine limiting pointwise point estimates).
c. Calculate limiting pointwise availability.
d. Determine system unavailability assuming all components have the same availability value as calculated in part (c).
e. Calculate 90% confidence limits for ROCOF using data in part (a) (assume no trend).

5.8 For the system represented by the fault tree below, if all the components are periodically tested, determine the average unavailability of the system.

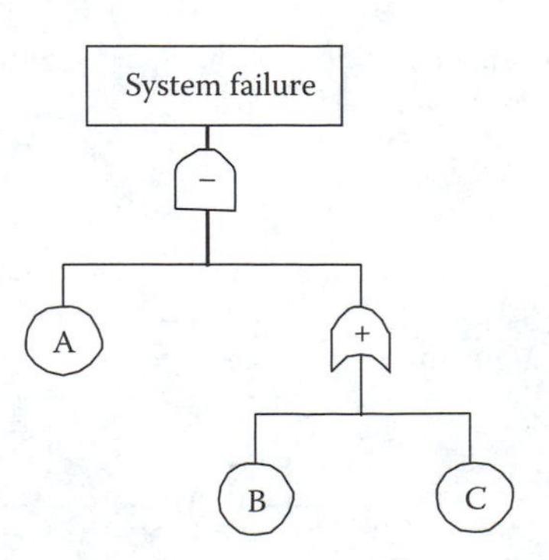

All components have a test interval of 1 month and average test duration of 1 h.

Components	Failure (Rate/h)	Repair (Rate/h)	Frequency of Repair
A	10^{-3}	10^{-1}	0.1
B	10^{-4}	10^{-2}	0.2
C	10^{-4}	10^{-2}	0.3

5.9 A repairable unit had the following failure characteristic in the past 5 years. (Assume the unit's life is currently at the beginning of its sixth year.)

Year	Number of Failure*** Required Maintenance
1	n
2	$0.9n$
3	$0.88n$
4	$0.72n$
5	$0.88n$

a. Is there any trend?
b. How many failures (in terms of fraction of n) do we expect to see in the sixth year?

5.10 The following data represent failure times and repair experiences from the life of a component.

Failure Number	t_0 (Age in Months)	Repair Duration (h)
1	4	3
2	7	4
3	15.5	7
4	20	3
5	26	7

Assume 30 days/months

Assuming that the current age of the unit is 30 months.

a. Is there a trend in the ROCOF and repairs?
b. Estimate the instantaneous availability at the present time of 30 months.
c. In a periodically tested system with a constant ROCOF of $10^{-4}\,\text{h}^{-1}$, average constant repair duration of 10 h, average test duration of 5 h and average constant frequency of repair of 0.05 per test, determine the best (optimum) inspection interval.

5.11 The failure time history of a component is shown below:

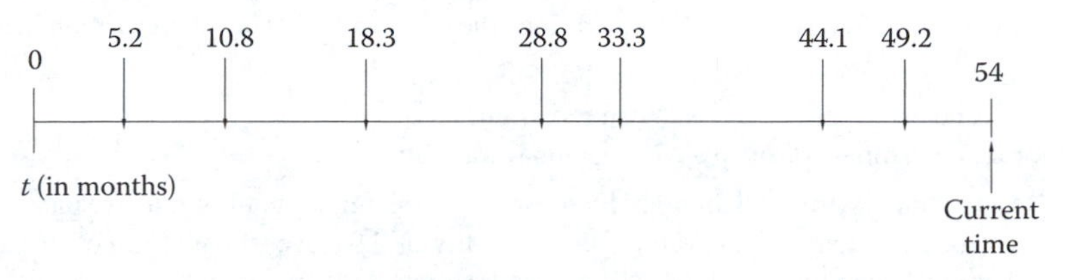

a. Is there practically a trend in ROCOF?
b. What is the conditional probability of no failure (i.e., reliability) in the next six months given the component is operating at the current accumulated component age of 54 months?
c. If the observed average downtime for repair is 14 h per repair and repair has no trend, what is the average conditional unavailability (given we are at the age 54 months and the component is operating) in the next 6 months?

d. By assuming no trend in ROCOF, how much error we would make in the unconditional average unavailability?
e. If we decide to adopt a periodically tested policy that reduces the failure rate by a factor of 5 (e.g., using a preventive maintenance program) with an average test duration of 5 min, average repair duration of 2 h and average frequency of repair of 0.01, what should we select as our inspection interval (T_0) to maintain the same unavailability value calculated in (c).

5.12 Consider the success tree for system S below. Determine the probability of failure by developing all mutually exclusive cut sets and path sets. Determine the unavailability of the top event "S-unavailable." Mean availabilities of the basic events are shown.

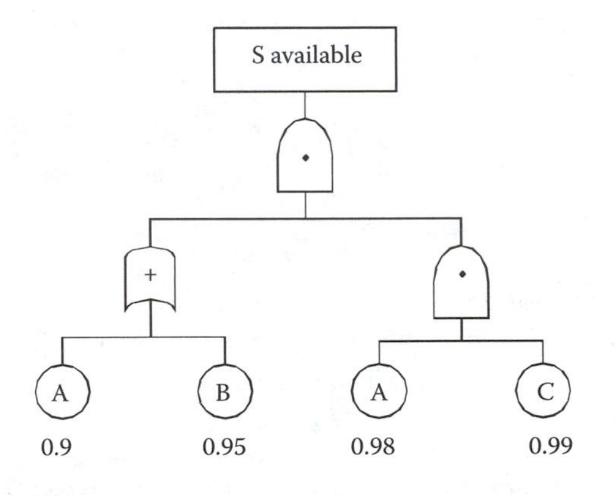

5.13 Failure and repair times of a periodically tested component have been reported as follows:

Number	Failure Time (Months)	Duration of Repair (Months)
1	73	0.1
2	132	0.5
3	193	0.05
4	148	0.15
5	212	0.2

Assuming that the present age of the component is 250 months and a monthly test with a duration of 2 h has been performed since the unit has been put into operation, determine

a. An estimate of the past mean ROCOF.
b. The estimate of the past mean repair rate.
c. The estimate of the past average unavailability.

5.14 Calculate system availability of Exercise 4.38 by using mutually exclusive path sets. Assume each component has an average unavailability of 0.01 over the mission of this system.

5.15 The following historical data for a periodically tested component are available. Duration of repair for the times that the component required repair (in hour) are 7, 8, 10, 9, and 8. Total number of tests is 50. Overall length of each test is 3.5 h. Length of operation: $T_0 = 1000$ h.

a. Is there any trend in the length of repair? What is the confidence level you can place on the trend conclusion?
b. Calculate average unavailability of the component assuming a constant rate of occurrence failure and constant duration of repair.

REFERENCES

1. Smith, W. and M. Leadbetter, On the renewal function for the Weibull distribution, *Technometrics*, 5, 243–302, 1963.
2. White, J. Weibull renewal analysis, *Proceedings of 3rd SAE Annual Reliability and Maintainability Conference*, New York, pp. 639–657, 1964.
3. Baxter, L. A., et al., On the tabulation of the renewal function, *Technometrics*, 24, 151–156, 1982.
4. Garg, A. and J. Kalagnanam, Approximations for the renewal functions, *IEEE Trans. Reliab.*, 47 (1), 66–72, 1998.
5. Blischke, W. R. and D. N. P. Murthy, *Warranty Cost Analysis*, Marcel Dekker, New York, 1994.
6. Kaminskiy, M. Simple bounds on cumulative intensity functions of renewal and G-renewal processes with increasing failure rate underlying distributions, *Risk Anal.: Int. J.*, 24 (5.33), 1035–1039, 2004.
7. Hoyland, A. and M. Rausand, *System Reliability Theory: Models and Statistical Methods*, Wiley, New York, 1994.
8. Crow, L. H., Reliability analysis for complex repairable systems, *Reliability and Biometry*, in F. Proschan and R. J. Serfling, (eds.), SIAM, Philadelphia, pp. 379–410, 1974.
9. Crow, L. H., Confidence interval procedures for the Weibull process with applications to reliability growth, *Technometrics*, 24, 67–72, 1982.
10. Duane, J. T., Learning curve approach to reliability monitoring, *IEEE Trans. Aerosp. Electron. Syst.*, 2, 563–566, 1964.
11. Cox, D. R. and P. A. W. Lewis, *The Statistical Analysis of Series of Events*, Chapman & Hall, London, 1966.
12. Vesely, W., Incorporating aging effects into probabilistic risk analysis using a taylor expansion approach, *Reliab. Eng. Syst. Saf.*, 32, 315–337, 1991.
13. Atwood, C., Parametric estimation of time-dependent failure rates for probabilistic risk assessment, *Reliab. Eng. Syst. Saf.*, 37, 181–194, 1992.
14. Rigdon, S. E. and A. P. Basu, *Statistical Methods for the Reliability of Repairable Systems,* Wiley, New York, 2000.
15. Crowder, M., A. Kimber, R. Smith, and T. Sweeting, *Statistical Analysis of Reliability Data*, Chapman & Hall, London, 1991.
16. Ascher, H. and H. Feingold, *Repairable Systems Reliability: Modeling and Inference, Misconception and Their Causes*, Marcel Dekker, New York, 1984.
17. Lindqvist, H., Statistical modeling and analysis of repairable systems, *Statistical and Probabilistic Models in Reliability*, in D. C. Ionescu and N. Limnios, (eds.), pp. 3–25, Birkhauser, Berlin, 1999.
18. Thompson, W. A. Jr. On the foundation of reliability, *Technometrics*, 23, 1–13, 1981.
19. Brown M. and F. Proschan, Imperfect maintenance, in J. Crowley and R. Johnson, eds, *Survival Analysis*, IMS Lecture Notes—Monograph Series, Hayward, Vol. 2, pp. 179–188, 1982.
20. Kijima, M. and N. Sumita, A useful generalization of renewal theory: Counting process governed by non-negative Markovian increments, *J. Appl. Probab.*, 23, 71–88, 1986.
21. Zhang, F. and. A. K. S. Jardine, Optimal maintenance models with minima repair, periodic overhaul and complete renewals, *IIE Trans*., 30, 1109–1119, 1998.
22. Kaminskiy, M. and V. Krivtsov, A Monte Carlo approach to estimation of G-renewal process in warranty data analysis, *Proceedings 2nd International Conference on Mathematical Methods in Reliability*, Bordeaux, France, pp. 583–586, 2000.
23. Hurtado, J. L., F. Joglar, and M. Modarres, Generalized renewal process: Models, parameter estimation and applications to maintenance problems, *Int. J. Performability Eng.*, 1, July 2005.
24. Kahle, W., Statistical models for the degree of repair in incomplete repair models, *International Symposium on Stochastic Models in Reliability, Safety, Security and Logistics,* Beer Sheva, Israel, pp. 178–181, 2005.
25. Filkenstein, M., The concealed age of distribution function and the problem of general repair, *J. Stat. Plann. Inference*, 65, 315–321, 1997.
26. Kaminskiy, M. and V. Krivtsov, A Monte Carlo approach to repairable system reliability analysis, *Probabilistic Safety Assessment and Management*, Springer-Verlag, London Ltd., New York, pp. 1063–1068, 1998.

27. Kijima, M., H. Morimura, and Y. Suzuki, Periodical replacement problem without assuming minimal repair, *European Journal of Operational Research*, 37, 194–203, 1988.
28. Barlow, R. E. and F. Proschan, *Mathematical Theory of Reliability*, SIAM, Philadelphia, 1996.
29. Barlow, R. and S. Gupta, Distribution-free life test sampling plans, *Technometrics*, 8, 591–603, 1966.
30. Kaminskiy, M., Simple bounds for counting processes with monotone rate of occurrence of failures. *Proceedings of International Symposium on Stochastic Models in Reliability, Safety, Security, and Logistics*, Beer Sheva, Israel, 182–185, 2005.
31. Leemis, L. M., *Reliability: Probabilistic Models and Statistical Methods*, Prentice-Hall, Englewood Cliffs, NJ, 1995.
32. Aalen, O. O., Non-parametric inference for a family of counting processes, *Ann. Statist.* 6, 701–726, 1978.
33. Nelson, W., Confidence limits for recurrence data: Applied to cost or number of repairs, *Technometrics*, 37, 147–157, 1995.
34. Nelson, W., *Recurrent Events Data Analysis for Product Repairs, Disease Recurrences, and Other Applications*, SIAM, ASA, Philadelphia, 2003.
35. Meeker, W. and L. Escobar, *Statistical Methods for Reliability Data*, Wiley, New York, 1998.
36. Lawless, J. and C. Nadeau, Some simple robust methods for the analysis of recurrent events, *Technometrics*, 37, 158–168, 1995.
37. Judge, G. G., R. C. Hill, W. E. Griffiths, H. Lutkepohl, and Lee, T.-Ch, *Introduction to the Theory and Practice of Econometrics*, Wiley, New York, 1988.
38. Lawless, J. F., *Statistical Models and Methods for Lifetime Data*, Wiley, New York, 1982.
39. Bassin, W. M., Increasing hazard functions and overhaul policy, *ARMS IEEE-69C 8-R*, pp. 173–180, 1969.
40. Bassin, W. M., A Bayesian optimal overhaul interval model for the Weibull restoration process, *J. Am. Stat. Soc.*, 68, 575–578, 1973.
41. Bain, L. J., *Statistical Analysis of Reliability and Life-Testing Models Theory and Methods*, Marcel Dekker, New York, 1978.
42. Crow, L. H., Evaluating the reliability of repairable systems, *Proceedings of the Annual Reliability and Maintenance Symposium*, IEEE, Orlando, FL, 1990.
43. Yanez, Modarres, M., et al., Generalized renewal process for analysis of repairable systems with limited failure experience, *Reliab. Eng. Syst. Saf.*, 77 (2), 167–180, 2002.
44. Hurtado, J. L., F. Joglar, and M. Modarres, Generalized renewal process: Models, parameter estimation and applications to maintenance problems, *International Journal on Performability Engineering*, 1, 67–79, 2005.
45. Mettas, A. and W. Zhao, Modeling and analysis of repairable systems with general repair, *Proceedings of the Annual Reliability and Maintainability Symposium*, Alexandria, Virginia, 2005.
46. Caldorola, G., Unavailability and failure intensity of components, *Nucl. Eng. Des. J.*, 44, 147, 1977.
47. Lofgren, E. *Probabilistic Risk Assessment Course Documentation*, U.S. Nuclear Regulatory Commission, NUREG/CR-4350, Vol. 5, System Reliability and Analysis Techniques, Washington, DC, 1985.
48. Vesely, W. E. and F. F. Goldberg, FRANTIC—A Computer Code for Time-Dependent Unavailability Analysis, NUREG-0193, 1977.
49. Vesely, W. E., F. F. Goldberg, J. T. Powers, J. M. Dickey, J. M. Smith, and E. Hall, *FRANTIC II—A Computer Code for Time-Dependent Unavailability Analysis*, U.S. Nuclear Regulatory Commission, NUREG/CR-1924, Washington, DC, 1981.
50. Barroeta, C. and M. Modarres, Risk and economic estimation of inspection policy for periodically tested repairable components, *Proceedings of PSAM-8*, New Orleans, 2006.
51. Ginzburg, T. and W. E. Vesely, *FRANTIC-ABC User's Manual: Time-Dependent Reliability Analysis and Risk Based Evaluation of Technical Specifications*, Applied Biomathematics, Inc., Setauket, New York, 1990.
52. Modarres, M., A method of predicting availability characteristics of series-parallel systems, *IEEE Trans. Reliab.*, R-33,(4), 309–312, 1984.

6 Selected Topics in Reliability Modeling

In this chapter, we discuss a number of topics important to reliability modeling. These topics are not significantly related to each other, nor are they presented in any particular order. Some of the topics are still the subject of current research; the methods presented represent a summary of the state of the art.

6.1 PROBABILISTIC PHYSICS-OF-FAILURE RELIABILITY MODELING

Reliability experts often deal with situations in which system and component reliability is better modeled and estimated by modeling the underlying processes and phenomena that lead to failure. This is particularly important in passive system failures. There are three types of models that can be used to assess reliability, capability, or probability of failure of the various items using physics-of-failure modeling. Originally derived from mechanics and metallurgy, they are called *mechanistic models of failure*. Physics of failure is an analysis of the nature of the underlying process that leads to the failure of items. Due to limitations in the understanding of such a process, at best physics-based reliability can be described by empirical models. Such models are universally involved with uncertainties and characterization of such uncertainties in reliability engineering would be necessary. Physics-based reliability engineering with consideration of uncertainties lead to probabilistic physics of failure. Three types of modeling are possible in physics of failure.

Stress–Strength Model: This model assumes that the item (usually a passive component or system) fails if applied stress (mechanical, thermal, etc.) exceeds its strength. The *stress* represents an aggregate of the challenges and external conditions. This failure model may depend on environmental conditions or the occurrence of critical events, rather than on the mere passage of time or cycles. Strength is often treated as a r.v. representing the effect of all conditions affecting the strength, or the lack of knowledge about the item's strength (e.g., the item's capability, mechanical strength, and dexterity). Two examples of this model are a steel bar in tension, and a transistor with a voltage applied across the emitter–collector.

Damage–Endurance Model: This model is similar to the stress–strength model, but the scenario of interest is when applied stress causes damage that accumulates irreversibly, as in corrosion, wear, and fatigue. The aggregate of challenges and external conditions leads to the metric represented as cumulative damage. The cumulative damage may not degrade performance. The item fails when and only when the cumulative damage exceeds the endurance, that is, the damage accumulates until the endurance of the item is reached. As such, the item's capacity is measured by its tolerance to the accumulated damage or endurance. Accumulated damage does not disappear when the stresses are removed, although sometimes treatments such as annealing are possible. Both damage and endurance are often treated as r.v.s. Similar to the stress–strength model, endurance is an aggregate measure for the effects of challenges and external conditions on the item's capability to withstand cumulative stresses. As such, endurance may remain constant or experience decline.

Performance–Requirement Model: A component or system's performance is its ability to deliver its intended function or efficiency, which may degrade over time (e.g., a compressor or pump's efficiency or output), eventually reaching a level that is unacceptable or violates a requirement. In safety engineering the difference between performance and efficiency is known as the *safety margin* (SM). In this case the item may still be reliable, but has fallen bellow a required level of performance.

The mathematical aspects of the first two modeling approaches will be discussed in the next two sections. The third model, mathematically speaking, is very similar to the second.

6.1.1 Stress–Strength Model

As discussed in Chapter 1, a failure occurs when the stress applied to an item exceeds its strength. The probability that no failure occurs is equal to the probability that the applied stress is less than the item's strength. That is,

$$R = \Pr(S > s), \tag{6.1}$$

where R is the reliability of the item, s is the r.v. representing applied stress, and S is the r.v. representing item's strength.

Examples of stress-related failures include the following:

1. Misalignment of a journal bearing, lack of lubricants, or incorrect lubricants generate an internal load (mechanical or thermal stress) that causes the bearing to fail.
2. The voltage applied to a transistor gate is too high, causing a high temperature that melts the transistor's semiconductor material.
3. Cavitation causes pump failure, which in turn causes a violent vibration that ultimately breaks the rotor.
4. Lack of heat removal from a feed pump in a power plant results in overheating of the pump seals, causing the seals to break.
5. Thermal shock causes a pressurized vessel to experience fracture due to crack growth.

Engineers need to ensure that the strength of an item exceeds the applied stress for all possible stress situations. Traditionally, in the deterministic design process, safety factors are used to cover the spectrum of possible applied stresses. This is generally a good engineering principle, but failures occur despite these safety factors. On the other hand, safety factors that are too stringent result in overdesign, high cost, and sometimes poor performance.

If the range of major stresses is known or can be estimated, a probabilistic approach can be used to address the problem. This approach eliminates overdesign, high costs, and failures caused by stresses that are not considered early in the design. If the distribution of r.v.s S and s can be estimated as $F(S)$ and $g(s)$, then

$$\begin{aligned} R &= \int_0^\infty f(S)\, \mathrm{d}g(s), \\ F &= \int_0^\infty f(S) \left[\int_S^\infty g(s)\, \mathrm{d}s \right] \mathrm{d}S. \end{aligned} \tag{6.2}$$

Figure 6.1 shows typical relation between $f(S)$ and $g(s)$ distributions.

The SM is defined as

$$\mathrm{SM} = \frac{E(S) - E(s)}{\sqrt{\mathrm{var}(S) + \mathrm{var}(s)}}. \tag{6.3}$$

The SM shows the relative difference between the mean values for stress and strength. The larger the SM, the more reliable the item will be. Use of Equation 6.3 is a more objective way to measure the safety of items. It also allows for calculation of reliability and probability of failure as compared with the traditional deterministic approach using mean safety factors. However, good data on the variability

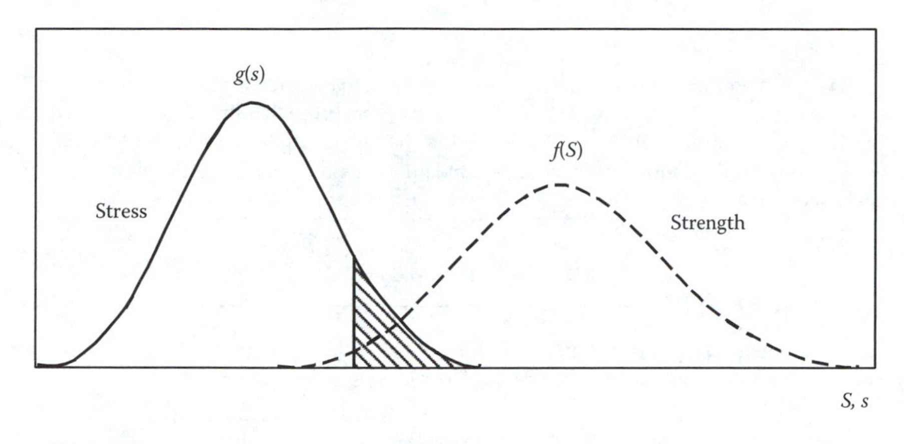

FIGURE 6.1 Stress–strength distributions.

of stress and strength are often not easily available. In these cases, engineering judgment can be used to obtain the distribution, including engineering uncertainty. The section on expert judgment explains methods for doing this in more detail.

The distribution of stress is highly influenced by the way the item is used and the internal and external operating environments. The design determines the strength distribution, and the degree of quality control in manufacturing primarily influences the strength variation.

For a normally distributed S and s,

$$R = \Phi(\mathrm{SM}), \tag{6.4}$$

where $\Phi(\mathrm{SM})$ is the cumulative standard normal distribution with $z = \mathrm{SM}$ (see Table A.1).

Example 6.1

Consider the stress and strength of a beam in a structure represented by the following normal distributions:

$$\mu_S = 420\,\text{kg/cm}^2 \text{ and } \sigma_S = 32\,\text{kg/cm}^2$$

$$\mu_s = 310\,\text{kg/cm}^2 \text{ and } \sigma_s = 72\,\text{kg/cm}^2$$

What is the reliability of this structure?

Solution:

$$\mathrm{SM} = \frac{420 - 310}{\sqrt{32^2 + 72^2}} = 1.4$$

with $z = 1.4$ and using Table A.1, $R = \Phi(1.4) = 0.91$.

Example 6.2

A r.v. representing the strength of a nuclear power plant containment building follows a lognormal distribution with a mean strength of 0.905 MPa and a standard deviation of 0.144 MPa. Three possible accident scenarios can lead to high-pressure conditions inside the containment building that may exceed its strength. The pressures cannot be calculated precisely but can be represented as another r.v. that follows a lognormal distribution.

continued

a. For a given accident scenario that causes a mean pressure load inside the containment of 0.575 MPa with a standard deviation of 0.117 MPa, calculate the probability that the containment fails.
b. If the four scenarios are equally probable, and each leads to high-pressure conditions inside the containment with the following mean and standard deviations, calculate the probability that the containment fails.

μ_L(MPa)	0.575	0.639	0.706	0.646
σ_L(MPa)	0.117	0.063	0.122	0.061

c. If the containment strength distribution is divided into the following failure mode contributors with the mean failure pressure and standard deviation indicated, repeat part a.

Failure Mode	**Mean Pressure, μ_S (MPa)**	**Standard Deviation, σ_S (MPa)**
Liner tear around personnel airlock	0.910	1.586×10^{-3}
Basemat shear	0.986	1.586×10^{-3}
Cylinder hoop membrane	1.089	9.653×10^{-4}
Wall-basemat junction shear	1.131	1.586×10^{-3}
Cylinder meridional membrane	1.241	1.034×10^{-3}
Dome membrane	1.806	9.653×10^{-4}
Personnel air lock door buckling	1.241	1.655×10^{-3}

Solution: If S is a lognormally distributed r.v. representing strength, and L is a lognormally distributed r.v. representing pressure stress (load), then the r.v., $Y = \ln(S) - \ln(L)$, is normally distributed.

For the lognormal distribution with a mean and standard deviation of μ_y and σ_y, the respective mean and standard deviation of the normal distribution, μ_t and σ_t, can be obtained using Equations 2.47 and 2.48. Then

a.

$$R = \Phi(\text{SM}) = \Phi\left(\frac{\mu_{S_t} - \mu_{L_t}}{\sqrt{\sigma_{S_t}^2 + \sigma_{L_t}^2}}\right).$$

$$R = \Phi\left(\frac{-0.112 - (-0.574)}{\sqrt{0.158^2 + 0.201}}\right) = 0.9649.$$

The probability of containment failure is

$$F = 1 - R = 0.0351.$$

b. Because the four scenarios are equally probable, the system is equivalent to a series system, such that $R = R_1 \times R_2 \times R_3 \times R_4$:

$$R_1 = \Phi(\text{SM1}) = \Phi(1.81) = 0.9649,$$

$$R_2 = \Phi(\text{SM2}) = \Phi(1.83) = 0.9664,$$

$$R_3 = \Phi(\text{SM3}) = \Phi(1.07) = 0.8577,$$

$$R_4 = \Phi(\text{SM4}) = \Phi(1.79) = 0.9633.$$

The probability of containment failure is

$$F = 1 - R = 1 - R_1 \times R_2 \times R_3 \times R_4 = 0.2296.$$

c. Because each failure mode may cause a system failure, this case can be treated as a series system. Because we know the median of the lognormal distribution instead of the mean, it takes several algebra steps to solve for the respective means and standard deviations:

$$R_a = \Phi(\text{SM}_a) = \Phi(2.38) = 0.9913$$

$$R_b = \Phi(\text{SM}_b) = \Phi(2.78) = 0.9973$$

$$R_c = \Phi(\text{SM}_c) = \Phi(3.27) = 0.9995$$

$$R_d = \Phi(\text{SM}_d) = \Phi(3.46) = 0.9997$$

$$R_e = \Phi(\text{SM}_e) = \Phi(3.92) \approx 1$$

$$R_f = \Phi(\text{SM}_f) = \Phi(5.78) \approx 1$$

$$R_g = \Phi(\text{SM}_g) = \Phi(3.92) \approx 1.$$

The probability of containment failure is

$$F = 1 - R = 1 - R_a \times R_b \times R_c \times R_d \times R_e \times R_f \times R_g = 0.0122.$$

If both the stress and strength distributions are exponential with parameters λ_s and λ_S, the reliability can be estimated as

$$R = \frac{\lambda_S}{\lambda_s + \lambda_S}. \tag{6.5}$$

For more information about stress–strength methods in reliability analysis, refer to [1] and [2].

6.1.2 Damage–Endurance Model

In this modeling technique failure occurs when applied stress causes permanent and irreversible damage as a function of time. When the level of damage reaches a level that the item is unable to endure, it fails. In practice, damage and endurance are not strict single values. Damage does not accumulate in the same way from sample to sample, and different initial conditions may result in drastically different levels of damage. Similarly, endurance limits of the samples are not the same, either because the strength of the material varies from sample to sample, or because the endurance limit is a characteristic of the material and varies from sample to sample. In any case, failure occurs only when accumulated damage exceeds the endurance limit of the material. The interference of variability, and the uncertainty in accumulated damage, endurance limit, and the occurrence of failure are illustrated in Figure 6.2. This figure shows how the interference of the accumulated damage and endurance results in a statistical distribution (describing uncertainty) of time to failure, which is the variable of interest in most reliability analyses.

Examples of cumulative damage include fatigue crack growth in a structure. When the crack size reaches a critical size, the structure fails. The critical size is the endurance limit. In nuclear reactors, the pressure vessel (a passive containment from the risk assessment point of view) can also be damaged by irradiation embrittlement; thus, its endurance will diminish over time to a point that it can no longer withstand possible anticipated or unanticipated thermal and mechanical applied stresses. This will cause the pressure vessel to rupture. It should be noted that time can affect both the damage level (i.e., increase it) and the endurance limit (i.e., decrease it). As such, as time passes the pdf of damage and the pdf of endurance approach each other.

From Figure 6.3, the problem is to establish a pdf of damage and endurance as a function of time or cycle of operation. Clearly, as we are unable to reduce all the uncertainties involved in measuring damage and endurance, they are often represented by probabilistic distribution functions as shown in Figure 6.3. The mathematical concept determining the reliability (and thus performance) of the item

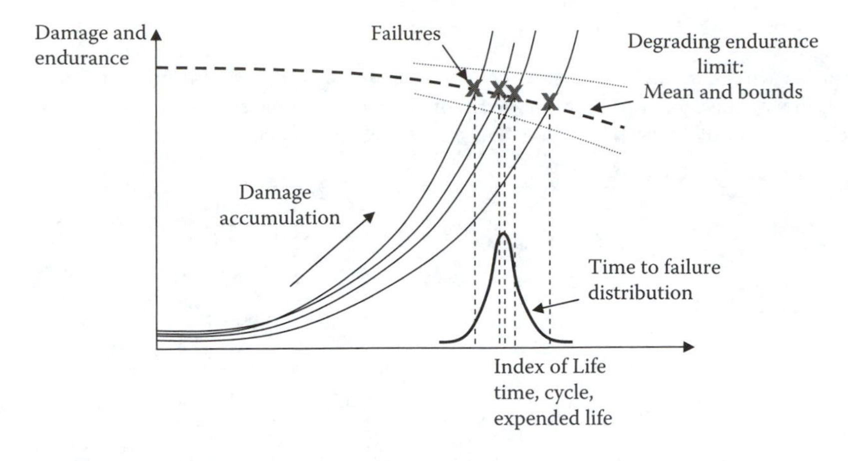

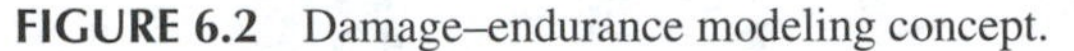
FIGURE 6.2 Damage–endurance modeling concept.

at a given time is similar to the strength–strength modeling approach, and Equations 6.1 through 6.4 apply to the damage–endurance model at a given age (time) or cycle of operation. Here, "stress" is now represented by "damage," and strength is represented by "endurance."

To estimate damage and endurance distributions, we need to understand the degradation processes leading to damage and reduction of endurance limits, and then perform degradation analysis. Degradation analysis (an element of physics of failure) involves the measurement and extrapolation of degradation or performance data that can be directly related to the presumed failure of the item in question. Many failure mechanisms can be directly linked to the degradation of parts on an item or a passive component, and degradation analysis allows the analyst to extrapolate to an assumed age or failure time based on the measurements of degradation or performance over time. To reduce testing or observation time further, tests and observations can be performed at elevated stress levels, and the degradation at these elevated levels can be measured. This type of analysis is known as *accelerated degradation*.

In some cases, it is possible to directly measure the degradation over time, as with the wear of a bearing or with the propagation of cracks in structures under random loading (causing fatigue-induced crack growth). In other cases, direct measurement of degradation might not be possible without invasive or destructive measurement techniques that would directly affect the subsequent performance of the item. In such cases, the degradation can be estimated through the measurement of certain performance characteristics, such as using resistance to gauge the degradation of a dielectric material. In both cases, it is necessary to define a level of degradation (endurance limit) at which a failure will occur. Once this endurance limit is established, it is simple to use basic mathematical

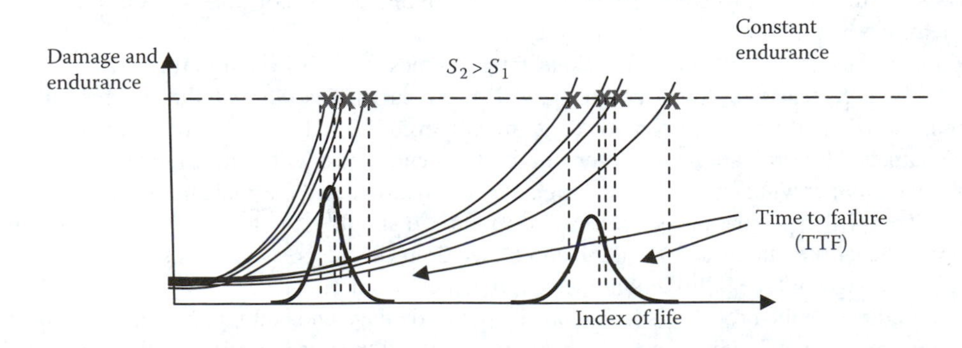

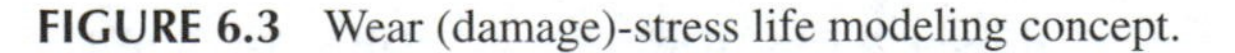
FIGURE 6.3 Wear (damage)-stress life modeling concept.

models (empirical models) to extrapolate failure probabilities (and thus, the performance measurements) over time to the point where the failure would occur. This is done at different damage levels (caused by applied stresses) and, therefore, each cycle to failure or time to failure is also associated with a corresponding stress level. Once the times to failure at the corresponding stress levels have been determined, it is merely a matter of analyzing the extrapolated failure times like conventional accelerated time-to-failure data.

Once the level of failure (or the degradation level that would constitute a failure) is defined, the degradation over time should be measured. The uncertainty in the results is directly related to the number of items and amount of data involved in the observations at each stress level, as well as to the amount of overstressing with respect to the normal operating conditions. The degradation of these items needs to be measured over time, either continuously or at predetermined intervals. Once this information has been recorded, the next task is to extrapolate the measurements to the defined failure level in order to estimate the failure time. Extrapolation is done by empirical models. These models have one or a combination of the following general forms:

$$\text{Linear: } y = ax + b, \tag{6.6}$$

$$\text{Exponential: } y = be^{ax}, \tag{6.7}$$

$$\text{Power: } y = bx^{a}, \tag{6.8}$$

$$\text{Logarithmic: } y = a\ \ln(x) + b, \tag{6.9}$$

where y represents the performance (e.g., the item's failure time, cumulative damage level, or endurance), x represents time, cycle, or stress applied, and a and b are model parameters to be estimated.

Once the model parameters a and b are estimated for each sample i, a time, cycle, or stress level, x_i, can be extrapolated that corresponds to the defined level of failure y. The computed x_i can now be used as our cycles or times to failure for subsequent life analysis. As with any sort of extrapolation, one must be careful not to extrapolate too far beyond the actual range of data in order to avoid large inaccuracies (modeling uncertainty, see [3]).

For example, consider developing the damage distribution in a damage–endurance model due to wear of a bearing in rotating machinery as a function of time. As expected from the physics of failure, the higher wear values lead to the lower life of the bearing. Note that a main assumption is that the mechanism of failure remains the same. Thus, the life distribution shape factor remains the same at all time as the damage progresses.

As illustrated in Figure 6.3, using the physics-of-failure model with one wear agent (i.e., τ_{max}), one significantly simplifies the wear problem and life estimation, because instead of dealing with many variables and conditions, only one value represents the aggregate effects of all others. In this figure, the degradation-life model in the form of Equation 6.8 is used and plotted in a log–log scale. The empirical model based on the physics of failure is shown in Equations 6.10 and 6.11.

Observations at higher stress levels (τ_{max1} and τ_{max2}) make the test time needed shorter and yield a better estimate of the probabilistic distribution of the life of the item. Typically, lognormal or Weibull distributions best represent the form of these distributions. Life distribution models in high stress levels are then extrapolated to the use level along the life acceleration model. Data from accelerated tests are then used to estimate the parameters of both the life-stress and distribution models. The wear agent is directly related to determining the distribution of damage of the bearing over time.

The maximum shear stress in the vicinity of the surface is an aggregate agent of failure responsible for material removal in the abrasive wear process. The wear rate (and thus the damage) can be expressed as a power relationship similar to Equation 6.8. Given the wear rate, the life of the bearing can be estimated by considering an endurance limit beyond which the bearing can be considered failed. This end state can be defined either based on the surface roughness or a certain amount of material removal on the bearing (e.g., if the coating thickness of bearing is removed). Based on the above description, the damage of the bearing can be related to the wear agent as

$$\dot{W} \propto \left[\frac{\tau_{max}}{\tau_{yp}}\right]^n, \quad (1/D) \propto L = \frac{C_0}{\dot{W}}. \tag{6.10}$$

Therefore,

$$L = C\left[\frac{\tau_{yp}}{\tau_{max}}\right]^n = \frac{K}{[\tau_{max}]^n}, \tag{6.11}$$

where $\dot{W}$ is the wear rate, C_0 is the end-state wear, C is a constant, L, D are life and damage of the bearing in terms of thickness removed (dependent variable), K, n are constants to be estimated from the accelerated test results, τ_{yp} is the material shear yield point, and τ_{max} is the maximum shear stress in the vicinity of the surface (independent variable).

Note that the presence of τ_{yp} (i.e., the shear yield point of the bearing material) enables this model also to consider the temperature degradation of the bearing material, which can be critical for coated-type bearings.

Assuming that the life is distributed lognormally, the distribution of life-stress can be obtained as

$$f(t \mid \mu, \sigma) = \frac{1}{t\sigma\sqrt{2\pi}} \exp\left[-\frac{1}{2}\frac{(\ln t - \mu)^2}{\sigma^2}\right]. \tag{6.12}$$

From Equations 6.11 and 6.12, μ = mean of the natural logarithms of the time to failure as a function of stress $= -\ln(K) - n\ln(\tau_{max})$. Therefore,

$$f(t \mid \mu, \sigma) = \frac{1}{t\sigma\sqrt{2\pi}} \exp\left[-\frac{1}{2}\frac{(\ln t + \ln(K) + n\ln(\tau_{max}))^2}{\sigma^2}\right]. \tag{6.13}$$

Example 6.3

A 2024-T3 Al component was tested to failure at two different stress values, 400 and 450 MPa, and the following data were obtained.

Estimate the number of cycles to failure at 300 MPa.

Solution: Assuming that the fatigue mechanism follows a lognormal life distribution, and using Equation 6.8 in the form of $L = K/[S]^n$ as the relationship for the stress-life model, when n and K are constants and S is the applied stress,

$$f(t \mid S) = \frac{1}{t\sigma_{T'}\sqrt{2\pi}} \exp\left[-\frac{1}{2}\frac{(\ln t + \ln(K) + n\ln(S))^2}{\sigma_{T'}^2}\right],$$

where t is cycles to failure, $\sigma_{T'}$ is the standard deviation of the natural logarithm of the cycles to failure, and K, n are the empirical model parameters.

The empirical model parameters are calculated through the maximum likelihood estimator approach as

$$K = 2.288\text{E–}18; n = 6.266; \text{ and } \sigma_{T'} = 0.195.$$

continued

Then the mean life using the model at 300 MPa is calculated from the expected value expression for a lognormal distribution,

$$\bar{t} = \exp\left[-\ln(K) - n\ln(S) + \frac{1}{2}\sigma_{T'}^2\right]$$
$$= \exp\left[-\ln(2.3\text{E–}18) - 6.3\ln(300) + \frac{1}{2}(0.2)^2\right] = 134.2\,\text{cycles},$$

clearly the same log standard derivation of 0.195 applies to the mean life.

Example 6.4

Time to failure of a journal bearing is believed to follow a lognormal distribution. Accelerated life (AL) tests revealed the median life of the bearing to be 2000 h in the first test, where a radial force of 750 pounds was applied on the bearing and 1500 h when the radial force was increased to 800 pounds in the second test. What is the expected median life for the bearing in normal operation where the radial force is 400 pounds?

Solution: Radial force on the bearing is the acceleration variable in these tests. Higher radial force results in higher pressure in the lubricant film and ultimately higher normal stress on the surface. The friction has also a linear relationship with the normal load being applied on the bearing. τ_{max}, which is the maximum shear stress in the vicinity of mating surfaces, is estimated from

$$\tau_{max} = ke\sqrt{\left(\frac{\sigma_n}{2}\right)^2 + \tau_f^2},$$

where τ_{max} is the maximum shearing stress, *ke* is the stress concentration factor, σ_n is the normal stress on the surface, τ_f is, $\mu\sigma_n$, the friction-generated shear stress, and μ is the friction factor.

Since normal stress σ_n is a linear function of bearing radial force, the τ_{max} becomes a linear function of applied radial force, and we have

$$\frac{T_1}{T_2} = \frac{2000}{1500} = \frac{(K/\tau_{max1}^n)}{(K/\tau_{max2}^n)} = \left(\frac{F_2}{F_1}\right)^n = \left(\frac{800}{750}\right)^n \Rightarrow n = 4.457,$$

$$\frac{T_{use}}{T_2} = \frac{T_{use}}{1500} = \frac{(K/\tau_{use}^n)}{(K/\tau_{max2}^n)} = \left(\frac{F_2}{F_{use}}\right)^n = \left(\frac{800}{400}\right)^{4.457} \Rightarrow T_{use} \approx 32,944\,\text{h}.$$

Example 6.5

An AISI 4130X cylindrical tank is subjected to degradation from the corrosion-fatigue mechanism. Calculate the crack growth resistance at 30 ksi-in$^{1/2}$, in a wet gas environment, according to the following data:

Solution: The crack growth behavior is generally modeled using the Paris equation,

$$\frac{da}{dN} = C(\Delta K)^m,$$

where a is the crack length, N is the number of cycles, ΔK is the stress intensity factor range, and C, m are material constants.

continued

The resistance of the material to crack growth can be written as the inverse of the crack growth rate,

$$\frac{1}{\mathrm{d}a/\mathrm{d}N} = \frac{1}{C(\Delta K)^m}.$$

Then, setting $1/\mathrm{d}a/\mathrm{d}N = L(V)$, where V represents the stress intensity factor and L the crack resistance,

$$L(V) = \frac{1}{C(V)^m}.$$

Assuming that the size of a crack induced by the corrosion-fatigue mechanism follows a lognormal life distribution, and an inverse power law relationship for the stress-life model is applicable,

$$f(a' \mid V) = \frac{1}{T\sigma_{a'}\sqrt{2\pi}} \exp\left[-\frac{1}{2}\frac{(\ln(a') + \ln(C) + m\ln(V))^2}{\sigma_{a'}^2}\right],$$

where a' is the natural logarithm of the crack growth resistance, $\sigma_{a'}$ is the standard deviation of the natural logarithms of the crack growth resistance, and C, m are the inverse power law (IPL) model parameters.

The IPL parameters are calculated through the maximum likelihood estimator approach as

$$C = 2.0496\text{E}11; m = -5.1591; \text{ and } \sigma_a = 0.8191.$$

Then, the mean life of the IPL-lognormal model at 30 ksi-in$^{1/2}$ is calculated from the expected value expression for a lognormal distribution,

$$\begin{aligned}\bar{a}' &= \exp\left[-\ln(C) - m\ln(V) + \frac{1}{2}\sigma_{a'}^2\right] \\ &= \exp\left[-\ln(2.1\text{E}11) + 5.2\ln(30) + \frac{1}{2}(0.8)^2\right] \\ &= 2.9\text{E} - 04 \text{ in/cycles}.\end{aligned}$$

6.1.3 Performance–Requirement Model

It is possible that the performance of an item, such as efficiency, properties (e.g., resistivity and emissivity), tolerance (accuracy in measurement), and output intensity (e.g., pressure, flow, and current), may change over the time or cycle of operation. Usually performance degrades and it is irreversible. The rate of performance decline is not only a function of both the material and physics of operation, but also the applied stresses and level of usage. As depicted in Figure 6.4, the performance

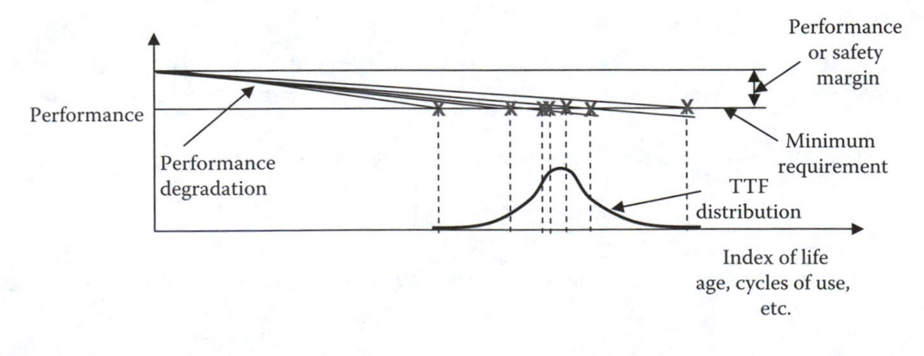

FIGURE 6.4 Performance–requirement model.

declines over time according to one of the empirical model in the form of Equations 6.6 through 6.9 (or their combination), where y would be the amount of performance at the time (or cycles of operation). When the performance level goes below a minimum level of performance (i.e., requirement), a failure occurs. While in engineering systems and components, performance usually declines over time or cycles of usage, it is equally possible that the level of requirement may change over time. This, however, may be a level set by the end users. The distance between the performance and requirement is often referred to as the "margin of operation." In safety systems and components this is particularly important and is known as the "SM." In such systems and components a failure may occur when the SM itself reaches a minimum level.

Example 6.6

The voltage provided by a specific type of battery is believed to degrade in time following a linear trend. The battery is considered failed when its voltage is less than 12 V.

Two measurements (6 months, 13.88 V) and (18 months, 13.63 V) became available from the field tests. If the variability of provided voltage can be characterized by a normal distribution with a constant standard deviation of 1 V, what percentage of batteries is expected to return due to low voltage failures within 2 years of guarantee period?

Solution: Using the available measurements one can estimate the parameters of the linear physical model of provided voltage as follows:

$$V = a \times t + b \Rightarrow \left. \begin{matrix} 13.88 = a \times 6 + b \\ 13.63 = a \times 18 + b \end{matrix} \right\} \Rightarrow \left\{ \begin{matrix} a = -0.0208 \\ b = 14.00 \end{matrix} \right. .$$

Using this physical model, the average provided voltage after 24 months (guarantee period) can be calculated as

$$V = a \times t + b \Rightarrow V(24) = -0.0208 \times 24 + 14 = 13.500\, V.$$

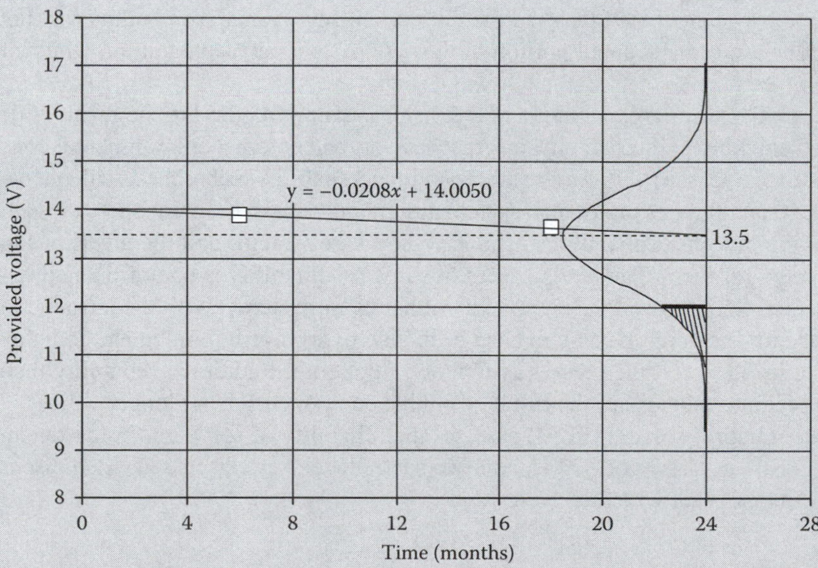

Since the voltage is normally distributed, the return percentage after 24 months is basically the cumulative distribution of voltage at 12 V as shown in the figure, and we have

$$\text{Return percentage} = F(24) = \Phi\left(\frac{12 - 13.500}{1}\right) = 0.067 \equiv 6.7\%.$$

6.2 SOFTWARE RELIABILITY ANALYSIS

6.2.1 Introduction

Many techniques have been developed for analyzing the reliability of physical systems. However, their extension to software has been problematic for two reasons. First, the software faults are design faults, while faults in physical systems are equipment breakage or human error. Second, software systems are more complex than physical systems, so the same reliability analysis methods may be impractical to use.

Software has deterministic behavior, whereas hardware behavior is both deterministic and stochastic. Indeed, once a set of inputs to the software has been selected, the software will either fail or execute correctly, provided that the computer and operating system with which the software will run is error-free. However, our knowledge of the inputs selected, of the computer, of the operating system, and of the nature and position of the fault may be uncertain. One may translate this uncertainty into probabilities. A software fault is a triggering event that causes software error. A software bug (error in the code [4–6]) is an example of a fault.

Accordingly, we can adopt a probabilistic definition for software reliability. Software reliability is the probability that the software product will not fail for a specified time under specified conditions. This probability is a function of the input to and use of the product, as well as a function of the existence of faults in the software. The inputs to the product will determine whether an existing fault is encountered or not.

Faults can be classified as design faults, operational faults, or transient faults. All software faults are design faults; however, hardware faults may occur in any of the three classes. Faults can also be classified by the source of the fault; software and hardware are two of the possible sources of the fault. Sources of faults are input data, system state, system topology, humans, environment, and unknown causes. For example, the source of many transient faults is unknown.

Failures in software are classified by mode and scope. A failure mode may be sudden or gradual, partial or complete. All four combinations of these are possible. The scope of failure describes the extent within the system of the effects of the failure. This may range from an internal failure, whose effect is confined to a single small portion of the system, to a pervasive failure, which affects much of the system (see [7]).

Software, unlike hardware, is unique in that its only failure modes are the result of design flaws, as opposed to any kind of internal physical mechanisms and external environmental conditions such as aging. For example, see [8,9]. As a result, traditional reliability techniques, which tend to focus on physical component failures rather than system design faults, have been unable to close the widening gap between the powerful capabilities of modern software systems and the levels of reliability that can be computed for them. The real problem of software reliability is one of managing complexity. There is a natural limitation on the complexity of hardware systems. With the introduction of digital computer systems, however, designers have been able to arbitrarily implement complex designs in software. The result is that the central assumption implicit in traditional reliability theory, that the design is correct and failures are the result of fallible components, is no longer valid.

A software reliability model (SRM) assesses the reliability of software. In the remainder of this section, details about classes of SRMs, and two such models, are discussed. Also discussed are the models used to assess software life cycles.

6.2.2 Software Reliability Models

Software failure may be defined as the probability of failure-free software operation for a specified period of time in a specified environment. Software failure may be due to errors, ambiguities, oversights, misinterpretation of specifications, and so on [10].

Several SRMs have been developed over the years. These techniques are variously referred to as "analyses" or "models," but there is a distinct difference between the two. An *analysis* (such as fault tree analysis, FTA) is carried out by creating a model (the fault tree) of a system, and then using that model to calculate properties of interest, such as reliability.

The standard reliability models such as FTA, event tree analysis (ETA), failure modes and effect analysis (FMEA), and Markov models discussed in this book are adequate for systems whose components remain unchanged for long periods of time. They are less flexible for systems that undergo frequent design changes. If, for example, the failure rate of a component is improved through design or system configuration changes, the reliability model must be re-evaluated. A reliability growth model (see Section 6.6) is more appropriate for these cases. In this model, software is tested for a period of time, during which failures may occur. These failures lead to modification to the design or manufacture of the component; the new version then goes back into test. This cycle is continued until design objectives are met. Software reliability growth is a very active research area today.

When these models are applied to software reliability, one can group them into two main categories: predictive models and assessment models [11]. Predictive models typically address the reliability of software early in the design cycle. Different elements of a life cycle development of software are discussed later. Predictive models are developed to assess the risks associated with the development of software under a given set of requirements and for specified personnel before the project truly starts. Predictive SRMs are few in number [11], and so the predictive models are not discussed in this section. Assessment models evaluate present and project future software reliability from failure data gathered when the integration of the software starts.

6.2.2.1 Classification

Most existing SRMs may be grouped into four categories:

1. Time-between-failure models
2. Fault seeding models
3. Input-domain-based models
4. Failure count models.

Each category of models is summarized here.

6.2.2.1.1 Time-between-Failure Models

This category includes models that provide an estimate of the times between failures in software. Key assumptions of this model are as follows:

- Independent times between successive failures
- Equal probability of exposure of each fault
- Embedded faults are independent of each other
- No new faults introduced during corrective actions.

Specific SRMs that estimate mean time between failures are

- Jelinski–Moranda's model [12]
- Schick and Wolverton's model [13]
- Littlewood—Verrall's Bayesian model [14,15]
- Goel and Okumoto's [16,17] imperfect debugging model.

6.2.2.1.2 Fault Seeding Model

This category of SRMs includes models that assess the number of faults in the software at time zero via seeding extraneous faults. Key assumptions of this model are as follows:

- Seeded faults are randomly distributed in the software.
- Indigenous and seeded faults have equal probabilities of being detected.

The specific SRM that falls into this category is the Mills Fault Seeding Model [18]. In this model, an estimate of the number of defects remaining in a program can be obtained by a seeding process that assumes a homogeneous distribution of a representative class of defects. The variables in this measure are the number of seed faults introduced, N_S; the number of intentional seed faults found, n_s; and the number of faults found, n_F, that were not intentionally seeded.

Before seeding, a fault analysis is needed to determine the types of faults expected in the code and their relative frequency of occurrence. An independent monitor inserts into the code N_S faults that are representative of the expected indigenous faults. During testing, both seeded and unseeded faults are identified. The number of seeded and indigenous faults discovered permits an estimate of the number of faults remaining for the fault type considered. The measure cannot be computed unless some seeded faults are found. The MLE of the unseeded faults is given by

$$\hat{N}_F = \frac{n_F N_S}{n_S}. \tag{6.14}$$

Example 6.7

Forty faults of a given type are seeded into a code and, subsequently, 80 faults of that type are uncovered, 32 seeded, and 48 unseeded. Calculate an estimate of unseeded faults. How many faults remain to be found?

Solution: Using Equation 6.14, $N_F = 60$, and the estimate of faults remaining is

$$N_F(\text{remaining}) = N_F - n_F = 60 - 48 = 12.$$

6.2.2.1.3 Input-Domain-Based Model

This category of SRMs includes models that assess the reliability of software when the test cases are sampled randomly from a well-known operational distribution of software inputs. The reliability estimate is obtained from the number of observed failures during execution. Key assumptions of these models are as follows:

- The input profile distribution is known.
- Random testing is used (inputs are selected randomly).
- The input domain can be partitioned into equivalent classes.

Specific models of this category are

- Nelson's model [19]
- Ramamoorthy and Bastani's model [20].

Nelson's model is typically used for systems with ultrahigh-reliability requirements, such as software used in nuclear power plants, and are limited to about 1000 lines of code. The model is applied to the validation phase of the software (acceptance test) to estimate the reliability. Nelson defines the reliability of a program that was run n times (for n test case) and failed n_f times as

$$R = 1 - \frac{n_f}{n}, \tag{6.15}$$

where n is the total number of test cases and n_f is the number of failures experienced out of these test cases.

6.2.2.1.4 *Failure Count Model*

This category of SRMs estimates the number of faults or failures experienced in specific intervals of time. Key assumptions of these models are as follows:

- Test intervals are independent of each other.
- Testing intervals are homogeneously distributed.
- The number of faults detected during nonoverlapping intervals is independent of each other.

The SRMs that fall into this category are

- Shooman's exponential model [21]
- Goel–Okumoto's [16,17] non-HPP
- Musa's [22] execution time model
- Goel's [23] generalized non-HPP model
- Musa–Okumoto's [24] logarithmic Poisson execution time model.

6.2.2.1.5 *Musa's Basic Execution Time Model (BETM)*

This model [22] assumes that failures occur in the form of a non-HPP. The unit of failure intensity is the number of failures per central process unit (CPU) time. This relates failure events to the processor time used by the software. In the BETM, the reduction in the failure intensity function remains constant, irrespective of whether any failure is being fixed.

The failure intensity, as a function of the number of failures experienced, is obtained from

$$\lambda(\mu) = \lambda_0 \left(1 - \frac{\mu}{\nu_0}\right), \tag{6.16}$$

where $\lambda(\mu)$ is the failure intensity (failures/CPU hour), λ_0 is the initial failure intensity at the start of execution, μ is the expected number of failures experienced up to a given point in time, and ν_0 is the total number of failures.

The number of failures that needs to be fixed to move from a present failure intensity to a target intensity is given by

$$\Delta\mu = \frac{\nu_0}{\lambda_0}(\lambda_p - \lambda_F). \tag{6.17}$$

The execution time required to reach this objective is

$$\Delta\tau = \frac{\nu_0}{\lambda_0} \ln\left(\frac{\lambda_p}{\lambda_F}\right). \tag{6.18}$$

In these equations, ν_0 and λ_0 can be estimated in different ways, see [25].

6.2.2.1.6 *Musa–Okumoto's Logarithmic Poisson Time Model*

According to the logarithmic Poisson time model (LPETM), the failure intensity is given by

$$\lambda(\mu) = \lambda_0 \exp(-\theta\mu), \tag{6.19}$$

where θ is the failure intensity decay parameter and λ, μ, λ_0 are the same as in the BETM. This model assumes that repair of the first failure has the greatest impact in reducing failure intensity and that the impact of each subsequent repair decreases exponentially.

TABLE 6.1
Execution Time Parameters

	Model	
Parameter	**Basic**	**Logarithmic Poisson**
Initial failure intensity	λ_0	λ_0
Failure intensity change		
Total failures	ν_0	—
Failure intensity decay parameter	—	θ

In the LPETM, no estimate of ν_0 is needed. The expected number of failures that must occur to move from a present failure intensity of λ_p to a target intensity of λ_F is

$$\Delta\mu = \frac{1}{\theta}\ln\left(\frac{\lambda_p}{\lambda_F}\right). \tag{6.20}$$

The execution time to reach this objective is given by

$$\Delta\tau = \frac{1}{\theta}\left(\frac{1}{\lambda_F} - \frac{1}{\lambda_p}\right). \tag{6.21}$$

As we have seen, the execution time components of these models are characterized by two parameters. These are listed in Table 6.1.

EXAMPLE 6.8

Assume that a software program will experience 200 failures in its lifetime. Suppose it has now experienced 100 of them. The initial failure intensity was 20 failures/CPU hour. Using BETM and LPETM, calculate the current failure intensity (assume that the failure intensity decay parameter is 0.02 per failure).

Solution: For BETM,

$$\lambda(\mu) = \lambda_0\left(1 - \frac{\mu}{\nu_0}\right) = 20\left(1 - \frac{100}{200}\right) = 10 \text{ failure per CPU hour.}$$

For LPETM,

$$\lambda(\mu) = \lambda_0 \exp(-\theta\mu) = 20\exp[-(0.02)(100)] = 2.7 \text{ failures per CPU hour.}$$

The most common approach to software reliability analysis is testing. Testing is often performed by feeding random inputs into the software and observing the output produced to discover incorrect behavior. Because of the complexity of today's modern computer systems, however, these techniques often result in the generation of an enormous number of test cases. For example, Petrella et al. [26] found that Ontario Hydro's validation testing of its Darlington Nuclear Generating Station's new computerized emergency reactor shutdown systems required a minimum of 7000 separate tests to demonstrate 99.99% reliability at 50% confidence.

Software reliability growth models have not had a great impact so far in reducing the quantity and cost of software testing necessary to achieve a reasonable level of reliability.

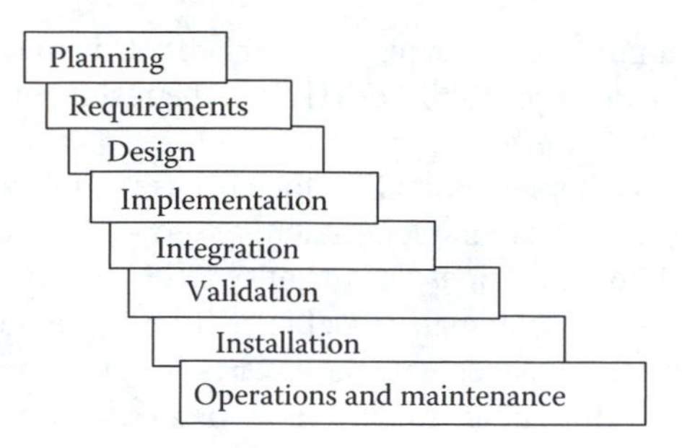

FIGURE 6.5 Waterfall model.

6.2.3 Software Life-Cycle Models

Many different life-cycle models exist for developing software systems. According to Boehm [27], the following types of process models can be distinguished:

- Sequential models
- Loop models (waterfall models)
- V-models (V stands for verification)
- Viewpoint models
- Spiral models.

These models have different motivations, strengths, and weaknesses. Many reliability, performance, and safety problems can be resolved only by careful design of the software. These must be addressed early in the life-cycle, no matter which life-cycle model is used. The life-cycle models generally require the same type of tasks to be carried out; they differ in the ordering of these tasks in time [7]. We will further elaborate on the Waterfall model.

6.2.3.1 Waterfall Model

This is a life-cycle model for software development. The classical waterfall model of software development assumes that each phase of the life cycle must be completed before the next phase can start [28]. The model permits the developer to return to previous phases. For example, if a requirement error is discovered during the implementation phase (Figure 6.5), the developer is expected to halt the development, return to the requirement phase, fix the problem, change the design accordingly, and then restart the implementation from the revised design. In practice, one may only stop the implementation affected by the newly discovered requirement. The waterfall model has been severely criticized as not being realistic to many software development situations [7]. However, despite this concern, it remains a useful model for situations where the requirements are known and stable before development begins, and where little change to the requirements is anticipated [7].

6.3 HUMAN RELIABILITY

It has long been recognized that human error has a substantial impact on the reliability of complex systems. Accidents at Three Mile Island and Chernobyl clearly show how human error can defeat engineered safeguards and play a dominant role in the progression of accidents. 70% of aviation accidents are caused by human malfunctions, and similar figures apply to the shipping and process industry. The Reactor Safety Study [29] revealed that more than 60% of the potential accidents in the nuclear industry are related to human errors. In general, the human contribution to overall system performance is at least as important as that of hardware reliability.

To obtain a precise and accurate measure of system reliability, human error must be taken into account. Analysis of system designs, procedures, and post-accident reports shows that human error can be an immediate accident initiator or can play a dominant role in the progress of undesired events. Without incorporating human error probabilities (HEPs), the results of a system reliability analysis are incomplete and often underestimated.

To estimate HEPs (and, thus, human reliability), one needs to understand human behavior. However, human behavior is very difficult to model. Literature shows that there is not a strong consensus on the best way to capture all human actions and quantify HEPs. The assumptions, mechanisms, and approaches used by any one specific human model cannot be applied to all human activities. Current human models need further advancement, particularly in capturing and quantifying intentional human errors. Limitations and difficulties in current human reliability analysis (HRA) include the following:

1. Human behavior is a complex subject that cannot be described as a simple component or system. Human performance can be affected by social, environmental, psychological, and physical factors that are difficult to quantify.
2. Human actions cannot be considered to have binary success and failure states, as in hardware failure. Furthermore, the full range of human interactions has not been fully analyzed by HRA methods.
3. The most difficult problem with HRA is the lack of appropriate data on human behavior in extreme situations.

Human error may occur in any phase of the design, manufacturing, construction, and operation of a complex system, resulting in many types of errors during system operation. The most notable errors are dependent failures whose occurrence can cause loss of system redundancy. These may be discovered during manufacturing and construction, or during system operation. Normally, quality assurance programs are designed and implemented to minimize the occurrence of these types of human error.

In this book, we are concerned with human reliability during system operation, where human operators are expected to maintain, supervise, and control complex systems. In the remainder of this section, human reliability models are reviewed, and important models are described in some detail. Emphasis is on the basic ideas, advantages, and disadvantages of each model, and their applicability to different situations. Then, we describe the most important area of data analysis in HRA. After the links between models and data are reviewed, the problems with human reliability data sources and respective data acquisition are addressed.

6.3.1 HRA Process

A comprehensive method of evaluating human reliability is the method called Systematic Human Action Reliability Procedure (SHARP) developed by Hannaman and Spurgin [30]. The SHARP defines seven steps of HRA. Each step consists of inputs, activities, rules, and outputs. The inputs are derived from prior steps, reliability studies, and other information sources, such as procedures and accident reports. The rules guide the activities that are needed to achieve the objectives of each step. The output is the product of the activities performed by analysts. The goals for each step are as follows:

1. *Definition*: Ensure that all different types of human interactions are considered.
2. *Screening*: Select the human interactions that are significant to system reliability.
3. *Qualitative Analysis*: Develop a detailed description of important human actions.
4. *Representation*: Select and apply techniques to model human errors in system logic structures, for example, fault trees, event trees, master logic diagram (MLD), or reliability block diagrams.
5. *Impact Assessment*: Explore the impact of significant human actions identified in the preceding step on the system reliability model.

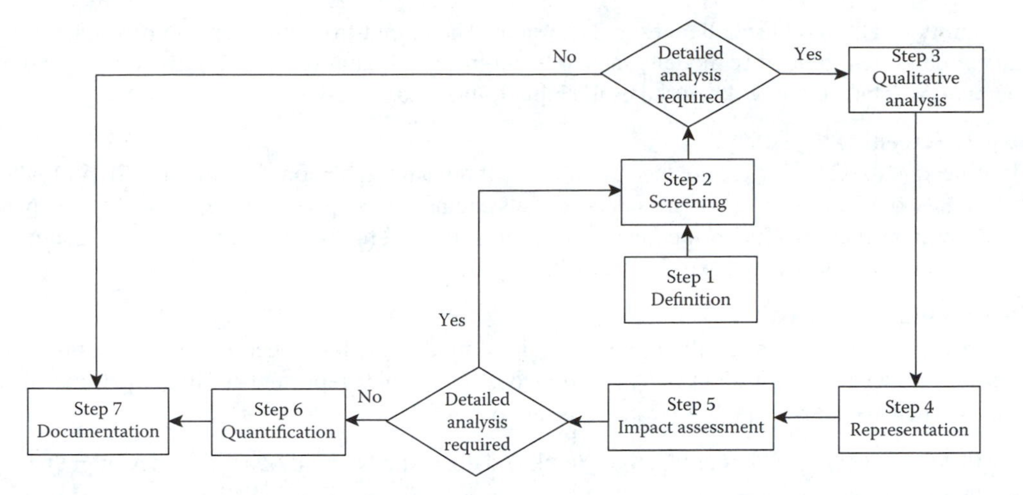

FIGURE 6.6 Systematic Human Action Reliability Procedure [30].

6. *Quantification*: Apply appropriate data to suitable human models to calculate probabilities for various interactions under consideration.
7. *Documentation*: Include all necessary information for the assessment to be understandable, reproducible, and traceable.

The relationships among these steps are shown in Figure 6.6.

These steps in human reliability consideration are described in more detail below.

Step 1: Definition

The objective of Step 1 is to ensure that key human interactions are included in the human reliability assessment. Any human actions with a potentially significant impact on system reliability must be identified at this step to guarantee the completeness of the analysis. Human activities can generally be classified according to the following types of behavior:

Type 1: Before any challenge to a system, an operator can affect availability, reliability, and safety by restoring safeguard functions during testing and maintenance.

Type 2: By committing an error, an operator can initiate a challenge to the system, causing the system to deviate from its normal operating envelope.

Type 3: By following procedures during the course of a challenge, an operator can operate redundant systems (or subsystems) and recover the systems to their normal operating envelope.

Type 4: By executing incorrect recovery plans, an operator can aggravate the situation or fail to terminate the challenge to the systems.

Type 5: By improvising, an operator can restore initially failed equipment to terminate a challenge.

As recommended by the SHARP, HRA should use these types of activities to investigate the system to reveal possible human interactions. Analysts can apply these types to different system levels as appropriate. For example, Type 1 interactions generally involve components, whereas Type 3 and Type 4 interactions are mainly operating actions that are performed at the system level. Type 5 interactions are recovery actions that may affect both systems and components. Type 2 interactions can generally be avoided by confirming that human-induced errors are included as contributors to the probability of all possible challenges to the system. The output from this step can be used to revise and enrich system reliability models, such as event trees and fault trees, to fully account for human interactions. This output will be used as the input to the next step.

Step 2: Screening

The objective of screening is to reduce the number of human interactions identified in Step 1 to those that might potentially challenge the safety of the system. This step provides analysts with a chance to concentrate their efforts on the key human interactions. This is generally done in a qualitative manner; the process is subjective.

Step 3: Qualitative Analysis

To incorporate human errors into equipment failure modes, analysts need more information about each key human interaction identified in the previous steps to help represent and quantify these human actions. The two goals of qualitative analysis are to

1. Postulate what operators are likely to think and do, and what kind of actions they might take in a given situation.
2. Postulate how an operator's performance may modify or trigger a challenge to the system.

This process of qualitative analysis may be broken down into four key stages:

1. Information gathering
2. Prediction of operator performance and possible human error modes
3. Validation of predictions
4. Representation of output in a form appropriate for the required function.

In summary, the qualitative analysis step requires a thorough understanding of what performance-shaping factors (PSFs) (e.g., task characteristics, experience level, environmental stress, and social-technical factors) affect human performance. Based on this information, analysts can predict the range of plausible human actions. The psychological model proposed by Rasmussen [31] is a useful way of conceptualizing the nature of human cognitive activities. The full spectrum of possible human actions following a misdiagnosis is typically very hard to recognize. Computer simulations of performance described by Woods et al. [32] and Amendola et al. [33] offer the potential to assist human reliability analysts in predicting the probability of human errors. Training of the human operators is also an important consideration in PSF evaluation as it has been shown in various accident investigations [34,35].

Step 4: Representation

To combine the HRA results with the system analysis models of Chapter 4, human error modes need to be transformed into appropriate representations. Representations are selected to indicate how human actions can affect the operation of a system.

Three basic representations have been used to delineate human interactions: the operator action tree (OAT) described by Wreathall [36], the confusion matrix described by Potash et al. [37], and the HRA event trees described by Swain and Guttman [38]. Figure 6.7 shows an example of OAT. The HRA tree is discussed in Section 6.3.2.

Step 5: Impact Assessment

Some human actions can introduce new impacts on the system response. This step provides an opportunity to evaluate the impact of the newly identified human actions on the system.

The human interactions represented in Step 4 are examined for their impact on challenges to the system, system reliability, and dependent failures. Screening techniques are applied to assess the importance of the impacts. Important human interactions are found, reviewed, and grouped into suitable categories. If the re-examination of human interactions identifies new human-induced challenges or behavior, the system analysis models (e.g., MLD, fault tree) are reconstructed to incorporate the results.

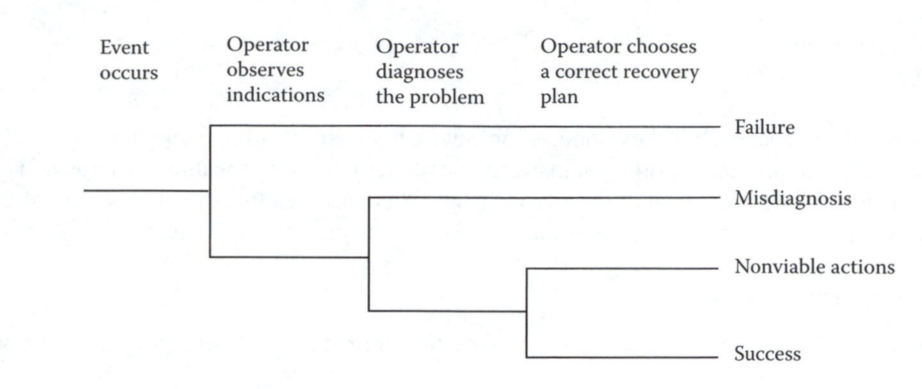

FIGURE 6.7 Operator action tree.

Step 6: Quantification

The purpose of this step is to assess the probabilities of success and failure for each human activity identified in the previous steps. In this step, analysts apply the most appropriate data or models to produce the final quantitative reliability analysis. Selection of the models should be based on the characteristics of each human interaction.

Guidance for choosing the appropriate data or models to be adopted is provided below:

- For procedural tasks, the data from Swain and Guttman [38] or the equivalent can be applied.
- For diagnostic tasks under time constrains, time–reliability curves from Hall et al. [39] or the human cognitive reliability (HCR) model from Hanamman et al. [40] can be used.
- For situations where suitable data are not available, expert opinion approaches, such as paired comparison by Hunns and Daniels [41] and the success likelihood index method (SLIM) by Embry et al. [42] can be used.
- For situations where multiple tasks are involved, the dependence rules discussed by Swain and Guttman [38] can be used to assess the quantitative impact.

Step 7: Documentation

The objective of Step 7 is to produce a traceable description of the process used to develop the quantitative assessments of human interactions. The assumptions, data sources, selected models, and criteria for eliminating unimportant human interactions should be carefully documented. The human impact on the system should be stated clearly.

6.3.2 HRA Models

The HRA models can be classified into the following categories. Representative models in each are also summarized below:

1. Simulation methods
   a. Maintenance personnel performance simulation (MAPPS)
   b. Cognitive environment simulation (CES)
2. Expert judgment methods
   a. Paired comparison
   b. Direct numerical estimation (absolute probability judgment)
   c. Success likelihood index method (SLIM)
3. Analytical methods
   a. Technique for human error rate prediction (THERP)

b. HRC correlation
c. Time–reliability correlation (TRC)

We will briefly discuss each of these models. Because human error is a complex subject, there is no single model that captures all important human errors and predicts their probabilities. Poucet [43,44] reports the results of a comparison of the HRA models. He concludes that the methods could yield substantially different results, and presents their suggested use in different contexts.

6.3.2.1 Simulation Methods

These methods primarily rely on computer models that mimic human behavior under different conditions.

6.3.2.1.1 Maintenance Personnel Performance Simulation

MAPPS, developed by Siegel et al. [45], is a computerized simulation model that provides human reliability estimations for testing and maintaining tasks. To perform the simulation, analysts must first determine the necessary tasks and subtasks that individuals must perform. Environmentally motivated tasks and organizational variables that influence personnel performance reliability are input into the program. Using the Monte Carlo simulation, the model can output the probability of success, time to completion, idle time, human load, and level of stress. The effects of a particular parameter or subtask performance can be investigated by changing the parameter and repeating the simulation.

The simulation output of task success is based on the difference between the ability of maintenance personnel and the difficulty of the subtask. The model used is

$$\Pr(\text{success}) = \frac{\exp(y)}{1 + \exp(y)}, \tag{6.22}$$

where $y > 0$ is the difference between personnel ability and task difficulty.

6.3.2.1.2 Cognitive Environment Simulation

Woods [32] has developed a model based on techniques from artificial intelligence (AI). The model is designed to simulate a limited-resources problem solver in a dynamic, uncertain, and complex situation. The main focus is on the formation of intentions, situations, and factors leading to intentional failures, forms of intentional failures, and the consequence of intentional failures.

Similar to the MAPPS model, the CES model is a simulation approach that mimics the human decision-making process during an emergency condition. But CES is a deterministic approach, which means the program will always obtain the same results [46] if the input is unchanged. The first step in CES is to identify the conditions leading to human intentional failures. CES provides numerous performance-adjusting factors to allow the analysts to test different working conditions. For example, analysts may change the number of people interacting with the system (e.g., the number of operators), the depth or breadth of their working knowledge, or the human–machine interface. Human error-prone points can be identified by running the CES for different conditions. The human failure probability is evaluated by knowing, *a priori*, the likelihood of occurrence of these error-prone points.

In general, CES is not a human rate quantification model. It is primarily a tool to analyze the interaction between problem-solving resources and task demands.

6.3.2.2 Expert Judgment Methods

The primary reason for using expert judgment in HRA is that there often exists little or no relevant or useful human error data. Expert judgment is discussed in more detail in Section 6.4. There are two requirements for selecting experts:

- They must have substantial expertise.
- They must be able to accurately translate this expertise into probabilities.

6.3.2.2.1 *Direct Numerical Estimation*

For the direct numerical estimation method described by Stillwell et al. [47], experts are asked to directly estimate the HEPs and the associated upper/lower bounds for each task. A consistency analysis might be performed to check for agreement among these judgments. Then, individual estimations are aggregated by either arithmetic or geometric average.

6.3.2.2.2 *Paired Comparison*

Paired comparison, described by Hunns and Daniel [41], is a scaling technique based on the idea that judges are better at making simple comparative judgments than making absolute judgments. An interval scaling is used to indicate the relative likelihood of occurrence of each task. Saaty [48] describes this general approach in the context of a decision analysis technique. The method is equally applicable to HRA.

6.3.2.2.3 *Success Likelihood Index Method*

The SLIM, developed by Embry et al. [42], is a structural method that uses expert opinion to estimate human error rates. The underlying assumption of SLIM is that the success likelihood of tasks for a given situation depends on the combination of effects from a small set of PSFs relevant to a group of tasks under consideration.

In this procedure, the experts are asked to assess the relative importance (weight) of each PSF with regard to its impact on the tasks of interest. An independent assessment is made on the level or the value of the PSFs in each task situation. After identifying and agreeing on the small set of PSFs, respective weights and ratings for each PSF are multiplied. These products are then summed to produce the SLI, varying from 0 to 100 after normalization. This value indicates the expert's belief regarding the positive or negative effects of PSFs on task success.

The SLIM approach assumes that the functional relationship between success probability and SLI is exponential. That is, $\log[\Pr(\text{Operator Success})] = a(\text{SLI}) + b$, where a and b are empirically estimated constants. To calibrate a and b, at least two human tasks of known reliability must be used in Equation 6.20, from which constants a and b are calculated.

This technique has been implemented as an interactive computer program. The first module, called Multi-Attribute Utility Decomposition (MAUD), analyzes a set of tasks to define their relative likelihood of success given the influence of PSFs. The second module, Systematic Approach to the Reliability Assessment of Humans (SARAH), is then used to calibrate these relative success likelihoods to generate absolute HEP. The SLIM technique has a good theoretical basis in decision theory. Once the initial data base has been established with the SARAH module, evaluations can be performed rapidly. This method does not require extensive decomposition of a task to an elemental level. For situations where no data are available, this approach enables HRA analysts to reasonably estimate human reliability. However, this method makes extensive use of expert judgment, which requires a team of experts to participate in the evaluation process. The resources required to set up the SLIM–MAUD data base are generally greater than other techniques.

6.3.2.3 Analytical Methods

These methods generally use a model based on some key parameters that form the value of human reliabilities.

6.3.2.3.1 *SPAR-H HRA Method*

Gertman et al. [49] have introduced the Standardized Plant Analysis Risk Model-Human (SPAR-H) analysis method to estimate HEPs associated with operator and crew actions (starting a machine, testing equipment) primarily of nuclear power plants. This approach decomposes the HEP into contributions from diagnosis failures and human action failures. Diagnosis tasks rely on human knowledge and experience to understand the actual conditions, planning, prioritizing, and determining the best course of actions. It also accounts for the context associated with the human failure event by considering the relevant PSFs and their assignment to correct for any PSF dependencies and appropriately adjust a base-case HEP. SPAR-H also provides predefined base-case HEPs. To account for uncertainties with probability estimates, the method employs a beta pdf. Finally it provides worksheets to ensure sensitivity in the analysis.

Eight PSFs account for the totality of human error context. These are

- Available time
- Stress and stressor
- Experience and training
- Complexity
- Ergonomics (including human–machine interface)
- Procedure
- Fitness for duty
- Work process.

A positive influence of PSF can reduce the base (nominal) failure probability whenever a negative influence increases. A good example of this is a highly experienced and trained operator who will do better than the base value. While SPAR-H uses the same approach as THERP (to be discussed next), it improves on it by accounting for dependency among PSFs.

The formula used to account for HEP, P_e is obtained from

$$P_e = \frac{P_{eb}\mathrm{PSF_c}}{P_{eb}(\mathrm{PSF_c} - 1) + 1}, \tag{6.23}$$

where P_{eb} is the base probability of human error taken as 0.01 for diagnosis tasks and 0.0001 for actions, and $\mathrm{PSF_c}$ is the composite PSF. SPAR-H suggests the PSFs. For example, the PSF (Avaiasa Time) offers the following PSF levels and PSF values:

1. Time available = time required = 10.
2. Nominal time = 1.
3. Time available $\geq 5 \times$ time required = 0.1.
4. Time available $> 50 \times$ time required = 0.01.

Once the PSF levels of all eight PSFs have been determined, the composite $\mathrm{PSF_c} = \prod_{i=1}^{8} \mathrm{PSF}_i$.

A qualitative method accounts for the dependency between PSFs and adjust $\mathrm{PSF_c}$.

6.3.2.3.2 Technique for Human Error Rate Prediction

The oldest and most widely used HRA technique is the THERP analysis developed by Swain and Guttman [38] and reported in the form of a handbook. The THERP approach uses conventional system reliability analysis modified to account for possible human error. Instead of generating equipment system states, THERP produces possible human task activities and the corresponding HEPs. THERP is carried out in the five steps described below:

1. *Define system failures of interest* Analysts examine system operation and analyze system safety to identify possible human interaction points and task characteristics and their impact on the systems. Then, analysts identify critical actions that require detailed analysis.

2. *List and analyze related human actions* The next step is to develop a detailed task analysis and human error analysis. The task analysis delineates the necessary task steps and the required human performance. The analyst then determines the errors that could possibly occur. The following human error categories are defined by THERP:
   - Errors of omission (omit a step or the entire task)
   - Errors of commission, including
     - Selection errors (select the wrong control, choose the wrong procedures)
     - Sequence errors (actions carried out in the wrong order)
     - Time errors (actions carried out too early/too late)
     - Qualitative error (action is done too little/too much).

   At this stage, opportunities for human recovery actions (recovery from an abnormal event or failure) should be identified. Without considering recovery possibilities, overall human reliability might be dramatically underestimated.

   The basic tool used to model tasks and task sequences is the HRA event tree. According to the time sequence or procedure order, the tree is built to represent possible alternative human actions. Therefore, if appropriate error probabilities of each subtask are known and the tree adequately depicts all human action sequences, the overall reliability of this task can be calculated. An example of an HRA event tree is shown in Figure 6.8.
3. *Estimate relevant error probabilities* As explained in the previous section, HEPs are required for the failure branches in the HRA event tree. Chapter 20 of Swain and Guttman [38] provides the following information:
   - Data tables containing nominal HEPs.

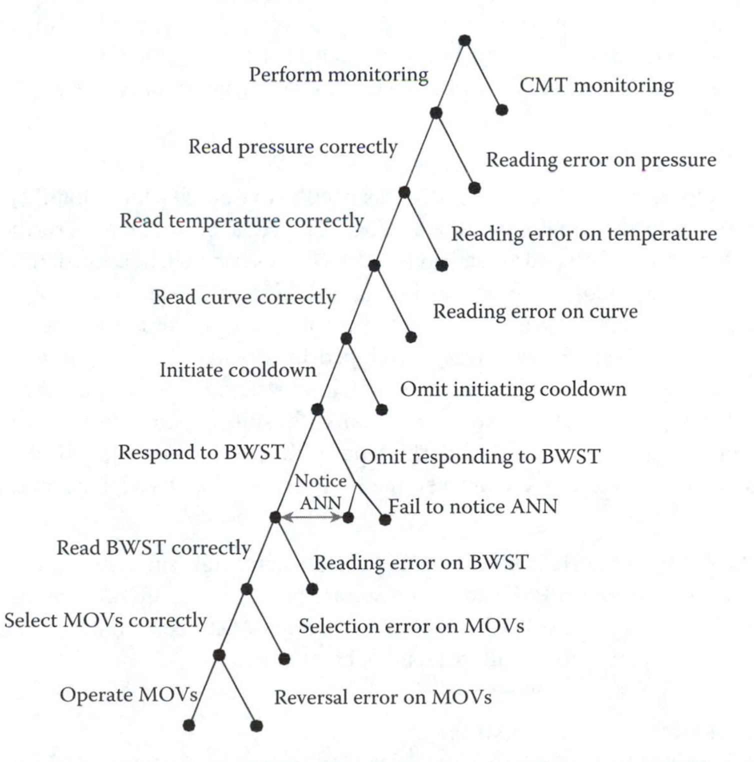

FIGURE 6.8 HRA event tree on operator actions during a small-break loss of coolant in nuclear plants. CMT, computer monitoring; ANN, annunciator; BWST, borated water storage tank; and MOV, motor-operated valve [30].

- Performance models explaining how to account for PSFs to modify the nominal error data.
- A simple model for converting independent failure probabilities into conditional failure probabilities.

In addition to the data source of THERP, analysts may use other data sources, such as the data from recorded incidents, trials from simulations, and subjective judgment data, if necessary.

4. *Estimate effects of error on system failure events* In the system reliability framework, the human error tasks are incorporated into the system model, such as a fault tree. In this way, the probabilities of undesired events can be evaluated, and the contribution of human errors to system reliability or availability can be estimated.
5. *Recommend changes to system design, and recalculate system reliability* A sensitivity analysis can be performed to identify dominant contributors to system unreliability. System performance can then be improved by reducing the sources of human error or redesigning the safeguard systems.

THERP's approach is very similar to the equipment reliability methods described in Chapter 4. The integration of HRA and equipment reliability analysis is straightforward using the THERP process, making it simple for system analysts to understand. In addition, compared with the data for other models, the data for THERP are much more complete and simple to use. The handbook contains guidance for modifying the listed data for different environments. The dependencies among subtasks are formally modeled, although subjective. Conditional probabilities are used to account for this kind of task dependence.

Very detailed THERP analysis can require a large amount of effort. In practice, by reducing the details of the THERP analysis to an appropriate level, the amount of work can be minimized. THERP is not appropriate for evaluating errors involving high-level decisions or diagnostic tasks. In addition, THERP does not model underlying psychological causes of errors. Since it is not an ergonomic tool, this method cannot produce explicit recommendations for design improvement.

6.3.2.3.3 HCR Correlation

During the development of SHARP, a model was needed to quantify the reliability of control room personnel responses to abnormal system operations. The HCR correlation, described by Hannaman et al. [40], was therefore developed as a normalized TRC (described in Section 6.3.2.3.4) whose shape is determined by the available time, stress, human-machine interface, and so on. Normalization is needed to reduce the number of curves required for a variety of situations. A set of three curves (skill-, rule-, and knowledge-based ideas, developed by Rasmussen [50]) represents all kinds of human decision behaviors. The application of HCR is straightforward, and the HCR correlation curves can be developed for different situations from the results of simulator experiments. Therefore, the validity can be verified continuously. This approach also has the capability of accounting for cognitive and environmental PSFs. However, the HCR correlation does have some disadvantages:

- The applicability of the HCR to all kinds of human activities is not verified.
- The relationships between PSFs and nonresponse probabilities are not well addressed.
- This approach does not explicitly address the details of human thinking processes. Thus, information about intentional failures cannot be obtained.

6.3.2.3.4 Time–Reliability Correlation

Hall et al. [39] concentrate on the diagnosis and decision errors of nuclear power plant operators after the initiation of an accident. They criticize the behavioral approach used by THERP and suggest that a more holistic approach be taken to analyze decision errors.

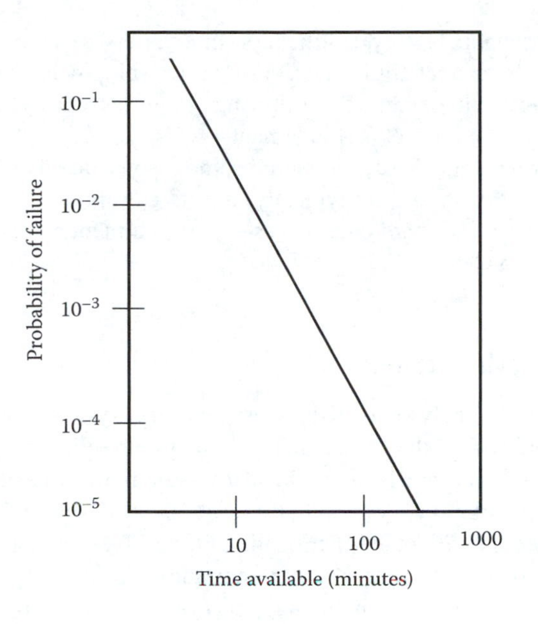

FIGURE 6.9 TRC for operators.

The major assumption of TRC is that the time available for diagnosis of a system fault is the dominant factor in determining the probability of failure. In other words, the longer people take to think, the more unlikely they are to make mistakes. The available time for decision and diagnosis is delimited by the operator's first awareness of an abnormal situation and the initiation of the selected response. Because no data were available when the TRC was developed, an interim relationship was obtained by consulting psychologists and system analysts. Recent reports confirm that the available time is an important factor in correctly performing cognitive tasks. A typical TRC is shown in Figure 6.9.

Dougherty and Fragola [51] is a good reference for TRC as well as other HRA methods. TRC is very easy and fast to use. However, TRC is still a preliminary approach: the exact relationship between time and reliability requires additional experimental and actual observations. Since this approach overlooks other important PSFs, such as experience level and task complexity, TRC focuses only on limited aspects of human performance in emergency conditions. The time available is the only variable in this model, so the estimation of the effect of this factor should be very accurate. However, TRC does not provide guidelines or information on how to reduce human error contributions.

6.3.3 Human Reliability Data

There is general agreement that a major problem for HRA is the scarcity of data on human performance that can be used to estimate human error rates and performance time. To estimate HEPs, one needs data on the relative frequency of the number of errors and/or the ratio of "near-misses" to the total number of attempts. Ideally, this information can be obtained from observing a large number of tasks performed in a given application. However, this is impractical for several reasons. First, error probabilities for many tasks, especially for rare emergency conditions, are very small. Therefore, it is very difficult to observe enough data within a reasonable amount of time to get statistically meaningful results. Second, possible penalties assessed against people who make errors in, for example, a nuclear power plant or an aircraft cockpit, discourage free reporting of all errors. Third, the costs of collecting and analyzing data could be unacceptably high. Moreover, estimation of performance times presents difficulties since data taken from different situations might not be applicable.

Data can be used to support HRA quantification in a variety of ways: for example, to confirm expert judgment, develop human reliability data, or support development of an HRA model. Currently, available data sources can be divided into the following categories: (1) actual data, (2) simulator data, (3) interpretive information, and (4) expert judgment.

Psychological scaling techniques, such as paired comparisons, direct estimation, SLIM, and other structured expert judgment methods, are typically used to extrapolate error probabilities. In many instances, the scarcity of relevant hard data makes expert judgment a very useful data source. This topic is discussed further in Chapter 7.

6.4 MEASURES OF IMPORTANCE

During the design reliability analysis, or risk assessment of a system, some components and their arrangement may be more critical than others in terms of system reliability. For example, a series set of components within a system is much more critical to a system (in terms of failure) than the same set of components in parallel within the system. In this section, we describe five methods of measuring the importance of components: Birnbaum, criticality, Fussell–Vesely, risk-reduction worth (RRW), and risk-achievement worth (RAW) measures of importance. Usually, importance measures are used in the failure space; however, in this book their application in the success space is also discussed.

6.4.1 Birnbaum Measure of Importance

Introduced by Birnbaum [52], this measure of component importance, $I_i^{\mathrm{B}}(t)$, for success space (as described by Sharirli [53]) is defined as

$$I_i^{\mathrm{B}}(t) = \frac{\partial R_{\mathrm{s}}[R(t)]}{\partial R_i(t)}, \tag{6.24}$$

where $R_{\mathrm{s}}[R(t)]$ is reliability of the system as a function of the reliability of its individual components, $R_i(t)$.

If, for a given component i, $I_i^{\mathrm{B}}(t)$ is large, it means that a small change in the reliability of component i, $R_i(t)$, will result in a large change in the system reliability $R_{\mathrm{s}}(t)$.

If system components are assumed to be independent, the Birnbaum measure of importance can be represented by [54]

$$I_i^{\mathrm{B}}(t) = R_{\mathrm{s}}[R(t)|R_i(t) = 1] - R_{\mathrm{s}}[R(t)|R_i(t) = 0], \tag{6.25}$$

where $R_{\mathrm{s}}[R(t)|R_i(t) = 1]$ and $R_{\mathrm{s}}[R(t)|R_i(t) = 0]$ are the values of reliability function of the system with the reliability of component i set to 1 and 0, respectively.

Equations 6.24 and 6.25 are often used in conjunction with the unreliability, unavailability, or risk function, $F_{\mathrm{s}}[Q_i(t)]$, given in terms of individual component unreliability or unavailability $Q_i(t)$. In this case, Equation 6.22 is replaced by

$$I_i^{\mathrm{B}}(t) = \frac{\partial F_{\mathrm{s}}[Q(t)]}{\partial Q_i(t)} = F_{\mathrm{s}}[Q(t)|Q_i(t) = 1] - F_{\mathrm{s}}[Q(t)|Q_i(t) = 0]. \tag{6.26}$$

Example 6.9

Consider the system shown below. Determine the Birnbaum importance of each component at $t = 720$ h. Assume an exponential time to failure.

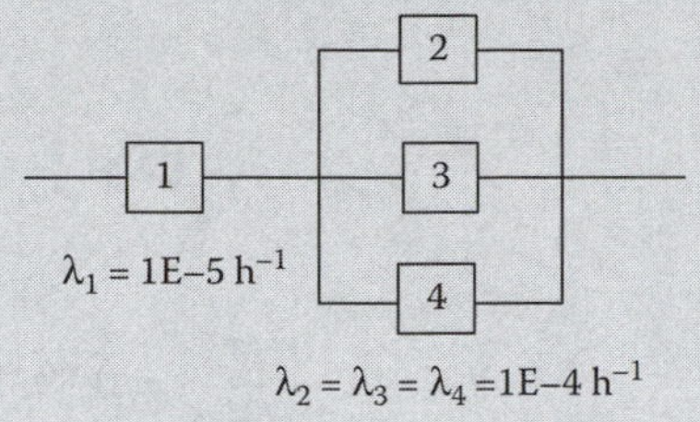

Solution:

$$R_1(t = 720) = 0.993, R_2(t = 720) = R_3(t = 720) = R_4(t = 720) = 0.9305.$$

The reliability function of the system is

$$R_S[R(t)] = R_1(t)\{1 - [1 - R_2(t)][1 - R_3(t)][1 - R_4(t)]\} = 0.9925.$$

Using Equation 6.25,

$$I_i^B(t) = R_S[R(t)|R_i(t) = 1] - R_S[R(t)|R_i(t) = 0],$$

$$I_1^B(t) = \{1 - [1 - R_2(t)]\}[1 - R_3(t)][1 - R_4(t)].$$

Therefore,

$$I_1^B(t = 720) \approx 0.999,$$

$$I_2^B(t) = R_1(t)[(1 - R_3(t))(1 - R_4(t))],$$

$$I_2^B(t = 720) \approx 0.005.$$

Similarly,

$$I_3^B(t = 720) = I_4^B(t = 720) \approx 0.005.$$

It can be concluded that the rate of improvement in component 1 has far more importance (impact) on system reliability than components 2, 3, and 4. For example, if the reliability of the parallel units increases by an order of magnitude, clearly the importance of components 2, 3, and 4 decreases (e.g., for $\lambda_2 = \lambda_3 = \lambda_4 = 10^{-4}\,\text{h}^{-1}$, $I_2^B = I_3^B = I_4^B \approx 0$, and $I_1^B = 1$). Similarly, if identical units are in parallel with component 1, the importance changes.

6.4.2 Criticality Importance

Birnbaum's importance for component i is independent of the reliability of component i itself. Therefore, I_i^B is not a function of $R_i(t)$. It would be more difficult and costly to further improve the more reliable components than to improve the less reliable ones. From this, the criticality importance of component i is defined as

$$I_i^{CR}(t) = \frac{\partial R_S[R(t)]}{\partial R_i(t)} \times \frac{R_i(t)}{R_S[R(t)]} \tag{6.27}$$

or

$$I_i^{CR}(t) = I_i^B(t) \times \frac{R_i(t)}{R_S[R(t)]}. \tag{6.28}$$

From Equation 6.28, we can see that the Birnbaum importance is corrected for reliability of the individual components relative to the reliability of the whole system. Therefore, if the Birnbaum importance of a component is high, but the reliability of the component is low with respect to the reliability of the system, then criticality importance assigns a low importance to this component. Similarly, Equation 6.28 can be represented by the unreliability or unavailability function:

$$I_i^{\mathrm{CR}}(t) = I_i^{\mathrm{B}}(t) \times \frac{Q_i(t)}{F_{\mathrm{s}}[Q(t)]}. \tag{6.29}$$

As such, in Example 6.9,

$$I_i^{\mathrm{CR}}(t) = 0.9925 \times \frac{0.993}{0.9925} = 0.993. \tag{6.30}$$

Since component 1 is more reliable, its contribution to the reliability of the system (i.e., its criticality importance) increases, whereas components 2, 3, and 4 will have a less important contribution to the overall system reliability.

A subset of the criticality importance measure is the inspection importance measure (I_i^W). This measure is defined as the product of the Birnbaum importance times the failure probability (unreliability or unavailability) of the component. Accordingly,

$$I^W(t) = I^{\mathrm{B}}(t) \times Q_i(t). \tag{6.31}$$

This measure is used to prioritize operability test activities to ensure high component readiness and performance.

6.4.3 Fussell–Vesely Importance

In cases where component i contributes to system reliability, but is not necessarily critical, the Fussell–Vesely importance measure can be used. This measure, introduced by W.E. Vesely and later applied by Fussell [55], is in the form of

$$I_i^{\mathrm{FV}}(t) = \frac{R_i[R(t)]}{R_{\mathrm{s}}[R(t)]}, \tag{6.32}$$

where $R_i[R(t)]$ is the contribution of component i to the reliability of the system. Similarly, using unreliability or unavailability functions,

$$I_i^{\mathrm{FV}}(t) = \frac{F_i[Q(t)]}{F_{\mathrm{s}}[Q(t)]}, \tag{6.33}$$

where $F_i[Q(t)]$ denotes the probability that component i is contributing to system failure or system risk.

The Fussell–Vesely importance measure has been applied to system cut sets to determine the importance of individual cut sets to the failure probability of the whole system. For example, consider the importance of I_k of the kth cut set representing a system failure. In that case, Equation 6.28 replaces

$$I_k^{\mathrm{FV}}(t) = \frac{Q_k(t)}{Q_{\mathrm{s}}(t)}, \tag{6.34}$$

where $Q_k(t)$ is the time-dependent probability that minimal cut set k occurs and $Q_{\mathrm{s}}(t)$ is the total time-dependent probability that the system fails (due to all cut sets).

Generally, the minimal cut sets with the largest values of I_k are the most important ones. Equation 6.34 is equally applicable to mutually exclusive cut sets. Consequently, system improvements should initially be directed toward the minimal cut sets with the largest importance values.

If the probability of all minimal cut sets or mutually exclusive cut sets is known, then the following approximate expression can be used to find the importance of individual components:

$$I_i^{\mathrm{FV}} = \frac{\sum_{j=1}^{m} Q_j(t)}{Q_{\mathrm{S}}(t)}, \tag{6.35}$$

where $Q_i(t)$ is the probability that the jth cut set that contains component i is failed and m is the number of minimal cut sets that contain component i.

Expression 6.35 is an approximation; the situation of two minimal cut sets containing component i failing at the same time is neglected since its probability is very small.

6.4.4 RRW Importance

The RRW importance is a measure of the change in unreliability (unavailability or risk) when an input variable, such as the unavailability of component, is set to zero—that is, by assuming that a component is "perfect" (or its failure probability is zero) and thus eliminating any postulated failure. This importance measure shows how much better the system can become as its components are improved.

This importance measure is used in failure domains, although it can also be used in the success domain. The calculation may be done either as a ratio or as a difference. Accordingly, as a ratio,

$$I_i^{\mathrm{RRW}} = \frac{F_{\mathrm{S}}[Q(t)]}{F_{\mathrm{S}}[Q(t) \mid Q_i(t) = 0]} \tag{6.36}$$

and as a difference,

$$I_i^{\mathrm{RRW}} = F_{\mathrm{S}}\,[Q(t)] - F_{\mathrm{S}}\,[Q(t) \mid Q_i(t) = 0], \tag{6.37}$$

where $F_{\mathrm{S}}[Q(t) \mid Q_i(t) = 0]$ is the system unreliability (unavailability or risk) when unreliability (or unavailability) of component i is set to zero.

In practice, this measure is used to identify elements of the system (such as components) that are the best candidates for improving system reliability (risk or unavailability).

6.4.5 RAW Importance

The RAW (increase) importance is the inverse of the RRW measure. In this method, the input variable (e.g., component unavailability) is set to one, and the effect of this change on system unreliability (unavailability or risk) is measured. Similar to RRW, the calculation may be done as a ratio or a difference. By setting component failure probability to one, RAW measures the increase in system failure probability assuming the worst case of failing the component. As a ratio, RAW measure is

$$I_i^{\mathrm{RAW}} = \frac{F_{\mathrm{S}}[Q(t) \mid Q_i(t) = 1]}{F_{\mathrm{S}}[Q(t)]} \tag{6.38}$$

and as a difference,

$$I_i^{\mathrm{RAW}} = F_{\mathrm{S}}[Q(t) \mid Q_i(t) = 1] - F_{\mathrm{S}}[Q(t)], \tag{6.39}$$

where $F_{\mathrm{S}}[Q(t) \mid Q_i(t) = 1]$ is the system unreliability (unavailability or risk) when unreliability (or unavailability) of component i is set to one.

The risk increase measure is useful for identifying elements of the system that are the most crucial for making the system unreliable (unavailable or increasing the risk). Therefore, components with high I^{RAW} are the ones that will have the most impact, should their failure probability unexpectedly rise.

Example 6.10

Consider the water-pumping system below. Determine the Birnbaum, criticality, and Fussell–Vesely importance measures of the valve (v), pump-1 (p-1), and pump-2 (p-2) using both reliability and unreliability versions of the importance measures.

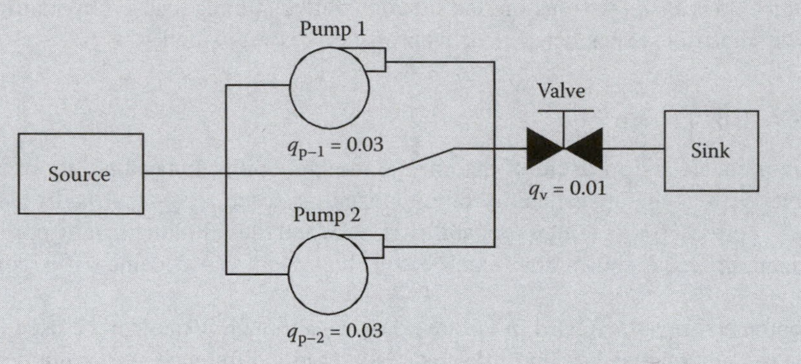

Solution: Because the component reliability, $R_{\text{p}-1} = R_{\text{p}-2} = 0.97, R_{\text{v}} = 0.99$, the reliability function is

$$R_{\text{S}}[R(t)] = R_{\text{v}}[R_{\text{p}-1} + R_{\text{p}-2} - R_{\text{p}-1} \times R_{\text{p}-2}] = 0.989.$$

Using the rare event approximation, the unreliability function is

$$F_{\text{S}}[Q(t)] = Q_{\text{p}-1} \times Q_{\text{p}-2} + Q_{\text{v}} = 0.011$$

1. Birnbaum's importance:

$$I_{\text{v}}^{\text{B}} = R_{\text{p}-1} + R_{\text{p}-2} - R_{\text{p}-1}R_{\text{p}-2} \approx 1,$$

$$I_{\text{p}-1}^{\text{B}} = R_{\text{v}} - R_{\text{v}}R_{\text{p}-2} \approx 0.03,$$

$$I_{\text{p}-2}^{\text{B}} = R_{\text{v}} - R_{\text{v}}R_{\text{p}-1} \approx 0.03.$$

Using the unreliability function,

$$I_{\text{v}}^{\text{B}} \approx 1,$$

$$I_{\text{p}-1}^{\text{B}} = Q_{\text{p}-2} \approx 0.03,$$

$$I_{\text{p}-2}^{\text{B}} = Q_{\text{p}-1} \approx 0.03.$$

2. Criticality importance:

$$I_{\text{v}}^{\text{CR}}(t) = 1 \times \frac{0.99}{0.989} \approx 1,$$

$$I_{\text{p}-1}^{\text{CR}}(t) = I_{\text{p}-2}^{\text{CR}}(t) = 0.03 \times \frac{0.97}{0.989} \approx 0.029.$$

The same criticality importance values are expected for the unreliability function. $R_i[R(t)]$ is obtained by retaining terms involving $R_i(t)$.

3. Fussell–Vesely importance:

$$R_{\mathrm{V}}[r(t)] = R_{\mathrm{S}}[R(t)] \approx 0.989,$$

$$R_{\mathrm{p-1}}[r(t)] = R_{\mathrm{V}} \times R_{\mathrm{p-1}} - R_{\mathrm{V}} \times R_{\mathrm{p-1}} \times R_{\mathrm{p-2}} \approx 0.029,$$

$$R_{\mathrm{p-2}}[r(t)] = R_{\mathrm{V}} \times R_{\mathrm{p-2}} - R_{\mathrm{V}} \times R_{\mathrm{p-1}} \times R_{\mathrm{p-2}} \approx 0.029,$$

$$I_{\mathrm{V}}^{\mathrm{FV}} = \frac{0.989}{0.989} = 1,$$

$$I_{\mathrm{p-1}}^{\mathrm{FV}} = I_{\mathrm{p-2}}^{\mathrm{FV}} = \frac{0.029}{0.989} \approx 0.029.$$

Using the unreliability function,

$$F_{\mathrm{V}}[R(t)] = Q_{\mathrm{V}} = 0.01,$$

$$F_{\mathrm{p-1}}[Q(t)] = Q_{\mathrm{p-1}} \times Q_{\mathrm{p-2}} = 0.0001,$$

$$F_{\mathrm{p-2}}[Q(t)] = Q_{\mathrm{p-2}} \times Q_{\mathrm{p-1}} = 0.0009.$$

Then,

$$I_{\mathrm{V}}^{\mathrm{FV}} = \frac{0.01}{0.011} = 0.9,$$

$$I_{\mathrm{p-1}}^{\mathrm{FV}} = I_{\mathrm{p-2}}^{\mathrm{FV}} = \frac{0.0009}{0.011} \approx 0.08.$$

EXAMPLE 6.11

Repeat Example 6.10 and calculate I^{RRW} and I^{RAW} for all components. Compare the results with I^{B}, I^{CR}, and I^{FV}.

Solution: The unreliability function is $F_{\mathrm{S}}[Q(t)] = Q_{\mathrm{p-1}}Q_{\mathrm{p-2}} + Q_{\mathrm{V}} = 0.011$.

1. For RRW,

$$F_{\mathrm{S}}[Q \mid Q_{\mathrm{V}} = 0] = Q_{\mathrm{p-1}} \times Q_{\mathrm{p-2}} = 0.03 \times 0.03 = 0.0009.$$

Therefore, for the ratio measure,

$$I_{\mathrm{V}}^{\mathrm{RRW}} = \frac{0.011}{0.0009} = 12.2.$$

For the difference measure,

$$I_{\mathrm{V}}^{\mathrm{RRW}} = 0.011 - 0.0009 = 0.01.$$

Similarly for pumps as ratio,

$$I_{\mathrm{p-1}}^{\mathrm{RRW}} = I_{\mathrm{p-2}}^{\mathrm{RRW}} = \frac{0.011}{0.01} = 1.1.$$

As difference,

$$I_{\mathrm{p-1}}^{\mathrm{RRW}} = I_{\mathrm{p-2}}^{\mathrm{RRW}} = 0.011 - 0.01 = 0.001.$$

continued

Note that in the ratio method, the larger numbers indicate increasing importance, whereas the reverse is true for the difference method. This is only a metric for identifying components when assured performance will highly affect system operation.

2. Similarly for RAW, the ratio method yields

$$I_{\mathrm{v}}^{\mathrm{RAW}} = \frac{1}{0.011} = 90.91.$$

For the difference method,

$$I_{\mathrm{v}}^{\mathrm{RAW}} \approx 1.$$

For the pumps, using the ratio method,

$$I_{\mathrm{p-1}}^{\mathrm{RAW}} = I_{\mathrm{p-2}}^{\mathrm{RAW}} = \frac{1 \times 0.03 + 0.01}{0.011} = 3.64.$$

For the difference method,

$$I_{\mathrm{p-1}}^{\mathrm{RAW}} = I_{\mathrm{p-2}}^{\mathrm{RAW}} = (1 \times 0.03 + 0.01) - 0.011 = 0.029.$$

The I_i^{RAW} shows the importance of component i with respect to system unreliability when component i fails. Clearly by comparing the results to $I^{\mathrm{B}}, I^{\mathrm{CR}}$, and I^{FV} with I^{RAW} and I^{RRW}, the relative importance value measured by I_i^{RAW} is consistent. This is expected since all other measures are related to the degradation of the component. I^{RAW} is related to the worth of improvement in component reliability. For a more in-depth discussion of importance measures used in reliability engineering and risk assessment, see [56].

6.4.6 Practical Aspects of Importance Measures

There are two principal factors that determine the importance of a component in a system: the structure of the system and the reliability or unreliability of the components. Depending on the measure selected, one or the other may be pertinent. Also, depending on whether we use reliability or unreliability, some of these measures behave differently. In Example 6.10, this is seen in $I_{\mathrm{p-1}}^{\mathrm{FV}}$ and $I_{\mathrm{p-2}}^{\mathrm{FV}}$, where their importance in the success space is almost 1 and in the failure space is 0.

The Birnbaum measure of importance completely depends on the structure of the system (e.g., whether the system is dominated by a parallel or series configuration). Therefore, it should only be used to determine the degree of redundancy and appropriateness of the system's logic.

The criticality importance is related to Birnbaum's. However, it is also affected by the reliability/unreliability of the components and the system. This measure allows for the evaluation of the importance of a component in light of its potential to improve system reliability. The effect of improvements on one component may result in changes in the importance of other components of the system.

Due to its simplicity, the Fussell–Vesely measure of importance has been widely used in practice, mostly for measuring importance in the failure space using unreliability/unavailability functions. The measure is influenced by the actual reliability/unreliability of the components and the systems as well as the logical structure of the system.

Generally, the importance of components should be used during design or evaluation of systems to determine which components or subsystems are important to the overall reliability of the system. Those with high importance could prove to be candidates for further improvements. In an operational context, items with high importance should be watched by the operators, since they are critical for the continuous operation of the system.

Some importance measures are calculated as dimensionless ratios, while others are absolute physical quantities or probabilities. The Birnbaum measure is an absolute measure, whereas the Fussell–Vesely is a relative one. Table 6.2 summarizes the importance measures discussed in this section.

It is widely believed that the relative measures (ratios) have the advantage of being more robust than the absolute measure: since many quantities appear in both the numerator and the denominator, it can be hoped that errors in their magnitudes will tend to divide out. This hope is realized in some models. On the other hand, either the denominator or the numerator in the relative measures may be dominated by terms that have nothing to do with the basic event of interest, so that errors or uncertainties in those terms may obscure the desired insights. It is thought that the risk-achievement ratio and the risk-reduction ratio are especially vulnerable to this kind of distortion. The *absolute* measures have the advantage of providing an immediate sense of whether a given event is negligible on an absolute scale. The relative measures do not provide this information; the user must obtain system failure probability and perform some arithmetic to obtain this information.

TABLE 6.2
Interpretation of Importance Measures

Name	Definition	Interpretation	Comments
Birnbaum	Pr(coefficient of component *i*)	How often component *i* is need to prevent system failure.	Absolute measure; directly measures sensitivity of probability of system failure (or risk) to probability of component *i* failure.
Criticality	Pr(coefficient of component *i*) × Pr(component *i* failure)/ Pr(system failure)	How often component *i* is needed to prevent system failure adjusted for relative probability of component *i* failure.	Absolute measure, measures the sensitivity of system failure probability with respect to failure probability of component *i*.
Fussell-Vesely	Pr[system failure (or risk) based on terms involving component *i*]/ Pr[total system failure (or risk)]	Fraction of system unavailability (or risk) involving failure of component *i*.	Dimensionless, relative measure; reflects how much relative improvement is theoretically available from improving performance of component *i*. Denominator may contain some terms having nothing to do with component *i* operation.
Risk reduction worth	Pr[system failure (or risk)]/Pr[system failure given component *i* operates]	Shows relative improvements in Pr (system failure) realizable by improving component *i*; how much relative harm component *i* does, by not being perfect.	Dimensionless, relative measure. Both the numerator and the denominator contain some terms having nothing to do with component *i* operation
Risk achievement worth	Pr[system failure given component *i* fails]/ Pr[total system failure (or risk)]	How much relative good is done by component *i*; factor by which Pr[system failure] would increase with no credit for component *i*.	Dimensionless, relative measure; Both the numerator and the denominator contain some terms having nothing to do with component *i* operation.

A number of other measures of importance have been introduced, as well as computer program importance calculations. For more information, the readers are referred to Lambert [57], Sharirli [53], NUREG/CR-4550 [58], and Zio and Podofillini [59]. Subject of uncertainty importance is used to identify contributors to output uncertainty attributed to the inputs of a model or a system. For more information on this subject, see [56,60,61].

6.5 RELIABILITY-CENTERED MAINTENANCE

6.5.1 History and Current Procedures

The reliability-centered maintenance (RCM) methodology is a systematic approach directed toward defining and developing applicable and effective failure management strategies. RCM finds its roots in the early 1960s. The initial development work was done by the North American civil aviation industry. It started when the airlines at that time noted that many of their maintenance practices were not only too expensive but also unsafe. This prompted the industry to put together a series of Maintenance Steering Groups (MSG) to review everything they were doing to keep their aircrafts airborne. These groups consisted of representatives of the aircraft manufacturers, the airlines, and the Federal Aviation Administration (FAA).

The first attempt at a rational, zero-based process for formulating maintenance strategies was promulgated by the Air Transport Association in Washington, DC, in 1968. The first attempt is now known as MSG-1. A refinement—now known as MSG-2—was promulgated in 1970.

In the mid-1970s, the U.S. Department of Defense became interested in the then-state of the art in aviation maintenance. It commissioned a report on the subject from the commercial aviation industry. This report, written by Stanley Nowlan and Howard Heap of United Airlines, was entitled "Reliability Centered Maintenance." The report was published in 1978, and it is still the leading document in physical asset management. RCM is a process used to decide what must be done to ensure that any item (e.g., system or process) continues to perform its function.

What users expect from their items is defined in terms of primary performance parameters such as output, throughput, speed, range, and carrying capacity. Where relevant, the RCM process also defines what users want in terms of risk (safety and environmental integrity), quality (precision, accuracy, consistency, and stability), control, comfort, containment, economy, and customer service.

The next step in the RCM process is to identify the ways the item can fail to live up to these expectations (failed states), followed by an FMEA, to identify all the events that are reasonably likely to cause each failed state.

Finally, the RCM process seeks to identify a suitable failure management policy for dealing with each failure mode in light of its consequences and technical characteristics. Failure management policy options include

- Predictive maintenance
- Preventive maintenance
- Failure-finding
- Changing the design or configuration of the system
- Changing the way the system is operated
- Run to failure.

The RCM process provides powerful rules for deciding whether any failure management policy is technically appropriate. It also provides adequate criteria for deciding how often routing tasks should be performed. The RCM methodology involves a systematic and logical step-by-step consideration of

1. The function(s) of a system or component.
2. The ways the function(s) can be lost.

3. The importance of the function and its failure.
4. A priority-based consideration that identifies those failure management activities that both reduce failure potential and are cost-effective.

The key steps of this process include

- *Definition of system boundaries*: Boundaries must be clearly identified and the level of detail for the analysis must be explained.
- *Determination of the functions of a system, its subsystems, or components*: Each component within the system or subsystem may have one or more functions. These should be explained. Inputs and outputs of functions across system boundaries must also be identified.
- *Determination of functional failures*: A functional failure occurs when a system or subsystem fails to provide its required function.
- *Determination of dominant failure modes*: One of the logical system analysis methods (e.g., fault tree or MLD) along with FMEA should be used to identify the modes that are the leading (highest probability) causes for functional failures.
- *Determination of corrective actions and optimal preventive maintenance schedules*: An applicable and effective course of action for each failure mode should be identified. This may involve implementing a preventive maintenance task, accepting the likelihood of failure, or initiating redesign.
- *Integration of the results*: The results of the failure management task, along with other specifics of implementation, are integrated into the maintenance plan.

From the above steps, it is clear that RCM methodology can be divided into two basic phases. First, the system and its boundaries are defined, and then the system is decomposed to subsystems and components, and their functions are identified along with those failures that are likely to cause loss of function. Second, each of the functional failures is examined to determine the associated failure mode and to determine whether or not there are effective failure management strategies (or tasks) that eliminate or minimize occurrence of the failure mode identified.

For those failure modes for which an effective failure management task is specified, further definition is necessary. Each task should be labeled as either time-directed, condition-monitoring, or failure-finding. Time-directed tasks are generally applicable when the probability of failure increases with time; that is, the failure mode has a positive trend, as discussed in Chapter 5. Time can be measured in several ways, including actual run-time or the number of startups (demands) or shutdowns of the component (with the given failure mode). Condition-monitoring tasks are generally applicable when one can efficiently correlate functional failures to detectable and measurable parameters of the system. For example, vibration of a pump can be measured to predict alignment problems. Failure-finding tasks are not preventive, but are intended to discover failures that are otherwise hidden. If no effective failure management task can be identified for a hidden failure, a scheduled functional failure-finding task may be devised.

To develop an optimal preventive maintenance program, an optimal schedule for the maintenance activities must be devised. In the following section, a reliability-based technique for optimizing a preventive maintenance schedule is discussed.

6.5.2 Optimal Preventive Maintenance Scheduling

In this section we consider a simple example of optimal preventive action scheduling, which minimizes the average total cost of system functioning per unit time. Denote the preventive maintenance interval by θ, the cost of failure that occurs during a system operation by c_1, and the cost of a preventive maintenance by c_2. Find the optimal value of θ that minimizes the average total cost per unit time. To find the mean length of the interval between two adjacent maintenance actions and the average cost of this interval per unit time, assume that

1. The interval between maintenance actions is constant and equals θ if there is no failure, and all maintenance actions are preventive.
2. If a failure has occurred, it is assumed that maintenance is instantly performed at a random time $\tau < \theta$. The mean length of this interval, Θ, is given by

$$\Theta = E(\tau) = \int_0^{\theta} R(t)\,\mathrm{d}t, \tag{6.40}$$

where $R(t)$ is the reliability function of the component.

The average cost per unit time, $C(\theta)$, can be written as

$$C(\theta) = \frac{c_1 F(\theta) + c_2 R(\theta)}{\Theta}, \tag{6.41}$$

where $F(Q) = 1 - R(Q)$ is the unreliability of the component.

The optimal value of θ can be found using the first-order condition (equating the first derivative to zero). That is,

$$\frac{\mathrm{d}C(\theta)}{\mathrm{d}\theta} = \frac{\mathrm{d}}{\mathrm{d}\theta}\left(\frac{c_1 F(\theta) + c_2 R(\theta)}{\Theta}\right) = 0, \tag{6.42}$$

which results in the following equation:

$$\lambda(\theta)\Theta - F(\theta) = \frac{c_2}{c_1 - c_2}, \tag{6.43}$$

where $\lambda(\theta) = f(\theta)/R(\theta)$. For the practical applications, it is better to rewrite this equation such that it expresses the relative cost dependence:

$$\lambda(\theta)\Theta - F(\theta) = \frac{c_2/c_1}{1 - c_2/c_1}. \tag{6.44}$$

In the general case, this equation can be solved only numerically.

The results of numerical solution of Equation 6.44 for the Weibull distribution with the shape parameter, β, equal to 2 (aging distribution), and the scale parameter, α, and some values of c_2/c_1 are given in the following table.

c_2/c_1	θ^*/α
0.9	5.078
0.8	2.274
0.7	1.529
0.6	1.243
0.5	1.080
0.4	0.968
0.3	0.885
0.2	0.816
0.1	0.759
0.05	0.733
0.025	0.720
0.01	0.713

6.5.2.1 Economic Benefit of Optimization

To estimate the economic benefit of the optimization considered, compare the average cost of failure per unit time without preventive maintenance, $E(c_1) = c_1/\text{MTTF}$, with the average cost of failure per unit time given by Equation 6.41 when $\theta = \theta^*$ (i.e., under the optimal schedule). That is, to calculate the ratio ε,

$$\varepsilon = \frac{\text{MTTF}}{\Theta^*}\left[F(\theta^*) + \frac{c_2}{c_1}R(\theta^*)\right]. \tag{6.45}$$

The results of calculations for the example considered are given in the following table.

c_2/c_1	ε
0.9	1.000
0.8	1.000
0.7	1.000
0.6	0.993
0.5	0.966
0.4	0.922
0.3	0.862
0.2	0.783
0.1	0.690
0.05	0.644
0.025	0.605
0.01	0.589

As one may anticipate, the less c_2/c_1 is, the greater the effect of optimization. For the values of $c_2/c_1 = 0.1$, it is about 30 cents/dollar.

6.6 RELIABILITY GROWTH

As the design cycle of a product progresses from concept to development, testing and manufacturing, implementing design changes along the way can improve the product's reliability to achieve a design goal. Typically, a formal Test Analyze And Fix (TAAF) program is implemented to discover design flaws and mitigate them. The gradual product improvement through the elimination of design deficiencies, which results in the increase of failure (inter)arrival times is known as *reliability growth*.

Generally speaking, reliability growth can be applicable to any level of design decomposition, ranging from a component to a complete system. The (nonrepairable) component-level reliability growth can be established by comparing a chosen reliability metric for the consecutive design iterations or the product development milestone. For further reading on multiple comparisons of component-level reliability data, see [62,63].

Most of the existing reliability growth models, however, are associated with repairable systems and, therefore, the basic reliability growth mathematics will be related to those considered in Section 5.1. The reliability growth methodology includes some new terms, which we have not formally defined so far. The first term is the *cumulative MTBF* (CMTBF), Θ_c, defined as the ratio of the expected cumulative number of failures $E[N(t)]$ to the total time on test t:

$$\Theta_c = \frac{t}{E[N(t)]}. \tag{6.46}$$

The second term is the *instantaneous* MTBF (I_{MTBF}), Θ_i, defined as the inverse of the ROCOF function considered in Section 5.1.1:

$$\Theta_i = \frac{1}{\lambda(t)}. \tag{6.47}$$

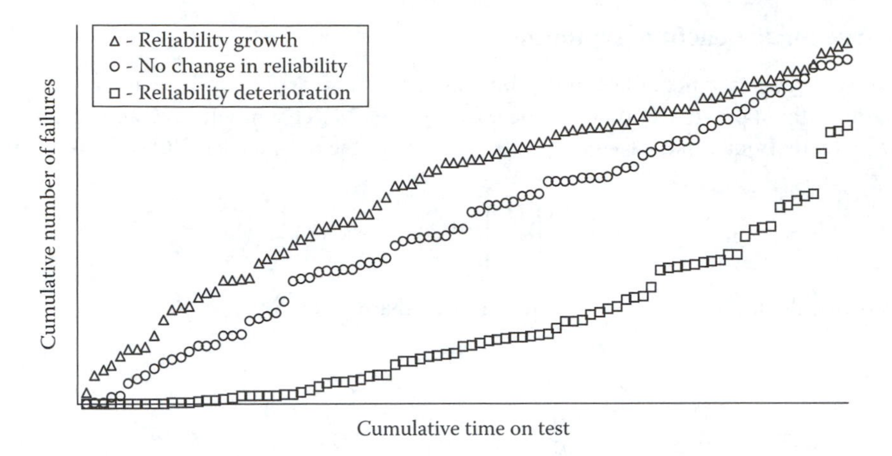

FIGURE 6.10 A nonparametric method of reliability growth evaluation.

The difference between the two is that Θ_c is a function of ROCOF integrated over the interval (0, t], whereas Θ_i is the inverse of ROCOF at a given point in time t.

6.6.1 Graphical Method

One of the easiest and most straightforward methods to assess reliability growth of a repairable system is to plot the cumulative number of failures versus cumulative time on test. Reliability growth is said to take place if the TAAF-based design changes lead to an incremental drop in the cumulative number of failures as a function of total time on test (Figure 6.10). Generally, the concave (convex) plot would indicate to a reliability growth (deterioration), while a straight line would be an indication of no change in reliability behavior.

6.6.2 Duane Method

Empirical studies conducted by Duane [64] on a number of repairable systems have shown that the CMTBF plotted against cumulative time on test in a log–log space exhibits an almost linear relationship. Duane postulated a reliability growth model that expresses the CMTBF as a function of total time on test in the following form:

$$\Theta_c(t) = \frac{t}{N} = \frac{t^{1-\beta}}{\lambda}, \tag{6.48}$$

where t is the total time on test and N is the cumulative number of failures.

Taking the log of both sides of Equation 6.48, one obtains

$$\ln[\Theta_c(t)] = (1-\beta)\ln(t) - \ln(\lambda), \quad t > 0, \tag{6.49}$$

which does indeed present an equation of a straight line. Parameters λ and β are referred to as the growth parameters and can be estimated using the linear regression technique discussed in Section 2.8. The inverse of parameter λ is sometimes referred to as the initial MTBF. The latter term becomes self-explanatory, if one sets t to one in Equation 6.48.

The instantaneous MTBF can be derived by differentiating Equation 6.48 with respect to t:

$$\Theta_i(t) = \left(\frac{\mathrm{d}N}{\mathrm{d}t}\right)^{-1} = \left[\frac{\mathrm{d}\,(t/\theta_c)}{\mathrm{d}t}\right]^{-1} = \frac{t^{1-\beta}}{\lambda\beta}. \tag{6.50}$$

Under the Duane model, the cumulative and instantaneous MTBFs are related to each other through parameter β:

$$\theta_c(t) = \beta\theta_i(t). \tag{6.51}$$

Keeping in mind that ROCOF, $\lambda(t)$, is the inverse of the instantaneous MTBF $\Theta_i(t)$, Equation 6.50 can be represented in the following form:

$$\lambda(t) = \beta\lambda t^{\beta-1}. \tag{6.52}$$

Note that this is the exact algebraic form of the NHPP ROCOF model discussed in Section 5.1.4.

Moreover, expression 6.52 formally coincides with the Weibull hazard rate function. As such, $\beta < 1$ represents reliability growth and $\beta > 1$ represents reliability degradation. O'Connor [1] has suggested the following engineering interpretation of the growth parameter β.

β	Interpretation
0.4–0.6	The program's top priority is the elimination of failure modes. The program uses accelerated tests and suggests immediate analysis and effective corrective action for all failures
0.3–0.4	The program gives priority to reliability improvement. The program uses normal environmental tests and well-managed analysis. Corrective action is taken for important failure modes
0.2–0.3	The program gives routine attention to reliability improvement. The program does not use applied environmental tests. Corrective action is taken for important failure modes
0.0–0.2	The program gives no priority to reliability improvement. Failure data are not analyzed. Corrective action is taken for important failure modes, but with low priority

Once the parameters of the Duane model are estimated, it becomes possible to determine the TAAF test time required to attain a given target instantaneous MTBF under a given rate of reliability growth β [65,66]:

$$t = (\theta_i\lambda\beta)^{1/(1-\beta)}. \tag{6.53}$$

EXAMPLE 6.12

The following are the miles between consecutive failures of a new automobile subsystem obtained through the TAAF program: 5940; 12,331; 21,010; 27,192; 19,910; 24,211; 26,422; 27,731; 26,862; and 29,271. Estimate the parameters of the Duane model and find the total test mileage required to attain the target MMBF (mean miles between failures) of 50,000 miles under the estimated rate of reliability growth.

continued

Solution:

Cumulative Failures (N)	Failure Interarrival Mileage (Δt)	Failure Arrival Mileage (t)	CMMBF ($\Theta_C = t/N$)	Log (t)	Log (Θ_C)
1	5940	5940	5940	3.77	3.77
2	12,331	18,271	9136	4.26	3.96
3	21,010	39,281	13,094	4.59	4.12
4	27,192	66,473	16,618	4.82	4.22
5	19,910	86,383	17,277	4.94	4.24
6	24,211	110,594	18,432	5.04	4.27
7	26,422	137,016	19,574	5.14	4.29
8	27,731	164,747	20,593	5.22	4.31
9	26,862	191,609	21,290	5.28	4.33
10	29,271	220,880	22,088	5.34	4.34

The plot of the two rightmost columns of the above table is shown in Figure 6.11.

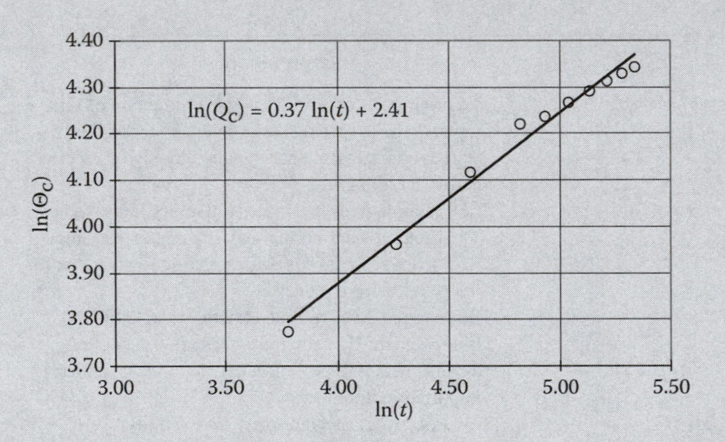

FIGURE 6.11 Duane model plot in Example 6.12.

Using Equation 2.139, the slope and intercept of the regression line in Figure 6.11 are estimated as 0.37 and 2.41, respectively. Then, the β parameter of the Duane model is

$$\beta = 1 - 0.37 = 0.63,$$

which corresponds to a program dedicated to the reduction of failures. The λ parameter is

$$\lambda = 10^{-2.41} = 0.0039\,\text{mile}^{-1}.$$

Using Equation 6.53, the test mileage required to attain the instantaneous MMBF of 50,000 miles is

$$t = (50,000 \times 0.0039 \times 0.63)^{1/(1-0.63)} = 443,568\ \text{miles}.$$

6.6.3 Army Material Systems Analysis Activity (AMSAA) Method

It is important to note that under Duane's assumption, Equation 6.50 and, consequently, Equation 6.52, are deterministic models. Crow [67] suggested that Equation 6.52 could be treated probabilistically as the ROCOF of a non-HPP (see Section 5.1.1). Such probabilistic interpretation of Equation 6.52 is known as the AMSAA model and offers two major advantages.

First, the model parameters can be estimated through the maximum likelihood method using Equations 5.30 through 5.31, and the confidence limits on these parameters can also be developed [67]. Second, the distribution of the number of failures $f[N(t)]$ can be obtained based on

$$\Pr[N(t) = n] = \frac{(\lambda t^{\beta})^n \exp(-\lambda t^{\beta})}{n!}, \quad n = 0, 1, 2, \ldots. \tag{6.54}$$

EXAMPLE 6.13

For the data in Example 6.12,

a. Find the MLEs of the AMSAA model parameters.
b. Determine the expected number of failures at 150,000 accumulated miles.
c. Find the probability that the actual number of failures at 150,000 miles will be greater than the expected value determined in (b).

Solution:

a. Using Equations 5.36 and 5.38, the estimates of the AMSAA model parameters are

$$\hat{\beta} = \frac{10}{\ln(220,880/5940) + \ln(220,880/12,331) + \cdots + \ln(220,880/26,862)} = 0.46,$$

$$\hat{\lambda} = \frac{10}{(220,880^{0.46})} = 0.035 \text{ mile}^{-1}.$$

b. The expected number of failures at 150,000 miles is

$$N(150,000) = \lambda t^{\beta} = 0.035(150{,}000^{0.46}) \approx 8.$$

c. The probabilities of the actual number of failures taking a value of less or equal to the expected number are provided in the table below.

n	$\Pr[N(150{,}000) > n] = \frac{8^n \exp(-8)}{n!}$
0	0.000335
1	0.002684
2	0.010735
3	0.028626
4	0.057252
5	0.091604
6	0.122138
7	0.139587
8	0.139587
Total	0.592547

Thus, the probability that the actual number of failures being greater than 7 is

$$\Pr[N(150,000) > 8] = 1 - 0.5925 = 0.4075.$$

See the software supplement for the automated reliability growth analysis.

The concepts of reliability growth are discussed by a number of authors. Balaban [68] presents the mathematical models of reliability growth; O'Connor [1] discusses general methods for sequential testing, reliability demonstration, and growth monitoring; and Fries and Sien [69] present a comprehensive survey of discrete reliability growth models.

EXERCISES

6.1 An engine crank shaft is a good example of a high reliability part of a car. Although it is pounded by each cylinder with every piston stroke, that single bar remains intact for a long time. Assume the strength of the shaft is normal with the mean S and standard deviation s, while the load per stroke is L with standard deviation l. A C-cylinder engine hits the shaft at C different places along it, so these events can be considered independent. Determine the reliability of the crank shaft.

a. Express the SM in terms of S, s, L, l, and C.
b. Estimate the reliability. Assume the motor turns at $X(t)$ revolutions.
c. Express the total number of reversals. $N(t)$ seen by each piston as a function of time.
d. If the shaft is subject to fatigue, express the reliability as a function of time. Metals fatigue, generally, follow Manson-Coffin Lae: $S(N) = SN^{(-1/q)}$. Assume also that the standard deviation, s, does not change with N. Also, q is a constant.
e. Determine the expected life (50% reliability) of the crank shaft turning at a constant rate, R (RPM).

6.2 Repeat Exercise 4.6 and calculate the Birnbaum and Vesely-Fussell importance measures for all events modeled in the fault tree.

6.3 The following data are given for a prototype of a system that undergoes design changes. A total of 10 failures have been observed since the beginning of the design. Estimate the Duane reliability growth model parameters. Discuss the results.

Failure Number	1	2	3	4	5	6	7	8	9	10
Cumulative Time on Test (h)	12	75	102	141	315	330	342	589	890	1007

6.4 In response to an RFQ, two vendors have provided a system proposal consisting of subsystem modules A, B, and C. Each vendor has provided failure rates and average corrective maintenance times for each module. Determine which vendor system has the best mean time to repair (MTTR) and which one you would recommend for purchase.

		Vendor 1		Vendor 2	
Module	No. in System	Failure Rate (per 10^4 h)	M_{ct} (min)	Failure Rate (per 104 h)	M_{ct} (min)
A	2	45	15	45	20
B	1	90	20	30	15
C	2	30	10	90	10

a. Describe the advantages of a preventive maintenance program.
b. Is it worth doing preventive maintenance if the failure rate is constant?

6.5 You are the project engineer for the development of a new airborne radar system with a design goal of 7000 h MTBF. The system has been undergoing development testing for the past six months, during which time eight failures have occurred in approximately 9000 test hours as follows.

Failure	Test Hours to Failure
1	1296
2	1582
3	1855
4	2310
5	3517
6	5188
7	6792
8	8902

How much time would you schedule for the balance of the test program, in order to have some confidence that your controller had met its goal? (*Note*: Each failure represents a different failure mode and corrective actions are being taken for each.)

6.6 A scenario is characterized by the following cutest:

$$S = ab + ac + d.$$

Probability of a and b can be obtained from an exponential model such as

$$\Pr(a) = 1 - \exp(-\lambda_a),$$

but probability of c and d can be obtained from a Weibull model such as

$$\Pr(c) = 1 - \exp\left[-\left(\frac{t}{\alpha}\right)^{\beta}\right].$$

If $\lambda_a = \lambda_b = 10^{-4}\,\text{h}^{-1}$ and $\alpha_c = \alpha_d = 2250\,\text{h}$ and $\beta_c = \beta_d = 1.4$. Determine the importance of parameters α, β, and λ in modeling this scenario. Use an importance measure of your choice, and justify your choice.

6.7 A system has the following event tree and the corresponding fault trees:

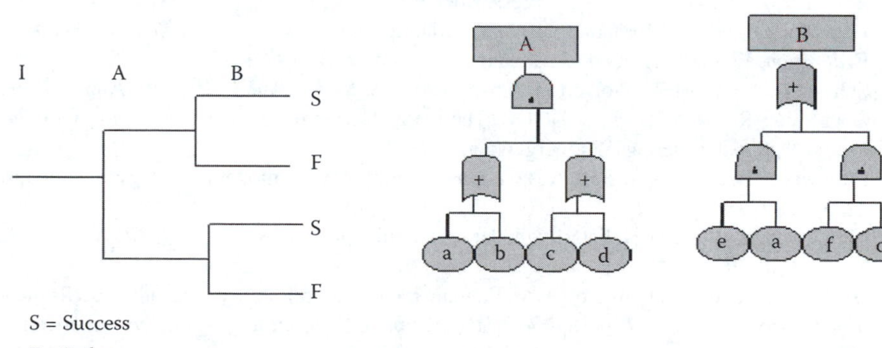

The failure data are listed below.

Component	MTTR (h)	MTBF (h)
a	12	17,500
b	17	18,200
c	20	27,500
d	17	18,200
e	7	12,900
f	15	19,000

Frequency of initiating event $I = 1.5\ \text{year}^{-1}$; components b and d have dependent (common-cause) failure with a β factor of 0.08.

a. Determine the minimal cut sets in terms of the basic events (I, a, b, c, etc.) for the event tree scenarios ending in failure (i.e., F).
b. Estimate the frequency of failure of the system.
c. Calculate the Fussell–Vesely importance measure of all components based on part B above.

REFERENCES

1. O'Connor, P. D. T., *Practical Reliability Engineering*, 3rd edition, Wiley, New York, 1991.
2. Kapur, K. C. and L. R. Lamberson, *Reliability in Engineering Design*, Wiley, New York, 1977.
3. Mosleh, A., N. Siu, C. Smidts, and C. Lui, *Model Uncertainty: Its Characterization and Quantification*, International Workshop Series on Advanced Topics in Reliability and Risk Analysis, Center for Reliability Engineering, University of Maryland, College Park, MD, 1995.
4. Martz, H. F., *A Comparison of Methods for Uncertainty Analysis of Nuclear Plant Safety System Fault Tree Models*, U.S. Nuclear Regulatory Commission and Los Alamos National Laboratory, NUREG/CR-3263, Los Alamos, NM, 1983.
5. Morchland, J. D. and G. G. Weber, A moments method for the calculation of confidence interval for the failure probability of a system, *Proceeding of the 1972 Annual Reliability and Maintainability Symposium*, pp. 505–572, 1972.
6. Morgan, M. G. and M. Henrion, *Uncertainty: A Guide to Dealing with Uncertainty in Quantitative Risk and Policy Analysis*, Cambridge University Press, Cambridge, 1990.
7. Lawrence, J. D., *Software Reliability and Safety in Nuclear Reactor Protection Systems*, Lawrence Livermore National Laboratory, NUREG/CR-6101, 1993.
8. McDermid, J. A., Issues in developing software for safety critical systems, *Reliab. Eng. Syst. Saf.*, 32, 1–24, 1991.
9. Keller, A. Z., *Reliability Aging and Growth Modeling in Reliability Modeling and Applications*, A. G. Colombo and A. Z. Keller, eds, ECSC, EEC, EAEC, Brussels and Luxembourg, 1987.
10. Lyu, M. R. *Handbook of Software Reliability Engineering*, McGraw-Hill, New York, 1995.
11. Smidts, C., Software reliability: Current approaches, new directions, *The Electronics Handbooks*, J. C. Whitaker ed., CRC Press & IEEE Press, Boca Raton, FL, 1996.
12. Jelinski, Z. and P. Moranda, Software reliability research, in *Statistical Computer Performance Evaluation*, W. Freiberger, ed., Academic Press, New York, 1972.
13. Schick, G. J. and R. W. Wolverton, *Assessment of Software Reliability*, 11th Annual Meeting German Operations Research Society, DGOR, Hamburg, Germany; also in *Proceedings of the Operations Research*, Physica-Verlag, Wirzberg-Wien, 1973.
14. Littlewood, B. and J. K.Verrall, A Bayesian reliability growth model for computer software, *Appl. Stat.*, 22, 332, 1973.
15. Littlewood, B., Software reliability model for modular program structure, *IEEE Trans. Reliab.*, R-28 (3), 241–246, 1979.
16. Goal, A. L. and K. Okumoto, A Markovian model for reliability and other performance measures of software systems, in *Proceedings of the National Computing Conference*, American Federation of Information Processing Societies, New York, Vol. 48, 1979.
17. Goal, A. L. and K. Okumoto, A time dependent error detection rate model for software reliability and other performance measures. *IEEE Trans. Reliab.*, R28, 206, 1979.
18. Mills, H. D., *On the Statistical Validation of Computer Programs*, IBM Federal Systems Division, Report 72-6015, Gaithersburg, MD, 1972.
19. Nelson, E., Estimating software reliability from test data, *Microelectron. Reliab.*, 17, 67, 1978.
20. Ramamoorthy, C. V. and F. B. Bastani, Software reliability: Status and perspectives, *IEEE Trans. Soft. Eng.*, SE-8, 359, 1982.
21. Shooman, M. L., Software reliability measurements models, in *Proceedings of the Annual Reliability and Maintainability Symposium*, American Society for Quality, Washington, DC, 1975.
22. Musa, J. D., A theory of software reliability and its application, *IEEE Trans. on Soft. Eng.*, September, 312–327, 1975.

23. Goel, A. L., *A Guidebook for Software Reliability Assessment*, Kept. RADC TR-83-176, 1983.
24. Musa, J. D. and K. Okumoto, A logarithmic Poisson execution time model for software reliability measurement, *ICSE*, 230–238, 1984.
25. Musa, J. D., A. Iannino, and K. Okumoto, *Software Reliability*, McGraw-Hill, New York, 1987.
26. Petrella, S., P. Michael, W. C. Bowman, and S. T. Lim, Random testing of reactor shutdown system software, in *Proceedings of the International Conference on Probabilistic Safety Assessment and Management*, G. Apostolakis, ed., Elsevier, New York, 1991.
27. Boehm W. B., A spiral model of software development and enhancement, *IEEE Comput.*, 21 (5), 61–72, 1988.
28. Pressman, R. S., *Software Engineering: A Practitioners Approach*, 2nd edition, McGraw-Hill, New York, 1987.
29. Reactor Safety Study, *An Assessment of Accidents in U.S. Commercial Nuclear Power Plants*, U.S. Regulatory Commission, WASH-1400, Washington, DC, 1975.
30. Hannaman, G. W. and A. J. Spurgin, *Systematic Human Action Reliability Procedure 6 (SHARP), Electric Power Research Institute*, NP-3583, Palo Alto, CA, 1984.
31. Rasmussen, J., Cognitive control and human error mechanisms, Chapter 6 in *New Technology and Human Error*, J. Rasmussen, K. Duncan, and J. LePlate, eds, Wiley, New York, 1987.
32. Woods, D. D., E. M. Roth, and H. Pole, Modeling human intention formation for human reliability assessment, *Reliab. Eng. Syst. Saf.*, 22, 169–200, 1988.
33. Amendola, A., U. Bersini, P. C. Cacciabue, and G. Mancini, Modeling operators in accident conditions: Advances and perspectives on cognitive model, *Int. J. Man Mach. Stud.*, 27, 599, 1987.
34. Kendrick, B., *Systematic Safety Training*, Marcel Dekker, New York, 1990.
35. Kendrick, B., *Investigating Accidents with STEP*, Marcel Dekker, New York, 1987.
36. Wreathall, J., *Operator Action Tree Method*, IEEE Standards Workshops on Human Factors and Nuclear Safety, Myrtle Beach, SC, 1981.
37. Potash, L., M. Stewart, P. E. Diets, C. M. Lewis, and E. M. Dougherty, Experience in integrating the operator contribution in the PRA of actual operating plants, *Proceedings of American Nuclear Society*, Topical Meeting on Probabilistic Risk Assessment, Port Chester, New York, 1981.
38. Swain, A. D. and H. E. Guttman, *Handbook of Human Reliability Analysis with Emphasis on Nuclear Power Applications*, U.S. Nuclear Regulatory Commission, NUREG/CR-1278, Washington, DC, 1983.
39. Hall, R. E., J. Wreathall, and J. R. Fragola, *Post Event Human Decision Errors: Operator Action/Time Reliability Correlation*, U.S. Nuclear Regulatory Commission, NUREG/CR-3010, Washington, DC, 1982.
40. Hannaman, G. W., A. J. Spurgin, and Y. D. Lukic, *Human Cognitive Reliability Model for PRA Analysis*, NUS Corporation, NUS-4531, San Diego, CA, 1984.
41. Hunns, D. M. and B. K. Daniels, *The Method of Paired Comparisons*, 3rd European Reliability Data Bank Seminar, University of Bradford, National Center of System Reliability, UK, 1980.
42. Embry, D. E., P. C. Humphreys, E. A. Rosa, B. Kirwan, and K. Rea, *SLIM–MAUD: An Approach to Assessing Human Error Probabilities Using Structured Expert Judgment*, U.S. Nuclear Regulatory Commission, NUREG/CR-3518, Washington, DC, 1984.
43. Poucet, A., Survey of methods used to assess human reliability in the human factors reliability benchmark exercise, *Reliab. Eng. Syst. Saf.*, 22, 257–268, 1988a.
44. Poucet, A., *State of the Art in PSA Reliability Modeling as Resulting from the International Benchmark Exercise Project*, NUCSAFE 88 Conference, Avignon, France., 1988b.
45. Siegel, A. I., N. D. Bartter, J. Wolf, H. E. Knee, and P. M. Haas, *Maintenance Personnel Performance Simulation (MAPPS) Model*, U.S. Nuclear Regulatory Commission, NUREG/CR-3626, Vols. I and II, Washington, DC, 1984.
46. Lichtenstein, S. B., B. Fischoff, and L. D. Phillips, Calibration of probabilities: The state of the art, in *Decision Making and Change in Human Affairs*, J. Jungerman and G. deZeeuw, eds, D. Reidel, Dordrecht, Holland, 1977.
47. Stillwell, W., D. A. Seaver, and J. P. Schwartz, *Expert Estimation of Human Error Problems in Nuclear Power Plant Operations: A Review of Probability Assessment and Scaling*, U.S. Nuclear Regulatory Commission, NUREG/CR-2255, Washington, DC, 1982.
48. Saaty, T. L., *The Analytic Hierarchy Process*, McGraw-Hill, New York, 1980.

49. Gertman, D., et al., *The SPAR-H Human Reliability Analysis Method.* U.S. Nuclear Regulatory Commission, NUREG/CR-6883, Washington, DC, 2005.
50. Rasmussen, J., Skills, rules and knowledge: Signals, signs and symbols and their distinctions in human performance models, *IEEE Trans. Syst. Man Cybern.*, SMC-3 (3), 257–268, 1982.
51. Dougherty, E. M. and J. R. Fragola, *Human Reliability Analysis: A System Engineering Approach with Nuclear Power Plant Applications*, Wiley, New York, 1988.
52. Birnbaum, Z. W., On the importance of different components in a multicomponent system, in *Multivariate Analysis-II*, P. R. Krishnaiah, ed., Academic Press, New York, 1969.
53. Sharirli, M., *Methodology for System Analysis Using Fault Trees, Success Trees and Importance Evaluations*, PhD dissertation, University of Maryland, Department of Chemical and Nuclear Engineering, College Park, MD, 1985.
54. Hoyland, A. and M. Rausand, *System Reliability Theory: Models and Statistical Methods*, Wiley, New York, 1994.
55. Fussell, J., How to hand calculate system reliability and safety characteristics, *IEEE Trans. Reliability*, R-24 (3), 169–174 1975.
56. Modarres, M., *Risk Analysis in Engineering: Techniques, Tools, and Trends*, CRC/Taylor & Francis, London, 2006.
57. Lambert, H. E. *Measures of Importance of Events and Cut Sets in Fault Trees in Reliability and Fault Tree Analysis*, R. Barlow, J. Fussell, and N. Singpurwalla, eds., SIAM, Philadelphia, PA, 1975.
58. U.S. Nuclear Regulatory Commission, Analysis of Core Damage Frequency: Sequoyah, Unit 1 Internal Events, NUREG/CR-4550, Vol. 5, 1990.
59. Zio, E. and L. Podofillini, Accounting for component interactions in the differential importance of a measure, *Reliab. Eng. Syst. Saf.*, 91 (2), 181–190, 2006.
60. Bier, V. M., *A Measure of Uncertainty Importance for Components in Fault Trees*, Transactions of the 1983 Winter Meeting of the American Nuclear Society, San Francisco, CA, 1983.
61. Severe Accident Risk, *An Assessment for Five U.S. Nuclear Power Plants*, U.S. Nuclear Regulatory Commission, NUREG-1150, Washington, DC, 1990.
62. Nelson, W., *Applied Life Data Analysis*, Wiley, New York, 1982.
63. Ascher, H. and H. Feingold, *Repairable Systems Reliability*, Marcel Dekker, New York, 1984.
64. Duane, J. T., Learning curve approach to reliability monitoring. *IEEE Trans. Aerosp.*, 2, 563–566, 1964.
65. Crowder, M. J., A. C. Kimber, R. L. Smith, and T. J. Sweeting, *Statistical Analysis of Reliability Data*, Chapman & Hall, London, 1993.
66. Gnedenko, B. and I. Ushakov, *Probabilistic Reliability Engineering*, Wiley, New York, 1995.
67. Crow, L. H., Reliability analysis for complex repairable systems, *Reliability and Biometry*, in F. Proschan and R. J. Serfling, eds., SIAM, Philadelphia, 1974.
68. Balaban, H. S., *Reliability Analysis for Complex Repairable Systems*, Reliability and Biometry, SIAM, 1978.
69. Fries, A. and A. Sen, Survey of discrete reliability growth models, *IEEE Trans. Reliab.*, 45 (4), 582–604, 1996.

7 Selected Topics in Reliability Data Analysis

7.1 ACCELERATED LIFE TESTING

The reliability models considered in the previous chapters are expressed in terms of time-to-failure distribution or in terms of probability of failure on demand. These models are not appropriate in cases, where one is interested in reliability dependence on stress factors such as ambient temperature, humidity, vibration, operator skill, and so on. This dependence is considered in reliability models with *explanatory variables* or *covariates* [1]. Such models are traditionally referred to as AL models, which may be confusing because applications of these models are not necessarily limited to accelerated life (AL) testing, as will be demonstrated in this section.

7.1.1 Basic AL Notions

A reliability model (AL reliability model) is defined as the relationship between the time-to-failure distribution of a device and its stress factors, such as load, cycling rate, temperature, humidity, and voltage. AL reliability models are based on physics-of-failure considerations.

Stress severity in terms of reliability (or time-to-failure distribution) is expressed as follows. Let $R_1(t; z_1)$ and $R_2(t; z_2)$ be the reliability functions of the item under constant stress conditions z_1 and z_2, respectively. It should be mentioned that stress condition, z, in general, is a vector of the stress factors. The stress condition z_2 is called more severe than z_1, if for all values of t the reliability of the item under stress condition z_2 is less than the reliability under stress condition z_1. That is,

$$R_2(t; z_2) < R_1(t; z_1). \tag{7.1}$$

7.1.1.1 Time Transformation Function for the Case of Constant Stress

For the monotonic cdfs $F_1(t; z_1)$ and $F_2(t; z_2)$, if constant stress condition z_1 is less severe than z_2, and t_1 and t_2 are the times at which $F_1(t; z_1) = F_2(t; z_2)$, there exists a function g (for all t_1 and t_2) such that $t_1 = g(t_2)$. Therefore,

$$F_2(t_2; z_2) < F_1[g(t_2), z_1]. \tag{7.2}$$

Because $F_1(t; z_1) < F_2(t; z_2)$, $g(t)$ must be an increasing function with $g(0) = 0$. The function $g(t)$ is called the *acceleration* or *the time transformation function.*

The AL reliability model is a deterministic transformation of time to failure. Two main time transformations are considered in reliability data analysis. These transformations are known as the *AL model* and *the proportional hazard* (PH) *model.*

7.1.1.2 AL Model

The AL model is the most popular type of reliability model with explanatory variables. For example, AT&T's Reliability Model [2] is based on the AL model.

It may be assumed that $z = 0$ for the normal (use) stress condition. Denote a time-to-failure cdf under normal stress condition by $F_0(\cdot)$. The AL time transformation is expressed in terms of $F(t; z)$ and $F_0(\cdot)$ and is given by the following relationship [3]:

$$F(t; z) = F_0[t\psi(z, A)], \tag{7.3a}$$

where $\psi(z, A)$ is a positive function connecting time to failure with a vector of stress factors z; and A is a vector of unknown parameters; for $z = 0$, $\psi(z, A)$ is assumed to be 1. The corresponding relationship for the pdf can be obtained from Equation 7.3a as

$$f(t; z) = f_0[t\psi(z, A)]\psi(z), \tag{7.3b}$$

where $f_0(\cdot)$ is the time-to-failure pdf under the normal stress condition. Relationship 7.3a is the scale transformation. It means that a change in stress does not result in a change of the shape of the distribution function, but changes its scale only. Relationship 7.3b can be written in terms of the acceleration function as follows:

$$g(t) = \psi(z, A)t. \tag{7.4}$$

Relationship 7.3a is equivalent to the linear with time acceleration function 7.4. The time-to-failure distributions of a device under the normal stress condition ($z = 0$) and the stress condition $z \neq 0$ are geometrically similar to each other. Such distributions are denoted as "belonging to the class of time-to-failure distribution functions, which is closed with respect to scale" [1].

The similarity property is widely used in physics and engineering. Because it is difficult to imagine that any change of failure mode or mechanism would not result in a change in the shape of the failure time distribution, relationship 7.3a can be also considered as a principle of failure mechanism conservation or a similarity principle, which states that the failure modes and mechanisms remain the same over the stress domain of interest. The analysis of sets of real life data often shows that the similarity of time-to-failure distributions really exists, so that a violation of the similarity can indicate a change in a failure mechanism.

The relationship for the 100pth percentile of time to failure, $t_p(z)$, can be obtained from Equation 7.3a as

$$t_p(z) = \frac{t_p^0}{\psi(z, A)}, \tag{7.5}$$

where t_p^0 is the 100pth percentile for the normal stress condition $z = 0$.

The relationship 7.5 is the percentile AL reliability model and is usually written in the form

$$t_p(z, B) = \eta(z, B), \tag{7.6}$$

where B is a vector of unknown parameters. Reliability models are briefly considered in Section 7.1.2.

The AL reliability model is related to the relationship for percentiles (Equation 7.5) as

$$t_p(z, B) = \frac{t_p^0}{\psi(z, A)}. \tag{7.7}$$

The corresponding relationship for failure rate can also be obtained from Equation 7.3a as

$$\lambda(t; z) = \psi(z, A)\lambda^0[t\psi(z, A)]. \tag{7.8}$$

Clearly, the relationship for percentiles (Equation 7.5) is the simplest one.

7.1.1.3 Cumulative Damage Models and AL Model

Some known cumulative damage models result in the similarity of time-to-failure distributions under quite reasonable restrictions. As an example, the Barlow and Proschan model [4] results in an aging (IFRA) time-to-failure distribution, introduced in Section 3.1.2.

Consider, for example, an item subjected to shocks occurring randomly in time. Let these shocks arrive according to the Poisson process with constant intensity. Each shock causes a random amount x_i of damage, where $x_1, x_2, \ldots, x_i$ are r.v.s distributed with a common cdf, $F(x)$, called a *damage distribution function*. The item fails when accumulated damage exceeds a threshold X. It has been shown by Barlow and Proschan [4] that for *any* damage distribution function $F(x)$, the time-to-failure distribution function is IFRA.

Now consider an item under the stress conditions characterized by different shock intensities λ_i and different damage distribution functions $F_i(x)$. It can be also shown that the similarity of the corresponding time-to-failure distribution functions will hold for all these stress conditions, $z_i(\lambda_i, F_i(x))$, if they have the same damage cdf, that is, $F_i(x) = F(x)$. A similar example from fracture mechanics is considered in [5].

7.1.1.4 PH Model

For the PH model the basic relationship analogous to Equation 7.3a is given by

$$F(t; z) = 1 - [1 - F_0(t)]^{\psi(z,A)} \tag{7.9a}$$

or, in terms of reliability function, $R(t)$, as

$$R(t; z) = R_0(t)^{\psi(z,A)}. \tag{7.9b}$$

The proper PH (Cox) model is known as the relationship for hazard rate [3], which can be obtained from Equations 7.9a or 7.9b as

$$\lambda(t; z) = \psi(z, A)\lambda^0(t), \tag{7.10}$$

where $\psi(z, A)$ is usually chosen as a log-linear function.

Note that the PH model time transformation does not normally retain the shape of the cdf, and the function $\psi(z)$ no longer has a simple relationship to the acceleration function, nor does it have a clear physical meaning. That is why the PH model is not as popular in reliability applications as the AL model. It should be mentioned that for the Weibull distribution (and only for the Weibull distribution) the PH model coincides with the AL model (Cox and Oaks [3]).

7.1.2 Some Popular AL (Reliability) Models

The most commonly used AL models for the percentiles (including median) of time-to-failure distributions are log-linear models. Two such models are the Power Rule Model and the Arrhenius Reaction Model [6]. The Power Rule model is given as

$$t_p(x) = \frac{a}{x^c}, \quad a > 0, \ c > 0, \ x > 0, \tag{7.11a}$$

where x is a mechanical or electrical stress, c is a unitless constant, and the unit of constant a being the product of time and the measure of x^c. In reliability of electrical insulation and capacitors, x is usually the applied voltage. To estimate fatigue life, the model is used as the analytical representation of the so-called S–N or Wöhler curve, where S is the stress amplitude and N is life in cycles to failure, such that

$$N = kS^{-b}, \tag{7.11b}$$

where b and k are material parameters estimated from test data. Because of the probabilistic nature of fatigue life at any given stress level, one must deal with not just one S–N curve, but with a family of S–N curves, so that each curve is related to a probability of failure as the parameter of the model. These curves are called S–N–P curves or curves of constant probability of failure on a stress versus life plot. It should be noted that relationship 7.11b is an empirical model.

Another popular model is the Arrhenius Reaction Rate Model:

$$t_p(T) = a \exp\left(\frac{E_a}{T}\right), \tag{7.12}$$

where T is the absolute temperature under which the unit is functioning, and E_a is the activation energy. This model is the one most widely used for expressing the effect of temperature on reliability. The application of the Arrhenius model for electronic component reliability estimation was briefly discussed in Section 3.7. Originally, the model was introduced as a chemical reaction rate model.

Another model is a combination of models (Equations 7.11 and 7.12):

$$t_p(x, T) = ax^{-c} \exp\left(\frac{E_a}{T}\right), \tag{7.13}$$

where x (as defined by Equation 7.11) is a mechanical or electrical stress. This model is used in fracture mechanics of polymers as well as a model for the electromigration failures in aluminum thin films of integrated circuits. In the last case, stress factor x is the current density.

Jurkov's model [6] is another popular AL reliability model:

$$t_p(x, T) = t_0 \exp\left(\frac{E_a - \gamma x}{T}\right). \tag{7.14}$$

This model is considered as an empirical relationship reflecting the thermal fluctuation character of long-term strength, that is, durability under constant stress [8]. For mechanical long-term strength, parameter t_0 is a constant, which is numerically close to the period of thermal atomic oscillations (10^{-11}–10^{-13} s); E_a is the effective activation energy, which is numerically close to vaporization energy for metals and to chemical bond energies for polymers; and γ is a structural coefficient. The model is widely used for reliability prediction problems of mechanical and electrical (insulation, capacitors) long-term strength.

The *a priori* choice of a model (or competing models) is based on physical considerations. Meanwhile, statistical data analysis of AL test results or collected field data, combined with FMEA, can be used to check the adequacy of the chosen model, or to determine the most appropriate model among the competing ones.

7.1.3 AL Data Analysis

7.1.3.1 Exploratory Data Analysis (Criteria of Linearity of Time Transformation Function for Constant Stress)

The experimental verification of the basic ALM assumption (Equation 7.3a) is not only important in failure mechanism study, but also has great practical importance because almost all statistical procedures for AL test planning and data analysis are based on this assumption. Several techniques can be used for verification of the linearity of the time transformation function. Some of them are briefly discussed below.

7.1.3.2 Two-Sample Criterion

Let us start with the historically first criterion, which makes clear the physical meaning of the idea of similarity of time-to-failure distributions. This criterion requires two special tests. During the first test, a sample is tested at constant stress level z_1 over time period t_1, at which z_1 is changed to a constant stress z_2 for time period t_2. Such loading pattern (load as function of time) can be called *stress profile* S_1. During the second test, another sample is first tested under z_2 during t_2 and then is tested under the stress level z_1 during time t_1 (stress profile S_2). The time transformation function will be a linear function of time if the reliability functions (or the corresponding failure probabilities) of the items after the first and the second tests are equal (i.e., a change of loading order does not change the cumulative damage).

The corresponding statistical procedure can be based on the analysis of the so-called 2×2 contingency tables [9]. This analysis was initially developed for comparing binomial proportions (probabilities).

The null hypothesis tested, H_0, is

$$H_0 : p_1(S_1) = p_2(S_2) = p, \tag{7.15}$$

where p is the failure probability during the test with stress profile $S_1(S_2)$, or, in terms of reliability functions, the null hypothesis is expressed as

$$H_0 : R_1(S_1) = [1 - p_1(S_1)] = R_2(S_2) = [1 - p_2(S_2)] = R = 1 - p. \tag{7.16}$$

The alternative hypothesis, H_1, is

$$H_1 : p_1(S_1) \neq p_2(S_2). \tag{7.17}$$

Let n_1 and n_2 be the sample sizes tested under stress profiles S_1 and S_2, respectively. Further, let $n_{1\text{f}}$ and $n_{2\text{f}}$ be the number of items failed during these tests. Denote the corresponding number of nonfailed items by $n_{1\text{s}}$ and $n_{2\text{s}}$. Obviously, $n_1 = n_{1\text{f}} + n_{1\text{s}}$ and $n_2 = n_{2\text{f}} + n_{2\text{s}}$. Finally, denote $N = n_1 + n_2$. These test data can be arranged in the following contingency table.

Stress Profile 1	**Stress Profile 2**
$n_{1\text{f}}$	$n_{2\text{f}}$
$n_{1\text{s}}$	$n_{2\text{s}}$

If H_0 is true,

1. the probability $p_1(S_1) = p_2(S_2) = p$ can be estimated as

$$\hat{p} = \frac{n_{1\text{f}} + n_{2\text{f}}}{n_1 + n_2} = \frac{n_{1\text{f}} + n_{2\text{f}}}{N}. \tag{7.18}$$

2. the reliability functions $R_1(S_1) = R_2(S_2) = R$ can be estimated as

$$\hat{R} = \frac{n_{1\text{s}} + n_{2\text{s}}}{n_1 + n_2} = \frac{n_{1\text{s}} + n_{2\text{s}}}{N}. \tag{7.19}$$

3. based on these estimates, the expected frequencies $n_{1\text{f}}$, $n_{2\text{f}}$, $n_{1\text{s}}$, and $n_{2\text{s}}$ can be estimated as

$$\begin{aligned} \hat{n}_{1\text{f}} &= \hat{p}n_1, \ \hat{n}_{2\text{f}} = \hat{p}n_2, \\ \hat{n}_{1\text{s}} &= \hat{R}n_1, \ \hat{n}_{2\text{s}} = \hat{R}n_2. \end{aligned} \tag{7.20}$$

The following measure of discrepancy between the observed and expected frequencies for the contingency table can be introduced as

$$W = \frac{(n_{1f} + \hat{n}_{1f})^2}{\hat{n}_{1f}} + \frac{(n_{1s} + \hat{n}_{1s})^2}{\hat{n}_{1s}} + \frac{(n_{2f} + \hat{n}_{2f})^2}{\hat{n}_{2f}} + \frac{(n_{2s} + \hat{n}_{2s})^2}{\hat{n}_{2s}}, \tag{7.21}$$

which, under the null hypothesis, follows an approximate χ^2 distribution with $(4 - 2 - 1) = 1$ degrees of freedom. Thus, for a significance level, α, the null hypothesis is rejected if the above sum W is greater than the critical value of $\chi^2_{1-\alpha}(1)$.

Example 7.1

Two samples of identical thin polymer film units were tested. The first sample of 48 units was tested under stress profile (S_1): for 1 h, the units were under a voltage of 50 V, then the voltage was instantaneously increased to 70 V, under which the sample was tested for another hour. The second sample of 52 units was tested under the backward stress profile (S_2): it was put under 70 V for the first hour, then the voltage was decreased to 50 V and the test was continued for another hour. The data obtained from the two sample tests are given in the table below. Test the null hypothesis: $p_1(S_1) = p_2(S_2) = p$.

Stress Profile 1	**Stress Profile 2**
$n_{1f} = 19$	$n_{2f} = 32$
$n_{1s} = 29$	$n_{2s} = 20$

Solution:
Find: $n_1 = n_{1f} + n_{1s} = 48$; $n_2 = n_{2f} + n_{2s} = 52$; $n_1 + n_2 = N = 100$.
The probability $p_1(S_1) = p_2(S_2) = p$ is estimated as

$$\hat{p} = \frac{n_{1f} + n_{2f}}{n_1 + n_2} = \frac{n_{1f} + n_{2f}}{N},$$

$$\hat{p} = \frac{19 + 32}{100} = \frac{51}{100}.$$

Similarly, the probability $R_1(S_1) = R_2(S_2) = R$ is estimated as

$$\hat{R} = \frac{n_{1s} + n_{2s}}{n_1 + n_2} = \frac{n_{1s} + n_{2s}}{N},$$

$$\hat{R} = \frac{29 + 20}{100} = \frac{49}{100}.$$

The corresponding expected frequencies are calculated as

$$\hat{n}_{1f} = \hat{p}n_1 = \frac{51(48)}{100} = 24.48,$$

$$\hat{n}_{2f} = \hat{p}n_2 = \frac{51(52)}{100} = 26.52,$$

$$\hat{n}_{1s} = \hat{R}n_1 = \frac{49(48)}{100} = 23.52,$$

$$\hat{n}_{2s} = \hat{R}n_2 = \frac{49(52)}{100} = 25.48.$$

Finally, find the value of χ^2 statistic as

$$W = \frac{(n_{1f} + \hat{n}_{1f})^2}{\hat{n}_{1f}} + \frac{(n_{1s} + \hat{n}_{1s})^2}{\hat{n}_{1s}} + \frac{(n_{2f} + \hat{n}_{2f})^2}{\hat{n}_{2f}} + \frac{(n_{2s} + \hat{n}_{2s})^2}{\hat{n}_{2s}}$$

$$= \frac{(19 + 24.48)^2}{24.48} + \frac{(32 + 26.52)^2}{26.52} + \frac{(29 + 23.52)^2}{23.52} + \frac{(20 + 25.48)^2}{25.48} = 4.81.$$

If α is chosen as 5%, $\chi^2_{0.95}(1) = 3.82$, then our null hypothesis is rejected, which means that AL model (Equation 7.3a) is not applicable for the polymer film specimens when the applied voltage is changed from 50 up to 70 V. This conclusion can indicate a change in failure mechanisms due to a voltage increase.

7.1.3.3 Checking the Coefficient of Variation

The second criterion is associated with the coefficient of variation (i.e., the standard deviation to mean ratio, σ/m). It is possible to show that if the time transformation function is linear with respect to time for some constant stress levels $z_1, z_2, \ldots, z_k$, the coefficient of variation of time to failure will be the same for all these stress levels.

7.1.3.4 Logarithm of Time-to-Failure Variance

It can also be shown that under the same assumption the variance of the logarithm of times to failure will be the same for the stress levels at which the AL model holds. For the lognormal time-to-failure distribution, Bartlett and Cochran's tests can be used for checking whether the variances are constant [6].

7.1.3.5 Quantile–Quantile Plots

The quantile–quantile plot is a curve such that the coordinates of every point are the time-to-failure quantiles (percentiles) for a pair of stress conditions of interest. If the time transformation function is linear in time (i.e., relationship 7.3a holds), the quantile–quantile plot will be a straight line going through the origin. A sample quantile, $\hat{t}_p$, of level p (i.e., an estimate of the respective true quantile, t_p) for a sample of size n is defined as

$$\hat{t}_p = \begin{cases} t_{([np])}, & \text{if } np \text{ is not integer, and} \\ \text{any value from the interval } \left(t_{([np])}, t_{([np]+1)}\right), & \text{if } np \text{ is integer,} \end{cases} \tag{7.22}$$

where $t_{(\cdot)}$ is the failure time (order statistic), and $[x]$ means *the greatest integer that does not exceed x*.

The corresponding data analysis procedure is realized in the following way. All the sample quantiles of a given constant stress condition are plotted on one axis and the sample quantiles of another stress condition are plotted on the other axis. If the sample sizes for two stress conditions are equal, the corresponding order statistics can be used as the sample quantiles. Using the points obtained (a pair of quantiles of the same level gives a point), a straight regression line can be fitted. The AL model will be applicable, provided that one obtains linear dependence between the sample quantiles, and that the hypothesis that the intercept of the fitted line is equal to zero is not rejected (for more details see [5]).

Example 7.2

For the data given in Example 2.33, verify the applicability of AL model (Equation 7.3) assumptions.

Solution:

The values of sample coefficients of variation (i.e., sample standard deviation to sample mean ratio) for the time-to-failure data obtained under the temperatures 50°C, 60°C, and 70°C as well as the corresponding logarithms of the time-to-failure variances are given in the following table. The values of sample coefficients of variation and the values of logarithms of the time-to-failure variances are very close to each other for the respective temperatures. Thus, the ALT model assumptions look realistic for the data given.

Temperature (°C)	Sample Coefficients of Variation	Logarithm Time-to-Failure Variances
50	0.678	0.632
60	0.573	0.302
70	0.626	0.521

The same conclusion could be drawn using the quantile–quantile plots for these data. They show strong linear dependence between the sample quantiles (all the correlation coefficients are greater than 0.95) and the respective intercepts of the fitted lines are reasonably insignificant. Figure 7.1 provides an example of the quantile–quantile plots for the temperatures 50°C and 70°C.

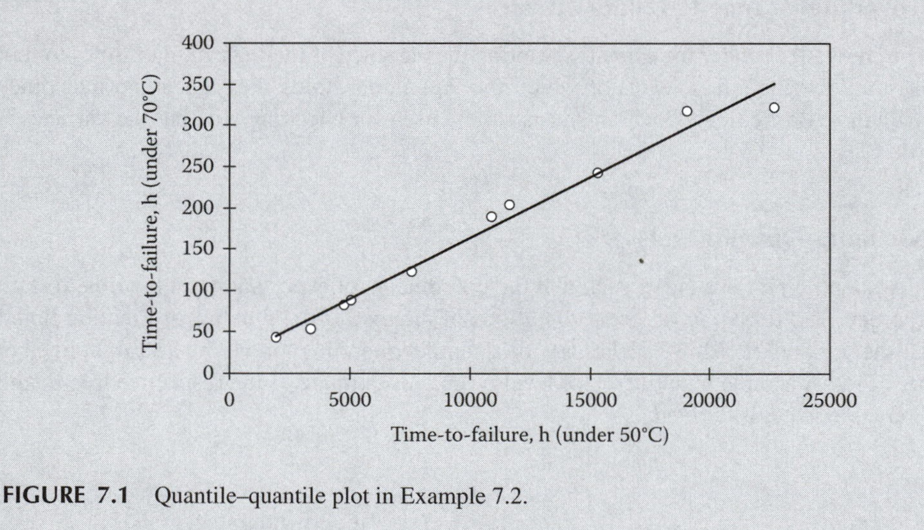

FIGURE 7.1 Quantile–quantile plot in Example 7.2.

7.1.3.6 Reliability Models Fitting: Constant Stress Case

Statistical methods for reliability model fitting on the basis of AL tests or field data collected can be divided into two groups—parametric and nonparametric. For the former, the time-to-failure distribution is assumed to be a specific parametric distribution, normal, exponential, or Weibull, whereas for the latter the only assumption is that the time-to-failure distribution belongs to a particular class of time-to-failure distribution, that is, continuous, IFR, or IFRA.

The most commonly used parametric methods are parametric regression (normal and lognormal, exponential, Weibull, and extreme value), the least squares method, and the maximum likelihood method. The following discusses the least-squares method for uncensored data [3,6].

The relationship for quantiles (Equation 7.5) can be written in terms of r.v.s as

$$t = \frac{t_0}{\psi(z)}, \tag{7.23}$$

where the time to failure, t_0, under normal stress has cdf, $F_0(\cdot)$.

Designate the expectation of log t_0 by μ_0. That is,

$$E(\log t_0) = \mu_0. \tag{7.24}$$

Using the equation above, one can write

$$\log t = \mu_0 - \log \psi(z) + \varepsilon, \tag{7.25}$$

where ε is a r.v. of zero mean with a distribution not depending on z. To make Equation 7.25 clear, note that any r.v. x having an expectation, $E(x)$, and a finite variance, var(x), can be represented as

$$x = E(x) + \varepsilon, \tag{7.26}$$

where $E(\varepsilon) = 0$, and $\text{var}(\varepsilon) = \text{var}(x)$.

If log $\psi(z)$ is a linear function with respect to parameter B's function (the case of log linear reliability model), that is,

$$\log \psi(Z, B) = ZB, \tag{7.27}$$

then Equation 7.25 can be written as

$$\log t = \mu_0 - ZB + \varepsilon, \tag{7.28}$$

which is a linear regression model with respect to parameter B.

When time-to-failure samples are uncensored, the regression equation for observations t_p, Z_i $(i = 1, 2, \ldots, n)$ is

$$\log t_i = \mu_0 - Z_iB + \varepsilon_i, \tag{7.29}$$

where for any time-to-failure distribution, $\varepsilon_i (i = 1, 2, \ldots, n)$ are independent and identically distributed r.v.s with an unknown variance and known distribution form (if the time-to-failure distribution is known). Thus, on the one hand, the least-squares technique (briefly considered in Section 2.8) for AL data analysis can be used as a nonparametric model. On the other hand, if time-to-failure distribution is known, one can use a parametric approach. The lognormal time-to-failure distribution is an example of the last case, which is reduced to standard normal regression. This is why the lognormal distribution is popular in AL practice. The respective example of a model parameter estimation problem for the Arrhenius model has already been considered in Chapter 2 (Example 2.33). The problems of optimal DoE for ALT are considered in Nelson [6].

7.1.4 AL Model for Time-Dependent Stress

The models considered in the previous sections are related to constant stress. The case of time-dependent stress is not only more general, but is also of more practical importance because its applications in reliability are not limited by AL testing problems. As an example, consider the time-dependent stress analog of the Power Rule model (Equation 7.11b).

The stress amplitude, S, experienced by a structural element often varies during its service life, so that the straightforward use of Equation 7.11b is not possible. In such situations, the Palmgren–Miner's Rule is widely used to estimate the fatigue life. The rule treats fatigue fracture as a result of a *linear accumulation* of partial fatigue damage fractions. According to the rule, the damage fraction Δ_i at any stress level S_i is linearly proportional to the ratio n_i/N_i, where n_i is the number of cycles of operation under stress level S_i, and N_i is the total number of cycles to failure (life) under the *constant* stress level S_i. That is,

$$\Delta_i(S_i) = \frac{n_i(S_i)}{N_i(S_i)}, \; n_1 \leq N_i. \tag{7.30}$$

The total accumulated damage, D, under different stress levels $S_i (i = 1, 2, \ldots, n)$ is defined as

$$D = \sum_i \Delta_i = \sum_i \frac{n_i}{N_i}. \tag{7.31}$$

It is assumed that failure occurs if the total accumulated damage $D \geq 1$.

AL tests with time-dependent stress such as step stress and ramp tests are also important. For example, one of the most common reliability tests of thin silicon dioxide films in metal-oxide semiconductor integrated circuits is the so-called ramp-voltage test. In this test, the oxide film is stressed to breakdown by a voltage that increases linearly with time [10].

Let $z(t)$ be a time-dependent stress such that $z(t)$ is integrable. In this case, the basic relationship 7.3a can be written in the form given by Cox and Oaks [3]:

$$F[t; z(t)] = F_0[\Psi(t)], \tag{7.32}$$

where

$$\Psi(t^{(z)}) = \int_0^{t^{(z)}} \Psi[z(s), \mathrm{A}]\, \mathrm{d}s$$

and $t^{(z)}$ is the time related to an item under the stress condition $z(t)$.

Based on Equation 7.32, the analogous relationships for the pdf and failure rate function can be obtained. The corresponding relationship for the 100pth percentile of time to failure $t_p[z(t)]$ for the time-dependent stress, $z(t)$, can be obtained from Equation 7.32 as

$$t_p^0 = \int_0^{t_p[z(t)]} \Psi[z(s), \mathrm{A}]\, \mathrm{d}s. \tag{7.33}$$

Using Equations 7.6 and 7.7, Equation 7.33 can be rewritten as

$$1 = \int_0^{t_p[z(t)]} \frac{1}{t_p^0\{\Psi[z(s), \mathrm{A}]\}^{-1}} \mathrm{d}s, \tag{7.34}$$

or, using Equation 7.7, in terms of the percentile reliability models, as

$$1 = \int_0^{t_p[z(t)]} \frac{1}{\eta[z(s), \mathrm{B}]} \mathrm{d}s. \tag{7.35}$$

7.1.4.1 AL Reliability Model for Time-Dependent Stress and Palmgren–Miner's Rule

It should be noted that relationship 7.35 is an exact nonparametric probabilistic continuous form of the Palmgren–Miner's rule. So, the problem of using AL tests with time-dependent stress is identical to the problem of cumulative damage addressed by the Palmgren–Miner's rule. Moreover, there exists a useful analogy between mechanical damage accumulation and electrical breakdown. For example, the power rule and Jurkov's models are used as the relationship for mechanical as well as for electrical long-term strength. There are two main applications of Equation 7.35:

1. Fitting an AL reliability model (estimating the vector of parameters, B, of percentile reliability model, $\eta(z, B)$, on the basis of AL tests with time-dependent stress).
2. Reliability (percentiles of time-to-failure) estimation (when reliability model is known).

EXAMPLE 7.3

The constant stress reliability model for a component is based on the Arrhenius model for the fifth percentile of time to failure given by the following equation:

$$t_{0.05} = 2.590 \exp\left(\frac{0.400}{0.862 \times 10^{-4}(273 + T)}\right),$$

where $t_{0.05}$ is fifth percentile in hours, and T is temperature in °C. Find the fifth percentile of time to failure for the following cycling temperature profile, $T(t)$:

$T(t) = 25°\text{C}$ for $0 < t \leq 24\,\text{h}$;
$T(t) = 35°\text{C}$ for $24 < t \leq 48\,\text{h}$;
$T(t) = 25°\text{C}$ for $48 < t \leq 72\,\text{h}$;
$T(t) = 35°\text{C}$ for $72 < t \leq 96\,\text{h}$; and so on.

Solution:
An exact solution can be found as a solution for the following equation (based on relationship 7.35):

$$\int_0^{t_p[T(s)]} \left[A \exp\left\{\frac{E_a}{b[T(s) + 273]}\right\}\right]^{-1} ds = 1.$$

Replacing the integral by the following sum, one obtains

$$\sum_{i=1}^{k(t)} \delta_i + \delta(t^*) = 1,$$

where $\delta_i = \delta$ is the damage accumulated in a complete cycle (48 h period), $\delta(t^*)$ is the damage accumulated during the last incomplete cycle, having duration t^*, k is the largest integer, for which $k\delta < 1$ and $\delta = \Delta_1 + \Delta_2$, where Δ_1 is the damage associated with the first 24 h of the cycle (under 25°C), and Δ_2 is the damage associated with the second part of the cycle (under 35°C). These damages can be calculated as

$$\Delta_1 = \frac{24}{A \exp\{E_a/b[T_1 + 273]\}}, \quad \Delta_2 = \frac{24}{A \exp\{E_a/b[T_2 + 273]\}},$$

where $T_1 = 25°\text{C}$ and $T_2 = 35°\text{C}$. The numerical calculations result in $\Delta_1 = 1.6003 \times 10^{-6}$ and $\Delta_2 = 2.6533 \times 10^{-6}$. Thus, $\delta = \Delta_1 + \Delta_2 = 4.2532 \times 10^{-6}$.

The integer k is calculated as $k = [1/\delta] = 235100$, where $[x]$ is the greatest integer that does not exceed x. Estimate the damage accumulated during the last incomplete cycle, $\delta(t^*)$, as

$$\delta(t^*) = 1 - k\delta = 1 - 235100(4.25 \times 10^{-6}) = 2.389 \times 10^{-6} > 1.6003 \times 10^{-6},$$

which means that the last temperature in the profile is 35°C. Find t^* as a solution of the following equation:

$$\int_0^{t^*-24} \left[A \exp\left\{\frac{E_a}{b[35 + 273]}\right\}\right]^{-1} ds = 2.15 \times 10^{-6} - 1.60 \times 10^{-6},$$

continued

which gives $t^* - 24 = 4.97$ (h). Finally, the exact solution is

$$t_p = 48k + 24 + 4.97 \approx 1.13 \times 10^7 \text{ (h)}.$$

The correction obtained is negligible, but in cases in which the cycle period is comparable with the anticipated life, the correction can be significant.

7.1.5 Exploratory Data Analysis for Time-Dependent Stress

Practically, the two sample criteria considered earlier are the criteria for the particular time-dependent stress. Generally speaking, the value of the integral in Equation 7.35 does not change when the stress history $z(t)$ is changed to $z(t_p - t)$, $t_p \geq t \geq 0$, which means that time is reversible under the AL model. Based on this property, it is not difficult to verify whether the AL model assumptions are applicable to a given problem. For example, each sample to be tested under time-dependent stress can be divided into two equal parts, so that the first subsample can be tested under the forward stress history, while the second subsample is tested under the backward stress.

7.1.5.1 Statistical Estimation of AL Reliability Models on the Basis of AL Tests with Time-Dependent Stress

Using Equation 7.33, the time-dependent percentile regression model can be obtained in the following form:

$$t_p^0 = \int_0^{\hat{t}_p[z(t)]} \Psi[z(s),\ \mathrm{A}]\,\mathrm{d}s, \tag{7.36}$$

where $\hat{t}_p[z(t)]$ is the sample percentile for an item under the stress condition (loading history) $z(t)$. The problem of estimating the vector A and t_p^0 in this case cannot be reduced to parameter estimation for a standard regression model, as in the case of constant stress.

Consider k different time-dependent stress conditions (loading histories), $z_i(t)$, $i = 1, 2, \ldots, k$, $[k > (\dim A) + 1]$, where the test results are complete or Type II censored samples and the number of uncensored failure times and the sample sizes are large enough to estimate the t_p as the sample percentile $\hat{t}_p$. In this situation, the parameter estimates for the AL reliability model (A and t_p^0) can be obtained using a least-squares method solution of the following system of integral equations:

$$t_p^0 = \int_0^{\hat{t}_p[z(t)]} \Psi[z_i(s),\ \mathrm{A}]\,\mathrm{d}s, \quad i = 1, 2, \ldots, k. \tag{7.37}$$

Example 7.4

Assume a model (Equation 7.13) for the 10th percentile of time to failure $t_{0.1}$ of a ceramic capacitor in the form

$$t_{0.1}(U, T) = aU^{-c} \exp\left(\frac{E_\mathrm{a}}{T}\right),$$

where U is the applied voltage and T is the absolute temperature. Consider a time-step-stress AL test plan using step-stress voltage in conjunction with constant temperature as accelerating stress factors. A test sample starts at a specified low voltage U_0 and is tested for a specified time Δt. Then the voltage is increased by ΔU, and the sample is tested at $U_0 + \Delta U$ during Δt. That is,

$$U(t) = U_0 + \Delta U \times En\left(\frac{t}{\Delta t}\right),$$

where $En(x)$ is the nearest integer not greater than x. The test will be terminated after the portion $p \geq 0.1$ of items fails. Thus, the test results are sample percentiles at each voltage-temperature combination. The test plan and simulated results with $\Delta U = 10\,\text{V}$, $\Delta t = 24\,\text{h}$ are given in Table 7.1. Estimate the model parameters a, c, and E_a.

Solution:

For this example, the system of integral Equations 7.37 takes the form

$$a = \int_0^{t_{0.1i}} \exp\left(-\frac{E_a}{T_i}\right)[U(s_i)]^c\, ds, \quad i = 1, 2, 3, 4$$

or

$$a = \int_0^{347.9} \exp\left(-\frac{E_a}{358}\right)[U(s_1)]^c ds \quad a = \int_0^{1688.5} \exp\left(-\frac{E_a}{398}\right)[U(s_2)]^c ds,$$

$$a = \int_0^{989.6} \exp\left(-\frac{E_a}{373}\right)[U(s_3)]^c ds \quad a = \int_0^{1078.6} \exp\left(-\frac{E_a}{373}\right)[U(s_4)]^c ds,$$

where $U(s_1) = U(s_2)$. Solving this system for the data above yields the following estimates for the model (Equation 7.13): $a = 2.23 \times 10^{-8}\,\text{hV}^{1.88}$, $E_a = 1.32 \times 10^{4}\,^\circ\text{K}$, and $c = 1.88$, which are close to the following values of the parameters used for simulating the data: $a = 2.43 \times 10^{-8}\,\text{H/V}^{1.87}$, $E_a = 1.32 \times 10^{4}\,^\circ\text{K}$, and $c = 1.87$. The values of the percentiles predicted using the model are given in the last column of Table 7.1.

TABLE 7.1
Ceramic Capacitors Test Results

Temperature (°K)	Voltage U_0 (V)	Sample Time-to-Failure Percentile (h)	Time-to-Failure Percentile (Predicted) (h)
398	100	347.9	361.5
358	150	1688.5	1747.8
373	100	989.6	1022.8
373	63	1078.6	1108.6

7.1.6 PH Model Data Analysis

In this section, we will consider a case study [12] on data analysis in the framework of the *PH model* discussed in Section 7.1.1.4.

7.1.6.1 Automotive Tire Reliability

By the design intention, an automobile tire should exhibit no failures during its useful life; however, some tires do fail prematurely. There are several kinds of failure modes observed in the field. This chapter focuses on a particular failure mode known as *tread and belt separation* (TBS). In the event of TBS, the whole (or a part of the) tread and the second (upper) steel belt leave the tire carcass and the first (lower) steel belt.

One could consider TBS as a sequence of two events: failure crack initiation in the wedge area (which usually starts as a "pocketing" at the edge of the second belt) followed by the crack propagation between the belts. Finite element analysis of tire geometry suggests that the largest strain occurs in the wedge area and is proportional to the wedge gauge.

While a small wedge encourages crack initiation, it is not itself a sufficient condition for TBS to occur. In order to propagate further, the crack must have favorable conditions, for example, low adhesion strength between belts and the proper energy input to separate the belts. These characteristics depend on physical, chemical, and mechanical properties of the rubber skim stock as well as the age of the tire. Hence, the following tire design characteristics represent *explanatory variables (covariates)* that could potentially affect the tire's life until the TBS failure:

- Tire age
- Wedge gauge (a geometric dimension of a tire)
- Interbelt gauge (a geometric dimension of a tire)
- End of the 2nd belt to buttress, EB2B (a geometric dimension of a tire)
- Peel force (adhesion force of rubber between steel belts, characterized as the force required to separate belts in the specimen of a given dimension)
- Percent of carbon black (a chemical ingredient of the rubber affecting its mechanical characteristics, such as tear resistance)

Field failure data turn out to be insufficient for the construction of a tire reliability model with explanatory variables. While the survival times *can* be estimated and even censoring can be properly accounted for, the data on the above-defined covariates are difficult to obtain because of the disintegration of the tire as a result of TBS. In order to overcome this problem and duplicate field failures in controlled conditions, a special laboratory test may be performed.

7.1.6.2 Reliability Test Procedure

Consider the test designed to reproduce the TBS failure mode observed in the field. To overcome the contradiction between the destructive nature of the test and the need to properly measure geometry and material properties of the tire, a special failure warning system may be considered. The main idea of this system is based on the fact that the internal crack must be developed inside the tire prior to its catastrophic failure. As a result of centrifugal forces at high speed, the large chunk of material lifts up at the crack location, which accompanied by further crack growth, then leads to a change of the tire eccentricity. Therefore, the vibration signature of the rotating tire can be used for early detection of the failure.

The failure warning system involved specially mounted accelerometers and a PC-based data acquisition system. A signal to stop the test is generated when the system senses the change in the tire's vibration pattern. If the test is not stopped, it leads to a full TBS separation. The associated time to *partial* TBS is thus equalized with that of *full* TBS for data analysis purposes. The partially disintegrated tire is then conveniently available for further tests to properly measure the mechanical, physical, and chemical covariates.

7.1.6.3 PH Model and Data Analysis

The PH model (Equation 7.10) offers a convenient and physically meaningful way of relating the life characteristic of an item to the vector of explanatory variables. The advantage of this model over parametric survival regression models is that it does not necessarily make any assumptions about the nature or the shape of the underlying survival distribution, thus reducing the uncertainty about model selection.

All the covariates included in the model should be checked for statistical independence and lack of autocorrelation. Failure times associated with competing failure modes (other than TBS) may be treated as censored responses.

TABLE 7.2
Tire Reliability Test Data

Tire Age	Wedge Gauge	Interbelt Gauge	EB2B	Peel Force	Percent of Carbon Black	Wedge Gauge *x* Peel Force	Time to Failure	Censoring (1—complete 0—censored)
1.22	0.81	0.88	1.07	0.63	1.02	0.46	1.02	0
1.19	0.69	0.77	0.92	0.68	1.02	0.43	1.05	1
0.93	0.77	1.01	1.11	0.72	0.99	0.49	1.22	0
0.85	0.80	0.57	0.98	0.75	1.00	0.42	1.17	1
0.85	0.85	1.26	1.03	0.70	1.02	0.64	1.09	0
0.91	0.89	0.94	1.00	0.77	1.03	0.59	1.09	1
0.93	0.98	0.84	0.92	0.72	1.00	0.55	1.17	1
1.10	0.76	0.94	1.01	0.84	0.98	0.55	1.10	0
0.95	0.53	0.96	0.91	0.58	1.00	0.27	1.00	1
0.94	0.87	1.11	0.88	0.72	0.99	0.65	1.15	1
1.08	1.13	1.12	0.93	0.75	0.96	0.79	0.98	1
0.89	1.03	1.28	0.97	0.68	1.02	0.53	1.24	0
1.41	0.79	0.83	0.91	1.00	1.00	1.00	0.98	1
1.50	0.72	0.76	0.97	0.76	0.96	0.35	1.15	1
1.21	0.54	0.70	0.95	0.59	1.00	0.30	0.65	1
2.01	0.76	0.94	1.01	0.53	1.00	0.35	0.97	1
1.49	0.64	0.70	1.02	0.71	0.97	0.41	0.85	0
1.55	0.63	0.71	1.13	0.66	1.00	0.40	0.98	0
1.23	0.84	1.09	1.04	0.76	0.98	0.57	1.02	0
2.60	1.05	1.21	1.07	1.06	0.99	1.05	1.14	0
2.26	0.98	1.34	1.02	0.87	1.00	0.89	1.18	0
1.66	1.13	0.68	1.18	1.02	0.98	0.86	1.18	0
2.03	0.96	1.12	1.11	0.57	1.01	0.47	0.91	0
0.38	1.15	1.01	0.97	0.81	1.00	0.86	0.75	0
0.45	1.23	1.01	0.96	0.74	1.00	0.91	0.79	0
0.38	0.89	1.03	0.99	0.84	0.99	0.74	0.87	0
0.09	1.37	1.29	1.06	2.27	1.00	2.61	0.87	0
0.09	1.35	1.44	0.95	2.33	1.00	3.00	0.87	0
0.09	1.49	1.13	0.91	2.15	1.00	2.75	0.90	0
0.15	1.32	1.11	0.91	1.90	1.00	2.18	0.91	0
0.17	1.68	1.12	1.05	1.74	1.02	2.44	0.79	0
0.17	1.71	0.98	1.05	1.68	1.02	2.42	0.83	0
0.17	1.63	1.05	1.02	1.44	1.03	2.16	0.84	0
1.05	1.04	1.06	1.02	1.03	1.03	0.93	1.28	0

Table 7.2 shows the test data set (coded for confidentiality). The results of the PH model estimation are shown in Table 7.3. The log-likelihood of the final solution is -16.008, whereas the log-likelihood of the null model (with all regression parameters being equal to zero) is -28.886. The likelihood ratio χ^2 statistic (the null model minus the final solution) is 25.757 with seven degrees of freedom and the associated *p*-value is 0.0005.

Table 7.4 shows the estimation results of the model that includes only statistically significant covariates. The log-likelihood of the final solution is -19.968, whereas the log-likelihood of the null model is -28.886. The likelihood ratio χ^2 statistic is 17.837 with four degrees of freedom and the associated *p*-value is 0.001. The exponential probability plot of Cox–Snell residuals was used to confirm the adequacy of the fitted model.

TABLE 7.3
Estimates of PH Model Based on Data in Table 7.2

Explanatory Variable	Coefficient	Standard Error	*t*-Value	*p*-Value
Tire age	2.109	1.393	1.514	0.130
Wedge gauge	−9.686	4.638	−2.088	0.037
Interbelt gauge	−10.677	4.617	−2.313	0.021
Belt2 to sidewall	−13.675	8.112	−1.686	0.092
Peel force	−34.293	13.651	−2.512	0.012
Percentage of carbon black	−48.349	33.448	−1.445	0.148
Wedge *x* peel force	20.839	8.860	2.352	0.019

TABLE 7.4
Estimates of PH Model with Statistically Significant Covariates Based on Data in Table 7.2

Explanatory Variable	Beta	Standard Error	*t*-Value	*p*-Value
Wedge gauge	−9.313	4.069	−2.289	0.022
Interbelt gauge	−7.069	2.867	−2.466	0.014
Peel force	−27.411	10.578	−2.591	0.010
Wedge A *x* peel force	18.105	7.057	2.566	0.010

Based on the obtained results, one may conclude that the wedge and interbelt gages along with the peel force are the major factors responsible for a tire's reliability in terms of TBS failure mode.

7.2 ANALYSIS OF DEPENDENT FAILURES

Dependent failures are extremely important in reliability analysis and must be given adequate treatment so as to minimize gross overestimation of reliability. In general, dependent failures are defined as events in which the probability of each failure is dependent on the occurrence of other failures. According to Equation 2.21, if a set of dependent events $\{E_1, E_2, \ldots, E_n\}$ exists, then the probability of each failure in the set depends on the occurrence of other failures in the set.

The probabilities of dependent events on the left-hand side of Equation 2.21 are usually, but not always, greater than the corresponding independent probabilities. Determining the conditional probabilities in Equation 2.21 is generally difficult. However, there are parametric methods that can take into account the conditionality and generate the probabilities directly. These methods are discussed later in this section.

Generally, dependence among various events, for example, failure events of two items, is due either to the internal environment of these systems or to the external environment (or events). The internal aspects can be divided into three categories: internal challenges, intersystem dependencies, and intercomponent dependencies. The external aspects are natural or human-made environmental events that make failures dependent. For example, the failure rates for items exposed to extreme heat, earthquakes, moisture, and flood will increase. The intersystem and intercomponent dependencies can be categorized into four broad categories: functional, shared equipment, physical, and human-caused dependencies. These are described in Table 7.5.

The major causes of dependence among a set of systems or components as described in Table 7.5 can be explicitly described and modeled, for example, by system reliability analysis models, such as fault trees. However, the rest of the causes can be collectively modeled using the concept of

TABLE 7.5
Types of Dependent Events

Dependent Event Type	Dependent Event Category	Subcategory	Example
Internal	1. Challenge	—	Internal transients or deviations from the normal operating envelope introduce a challenge to a number of items
	2. Intersystem (failure between two or more systems)	1. Functional	Power to several independent systems is from the same source
		2. Shared equipment	The same equipment, e.g., a valve, is shared between otherwise independent systems
		3. Physical	The extreme environment, e.g., high temperature causes dependencies between independent systems
		4. Human	Operator error causes failure of two or more independent systems
	3. Intercomponent	1. Functional	A component in a system provides multiple functions
		2. Shared equipment	Two independent trains in a hydraulic system share the same common header
		3. Physical	Same as system interdependency above
		4. Human	Design errors in redundant pump controls introduces a dependency in the system
External	—	—	Earthquake or fire fails a number of independent systems or components

common-cause failures (CCFs). CCFs are the collection of all sources of dependencies described in Table 7.5 (especially between components) that are not known or that are difficult to explicitly model in the system or component reliability analysis. For example, functional and shared equipment dependencies are explicitly modeled in the system analysis, but other dependencies are considered collectively using CCF.

CCFs have been shown by many reliability studies to contribute significantly to the overall unavailability or unreliability of complex systems. There is no unique and universal definition for CCFs. However, a fairly general definition of a CCF is given by Mosleh et al. [13] as "a subset of dependent events in which two or more component fault states exist at the same time, or in a short time interval, and are direct results of a shared cause."

To better understand CCFs, consider a system with three redundant components A, B, and C. The total failure probability of A can be expressed in terms of its independent failure A_I and dependent failures as follows:

- C_{AB} is the failure of components A and B (and not C) from common causes.
- C_{AC} is the failure of components A and C (and not B) from common causes.
- C_{ABC} is the failure of components A, B, and C from common causes.

Component A fails if any of the above events occur. The equivalent Boolean representation of the total failure of component A is $A_T = A_I + C_{AB} + C_{AC} + C_{ABC}$. Similar expressions can be developed for components B and C.

Now, suppose that the success criteria for the system is 2-out-of-3 for components A, B, and C. Accordingly, the failure of the system can be represented by the following events (cut sets):

$(A_I \cdot B_I)$, $(A_I \cdot C_I)$, $(B_I \cdot C_I)$, C_{AB}, C_{AC}, C_{BC}, and C_{ABC}. Thus, the Boolean representation of the system failure will be

$$S = (A_I \times B_I) + (A_I \times C_I) + (B_I \times C_I) + C_{AB} + C_{AC} + C_{BC} + C_{ABC}. \tag{7.38}$$

If only independence is assumed, the first three terms of the above Boolean expression are used, and the remaining terms are neglected. Applying the rare event approximation, the system failure probability Q_S is given by

$$Q_S \approx \Pr(A_I)\Pr(B_I) + \Pr(A_I)\Pr(C_I) + \Pr(B_I)\Pr(C_I) + \Pr(C_{AB}) + \Pr(C_{AC}) + \Pr(C_{BC}) + \Pr(C_{ABC}). \tag{7.39}$$

If components A, B, and C are similar (which is often the case since common causes among different components have a much lower probability), then

$$\Pr(A_I) = \Pr(B_I) = \Pr(C_I) = Q_1, \quad \Pr(C_{AB}) = \Pr(C_{AC}) = \Pr(C_{BC}) = Q_2, \quad \text{and} \quad \Pr(C_{ABC}) = Q_3. \tag{7.40}$$

Therefore,

$$Q_s = 3(Q_1)^2 + 3Q_2 + Q_3. \tag{7.41}$$

In general, one can introduce the probability Q_k representing the probability of CCF among k specific components in a component group of size m, such that $1 \leq k \leq m$. The CCF models for calculating

TABLE 7.6
Key Characteristics of the CCF Parametric Models

Estimation Approach	Model	Model Parameters	General Form for Multiple Component Failure Probabilities
Nonshock models single parameter	β-Factor	β	$Q_k \begin{cases} (1-\beta)Q_t, & k=1; \\ 0, & l<k<m; \\ \beta Q_t, & k=m \end{cases}$
Nonshock models multi parameter	MGL	β, γ, δ	$Q_k = \frac{1}{\binom{m-1}{k-1}}(1-\rho_{k+1})\left(\prod_{i=1}^{k}\rho_i\right)Q_t$, $k = 1, 2, \ldots, m$ $\rho_1 = 1$, $\rho_2 = \beta, \ldots, \rho_{k+1} = 0$
	α-Factor	$\alpha_1, \alpha_2, \ldots, \alpha_m$	$Q_k = \frac{k}{\binom{m-1}{k-1}}\frac{\alpha_k}{\alpha_t}Q_t$, $k = 1, 2, \ldots, m$, $\alpha_t = \sum_{k=1}^{m} k\alpha_k$
Shock models	BFR	μ, ρ, ω	$Q_k = \begin{cases} \mu\rho^k(1-\rho)^{m-k}, & k=1; \\ \mu\rho^m + \omega, & k=m \end{cases}$

Source: Mosleh, A., *Reliab. Eng. Syst. Saf.*, 34, 249–292, 1991.

Q_k are summarized in Table 7.6. In this table, Q_t is the total probability of failure accounting both for common cause and independent failures, and $\alpha, \beta, \gamma, \delta, \mu, \rho$, and ω are the parameters estimated from the failure data on these components.

CCF parametric models can be divided into two categories: single-parameter models and multiple-parameter models. The remainder of this section discusses these two categories in more detail and elaborates on the parameter estimation of the CCF models.

7.2.1 Single-Parameter Models

Single-parameter models are those that use one parameter in addition to the total component failure probability to calculate the CCF probabilities. One of the most commonly used single-parameter models defined by Fleming [14] is called the β-factor model. It is the first parametric model applied to CCF events in risk and reliability analysis. The sole parameter of the model, β, can be associated with that fraction of the component failure rate that is due to the CCFs experienced by the other components in the system. That is,

$$\beta = \frac{\lambda_c}{\lambda_c + \lambda_I} = \frac{\lambda_c}{\lambda_t}, \tag{7.42}$$

where λ_c is a failure rate due to CCFs, λ_I is a failure rate due to independent failures, and $\lambda_t = \lambda_c + \lambda_I$.

An important assumption of this model is that whenever a common-cause event occurs, all components of a redundant component system fail. In other words, if a CCF shock strikes a redundant system, all components are assumed to fail instantaneously.

Based on the β-factor model, for a system of m components, the probabilities of basic events involving k specific components (Q_k), where $1 \leq k \leq m$, are equal to zero, except Q_1 and Q_m. These quantities are given as

$$\begin{aligned} Q_1 &= (1-\beta)Q_t, \\ Q_2 &= 0, \\ &\vdots \\ Q_{m-1} &= 0, \\ Q_m &= \beta Q_t, \end{aligned} \tag{7.43}$$

with $m = 1, 2, \ldots$.

In general, the estimate for the total component failure rate is generated from generic sources of failure data, while the estimators of the corresponding β-factor do not explicitly depend on generic failure data, but rather rely on specific assumptions concerning data interpretation. The point estimator of β is discussed in Section 7.2.3. In addition, some recommended values of β are given in Mosleh et al. [13]. It should be noted that although this model can be used with a certain degree of accuracy for two-component redundancy, the results tend to be conservative for a higher level of redundancy. However, due to its simplicity, this model has been widely used in risk and reliability studies. To obtain more reasonable results for a higher level of redundancy, more generic parametric models should be used.

Example 7.5

Consider the following system with two redundant trains. Suppose each train is composed of a valve and a pump (each driven by a motor). The pump failure modes are "failure to start" (PS) and "failure to run following a successful start" (PR). The valve failure mode is "failure to open" (VO). Develop an expression for the probability of system failure.

continued

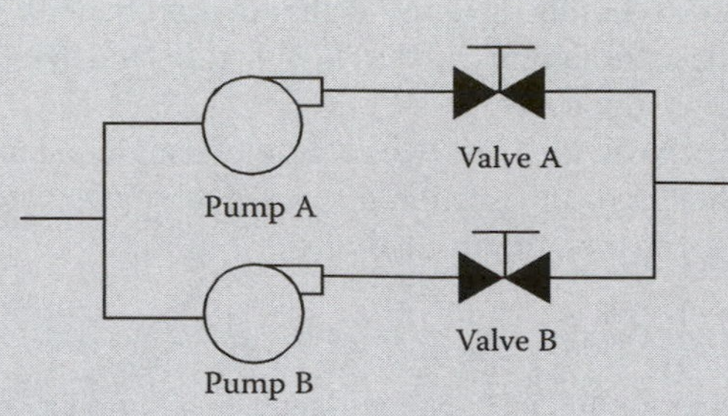

Solution:
Develop a system fault tree to include both independent and CCFs of the components, where

P_A is the independent failure of pump *A*
P_B is the independent failure of pump *B*
P_{AB} is the dependent failure of pumps *A* and *B*
V_A is the independent failure of valve *A*
V_B is the independent failure of valve *B*
V_{AB} is the dependent failure of valves *A* and *B*.

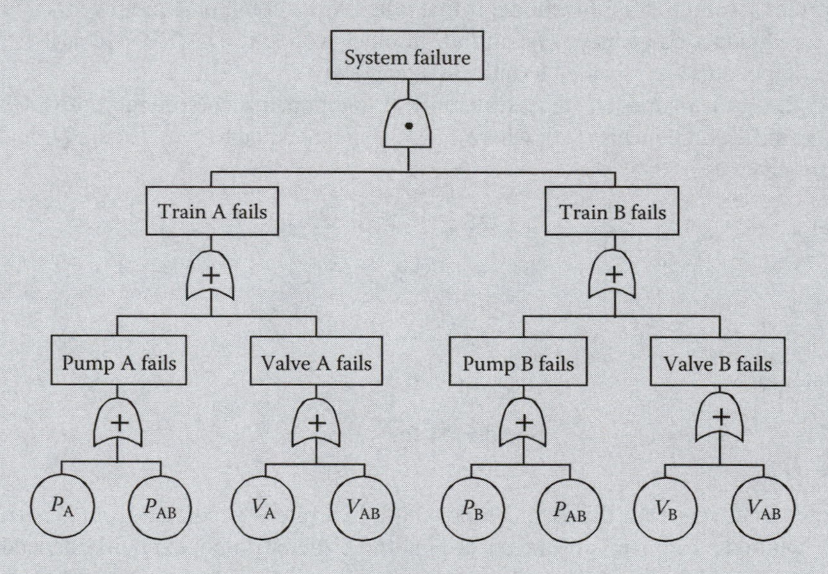

By solving the fault tree, the following cut sets can be identified:

$$C_1 = (P_A, P_B), C_2 = (P_{AB}),$$
$$C_3 = (V_A, V_B), C_4 = (V_{AB}),$$
$$C_5 = (P_A, V_B), C_6 = (P_B, V_A),$$

where

$$P_A = P_B = (P_{PS} + P_{PR}).$$

Use the β-factor method to calculate the probability of each cut set:

$$\Pr(C_1) = (1 - \beta_{PS})^2(q_{PS})^2 + (1 - \beta_{PR})^2(\lambda_{PR}\, t)^2 + 2(1 - \beta_{PS})(q_{PS})(1 - \beta_{PR})(\lambda_{PR}\, t),$$
$$\Pr(C_2) = \beta_{PS}(q_{PS}) + \beta_{PR}(\lambda_{PR}\, t),$$

$$\Pr(C_3) = (1 - \beta_{VO})^2 (q_{VO})^2,$$

$$\Pr(C_4) = \beta_{VO}(q_{VO}),$$

$$\Pr(C_5) = \Pr(C_6) = [(1 - \beta_{PS})(q_{PS}) + (1 - \beta_{PR})(\lambda_{PR}\, t)][(1 - \beta_{VO})(q_{VO})],$$

where q is the probability of failure rate on demand, λ is the failure rate to run, and t is the mission time. System failure probability is calculated using rare event approximation, as follows:

$$Q_s = \sum_{i=1}^{6} \Pr(C_i).$$

7.2.2 Multiple-Parameter Models

Multiple-parameter models are used to obtain a more accurate assessment of CCF probabilities in systems with a higher level of redundancy. These models have several parameters that are usually associated with different event characteristics. This category of models can be further divided into two subcategories, namely, shock and nonshock models. The multiple Greek model and the α-factor model are nonshock models, whereas the binomial failure rate (BFR) model is a shock model. These models are further discussed below.

7.2.2.1 Multiple Greek Letter Model

The multiple Greek letter (MGL) model introduced by Fleming et al. [15] is a generalization of the β-factor model. New parameters, such as γ and δ, are used in addition to β to distinguish among common-cause events affecting different numbers of components at a higher level of redundancy. For a system of m redundant components, $m - 1$ different parameters are defined. For example, for $m = 4$ the model includes the following three parameters (see Table 7.6).

1. Conditional probability that the common cause of an item failure will be shared by one or more additional items, β.
2. Conditional probability that the common cause of an item failure that is shared by one or more items will be shared by two or more items in addition to the first, γ.
3. Conditional probability that the common cause of an item failure shared by two or more items will be shared by three or more items in addition to the first, δ.

It should be noted that the β-factor model is a special case of the MGL model in which all other parameters excluding β are equal to 1.

The following estimates of the MGL model parameters are used as generic values:

Number of Components (m)	MGL Parameters		
	β	γ	δ
2	0.10	X	X
3	0.10	0.27	X
4	0.11	0.42	0.4

Consider the 2-out-of-3 success model described before. If we were to use the MGL model, then equivalent equations for Equation 7.43 for $m = 3$ (see Table 7.6) take the form

$$Q_1 = \frac{1}{\binom{3-1}{1-1}}(1-\rho_{1+1})\left(\prod_{i=1}^{1}\rho_i\right)Q_t = \frac{1}{2}(1-\rho_2)(\rho_1)Q_t. \tag{7.44}$$

Since $\rho_1 = 1$ and $\rho_2 = \beta$,

$$Q_1 = \frac{1}{2}(1-\beta)Q_t. \tag{7.45}$$

Similarly,

$$Q_2 = \frac{1}{2}(1-\rho_3)(\rho_1\rho_2)Q_t. \tag{7.46}$$

With $\rho_1 = 1, \rho_2 = \beta$ and $\rho_3 = \gamma$,

$$Q_2 = \frac{1}{2}\beta(1-\gamma)Q_t \tag{7.47}$$

Also,

$$Q_3 = \frac{1}{2}(1-\rho_4)(\rho_1\rho_2\rho_3)Q_t. \tag{7.48}$$

With $\rho_1 = 1$, $\rho_2 = \beta$, $\rho_3 = \gamma$ and $\rho_4 = 0$,

$$Q_3 = \beta\gamma Q_t. \tag{7.49}$$

To compare the result of the β-factor and the MGL, consider a case where the total failure probability of each component (accounting for both dependent and independent failures) is 8×10^{-3}. According to the β-factor model, the failure probability of the system including CCFs, if $\beta = 0.1$, would be

$$\begin{aligned} Q_s &= 3(1-\beta)^2Q_t^2 + \beta Q_t \\ &= 3(1-0.1)^2(8\times10^{-3})^2 + 0.1(8\times10^{-3}) \\ &= 9.6\times10^{-4}. \end{aligned} \tag{7.50}$$

However, the MGL model with $\beta = 0.1$ and $\gamma = 0.27$ will predict the system failure probability as

$$\begin{aligned} Q_s &= \frac{3}{4}(1-\beta)^2Q_t^2 + \frac{3}{2}\beta(1-\gamma)Q_t + \beta\gamma Q_t \\ &= \frac{3}{4}(1-0.1)^2(8\times10^{-3})^2 + \frac{3}{2}0.1(1-0.27)(8\times10^{-3}) + 0.1(0.27)(8\times10^{-3}) \\ &= 1.13\times10^{-3}. \end{aligned} \tag{7.51}$$

The difference is small, but the MGL model is more accurate than the β-factor model.

7.2.2.2 α-Factor Model

The α-factor model discussed by Mosleh and Siu [16] develops CCF failure probabilities from a set of failure ratios and the total component failure rate. The parameters of the model are the fractions of the total probability of failure in the system that involves the failure of k components due to a common cause, a_k.

The probability of a common-cause basic event involving the failure of k components in a system of m components is calculated according to the equation given in Table 7.6. For example, the probabilities of the basic events of the three-component system described earlier will be

$$\begin{aligned} Q_1 &= (\alpha_1/\alpha_t)Q_t, \\ Q_2 &= (2\alpha_2/\alpha_t)Q_t, \\ Q_3 &= (3\alpha_3/\alpha_t)Q_t, \end{aligned} \tag{7.52}$$

where $\alpha_t = \alpha_1 + 2\alpha_2 + 3\alpha_3$. The table below [17] provides generic values of α factors.

Number of Items (m)	α-Factor			
	α_1	α_2	α_3	α_4
2	0.95	0.050	—	—
3	0.95	0.040	0.01	—
4	0.95	0.035	0.01	0.005

Therefore, the system failure probability for the three redundant components discussed earlier can now be written as

$$Q_s = 3\left(\frac{\alpha_1}{\alpha_t}Q_t\right)^2 + 3\left(\frac{\alpha_2}{\alpha_t}Q_t\right) + 3\left(\frac{\alpha_3}{\alpha_t}Q_t\right). \tag{7.53}$$

Accordingly, using the generic values for the 2-out-of-3 success $\alpha_t = 0.95 + 0.08 + 0.03 = 1.06$. Thus,

$$\begin{aligned} Q_s &= 3\left[\frac{0.95}{1.06}(8 \times 10^{-3})\right]^2 + 3\left[\frac{0.04}{1.06}(8 \times 10^{-3})\right] + 3\left[\frac{0.01}{1.06}(8 \times 10^{-3})\right] \\ &= 1.28 \times 10^{-3}. \end{aligned} \tag{7.54}$$

which is closely consistent with the MGL model results.

7.2.2.3 BFR Model

The BFR model discussed by Atwood [18], unlike the α-factor model and MGL model, is a shock-dependent model. It estimates the failure frequency of two or more components in a redundant system as the product of the CCF shock arrival rate and the conditional failure probability of components, given the shock has occurred. This model considers two types of shock: lethal and nonlethal. The assumption is that, given a nonlethal shock, components fail independently, each with a probability of ρ, whereas in the case of a lethal shock, all components fail with a probability of 1. The expansion of this model is called the multinomial failure rate (MFR) model. In this model, the conditional probability of failure of k components is calculated directly from component failure data without any further assumptions. Therefore, the MFR model becomes essentially the same as the nonshock models because separating the CCF frequency into the shock arrival rate and conditional probability of failure given shock has occurred is, in general, a statistical rather than a physical modeling step. The parameters of the BFR model generally include

- Nonlethal shock arrival rate, μ.
- Conditional probability of failure of each component, given the occurrence of a nonlethal shock, ρ.
- Lethal shock arrival rate, ω.

It should be noted that due to the BFR model complexity and the lack of data to estimate its parameters, it is not widely used in practice.

7.2.3 Data Analysis for CCFs

Despite the difference among the models described in Section 7.2.1, they all have similar data requirements in terms of parameter estimation. One should not expect any significant difference among the numerical results provided by these models. The relative difference in the results may be attributed to the statistical aspects of the parameter estimation, which has to do with the assumptions made in developing a parameter estimator and the dependencies assumed in CCF probability quantification.

The most important steps in quantifying CCFs are collecting information from the raw data and selecting a model that can use most of this information. Statistical estimation procedures discussed in Chapters 2 and 3 can be applied to estimate the CCF model parameters. If separate models rely on the same type of information in estimating the CCF probabilities, and similar assumptions regarding the mechanism of CCFs are used, comparable numerical results can be expected. Table 7.7 summarizes simple point estimators for parameters of various nonshock CCF models. In this table, n_k is the total number of observed failure events involving failure of k similar components due to a common cause; m is the total number of redundant items considered; and N_D is the total number of system demands. If the item is normally operating (not on a standby), then N_D can be replaced by the total test (operation) time T. The estimators in Table 7.7 are based on the assumption that in every system demand, all components and possible combinations of components are challenged. Therefore, the estimators apply to systems whose tests are nonstaggered.

TABLE 7.7
Simple Point Estimators for Various Parametric Models

Model	Point Estimator
β-Factor	$\hat{Q}_t = \frac{1}{mN_D}\sum_{k=1}^{m} kn_k$
	$\hat{\beta} = \sum_{k=2}^{m} kn_k \Big/ \sum_{k=1}^{m} kn_k$
MGL	$\hat{Q}_t = \frac{1}{mN_D}\sum_{k=1}^{m} kn_k \quad \hat{\beta} = \sum_{k=2}^{m} kn_k \Big/ \sum_{k=1}^{m} kn_k$
	$\hat{\gamma} = \sum_{k=3}^{m} kn_k \Big/ \sum_{k=1}^{m} kn_k \quad \hat{\delta} = \sum_{k=4}^{m} kn_k \Big/ \sum_{k=1}^{m} kn_k$
α-Factor	$\hat{Q}_t = \frac{1}{mN_D}\sum_{k=1}^{m} kn_k$
	$\hat{\alpha}_k = n_k / \sum_{k=1}^{m} kn_k, \quad k = 1, 2, \ldots, m$

EXAMPLE 7.6

For the system described in Example 7.5, estimate the β parameters, λ and q, for the valves and pumps based on the following failure data:

	Event Statistic		
Failure Mode	n_1	n_2	Total (h) or N_D
Pump fails to start (PS)	10	1	500 (demands)
Pump fails to run (PR)	50	2	10,000 (hours)
Valve fails to open (VO)	10	1	10,000 (demands)

In the above table, n_1 is the number of observed independent failures, and n_2 is the number of observed events involving double CCF. Calculate the system unreliability for a mission of 10 h.

Solution:

From Table 7.7,

$$\hat{\beta} = \frac{2n_2}{n_1 + 2n_2}.$$

Apply this formula to β_{PR}, β_{PS}, and β_{VO} using appropriate values for n_1 and n_2:

$$n_{PS} = n_1 + 2n_2 = 12,$$

$$n_{PR} = n_1 + 2n_2 = 54,$$

$$n_{VO} = n_1 + 2n_2 = 17.$$

Accordingly, use Equations 3.65 and 3.97 for estimating λ and q, respectively:

$$q_{PS} = \frac{12}{500} = 2.4 \times 10^{-2}\text{D}^{-1} \quad \beta_{PS} = \frac{2}{12} = 0.17,$$

$$\lambda_{PR} = \frac{54}{10,000} = 5.4 \times 10^{-3}\,\text{h}^{-1} \quad \beta_{PR} = \frac{4}{54} = 0.07,$$

$$q_{VO} = \frac{12}{10,000} = 1.7 \times 10^{-3}\text{D}^{-1} \quad \beta_{VO} = \frac{2}{17} = 0.12.$$

Therefore, using the cut-set probability equations developed in Example 7.5, the estimates of the failure probabilities at 10 h of operation for each cut set are

$$\Pr(C_1) = (1-0.17)^2(2.4\times10^{-2})^2 + (1-0.07)^2(5.4\times10^{-3}\times10)^2$$
$$+\,2(1-0.17)(2.4\times10^{-2})(1-0.07)(5.4\times10^{-3}\times10) = 4.9\times10^{-3},$$

$$\Pr(C_2) = 0.17(2.4\times10^{-2}) + 0.07(5.4\times10^{-3})(10) = 7.9\times10^{-3},$$

$$\Pr(C_3) = (1-0.12)^2(1.7\times10^{-3})^2 = 2.2\times10^{-6},$$

$$\Pr(C_4) = 0.12(1.7\times10^{-3}) = 2.0\times10^{-4},$$

$$\Pr(C_5) = \left[(1-0.17)(2.4\times10^{-2}) + (1-0.07)(5.4\times10^{-3}\times10)\right]\left[(1-0.12)(1.7\times10^{-3})\right]$$
$$= 1.1\times10^{-4},$$

$$\Pr(C_6) = \Pr(C_5) = 1.1\times10^{-4}.$$

continued

Thus, the system failure probability is

$$Q_S \cong \sum_{i=1}^{6} \Pr(C_i) = 1.3 \times 10^{-2}.$$

7.3 UNCERTAINTY ANALYSIS

Uncertainty arises primarily due to a lack of reliable information, such as a lack of information about the ways a given system may fail. Uncertainty may also arise due to linguistic imprecision, such as the expression "System A is highly reliable." Furthermore, uncertainty may be divided into two kinds: the aleatory models of the world and epistemic uncertainty. For example, the Poisson model for modeling the inherent randomness in the occurrence of an event (e.g., a failure event) can be considered the "world model" of the occurrence of failure. The variability associated with the results obtained from this model represents the aleatory uncertainty. The epistemic uncertainty, on the other hand, describes our state of knowledge about this model. For example, the uncertainty associated with the choice of the Poisson model itself and its parameter λ is considered epistemic.

Consider a Weibull distribution used to represent the time to failure. The choice of the distribution model itself involves some modeling uncertainty (epistemic); however, the variability of time to failure is the aleatory uncertainty. We may even be uncertain about the way we construct the failure model. For example, our uncertainty about parameters α and β of the Weibull distribution representing time-to-failure distribution may be depicted by another distribution—for example, a lognormal distribution. In this case, the lognormal distribution models represent the epistemic uncertainty about the Weibull distribution model.

The most common practice in measuring uncertainty is the use of the probability concept. In this book, we have used this measure of uncertainty only. As we discussed in Chapter 2, there are different interpretations of probability. This also affects the way uncertainty analysis is performed. In this section, we first briefly discuss uncertainty in choice of models and then present methods of measuring the uncertainty about the parameters of the model. Then we discuss methods of propagating uncertainty in a complex model. For example, in a fault tree model representing a complex system, the uncertainty assigned to each leaf of the tree can be propagated to obtain a distribution of the top event probability.

The simplest way to measure uncertainty is to use sample mean $\bar{x}$ and variance S^2, described by Equations 2.112 and 2.114. We have discussed earlier in Chapter 2 that estimations of $\bar{x}$ and S^2 are, themselves, subject to some uncertainty, and it is important to describe this uncertainty by confidence intervals of $\bar{x}$ and S^2, for example, by using Equation 2.121. This brings another level of uncertainty. The confidence intervals associated with different types of distributions were discussed in Chapter 3. For a binomial model, the confidence intervals can be obtained from Equations 3.98 and 3.99. Similarly, if the data are insufficient, then the subjectivist definition of probability can be used, and different Bayesian probability intervals can be obtained (see Section 3.6).

Generally, the problem of finding the distribution of a function of r.v.s is difficult, which is why for most reliability and risk assessment applications, the problem is reduced to an estimation of the mean and variance (or standard deviation) of function of r.v.s. Such techniques are considered in the following sections. It should be mentioned that the uses of these techniques are, by no means, limited to reliability and risk assessment problems; they are also widely used in engineering.

7.3.1 Types of Uncertainty

Because different types of uncertainties are generally characterized and treated differently, it is useful to identify three types of uncertainty: *parameter uncertainty*, *model uncertainty*, and *completeness uncertainty*.

7.3.1.1 Parameter Uncertainties

Parameter uncertainties are those associated with the values of the fundamental parameters of the reliability or risk model, such as failure rates, or event probabilities including HEPs. They are typically characterized by establishing probability distributions on the parameter values.

Parameter uncertainties can be explicitly represented and propagated through the reliability or risk model, and the probability distribution of the relevant metrics (e.g., reliability, unavailability, and risk) can be generated. Various measures of central tendency, such as the mean, median, and mode, can be evaluated. For example, the distribution can be used to assess the confidence with which reliability targets are met. The results are also useful for studying the contributions from various elements of a model and for determining whether the tails of the distributions are being determined by uncertainties of a few significant elements of the reliability or risk model. If so, these elements can be identified as candidates for compensatory measures and/or monitoring.

In Chapter 3, we discussed measures for quantifying uncertainties of parameter values of distribution models for both frequentist and subjectivist (Bayesian) methods. Examples of these parameters are MTTF, μ; failure rate, λ; and probability of failure on demand, p, of a component. Uncertainty of the parameters is primarily governed by the amount of field data available about failures and repairs of the items. Because of these factors, a parameter does not take a fixed and known value, and has some random variability. In Section 7.3.2, we discuss how the parameter uncertainty is propagated in a system to obtain an overall uncertainty about the system failure.

7.3.1.2 Model Uncertainties

There are also uncertainties as to how to model specific elements of the reliability or risk. Model uncertainty may be analyzed in different ways. It is possible to include some model uncertainty by incorporating with the reliability/risk model a discrete probability distribution over a set of models for a particular issue (e.g., various models for reliability growth or human reliability). In principle, uncertainty in choosing a model can be handled in the same way as parameter uncertainty. For example, if a set of candidate models are available, one could construct a discrete probability distribution (M_i, p_i), where p_i is the degree of belief (in subjectivist terms) in model M_i as being the most appropriate representation. This has been done for the modeling of seismic hazard, for example, where the result is a discrete probability distribution on the frequencies of earthquakes. This uncertainty can then be propagated in the same way as the parameter uncertainties. Other methods are also available. For example, see [19].

It is often instructive to understand the impact of a specific assumption on the prediction of the model. The impact of using alternate assumptions or models may be addressed by performing appropriate sensitivity studies, or they may be addressed using qualitative arguments. This may be a part of the model uncertainty evaluation.

There are two aspects of modeling uncertainty at the component level or system level. In estimating uncertainty associated with unreliability or unavailability of a basic component, a modeling error can occur as a result of using an incorrect distribution model. Generally, it is very difficult to estimate an uncertainty measure for these cases. However, in a classical (frequentist) approach, the confidence level associated with a goodness-of-fit test can be used as a measure of uncertainty. For the reliability analysis of a system, one can say that a model describes the behavior of a system as viewed by the analyst. However, analysts can make mistakes due to a number of constraints, namely, their degree of knowledge and understanding of the system design and their assumptions about the system, as reflected in the reliability model (e.g., a fault tree).

Of course, one can minimize these sources of uncertainty, but one cannot eliminate them. For example, a fault tree based on an analyst's understanding of the success criteria of the system can be incorrect if the success criteria used are in error. For this reason, a more accurate dynamic analysis of the system may be needed to obtain correct success criteria.

Defining and quantifying the uncertainty associated with a model is very complex and cannot easily be associated with a quantitative representation (e.g., probabilistic representation). The reader is referred to Morgan and Henrion [20] for more discussion on this topic.

7.3.1.3 Completeness Uncertainty

Completeness is not in itself an uncertainty, but a reflection of the scope of reliability and risk analysis limitations. The result is, however, an uncertainty about where the true reliability or risk lies. The problem with completeness uncertainty is that, because it reflects unanalyzed contributions (e.g., contribution due to exclusion of certain failure modes in a FTA), it is difficult, if not impossible, to estimate the uncertainty magnitude. Thus, for example, the impact on actual reliability/risk from unanalyzed issues, such as the influences of the organization factor on equipment performance (e.g., reliability) and on quality assurance, cannot be explicitly assessed.

7.3.2 Uncertainty Propagation Methods

Consider a general case of a system performance characteristic Y (e.g., system reliability or unavailability). Based on an aleatory model of the system, a general function of uncertain quantities x_i and uncertain parameters θ_i can describe this system performance characteristic as

$$Y = f(x_1, x_2, \ldots, x_n, \theta_1, \theta_2, \ldots, \theta_m). \tag{7.55}$$

A simple example is a system composed of elements having the exponential time-to-failure distributions. In this case, Y can be the MTTF of the system; $x_i (i = 1, 2, \ldots, n)$ are the estimates of the MTTFs of the system components; and $\theta_i (i = 1, 2, \ldots, m)$ are the standard deviations (errors) of these estimates. The system performance characteristic, Y, can also be the probability of the top event of a fault tree, in which case x_i will be the failure probability (unavailability) of each component represented in the fault tree, and θ_i's will be the parameters of the distribution models representing x_i.

The variability of Y as a result of the variability of the basic parameters x_i and θ_i is estimated by the methods of propagation. These methods are discussed below.

7.3.2.1 Method of Moments

Write the function (Equation 7.55) in the following form:

$$Y = f(x_1, x_2, \ldots, x_n, S_1, S_2, \ldots, S_n), \tag{7.56}$$

where $x_i (i = 1, 2, \ldots, n)$ are the estimates of reliability parameters (e.g., MTTF, failure rate, and probability of failure on demand) of a system component, and $S_i (i = 1, 2, \ldots, n)$ are the respective standard deviations (errors).

Assume that

- $f(x_1, x_2, \ldots, x_n, S_1, S_2, \ldots, S_n) \equiv f(X, S)$ satisfies the conditions of Taylor's theorem.
- the estimates $x_i (i = 1, 2, \ldots, n)$ are independent and unbiased with expectations (true values) $\mu_i (i = 1, 2, \ldots, n)$.

Using Taylor's series expansion about μ_i, and denoting $(x_1, x_2, \ldots, x_n)$ by X and $(S_1, S_2, \ldots, S_n)$ by S, we can write

$$\begin{aligned} Y &= f(X; S) \\ &= f(\mu_1, \mu_2, \ldots, \mu_n; S) + \sum_{i=1}^{n} \left[\frac{\partial f(X)}{\partial X} \right]_{x_i=\mu_i} (x_i - \mu_i) \\ &+ \frac{1}{2!} \sum_{j=1}^{n} \sum_{i=1}^{n} \left[\frac{\partial^2 f(X)}{\partial x_i \partial x_j} \right]_{x_i=\mu_i, x_j=\mu_j} (x_i - \mu_i)(x_j - \mu_j) + R, \end{aligned} \tag{7.57}$$

where R represents the residual terms.

Taking the expectation of Equation 7.57 (using the algebra of expectations given in Table 2.2), one obtains

$$\begin{aligned} E(Y) &= f(\mu_1, \mu_2, \ldots, \mu_n; S) + \sum_{i=1}^{n} \left[\frac{\partial f(X)}{\partial X} \right]_{x_i=\mu_i} E(x_i - \mu_i) \\ &+ \frac{1}{2!} \sum_{j=1}^{n} \sum_{i=1}^{n} \left[\frac{\partial^2 f(X)}{\partial f x_i \partial f x_j} \right]_{x_i=\mu_i, x_j=\mu_j} E[(x_i - \mu_i)(x_j - \mu_j)] + E(R). \end{aligned} \tag{7.58}$$

Because the estimates $x_i (i = 1, 2, \ldots, n)$ are unbiased with expectations (true values) μ_i, the second term in the above equation is canceled. Dropping the residual term, $E(R)$, and assuming that the estimates x_i are independent, one obtains the following approximation:

$$E(Y) \approx f(\mu_1, \mu_2, \ldots, \mu_n; S) + \frac{1}{2} \sum_{i=1}^{n} \left[\frac{\partial^2 f(X)}{\partial x_i^2} \right]_{x_i=\mu_i} S^2(x_i). \tag{7.59}$$

For the more general and practical applications of the method of moments, we need to obtain the point estimate $\hat{Y}$ and its variance $\text{var}(\hat{Y})$. Replacing μ_i with $x_i (i = 1, 2, \ldots, n)$, we obtain

$$\hat{Y} \approx f(x_1, x_2, \ldots, x_n; S) + \frac{1}{2} \sum_{i=1}^{n} \left[\frac{\partial^2 f(X)}{\partial x_i^2} \right]_{x=x_i} S^2(x_i). \tag{7.60}$$

If, for a given uncertainty analysis problem, the second term can be neglected, the estimate (Equation 7.28) is reduced to the following simple form, which can be used as the point estimate:

$$\hat{Y} \approx f(x_1, x_2, \ldots, x_n). \tag{7.61}$$

To obtain a simple approximation for the variance (as a measure of uncertainty) of the system performance characteristic Y, consider the first two-term approximation for Equation 7.57. Taking the variance and treating the first term as constant, one obtains

$$\text{var}(\hat{Y}) = \text{var} \sum_{i=1}^{n} \left[\frac{\partial f(X)}{\partial x_i} \right]_{x_i=\mu_i} (x_i - \mu_i) = \sum_{i=1}^{n} \left[\frac{\partial f(X)}{\partial x_i} \right]_{x_i=\mu_i}^{2} S^2(x_i). \tag{7.62}$$

Example 7.7

For the system shown below, the constant failure rate of each component has a mean value of 5×10^{-3} per hour. If the failure rate can be represented by a r.v. that follows a lognormal distribution with a coefficient of variation of 2, calculate the mean and standard derivation of the system unreliability at $t = 1$, 10, and 100 h.

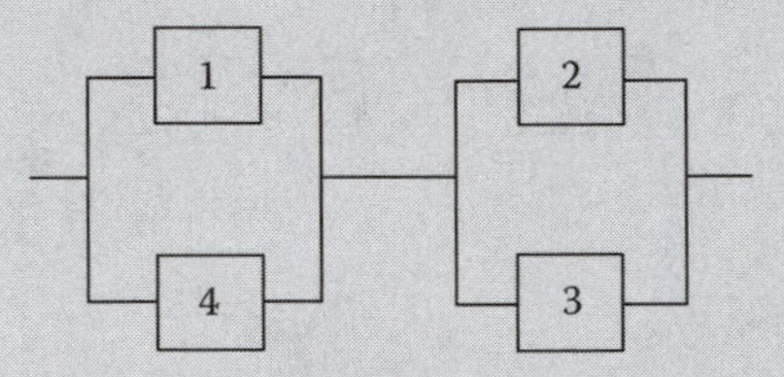

Solution:
System unreliability can be obtained from the following expression:

$$Q = q_1 \times q_4 + q_3 \times q_2 - q_1 \times q_2 \times q_3 \times q_4;$$

since $q_i = 1 - e^{-\lambda t}$, then

$$Q = (1 - e^{-\lambda_1 t})(1 - e^{-\lambda_2 t}) + (1 - e^{-\lambda_3 t})(1 - e^{-\lambda_4 t})$$
$$- (1 - e^{-\lambda_1 t})(1 - e^{-\lambda_2 t})(1 - e^{-\lambda_3 t})(1 - e^{-\lambda_4 t}).$$

Note that $\hat{\lambda}_1 = \hat{\lambda}_2 = \hat{\lambda}_3 = \hat{\lambda}_4 = \hat{\lambda} = 5 \times 10^{-3}\ \text{h}^{-1}$. Using Equation 7.60 and neglecting the second term (due to its insignificance),

$$\begin{aligned}\hat{Q} &= (1 - e^{-\hat{\lambda}_1 t})(1 - e^{-\hat{\lambda}_2 t}) + (1 - e^{-\hat{\lambda}_3 t})(1 - e^{-\hat{\lambda}_4 t}) \\ &\quad - (1 - e^{-\hat{\lambda}_1 t})(1 - e^{-\hat{\lambda}_2 t})(1 - e^{-\hat{\lambda}_3 t})(1 - e^{-\hat{\lambda}_4 t}) \\ &= 2(1 - e^{-\hat{\lambda} t})^2 - (1 - e^{-\hat{\lambda} t})^4 \\ &= 1 - 4e^{-2\hat{\lambda} t} + 4e^{-3\hat{\lambda} t} - e^{-4\hat{\lambda} t}.\end{aligned}$$

$$\hat{Q}_{1\,\text{h}} = 4.97 \times 10^{-5},$$
$$\hat{Q}_{10\,\text{h}} = 4.75 \times 10^{-3},$$
$$\hat{Q}_{100\,\text{h}} = 0.286.$$

calculate the derivatives.
For example,

$$\frac{\partial Q}{\partial \lambda_1} = te^{-\lambda_1 t}(1 - e^{-\lambda_2 t}) - te^{-\lambda_1 t}(1 - e^{-\lambda_2 t})(1 - e^{-\lambda_3 t})(1 - e^{-\lambda_4 t}).$$

Repeating for other derivatives of Q with respect to λ_2, λ_3, and λ_4 yields

$$\frac{\partial Q}{\partial \lambda_i} = 4t\,e^{-\lambda_i t}\left[(1 - e^{-\lambda t}) - (1 - e^{-\lambda t})^3\right],$$

and by Equation 7.62,

$$
\begin{aligned}
S^2(Q) &= 4\,\mathrm{var}(\lambda_i)\{t\,\mathrm{e}^{-\hat{\lambda}_i t}[(1-\mathrm{e}^{-\hat{\lambda}t}) - (1-\mathrm{e}^{-\hat{\lambda}t})^3]\}^2 \\
&= (2\hat{\lambda}_i)^2\{t\,\mathrm{e}^{-\hat{\lambda}_i t}[(1-\mathrm{e}^{-\hat{\lambda}t}) - (1-\mathrm{e}^{-\hat{\lambda}t})^3]\}^2, \\
S^2(Q)_{1\,\mathrm{h}} &= 2.51 \times 10^{-13}, \\
S^2(Q)_{10\,\mathrm{h}} &= 2.14 \times 10^{-11}, \\
S^2(Q)_{100\,\mathrm{h}} &= 4.07 \times 10^{-10}
\end{aligned}
$$

using $S(\lambda_i) = 2\hat{\lambda} = 2(5 \times 10^{-3}) = 0.01$. Therefore, $\mathrm{var}(\lambda_i) = S2\lambda_i = 10^{-4}$. It is now possible to calculate the coefficient of variation for system unreliability as

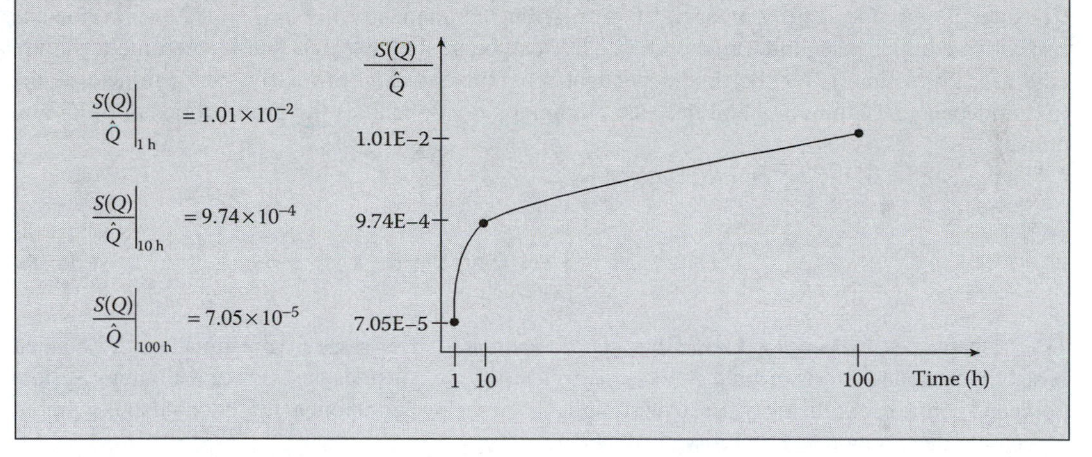

For a more detailed consideration of the reliability applications of the method of moments, the reader is referred to Morchland and Weber [21]. Apostolakis and Lee [22] propagate the uncertainty associated with parameters x_i by generating lower-order moments, such as the mean and variance for Y, from the lower-order moments of the distribution for x_i. A detailed treatment of this method is covered in a comparison study of the uncertainty analysis method by Martz et al. [23].

For a special case when $Y = \sum_{i=1}^{n} x_i$ (e.g., a series composed of components having the exponential time-to-failure distributions with failure rates x_i), and dependent x_i's, the variance of $\hat{Y}$ is given by

$$\mathrm{var}[\hat{Y}] = \sum_{i=1}^{n} \mathrm{var}[x_i] + 2\sum_{i=1}^{n-1}\sum_{j=i+1}^{n} \mathrm{cov}[x_i, x_j]. \tag{7.63}$$

In the case where $Y = \prod_{i=1}^{n} x_i$, and x_i's are independent (a parallel system composed of components having reliability functions, $x_i[i = 1, 2, \ldots, n]$),

$$E(Y) = \prod_{i=1}^{n} E(x_i) \tag{7.64}$$

and

$$\mathrm{var}[Y] \approx \left[\sum_{i=1}^{n} \frac{\mathrm{var}(x_i)}{E^2(x_i)}\right] \times E^2(Y). \tag{7.65}$$

Dezfuli and Modarres [24] have expanded this approach to efficiently estimate a distribution fit for Y when x_i's are highly dependent. The method of moments provides a quick and accurate estimation of lower moments of Y based on the moments of x_i, and the process is simple. However, for highly nonlinear expressions of Y, the use of only low-order moments can lead to significant inaccuracies, and the use of higher moments is complex.

7.3.3 System Reliability Confidence Limits Based on Component Failure Data

Estimation of system reliability usually involves incorporation of component failure model uncertainties. In this section, we consider some practical approaches to eliminating this type of uncertainty for series systems.

7.3.3.1 Lloyd–Lipow Method

Consider a series system composed of m different components. Let $p_i (i = 1, 2, \ldots, m)$ be the respective component failure probabilities. They can be treated as $F_i(t)$: that is, the time-to-failure cdfs at a given time t, for the time-dependent reliability models. Similarly, they can also be the time-independent failure probabilities (the binomial model), such as the probabilities of failure on demand.

The reliability of the system, R_s, is given by

$$R_s = \prod_{i=1}^{m}(1 - p_i). \tag{7.66}$$

The probabilities, p_i, are not known but can be estimated. The respective estimates are obtained based on component tests or field data. In the following, we consider methods of estimating system point and confidence reliability based on straightforward use of component test data—that is, without estimating the component reliability characteristics.

We start with the Lindstrom–Madden method, which is more frequently referred to as the Lloyd–Lipow method, due to their book [25] in which the method was first described. Note that the Lindstrom–Madden method is a heuristic one.

To simplify our consideration, let us limit ourselves to the case of a two-component series system. Assume that the test results for the components are given in the following form:

- N_1 is the number of the first components tested, and d_1 is the number of failures observed during the test.
- N_2 is the number of the second components tested, and d_2 is the number of respective failures observed during the test.

Without loss of generality, suppose that $N_2 > N_1$. These test results can be represented by the following two sets:

$$\begin{aligned} &x_{1i} (i = 1, 2, \ldots, N_1), \\ &x_{2i} (i = 1, 2, \ldots, N_2), \end{aligned} \tag{7.67}$$

where x_{1i} and x_{2i} take on the value 1, if the respective component failed during the test. They take zero values if the respective component did not fail during the test. Let us have $N_1 - d_1$ survived units among N_1 first components tested, and $N_2 - d_2$ survived units among N_2 second components tested.

Select randomly N_1 elements from the set x_{2i}. Randomly combining each of these elements with elements from the set $x_{1i} (i = 1, 2, \ldots, N_1)$, obtain N_1 pairs (x_{1j}, x_{2j}) with $j = 1, 2, \ldots, N_1$. The idea of the Lindstrom–Madden method is to treat these pairs as fictitious test results of N_1 series systems composed of the first and the second components. The number of fictitious series systems expected to fail (i.e., having at least one failed component) D_s, is given by

$$D_s = N_1(1 - \hat{R}_s), \tag{7.68}$$

where

$$\hat{R}_s = \left(1 - \frac{d_1}{N_1}\right)\left(1 - \frac{d_2}{N_2}\right) \tag{7.69}$$

is the point estimate of the series system reliability function. The value of D_s is the "equivalent" number of failures for a sample of N_1 series systems of interest [26]. Note that, similar to Bayes's approach, D_s is not necessarily an integer. To obtain confidence limits for the system reliability, one must use the Clopper–Pearson procedure, discussed in Chapter 3.

In the general case of a system composed of k components, the number of the fictitious series systems expected to fail, D_s, is given by

$$D_s = N_{1s}(1 - \hat{R}_s), \tag{7.70}$$

where $N_{1s} = \min(N_1, N_2, \ldots, N_k)$ and

$$\hat{R}_s = \prod_{i=1}^{k}\left(1 - \frac{d_i}{N_i}\right). \tag{7.71}$$

Based on D_s and N_{1s}, the respective confidence limits for the system reliability are constructed in a similar way, using the Clopper–Pearson procedure.

Example 7.8

Two components were tested under the following time-terminated test plans. A sample of 110 units of the first component was tested during 2000 h. The failures were observed at 3, 7, 58, 145, 155, 273, 577, 1104, 1709, and 1999 h. A sample of 100 units of the second component was tested during 1000 h. The failures were observed at 50, 70, 216, 235, 295, 349, 368, and 808 h. Find the point estimate and 90% lower confidence limit for the reliability function at 1000 h for the two-component series system composed of these components.

Solution:
Find the number of the series systems "tested" as

$$N_{1s} = \min(N_1, N_2) = \min(110, 100) = 100.$$

For the 1000 h interval we have $d_1 = 7$ and $d_2 = 8$.
Using Equation 7.69, find

$$\hat{R}_s(1000) = \left(1 - \frac{7}{110}\right)\left(1 - \frac{8}{100}\right) = 0.861,$$

$$D_s = 100(1 - 0.861) = 13.9.$$

continued

Using the Clopper–Pearson procedure in the form Equation 3.104 with $n = 100$ and $r = 13.9$, find the 90% upper confidence limit for the system reliability function at 1000 h, $R_1(1000)$, as a solution of the following equation:

$$I_{R_1}(100 - 13.9, 13.9 + 1) \leq 0.1,$$

which gives $R_1(1000) \approx 0.806$.

Note that the solution to the problem does not depend on particular time-to-failure distributions of the components, which shows that the Lindstrom–Madden method is nonparametric.

7.3.4 Maximus Method

The Lindstrom–Madden method considered in the previous section can be applied to series systems only. The Maximus method, which we briefly discuss below, is a generalization of the Lindstrom–Madden method for series–parallel arrangement of subsystems of components [27].

Under this method, the basic steps for constructing the lower confidence limit for system reliability based on component failure data are as follows:

1. Reduce each subsystem to an equivalent component. Treat the components of the reduced system as each having its equivalent failure data obtained from the reduction performed.
2. Obtain the maximum likelihood point estimate of system reliability, $\hat{R}_S$, based on the system configuration and component equivalent failure data.
3. Calculate the equivalent system sample size, N_S, according to the reduced system configuration and the respective equivalent component failure data from step 1.
4. Calculate the equivalent number of system failures, D_S, as

$$D_S = N_S(1 - \hat{R}_S). \qquad (7.72)$$

5. Using the Clopper–Pearson procedure (Equation 3.98) with N_S, D_S, and a chosen confidence probability, calculate the lower confidence limit for the system reliability.

7.3.4.1 Classical Monte Carlo Simulation

There are three techniques to estimate system reliability confidence based on the Monte Carlo simulation: classical Monte Carlo simulation, Bayes's Monte Carlo method, and the bootstrap method.

The classical Monte Carlo method is based on classical component probabilistic models (failure distributions) that are obtained using failure data only. In other words, each component of the system analyzed is provided with a failure (time-to-failure or failure-on-demand) distribution, fitted using real failure data.

If we knew the *exact* values of reliability characteristic of the system components, in principle, we would be able to calculate the system reliability using the system reliability function using, for example, Equations 4.1 and 4.7. However, instead of exact component reliability characteristics, we have only their estimates, which are r.v.s. Since there are no failure data for the system as a whole, we have to treat any system reliability characteristic as a r.v.s transformation result, obtained using the system reliability function. As mentioned in Section 3.1, it is generally not easy to find the distribution of transformed r.v.s, which is why the Monte Carlo approach turns out to be a practical tool for solving many problems associated with estimating complex system reliability.

In the framework of the classical Monte Carlo approach, there can be different algorithms for system reliability estimation. The following example illustrates the general steps for constructing the lower confidence limit for system reliability using this method. These steps are

1. For each component of the system given, obtain a classical estimate (e.g., the MLE) of component reliability, R_i ($i = 1, 2, \ldots, n$, where n is the number of components in the system), generating it from the respective estimate distribution.
2. Calculate the corresponding classical estimate of the system reliability,

$$\hat{R}_S = f(R_1, R_2, \ldots, R_n), \tag{7.73}$$

where $f(\cdot)$ is the system reliability function.
3. Repeat steps 1 and 2 a sufficiently large number of times, n, (e.g., 10000) to obtain a large sample of $\hat{R}_S$.
4. Using the sample obtained, and a chosen confidence level $(1 - \alpha)$, construct the respective lower confidence limit for the system reliability of interest as a sample percentile of level (discussed in Section 7.1):

$$\hat{R}_{S_p} = \begin{cases} \hat{R}_{S_p}, & \text{if } np \text{ is not integer and any value from the interval;} \\ [\hat{R}_{S(np)}, \hat{R}_{S(np+1)}], & \text{if } np \text{ is integer.} \end{cases} \tag{7.74}$$

7.3.4.2 Bayes's Monte Carlo Simulation

The only difference between the classical Monte Carlo and the Bayesian approaches is related to component reliability estimation. Under the Bayes approach, we need to provide prior information and respective prior distributions for each unique component in the given system. Then we need to obtain the corresponding posterior distributions. Once these distributions are obtained, the same steps used in the classical Monte Carlo approach are performed.

In the absence of prior information about reliability of the system components and binomial data with moderate sample size, Martz and Duran [27] recommend using the beta distribution having parameters 0.5 and 0.5 as an appropriate prior distribution, which they call *noninformative* prior. Note that such noninformative prior has a mean of 0.5 and a coefficient of variation very close to the coefficient of variation of the standard uniform distribution (0, 1). Also recall that the standard uniform distribution is a particular case of the beta distribution with parameters $\alpha = 1, \beta = 1$ (see Section 2.3).

7.3.4.3 Bootstrap Method

The bootstrap method, introduced by Efron and Tibshirani in [28], is a Monte Carlo simulation technique in which new samples are generated from the data of an original sample. The method's name, derived from the old saying about pulling yourself up by your own bootstraps, reflects the fact that one available sample gives rise to many others.

Unlike the classical and the Bayes Monte Carlo techniques, the bootstrap method is a universal nonparametric method. To illustrate the basic idea of this method, consider the following simple example, in which the standard error of a median is estimated [29].

Consider an original sample, $x_1, x_2, \ldots, x_n$, from an unknown distribution. The respective *bootstrap sample*, $X^b = x_1^b, x_2^b, \ldots, x_n^b$, is obtained by randomly sampling n times with replacement from the original sample $x_1, x_2, \ldots, x_n$. The bootstrap procedure consists of the following steps:

- Generate a large number, N, of bootstrap samples X_i^b ($i = 1, 2, \ldots, N$).
- Evaluate the sample median, $x_{0.5}(X_i^b)$, for each bootstrap sample obtained. This is called the *bootstrap replication.*
- Calculate the *bootstrap estimate* of standard error of the median of interest:

$$\hat{S}_{x_{0.5}} = \sqrt{\frac{\sum_{i=1}^{N} [x_{0.5}(X_i^b) - \hat{x}_{0.5}]^2}{N - 1}}, \tag{7.75}$$

where

$$\hat{x}_{0.5} = \sum_{i=1}^{N} \frac{x_{0.5}(X_i^b)}{N}. \tag{7.76}$$

Note that no assumption about the distribution of r.v. x was introduced.

For some estimation problems, the results obtained using the bootstrap approach coincide with respective known classical ones. This can be illustrated by the following example related to binomial data [27].

Assume that for each component of the system of interest, we have the data collected in the form $\{S_i, N_i\}$ ($i = 1, 2, \ldots, n$, where n is the number of components in the system), where N_i is the number of units of ith component tested (or observed) during a fixed time interval (the same for all n components of the system), and S_i is the respective number of units survived. The basic steps of the corresponding bootstrap simulation procedure are as follows:

1. For each component of the system given, obtain the bootstrap estimate of component reliability, R_i, ($i = 1, 2, \ldots, n$, where n is the number of component in the system), generating it from the binomial distribution with parameters N_i and $p = S_i/N_i$. In the case when $S_i = N_i$, that is, $p = 1$, one needs to smooth the bootstrap, replacing p by $(1-\varepsilon)$ where $\varepsilon \ll 1$. This procedure is discussed in Efron and Tibshirani [29].
2. Calculate the corresponding classical estimate of the system reliability, using Equation 7.73 with R_i, ($i = 1, 2, \ldots, n$) obtained from the results of step 1.
3. Repeat steps 1–2 a sufficiently large number of times, n, (e.g., 10,000) to obtain a large sample of $\hat{R}_S$.
4. Based on the sample obtained and a chosen confidence level $(1-\alpha)$, construct the respective lower confidence limit for the system reliability of interest as a sample percentile of level α, using Equation 7.74.

Example 7.9

Consider a fault tree, the top event, T, of which is described by the following expression:

$$T = C_1 + C_2C_3,$$

where C_1, C_2, and C_3 are the cut sets of the system modeled by the fault tree. If the following data are reported for the components representing the respective cut sets, determine a point estimate and 95% confidence interval for the system reliability $\hat{R}_S = 1 - \Pr(T)$ using (a) the system reduction methods and (b) the bootstrap method.

Component	Number of Failures d	Number of Trials N
C_1	1	1785
C_2	8	492
C_3	4	371

Solution:

a. The second cut set can be considered as a parallel subsystem containing components C_2 and C_3. Therefore, we shall apply the Maximus method to reduce this subsystem to an equivalent component C_{23}. The maximum likelihood point estimate of the equivalent component reliability can be obtained as

$$\hat{R}_{C_{23}} = 1 - \Pr(C_{23}) = 1 - \left(\frac{8}{492}\right)\left(\frac{4}{371}\right) = 0.99982.$$

The equivalent number of trials for C_{23} is

$$N_{C_{23}} = \min(N_{C_2}, N_{C_3}) = \min(492, 371) = 371$$

and using Equation 7.70, the equivalent number of failures is

$$D_{C_{23}} = N_{C_{23}}(1 - \hat{R}_{C_{23}}) = 371(1 - 0.99982) = 0.06678.$$

Now we can treat C_1 and C_{23} as a series system and apply the Lloyd–Lipow method to reduce it. Using Equation 7.71, the estimate of system reliability is

$$\hat{R}_S = \left(1 - \frac{1}{1785}\right)\left(1 - \frac{0.067}{371}\right) = 0.99926.$$

Then, keeping in mind the equivalent number of system trials of

$$N_S = \min(N_{C_1}, N_{C_{23}}) = \min(1785, 371) = 371.$$

From Equation 7.70, the fictitious number of system failures is

$$D_S = N_S(1 - \hat{R}_S) = 371(1 - 0.99926) = 0.27454.$$

Using (Equations 3.98 and 3.99) with $n = 371$ and $r = 0.27$, the 95% lower and upper confidence limits for the system reliability estimate are found to be

$$0.99758 \leq \hat{R}_S \leq 0.99988.$$

TABLE 7.8
The Bootstrap Solution in Example 7.9

Monte Carlo Run Number	Component	C_1	C_2	C_3	Estimate of System Reliability R_S
	Number of failures d	1	8	4	
	Number of trials N	1785	492	371	
	Binomial probability of Failure $p = d/N$	0.00056	0.01626	0.01078	
1	Observed number of failures in N binomial trials with parameter p, d_i	0	7	4	0.99985
	Bootsrap replication, $p_i^b = d_i/N$	0.00000	0.01423	0.01078	
2	Observed number of failures in N binomial trials with parameter p, d_i	0	7	6	0.99977
	Bootsrap replication, $p_i^b = d_i/N$	0.00000	0.01423	0.01617	
3	Observed number of failures in N binomial trials with parameter p, d_i	1	9	4	0.99924
	Bootsrap replication, $p_i^b = d/N$	0.00056	0.01829	0.01078	
...	...	...	...	...	...
10,000	Observed number of failures in N binomial trials with parameter p, d_i	2	10	5	0.99861
	Bootsrap replication, $p_i^b = d_i/N$	0.00112	0.02033	0.01348	
				$E(R_S)$	9.9926×10^{-1}
				$\mathrm{Var}(R_S)$	3.4500×10^{-7}

continued

b. The bootstrap estimation can be obtained as follows:

1. Using the failure data for each component, compute the estimate of the binomial probability of failure and treat it as a nonrandom parameter p.
2. Simulate N binomial trials of a component and count the observed number of failures.
3. Obtain a bootstrap replication of p by dividing the observed number of failures by the number of trials. Once the bootstrap replications are computed for each component, find the estimate of system reliability using Equation 7.73.
4. Repeat steps 2 and 3 a sufficiently large number of times, and use Equation 7.74 to obtain the interval estimates of system reliability.

The procedure and results of the bootstrap solution are summarized in Table 7.8. As seen there, the point estimate of system reliability very closely coincides with the one obtained in Part (a).

From the distribution of the system reliability estimates, the 95% confidence bounds can be obtained as the 2.5% and 97.5% sample percentiles—see Equation 7.74.

Example 7.10

The system reliability confidence bounds can also be estimated through the use of Clopper–Pearson procedure. The fictitious number of system trials is

$$N_S = \frac{\hat{R}_S(1-\hat{R}_S)}{\mathrm{var}(\hat{R}_S)} = \frac{0.99926[1-0.99926]}{3.45\times 10^{-7}} = 2143.3.$$

Then, the fictitious number of system failures is

$$D_S = N_S(1-\hat{R}_S) = 2143.3(1-0.99926) = 1.6.$$

Using Equations 3.98 and 3.99, with $n = 2143.3$ and $r = 1.6$, the 95% lower and upper confidence limits for the system reliability estimate are found to be

$$0.99899 \le \hat{R}_S \le 0.99944,$$

which is quite consistent with the results obtained from the other two methods.

Martz and Duran [27] performed some numerical comparisons of the Maximus, bootstrap, and Bayes Monte Carlo methods applied to 20 simple and moderately complex system configurations and simulated binomial data for the system components. They made the following conclusions about the regions of superior performance of the methods.

1. The Maximus method is, generally, superior for (a) moderate to large series systems with small quantities of test data per component and (b) small series systems composed of repeated components.

2. The bootstrap method is recommended for highly reliable and redundant systems.
3. The Bayes Monte Carlo method is, generally, superior for (a) moderate to large series systems of reliable components with moderate to large samples of test data and (b) small series systems composed of reliable nonrepeated components.

7.3.5 Graphic Representation of Uncertainty

The results of a probabilistic uncertainty analysis should be presented in a clear manner that aids analysts in developing appropriate qualitative insights. Generally, we will discuss three different ways of presenting probability distributions so that their use is not limited by uncertainty analysis: plotting the pdf, plotting the cdf, or displaying selected percentiles, as in a Tukey [30] box plot. Figure 7.2 shows examples.

Plotting the pdf shows the relative probabilities of different values of the parameters. One can see the areas or ranges where high densities (occurrences) of r.v. occur (e.g., the modes). One can judge

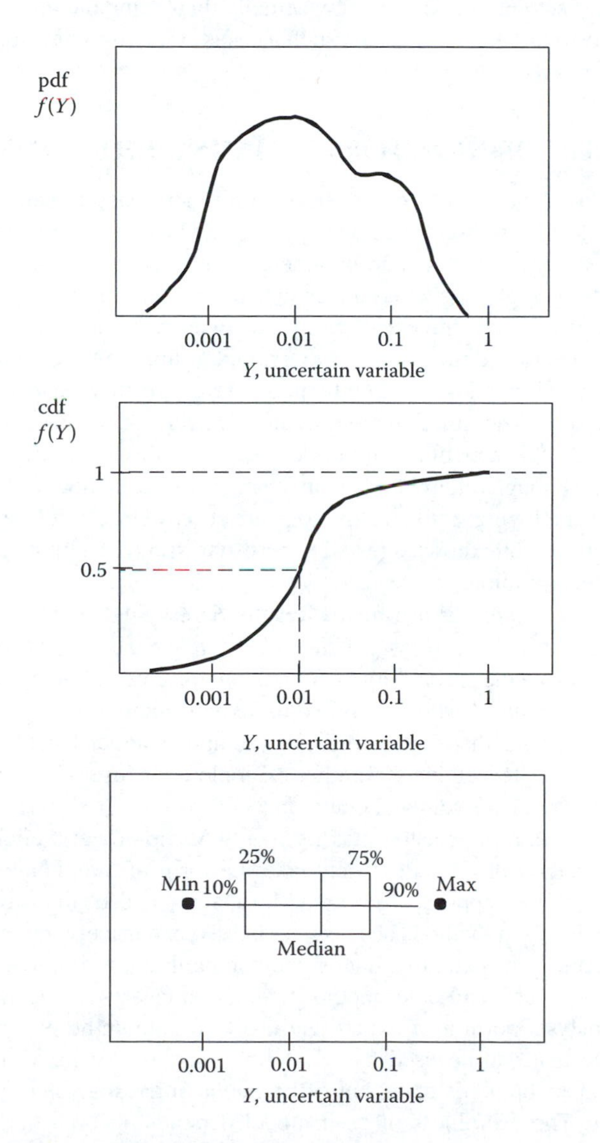

FIGURE 7.2 Three conventional methods of displaying distribution.

symmetry and skewness and the general shape of the distribution (e.g., bell-shaped versus J-shaped). The cdf is best for displaying percentiles (e.g., median) and the respective confidence intervals. It can be easily used for both continuous and discrete distributions.

The standard Tukey box shows a horizontal line from the 10th to 90th percentiles, a box between the lower percentiles (e.g., from the 25th to the 75th percentiles), and a vertical line at the median, and points at the minimum and maximum observed values. This method clearly shows the important quantities of the r.v.

In cases where statistical uncertainty limits are estimated, the Tukey box can be used to describe the confidence intervals. Consider a case where the distribution of a variable Y is estimated and described by a pdf. For example, a pdf of time to failure (the aleatory model) can be represented by an exponential distribution. The value of λ for this exponential distribution is represented, under the Bayes approach, by a lognormal distribution. Then, $f(Y|\lambda)$ for various values of λ can be plotted, and families of curves can be developed to show an aggregate effect of both kinds of uncertainty. This epistemic uncertainty is shown in Figure 7.3.

In general, a fourth method can be shown by actually displaying the probability densities of λ in a multidimensional form. Figure 7.4 presents such a case for a two-dimensional distribution. In this figure $f(Y|\lambda)$ is shown for various values of λ.

7.4 USE OF EXPERT OPINION FOR ESTIMATING RELIABILITY PARAMETERS

It is often necessary to solicit expert opinions in reliability analysis; in many cases, the use of these opinions is unavoidable. One reason for using experts is the lack of a statistically significant amount of empirical data necessary to estimate new parameters. Another reason for using experts is to assess the likelihood of a one-time event, such as the chance of rain tomorrow.

However, the use of expert judgment that requires extensive knowledge and experience in the subject field is not limited to one-time events. For example, suppose we are interested in using a new and highly capable microcircuit device currently under development by a manufacturer and expected to be available for use soon. The situation requires an immediate decision on whether or not to design an electronic box around this new microcircuit device. Reliability is a critical decision criterion for the use of this device. Although reliability data on the new device are not available, reliability data on other types of devices employing similar technology are accessible. Therefore, reliability assessment of the new device requires both knowledge and expertise in similar technology, and can be achieved through the use of expert opinion.

Some specific examples of expert use are the Reactor Safety Study [31]; IEEE-Standard 500 [32]; and *Severe Accident Risk: An Assessment for Five U.S. Nuclear Power Plants* [33], in which expert opinion was used to estimate the probability of component failures and other rare events. The Electric Power Research Institute study (1986) has relied on expert opinion to assess seismic hazard rates. Other applications include weather forecasting. For example, Clemens and Winkler [34] discuss the use of expert opinion by meteorologists. Another example is the use of expert opinion in assessing human error rates discussed by Swain and Guttman [35].

The use of expert opinion in decision making is a two-step process: elicitation and analysis of expert opinion. The method of elicitation may take the form of individual interviews, interactive group sessions, or the Delphi approach discussed by Dalkey and Helmer [36]. The relative effectiveness of different elicitation methods has been addressed extensively in the literature. Techniques for improving the accuracy of expert estimates include calibration, improvement in questionnaire design, motivation techniques, and other methods, although clearly no technique can be applied to all situations. The analysis portion of expert use involves combining expert opinions to produce an aggregate estimate that can be used by reliability analysts. Again, various aggregation techniques for pooling expert opinions exist, but of particular interest are those adopting the form of mathematical models. The usefulness of each model depends on both the reasonableness of the assumptions (implicit and explicit) carried by the model as it mimics the real-world situation, and

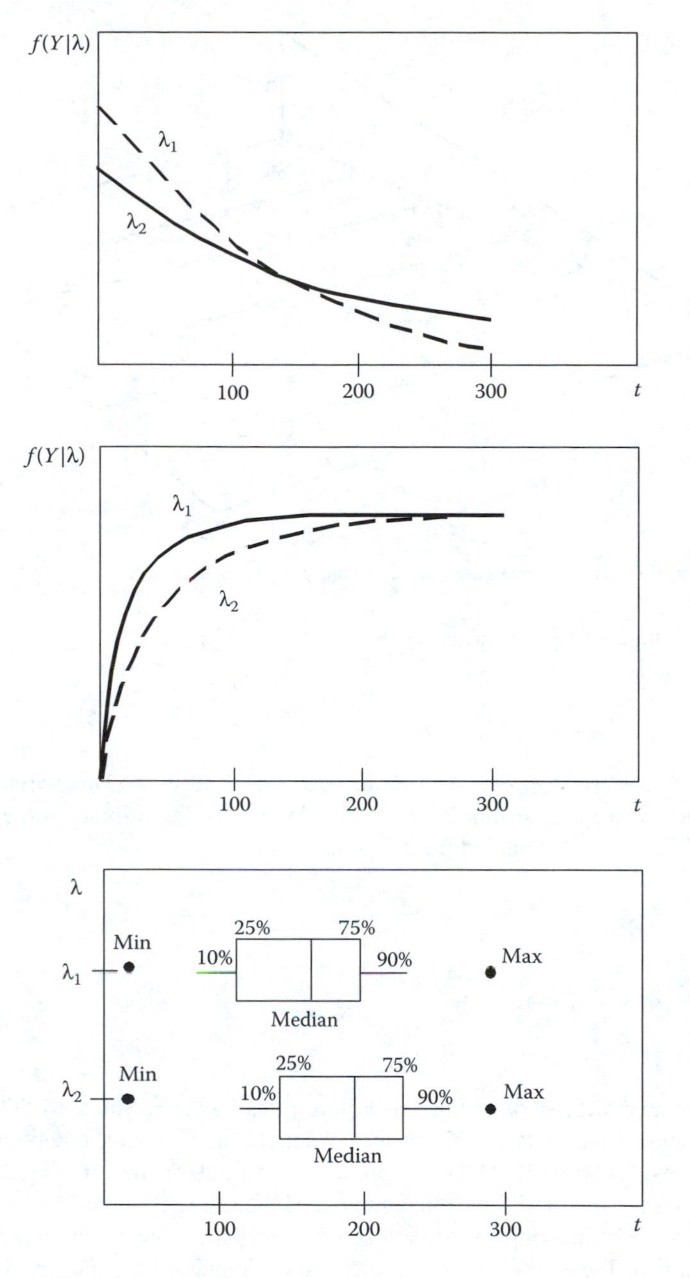

FIGURE 7.3 Representation of uncertainties.

the ease of implementation from the user's perspective. The term "expert" generally refers to any source of information that provides an estimate and includes human experts, measuring instruments, and models.

Once the need for expert opinion is determined and the opinion is elicited, the next step is to establish the method of opinion analysis and application. This is a decision task for the analysts, who may simply decide that the single best estimate of the value of interest is the estimate provided by the arithmetic average of all estimates, or an aggregate from a nonlinear pooling method, or some other opinion. Two methods of aggregating expert opinion, the geometric averaging technique and the Bayesian technique, are discussed in more detail here.

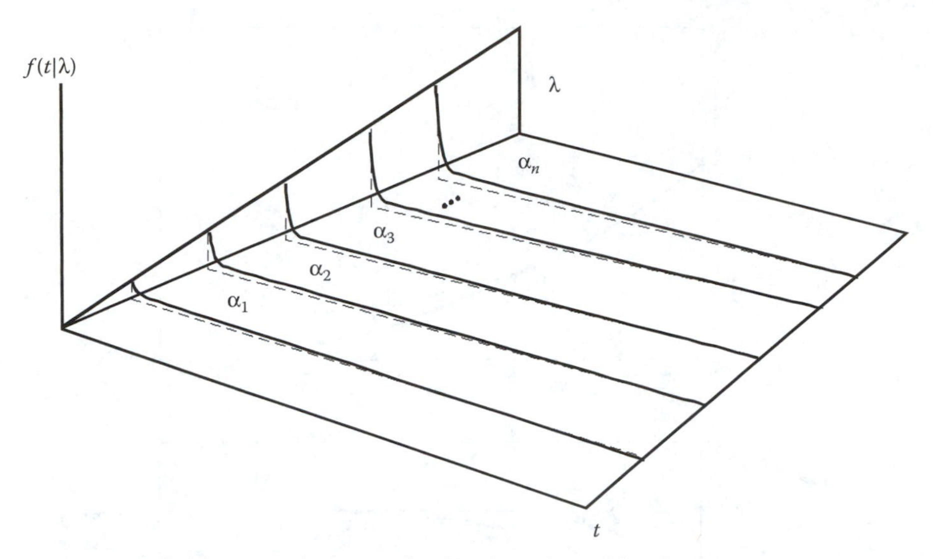

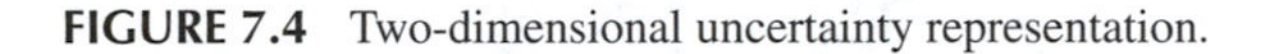
FIGURE 7.4 Two-dimensional uncertainty representation.

7.4.1 Geometric Averaging Technique

Suppose n experts are asked to estimate the failure rate of an item. The estimates can be pooled using the geometric averaging technique. For example, if λ_i is the estimate of the ith expert, then an estimate of the failure rate is obtained from

$$\hat{\lambda} = \sqrt[n]{\prod_{i=1}^{n} \lambda_i}. \tag{7.77}$$

This was the primary method of estimating failure rates in IEEE-Standard 500 [32]. The IEEE-Standard 500 contains rate data for electronic, electrical, and sensing components. The reported values were synthesized primarily from the opinions of some 200 experts (using a form of the Delphi procedure). Each expert reported "low," "recommended," and "high" values for each failure rate under normal conditions, and a "maximum" value that would be applicable under all conditions (including abnormal conditions). The estimates were pooled using Equation 7.77. For example, for maximum values

$$\hat{\lambda}_{\max} = \sqrt[n]{\prod_{i=1}^{n} \lambda_{\max,i}}. \tag{7.78}$$

As discussed by Mosleh and Apostolakis [37], the use of geometric averaging implies that (1) all the experts are equally competent, (2) the experts do not have any systematic biases, (3) experts are independent, and (4) the preceding three assumptions are valid regardless of which value the experts are estimating, such as high, low, or recommended.

The estimates can be represented in the form of a distribution. Apostolakis et al. [38] suggest the use of a lognormal distribution for this purpose. In this approach, the "recommended" value is taken as the median of the distribution, and the error factor (EF) is defined as

$$\mathrm{EF} = \sqrt{\frac{\hat{\lambda}_{0.95}}{\hat{\lambda}_{0.05}}}. \tag{7.79}$$

7.4.2 Bayesian Approach

As discussed by Mosleh and Apostolakis [37], the challenge of basing estimates on expert opinion is maintaining coherence throughout the process of formulating a single best estimate based on the experts' actual estimates and their credibilities. Coherence is the notion of internal consistency within a person's state of belief. In the subjectivist school of thought, a probability is defined as a measure of personal uncertainty. This definition assumes that a coherent person will provide his or her probabilistic judgments in compliance with the axioms of probability theory.

An analyst often desires a modeling tool that can aid him or her in formulating a single best estimate from expert opinion(s) in a coherent manner. Informal methods such as simple averaging will not guarantee this coherence. Bayes's theorem, however, provides a framework to model expert belief and ensures coherence of the analysts in arriving at a new degree of belief in light of expert opinion. According to the general form of the model given by Mosleh and Apostolakis, the state-of-knowledge distribution of a failure rate λ, after receiving an expert estimate $\hat{\lambda}$, can be obtained by using Bayes's theorem in the following form:

$$\prod(\lambda|\hat{\lambda}) = \frac{1}{k} L(\hat{\lambda}|\lambda)\pi_0(\lambda), \tag{7.80}$$

where $\pi_0(\lambda)$ is the prior distribution of λ; $\Pi(\lambda|\hat{\lambda})$ is the posterior distribution of λ; $L(\hat{\lambda}|\lambda)$ is the likelihood of receiving the estimate $\hat{\lambda}$, given the true failure rate λ; and k is a normalizing factor.

One of the models suggested for the likelihood of observing $\hat{\lambda}$ given λ is based on the lognormal distribution in the following form:

$$L(\hat{\lambda}|\lambda) = \frac{1}{\sqrt{2\pi}\sigma\hat{\lambda}} \exp\left[-\frac{1}{2}\left(\frac{\ln\hat{\lambda} - \ln\lambda - \ln b}{\sigma}\right)^2\right], \tag{7.81}$$

where b is a bias factor ($b = 1$ when no bias is assumed) and σ is the standard deviation of logarithm of $\hat{\lambda}$, given λ. When the analyst believes no bias exists among the experts, she or he can set $b = 1$. The quantity σ, therefore, represents the degree of accuracy of the experts' estimate as viewed by the analyst. The work by Kim [39], which includes a Bayesian model for a relative ranking of experts, is an extension of the works by Mosleh and Apostolakis.

7.4.3 Statistical Evidence on the Accuracy of Expert Estimates

Among the attempts to verify the accuracy of expert estimates, two types of expert estimates are studied here: assessment of single values and assessment of distributions.

Notable among the studies on the accuracy of expert assessments of a single estimate is Snaith's study [40]. In this study, observed and predicted reliability parameters for some 130 pieces of different equipment and systems used in nuclear power plants were evaluated. The predicted values included both direct assessments by experts and the results of analysis. The objective was to determine correlations between the predicted and observed values. Figure 7.5 shows the ratio ($R = \lambda/\hat{\lambda}$) of observed to predicted values plotted against their cumulative frequency. As shown, the majority

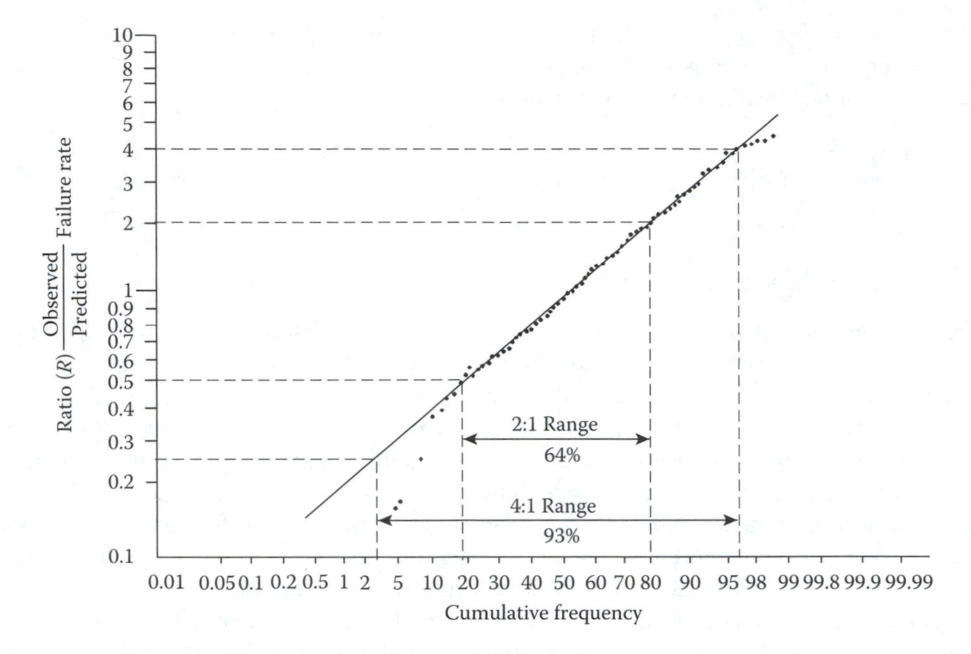

FIGURE 7.5 Frequency distribution of the failure rate ratio [40].

of the points lie within the dashed boundary lines. Predicted values are within a factor of 2 from the observed values, and 93% are within a factor of 4. The figure also shows that $R = 1$ is the median value, indicating that there is no systematic bias in either direction. Finally, the linear nature of the curve shows that R tends to be lognormally distributed, at least within the central region. This study clearly supports the use and accuracy of expert estimation.

Among the studies of expert estimation are works by cognitive psychologists. For example, Lichtenstein et al. [41] tested the adequacy of probability assessments and concluded that "the overwhelming evidence from research on uncertain quantities is that people's probability distributions tend to be biased." Commenting on judgmental biases in risk perception, Slovic et al. [42] stated: "A typical task in estimating uncertain quantities like failure rates is to set upper and lower bounds such that there is a 98% chance that the true value lies between them. Experiments with diverse groups of people making different kinds of judgments have shown that, rather than 2% of true values falling outside the 98% confidence bounds, 20% to 50% do so. Thus, people think that they can estimate such values with much greater precision than is actually the case."

Based on the above conclusion, Apostolakis [43] has suggested the use of the 20th and 80th percentiles of lognormal distributions instead of the 5th and 95th when using Equation 7.79, to avoid a bias toward low values, overconfidence of experts, or both. When using the Bayesian estimation method based on Equation 7.81, the bias can be accounted for by using larger values of σ and b in Equation 7.81.

7.5 PROBABILISTIC FAILURE ANALYSIS

Statistical, probabilistic, or deterministic methods are used to analyze failures. While all three methods or combinations of them can be used, in this section we rely primarily on the statistical methods for the analysis of failures. However, for evaluating the results of the analysis, we mainly use deterministic techniques. Probabilistic (Bayesian) and deterministic techniques are equally applicable. However, since Bayesian techniques may require expert or prior knowledge about equipment failures, they should be used only when observed failure data are sparse.

The statistical methods described in this section are based on the classical inference methods discussed in Chapters 3 and 5. That is, the history of failure or event occurrences is first studied to determine whether a statistically significant trend can be detected. If not, the traditional maximum likelihood parameter estimating method is used to determine the failure characteristic of the item—for example, to determine the failure rate or demand failure probability of an item.

If the trend analysis method discussed in Chapter 5 shows a significant trend in the data, it is important to determine the nature and degree to which the failure characteristic of the item is changed. Classical statistics methods are used to determine the failure characteristics of equipment if no trends are exhibited.

When the failure characteristics of an item with or without trend are determined, they must be evaluated to determine whether they show any change in the capability of the item. Both statistical and nonstatistical techniques can be used to detect changes.

When significant changes are detected, it is important to search for possible reasons for such changes. This may require an analysis to determine the root causes of the detected changes or observed failure events. No standard practice exists for determining root-cause failures. Engineers often use ad hoc techniques for this purpose.

Figure 7.6 shows the overall approach employed in this section. The basis for the statistical methods used in this document is explained in the remainder of this section.

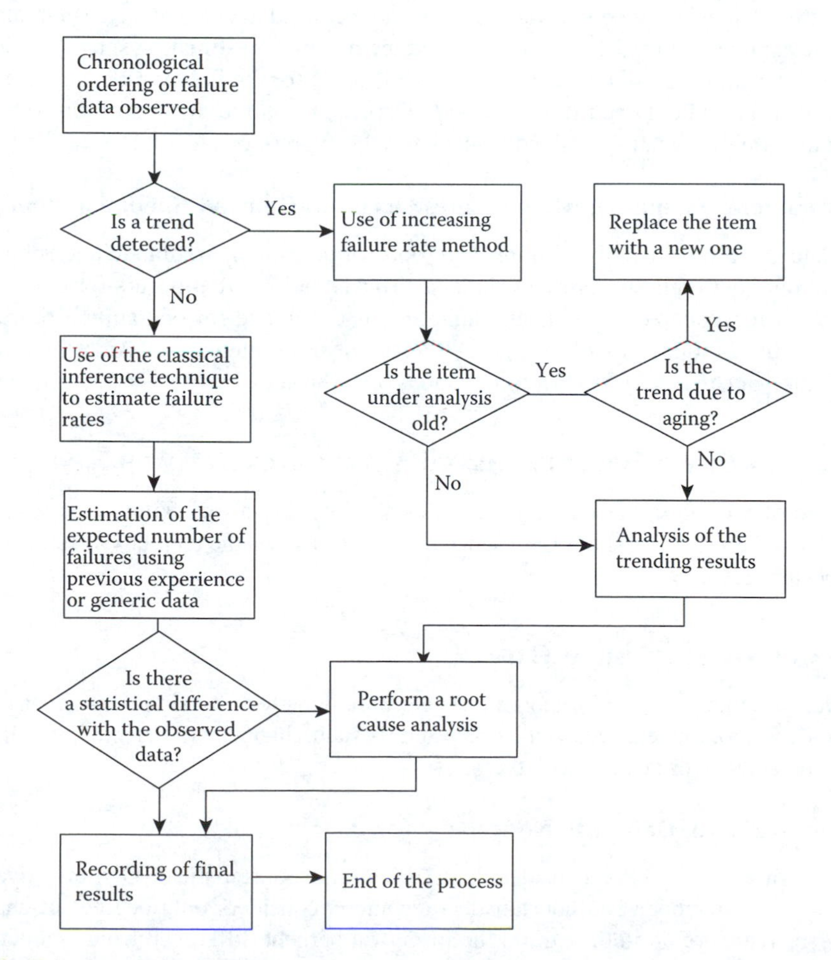

FIGURE 7.6 Failure analysis process.

7.5.1 Detecting Trends in Observed Failure Events

Statistical estimators for equipment failure characteristics should be used only after it has been determined that the failure occurrence is reasonably constant—that is, there is no evidence of an increasing or decreasing trend. In Chapter 5, we described the Centroid method to test for the possibility of a trend. In Equation 5.7, since statistic U is a sensitive measure, one could use the following practical criteria to ensure detection of trends, especially when the amount of data is limited.

1. When $U > 0.5$ or $U < -0.5$, assume a reasonable trend exists.
2. Otherwise, depending on the age and the item's recent failure history, assume a mostly constant failure rate (or failure probability) or a mild trend exists.

7.5.2 Failure Rate and Failure Probability Estimation for Data with No Trend

Sections 3.4 and 3.5 dealt with statistical methods for estimating failure rate and failure probability parameters of components when there is no trend in failures. The objective here is to find a point estimate and a confidence interval for the parameters of interest.

7.5.2.1 Parameter Estimation when Failures Occur by Time

When failures of equipment occur by time (r failures in T hours), the exponential distribution is most commonly used. Therefore, when the failure events are believed to occur at a constant rate (i.e., with no trend), the exponential model is reasonable and the parameter estimation should proceed. In this case, λ parameter must be estimated. The point estimator is for the failure rate parameter (λ) of the exponential distribution obtained from $\hat{\lambda} = r/T$. Depending on the method of observing data, the confidence interval of λ can be obtained from one of the expressions in Table 3.2.

7.5.2.2 Parameter Estimation when Failures Occur on Demand (Binomial Model)

When the data are in the form of X failures in n trials (or demands), no time relationship exists, and the binomial distribution best represents the data. This situation often occurs for equipment in the standby mode—for example, a redundant pump that is demanded for operation n times in a fixed period of time. In a binomial distribution, the only parameter of interest is p. An estimate of p and its confidence interval can be obtained from Equations 3.97 through 3.99.

7.5.3 Failure Rate and Failure Probability Estimation for Data with Trend

The existence of a trend in the data indicates that the interarrivals of failures are not statistically similar, and thus Equation 5.8 should be used. Chapter 5 describes the methods of estimating the rate of failure occurrence $\lambda(t)$.

7.5.4 Evaluation of Statistical Data

After the data are analyzed, it is important to determine whether any significant changes between the past data and more recent data can be detected. If such changes are detected, it is important to formulate a procedure for dealing with them.

7.5.4.1 Evaluation of Data with No Trend

Two methods of evaluation are considered, statistical and nonstatistical. One effective statistical technique is the χ^2 method. The nonstatistical technique considers only degrees of change in the failure characteristics of an item (e.g., in the form of a percent difference from a generic value or prior experience). Proper action is suggested based on a predefined criterion.

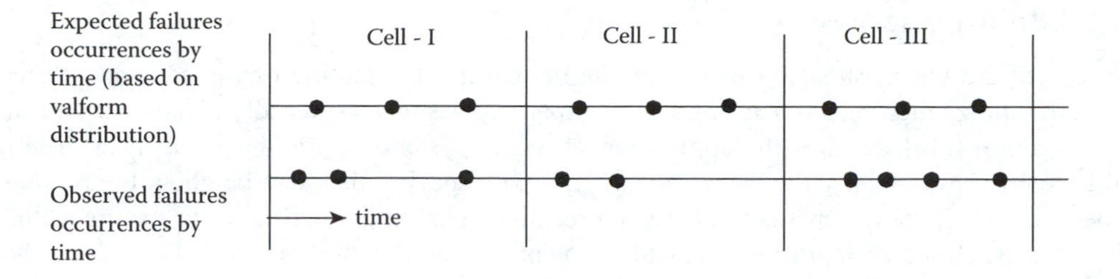

FIGURE 7.7 Comparison of expected and observed failure occurrences by an item.

As mentioned earlier, the χ^2 method can be adapted to the type of problems considered here. The χ^2 method was described in Chapter 2. In failure analysis, the χ^2 test can be used to determine whether the observed failure data are statistically different from generic data or from past history of the same or a similar item. For example, consider Figure 7.7.

If the expected number of failures, based on generic failure data or previously calculated values (e.g., using statistical analysis), are determined and compared with the observed failures, one can statistically measure the difference.

It is easy to divide the time line (or in a demand-type item, the number of demands) into equal time demand intervals (e.g., three intervals as in Figure 7.7) and compare them to see whether the observed and expected failures in each interval are statistically different.

For example, for data in Figure 7.7, the following χ^2 statistic can be calculated:

$$W = (3 - 3)^{2/3} + (3 - 2)^{2/3} + (3 - 4)^{2/3} = 2/3. \tag{7.82}$$

This shows that there is a slight difference between the observed and expected data, but depending on the desired level of confidence, this may or may not be acceptable.

The nonstatistical technique uses only a percent difference between the estimated failure rate $\hat{\lambda}$ and the generic failure rate λ_g. For example,

$$e = \left| \frac{\hat{\lambda} - \lambda_g}{\lambda_g} \right| \times 100. \tag{7.83}$$

If the difference is large (more than 100), one can assume the data are different, and further root-cause analysis is required.

7.5.4.2 Evaluation of Data with Trend

Generally, there is no set rule for this purpose. One approach is to use the doubling failure concept. If two consecutive intervals of (t_1, t_2) and (t_2, t_3) are such that $t_2 - t_1 = t_3 - t_2$, and the expected number of failures in each interval (N_1 and N_2, respectively) are such that $N_2/N_1 = 2$, then it is easy to prove, using Equation 5.74 that $\beta = 1.58$. Accordingly, for $N_2/N_1 = 5$, $\beta = 2.58$. These can be used as guidelines for determining the severity of the trend. For example, one can assume the following:

- If $1 \leq \beta \leq 1.58$, the trend is mildly increasing. Suggest a root-cause analysis and implement a careful monitoring system.
- If $1.58 < \beta \leq 2.58$, the trend is major. Suggest replacement or root-cause analysis.
- If $\beta > 2.58$, the trend is significant. Cease operation of the item and determine the root cause of the trend.

7.5.5 Root-Cause Analysis

Root causes are the most basic causes that can be reasonably identified by experts and corrected so as to minimize their recurrence. A group of experts (investigators) usually identifies root causes. Modarres et al. [44] explains the application of expert systems in root-cause analysis. The goal of the experts is to identify the basic causes. The more specific they can be about the reasons an incident occurred, the easier it is to arrive at a recommendation that will prevent recurrence of the failure events. However, root causes should not be investigated to the extreme. The analysis should yield the most out of the time spent and identify only those root causes for which a reasonable corrective action exists. Therefore, very complex and specific mechanisms of failure do not need to be identified, especially when corrective actions can be determined at a higher level of abstraction. The recommended corrective actions should be specific and should directly address the root causes identified during the analysis. Root-cause analysis involves three steps.

1. Determining events and causal factors.
2. Coding and documenting root causes.
3. Generating recommendations.

Charting the event and causal factors provides a road map for experts to organize and analyze the information that they gather, identify their findings, and highlight gaps in knowledge as the investigation progresses. For example, a sequence diagram similar to that in Figure 7.8 may be developed, showing the events leading up to and following an occurrence as well as the conditions and their causes surrounding the failure event. The process is performed inductively and in progressively more detail.

Figure 7.8a shows the causal relations leading to a "failure event," including the conditions, events, and causal factors.

Following this step, the causal factors and events should be documented. One method suggested by the *Root-Cause Analysis Handbook* (1991) uses a root-cause tree involving six levels. From the event and causal factors chart, these levels are described and documented. Figure 7.8b shows an example of the levels used, and Figure 7.8c shows an example of a report based on this classification.

The final and most important step in this process is to generate a list of recommendations. This process is based on the experience of the experts. However, as a general guideline, the following items should be considered when recommending corrective actions.

1. At least one corrective action should be identified for each root cause.
2. The corrective action should directly and unambiguously address the root cause.
3. The corrective action should not have secondary degrading effects.
4. The consequences of the recommended (or not recommended) corrective actions should be identifiable.
5. The cost associated with implementation of the corrective action should be estimated.
6. The need for special resources and training for implementation of the action should be identified.
7. The effect on the frequency of item failure should be estimated.
8. The impact the corrective action is expected to have on other items or on workers should be addressed.
9. The effect of the corrective action should be easily measurable.
10. Other possible corrective actions that are more resource intensive but more effective should be listed.

Root-cause analysis is a major field of study. For further reading in this subject, see [45–48].

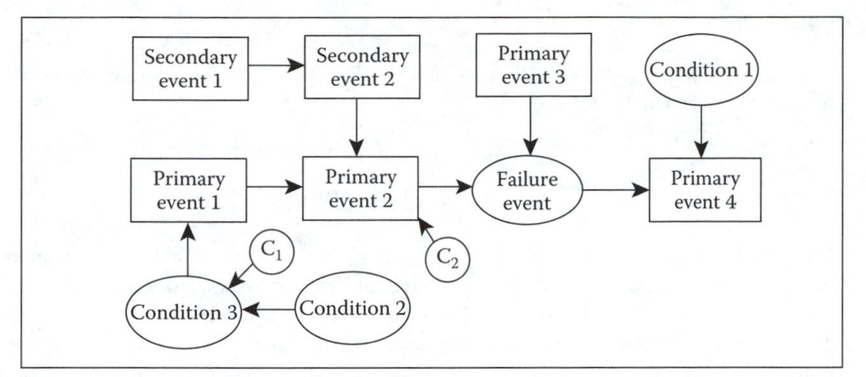

Levels of the root cause tree

Level	Shape	Description	Examples
A		Primary difficulty source	• Equipment difficulty • Operations difficulty • Technical difficulty
B		Area of responsibility	• Equipment reliability/ design • Production organization • Technical support organization
C		Equipment problem category	• Design • Installation/corrective/ preventive maintenance difficulty • Fabrication difficulty
D		Major root cause category	• Design review/ verification • Training • Management systems
E		Near root cause	• Procedures followed incorrectly • Workplace layout • Supervision during work
F		Root cause	• More than one action per step • Conflicting layouts • No supervision

Causal factor	Path through root cause tree	Recommendations
Operator had not previously been to motor control center Background: The Operator who went to verify the position of the Motor-Generator (M-G) switchgear had not been required to use this particular switchgear in the past. If he had been shown the switchgear as part of the training, it is unlikely that he would have forgotten its location.	• Operations difficulty • Production organization • Training The Operator had never been trained on location of equipment (including switchgear) in motor control center	• Include a tour of the Motor Control Centers as part of on-the-job training. Provide specific instructions on how to use drawings for verifying the positons of important places of equipment. (Production training department.)

FIGURE 7.8 Events and causal factors chart.

EXERCISES

7.1 Consider two resistors in parallel configuration. The mean and standard deviation for the resistance of each are as follows:

$$\mu_{R_1} = 25\Omega \quad \sigma_{R_1} = 0.1\mu_{R_1},$$
$$\mu_{R_2} = 50\Omega \quad \sigma_{R_2} = 0.1\mu_{R_2}.$$

Using one of the statistical uncertainty techniques, obtain the mean and standard deviation of the equivalent resistor. In what ways is the uncertainty associated with the equivalent resistance different from the individual resistor? Discuss the results.

7.2 The results of a bootstrap evaluation give $\mu = 1 \times 10^{-4}$ and $\sigma = 1 \times 10^{-3}$. Evaluate the number of pseudofailures, F, in N trials for an equivalent binomial distribution. Estimate the 95% confidence limits of μ.

7.3 Repeat Exercise 4.6 and assume that a CCF between the valves and the pumps exists. Using the generic data in Table B.1, calculate the probability that the top event occurs. Use a β-factor method with $\beta = 0.1$ for valves and pumps. Discuss whether the selection of $\beta = 0.1$ is sensitive to the end result.

7.4 A class of components is temperature sensitive in that the components will fail if the temperature is raised too high. Uncertainty associated with a component's failure temperature is characterized by a continuous uniform distribution such as shown below:

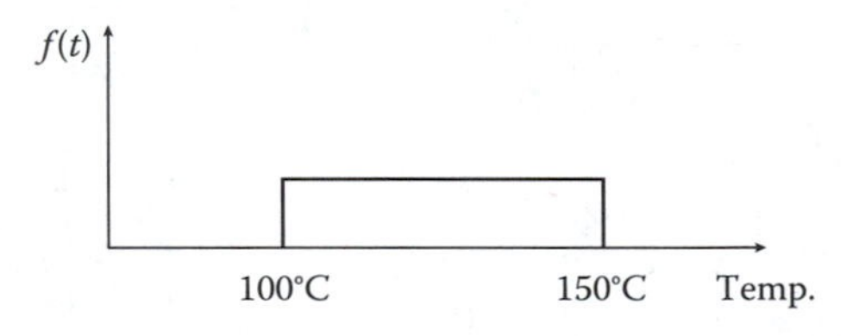

If the temperature for a particular component is uncertain but can be characterized by an exponential distribution with $\lambda = 0.05$ per degree Celsius, calculate reliability of this component.

7.5 Consider the cut sets below describing the failure of a simple system: $F = AB + BC$. The following data have been found for components A, B, and C:

Components	***A***	***B***	***C***
Number of failure	5	12	1
Total test time (h)	1250	4315	2012

a. Use the system reduction methods to calculate equivalent number of failures and total test time for failure of the system.
b. Given the results of (a), calculate the 90% confidence limits for the unreliability of this system.

7.6 The risk R of scenario I may be evaluated by using the expression $R_i = f_i \times C_i$, where f_i is the frequency and C_i is the consequence of this scenario. For a given scenario, if the mean frequency is 1×10^{-8} per year with a standard deviation of 1×10^{-8} per year, and mean consequences is 1000 injuries with a standard deviation of 100, determine the *mean risk* of the scenario and its associated standard deviation.

7.7 Consider the system below

If all components are identical and subject to dependent failures (CCF), calculate the failure probability of the system using the α-Factor Model for parametric CCF probability assessment (use generic data discussed in Section 7.2). Total failure probability of one unit is 0.001.

7.8 Consider the reliability function for a Weibull distribution time to failure:

$$R(t) = e^{-(t/\alpha)^{\beta}}$$

Suppose we are uncertain about the value of β. We have estimated the mean value of β, $\mu\beta = 1.5$ with a coefficient variation of 0.5. Assuming $\alpha = 1400$ h (with no uncertainty), compute the mean and coefficient of variation of $R(t)$ for $t = 1000$ h.

7.9 Consider the system as represented by the fault tree in Problem 5.8. If each component is replaceable and can be represented by an exponential time-to-failure model.

a. Determine the MTTF of the whole system.
b. Estimate MTTF's coefficient variation if the following data are applied.

Component	Failure Rate (λ)	Coefficient of Variation
A	10^{-3}	0.2
B	10^{-4}	0.1
C	10^{-4}	0.1

7.10 Consider the following reliability block diagram of a system

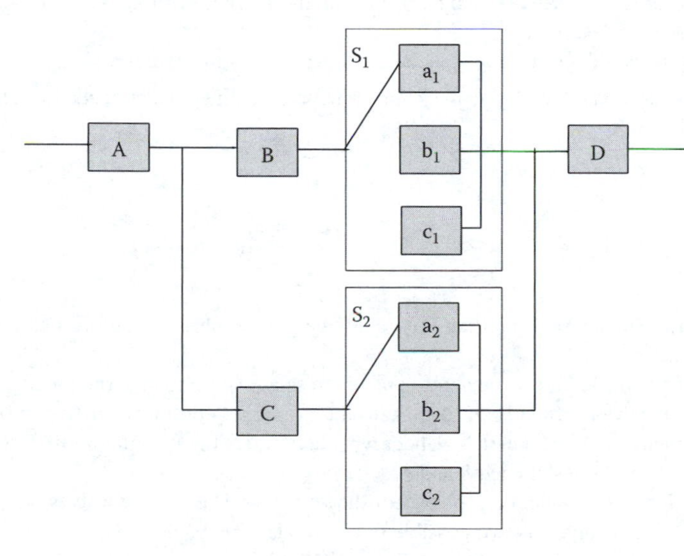

If S_1 and S_2 are standby subsystems with perfect switching, no standby failure and identical components (i.e., a_1, b_1, and c_1 are identical and so are a_2, b_2, and c_2)

a. Determine reliability of the system as a function of time. (Assume constant failure rate: λ_A, λ_B, λ_C, λ_D and $\lambda_{a1} = \lambda_{b1} = \lambda_{c1} = \lambda_1$, and $\lambda_{a2} = \lambda_{b2} = \lambda_{c2} = \lambda_2$.
b. If the units (i.e., the blocks) are "revealed fault repairable" units with constant failure rate λ and repair rate μ, determine the availability of the system. (Note that λ_A, λ_B, λ_C, λ_D and $\lambda_{a1} = \lambda_{b1} = \lambda_{c1} = \lambda_1$, and $\lambda_{a2} = \lambda_{b2} = \lambda_{c2} = \lambda_2$ and μ_A, μ_B, μ_C, μ_D and $\mu_{a1} = \mu_{b1} = \mu_{c1} = \mu_1$, and $\mu_{a2} = \mu_{b2} = \mu_{c2} = \mu_2$ are constants.)
c. Develop a corresponding fault tree for this diagram with the top event of "no flow out of D" in terms of units A, B, C, and D and subsystems S_1 and S_2.
d. Determine minimal cut sets of the system for the fault tree top event.
e. If components B and C are identical and subject to CCF, using the cut sets in (d) determine the probability of failure of the system (use an α-factor model with generic data for dependent failure analysis).

7.11 A reliability engineering test of 10 replaceable units has resulted in the following data (with some censored units).

i	t_i (h)
1	172
2	375+
3	673
4	842
5	1211+
6	1920
7	2711+
8	3922+
9	4931
10	6111

a. Is a Weibull distribution a proper fit? Use Kaplan–Meier and Rank-Increment rank statistics to plot and compare results.
b. What is your estimate of the reliability at 2500 h, using each method in (a)?
c. What is the mean residual life if the unit has worked without failure for 2500 h using each method in (a)?
d. What is the uncertainty associated with estimates in (b) and (c)?

7.12 In a structural reliability analysis for a pipeline specimen under cyclic load, the model often used is

$$NS^m = c,$$

where S = stress amplitude, N = number of cycles to failure, and m and c are constants.

If the mean and variance of applied stress are (μ_S and σ_S^2), the parameter m is known without any uncertainty, and the mean and variance of constant c are (μ_c and σ_c^2). Determine the mean and variance of N. Under what conditions the uncertainty in N would be higher than uncertainties of each S and c?

7.13 Consider a gas boiler control system shown below. The two gas valves are identical and in series; each valve has two possible failure modes.

Valve stuck open (probability $F_1 = 10^{-4}$).

Major leakage flow through valve (probability $F_2 = 10^{-3}$).

The three switches are identical and in series; each switch has three possible failure modes.

Stuck closed (probability $F_3 = 2 \times 10^{-3}$).

Short circuit (probability $F_4 = 2 \times 10^{-3}$).

Earth fault (probability $F_5 = 10^{-3}$).

If a current flows through the circuit, the solenoids are energized and both valves V_1 and V_2 are open.

The probability of the pilot flame failing $F_6 = 10^{-1}$.

The probability of a source of ignition existing $F_7 = 1$.

a. Draw a fault tree associated with the top event of "hazard of a gas explosion."
b. Use the data given to calculate the probability of a gas explosion.
c. Recalculate (b) assuming CCFs between the valves. Use the β-factor method with $\beta = 0.05$.

Schematic diagram of the control system for the gas boiler is shown below. A system failure corresponds to a gas explosion due to unburned gas being ignited

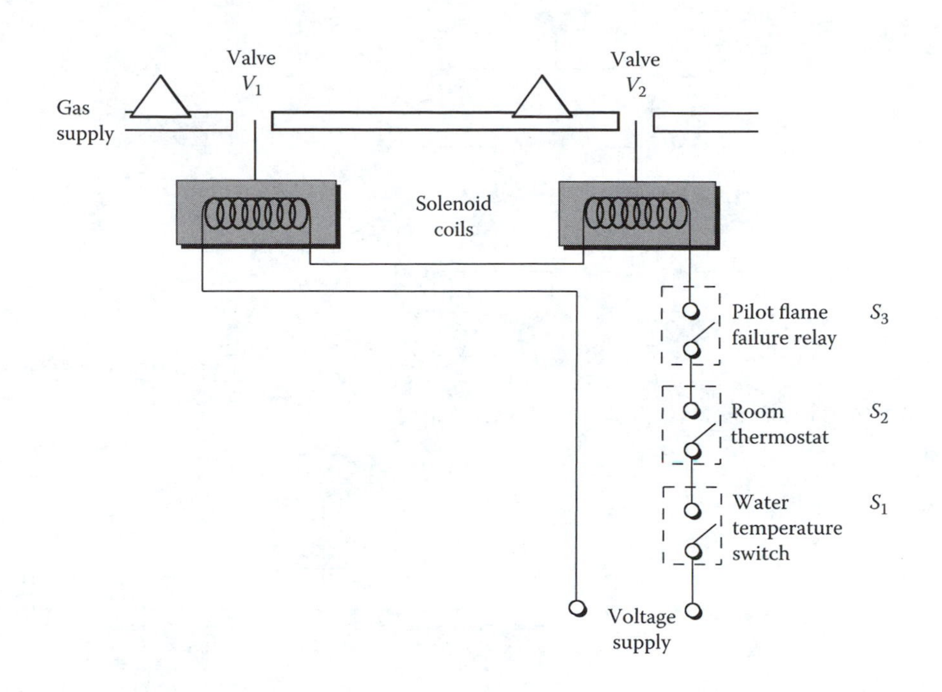

7.14 A decision has to be made whether to buy two, three, or four diesel generators for the electrical power system for an oil platform. Each generator will normally be working and can supply up to 50% of the total power demand. The reliability of each generator can be specified by a constant failure rate of 0.21 per year. The generators are to be simultaneously tested (the test and repair time is negligible) at 6-month intervals.

a. The required system availability must be at least 0.99. How many generators must be bought?
b. Calculate the MTBF for the chosen system.
c. Explain how the average availability of a system with three generators is affected if we assume the probability of generator failures are dependent. Use an α-factor method with generic data.

7.15 For the fault tree below, determine

1. The minimal cut sets.
2. The pointwise average *unavailability* for one test cycle (assume independence).
3. Repeat 2. Assuming CCF between components "*A* and *C*" and "*A* and *D*" using the β-factor model with $\beta = 0.05$.

Assume *A*, *B*, *C*, and *D* are periodically tested (no test staggering) and with the following data:

Component	λ	T_t	T_o	f	T_r
A	0.0001	2	720	0.1	10
B	0.0005	2	720	0.1	10
C	0.0005	2	720	0.1	10
D	0.0005	2	720	0.1	10

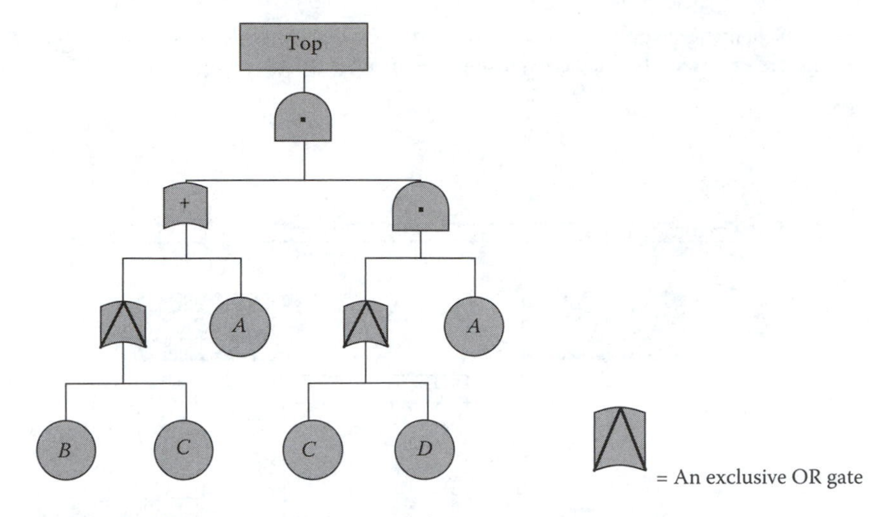

7.16 Consider the MLD (in the failed space) below.

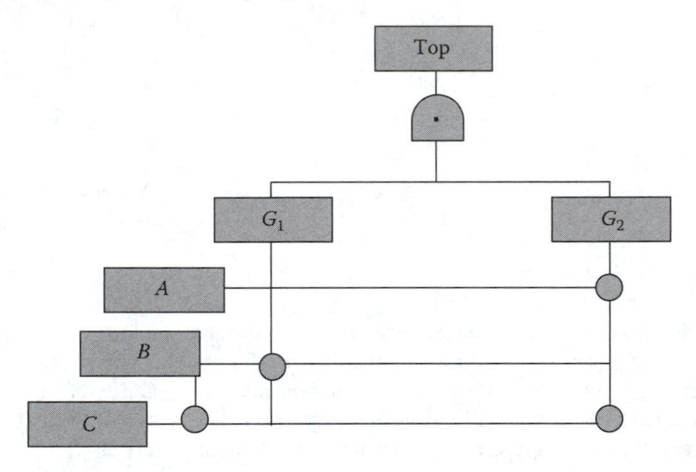

1. Write the *mutually excusive* cut sets.
2. If each component is distributed lognormally according to the table below, determine the *mean and standard deviation* of each mutually exclusive cut set (G_1 and G_2 are intermediate events).
3. Determine the probability that event T occurs.

Component	Mean	Standard Deviation
A	0.001	0.001
B	0.005	0.005
C	0.0001	0.0001

7.17 What is the MTBF in the future of a repairable component with the following observed times of failure (current time coincides with the occurrence of last failure)?

Failure Number	1	2	3	4	5	6
Interarrival of failures (h)	832	742	848	220	975	672

7.18 Time to failure of incandescent light bulbs can be described accurately with a lognormal distribution. A test engineer claims that a 10% increase in voltage decreases life by approximately 50%. A particular brand of 100-W bulb has a median life of 1200 h at 110 V.

a. Give an inverse power relationship expression for the life of such light bulbs as a function of voltage.
b. Calculate the time-acceleration factors for operating the light bulb at 140 volts.

7.19 A scenario is characterized by the following cut sets: $S = ab + ac + d$. Probability of a and b can be obtained from an exponential model such as $\Pr_a(t) = 1 - \exp(-\lambda_a t)$, but probability of c and d can be obtained from a Weibull model such as

$$\Pr_c(t) = 1 - \exp\left[-\left(\frac{t}{\alpha}\right)^{\beta}\right].$$

If $\lambda_a = \lambda_b = 10^{-4}\,\text{h}^{-1}$, $\alpha_c = \alpha_d = 2250\,\text{h}$, and $\beta_c = \beta_d = 1.4$, determine the importance of parameters α, β, and λ at a given time T in modeling this scenario. Use an importance measure of your choice, and justify your choice.

REFERENCES

1. Leemis, L. M., *Reliability: Probabilistic Models and Statistical Methods*, Prentice-Hall, Englewood Cliffs, NJ, 1995.
2. Klinger, D. J., Y. Nakada, and M. Menendez, eds, *AT&T Reliability Manual*, Van Nostrand Reinhold, New York, 1990.
3. Cox, D. R. and D. Oaks, *The Analysis of Survival Data*, Chapman & Hall, London, New York, 1984.
4. Barlow, R. E. and F. Proschan, *Statistical Theory of Reliability and Life Testing: Probability Models*, To Begin With, Silver Spring, MD, 1981.
5. Crowder, M. J., A. C. Kimber, R. L. Smith, and T. J. Sweeting, *Statistical Analysis of Reliability Data*, Chapman & Hall, London, New York, 1991.
6. Nelson, W., *Accelerated Testing: Statistical Models, Test Plans and Data Analysis*, Wiley, New York, 1990.
7. Sobczyk, K. and B. F. Spencer, Jr., *Random Fatigue: From Data to Theory*, Academic Press, New York, 1992.
8. Goldman, A. Ya, *Prediction of the Deformation Properties of Polymeric and Composite Materials*, American Chemical Society, Washington, DC, 1994.
9. Nelson, W., *Applied Life Data Analysis*, Wiley, New York, 1982.
10. Chan, C. K., A proportional hazard approach to SiO_2 breakdown voltage, *IEEE Trans. Reliab.*, R-39, 147–150, 1990.
11. Kaminskiy, M., I. Ushakov, and J. Hu, Statistical inference concepts, in *Product Reliability, Maintainability, and Supportability Handbook*, M. Pecht, ed., CRC Press, New York, 1995.
12. Krivtsov, V. V., D. E. Tananko, and T. P. Davis, A regression approach to tire reliability analysis, *Reliab. Eng. Syst. Saf.*, 78 (3), 267–273, 2002.
13. Mosleh, A. et al., *Procedure for Treating Common Cause Failures in Safety and Reliability Studies*, U.S. Nuclear Regulatory Commission, NUREG/CR-4780, Vol. I and II, Washington, DC, 1988.
14. Fleming, K. N., A reliability model for common mode failures in redundant safety systems, *Proceeding of the 6th Annual Pittsburgh Conference on Modeling and Simulations*, Instrument Society of America, Pittsburgh, PA, 1975.
15. Fleming K. N., A. Mosleh, and R. K. Deremer, A systematic procedure for the incorporation of common cause event, into risk and reliability models, *Nucl. Eng. Des.*, 58, 415–424, 1986.
16. Mosleh, A. and N. O. Siu, A Multi-parameter, event-based common-cause failure model, *Proceedings of the 9th International Conference on Structural Mechanics in Reactor Technology*, Lausanne, Switzerland, 1987.
17. Mosleh, A., Common cause failures: An analysis methodology and examples, *Reliab. Eng. Syst. Saf.*, 34, 249–292, 1991.
18. Atwood, C. L., *Common Cause Failure Rates for Pumps*, NUREG/CR-2098, U.S. Nuclear Regulatory Commission, Washington, DC, 1983.
19. Mosleh, A., N. Siu, C. Smidts, and C. Lui, *Model Uncertainty: Its Characterization and Quantification*, International Workshop Series on Advanced Topics in Reliability and Risk Analysis, Center for Reliability Engineering, University of Maryland, College Park, MD, 1995.

20. Morgan, M. G. and M. Henrion, *Uncertainty: A Guide to Dealing with Uncertainty in Quantitative Risk and Policy Analysis*, Cambridge University Press, Cambridge, 1990.
21. Morchland, J. D. and G. G. Weber, A moments method for the calculation of confidence interval for the failure probability of a system, *Proceeding of the 1972 Annual Reliability and Maintainability Symposium*, pp. 505–572, 1972.
22. Apostolakis, G. and V. T. Lee, Methods for the estimation of confidence bounds for the top event unavailability of fault trees, *Nucl. Eng. Des.*, 41, 411–419, 1977.
23. Martz, H. F., *A Comparison of Methods for Uncertainty Analysis of Nuclear Plant Safety System Fault Tree Models*, U.S. Nuclear Regulatory Commission and Los Alamos National Laboratory, NUREG/CR-3263, Los Alamos, NM, 1983.
24. Dezfuli, H. and M. Modarres, Uncertainty analysis of reactor safety systems with statistically correlated failure data, *Reliab. Eng. J.*, 11 (1), 47–64, 1984.
25. Lloyd, D. K. and M. Lipow, *Reliability: Management, Methods and Mathematics*, Prentice Hall, Englewood Cliff, NJ, 1962.
26. Ushakov, I. A., ed., *Handbook of Reliability Engineering*, Wiley, New York, 1994.
27. Marz, H. F. and B. S. Duran, A comparison of three methods for calculating lower confidence limits on system reliability using binomial component data, *IEEE Trans. Reliab.*, R-34 (2), 113–121, 1985.
28. Efron, B. A. and R. J. Tibshirani, *An Introduction to the Bootstrap*, Chapman & Hall, London, New York, 1979.
29. Efron, B. A. and R. J. Tibshirani, *An Introduction to the Bootstrap*, Chapman and Hall, New York, 1993.
30. Tukey, *Protection Against Depletion of Stratospheric Ozone by Chlorofluorocarbons*, Report by the Committee on Impacts of Stratospheric Change and the Committee on Alternative for the Reduction of Chlorofluorocarbon Emission, National Research Council, Washington, DC, 1979.
31. Reactor Safety Study, *An Assessment of Accidents in U.S. Commercial Nuclear Power Plants*, U.S. Regulatory Commission, WASH-1400, Washington, DC, 1975.
32. IEEE Standard-500, *IEEE Guide to the Collection and Presentation of Electrical, Electronic and Sensing Component Reliability Data for Nuclear Powered Generation Stations*, Institute of Electrical and Electronic Engineers, Piscataway, NJ, 1984.
33. Severe Accident Risk, *An Assessment for Five U.S. Nuclear Power Plants*, U.S. Nuclear Regulatory Commission, NUREG-1150, Washington, DC, 1990.
34. Clemens, R. J. and R. L. Winkler, Unanimity and compromise among probability forecasters, *Manage. Sci.*, 36, 767–779, 1990.
35. Swain, A. D. and H. E. Guttman, *Handbook of Human Reliability Analysis with Emphasis on Nuclear Power Applications*, U.S. Nuclear regulatory Commission, NUREG/CR-1278, Washington, DC, 1983.
36. Dalkey, N. and O. Helmer, An experimental application of the Delphi method to the use of experts, *Manage. Sci.*, 9, 458–467, 1963.
37. Mosleh, A. and G. Apostolakis, *Combining Various Types of Data in Estimating Failure Rates*, Transaction of the 1983 Winter Meeting of the American Nuclear Society, San Francisco, CA, 1983.
38. Apostolakis, G., S. Kaplan, B. Garrick, and R. Duphily, Data specialization for plant specific risk studies, *Nucl. Eng. Des.*, 56, 321–329, 1980.
39. Kim, J. H., *A Bayesian Model for Aggregating Expert Opinions*, PhD Dissertation, University of Maryland, Department of Materials and Nuclear Engineering, College Park, MD, 1991.
40. Snaith, E. R., *The Correlation Between the Predicted and Observed Reliabilities of Components, Equipment and Systems*, National Center of Systems Reliability, U.K. Atomic Energy Authority, NCSR-R18, 1981.
41. Lichtenstein, S. B., B. Fischoff, and L. D. Phillips, Calibration of probabilities: The state of the art, in *Decision Making and Change in Human Affairs*, J. Jungerman and G. de Zeeuw, eds, D. Reidel, Dordrecht, Holland, 1977.
42. Slovic, P., B. Fischhoff, and S. Lichtenstein, Facts versus fears: Understanding perceived risk, in *Societal Risk Assessment*, R. C. Schwing and W. A. Albers, Jr, eds, Plenum Press, New York, 1980.
43. Apostolakis, G., Data analysis in risk assessment, *Nucl. Eng. Des.*, 71, 375–381, 1982.
44. Modarres, M., L. Chen, and M. Danner, A Knowledge-based approach to root-cause failure analysis, *Proceedings of the Expert Systems Applications for the Electric Power industry Conference*, Orlando, FL, 1989.

45. Chu, C., *Root Cause Guidebook: Investigation and Resolution of Power Plant Problems*, Failure Prevention, Inc., San Clemente, CA, 1989.
46. Hendrick, K. and L. Benner, *Investigating Accidents with STEP*, CRC Press, New York, 1986.
47. Hendrick, K., *Systematic Safety Training*, CRC Press, New York, 1990.
48. Ferry, T. S., *Modern Accident Investigation and Analysis*, 2nd Ed, Wiley, New York, 1988.

8 Risk Analysis

Risk analysis has three elements: risk assessment, risk management, and risk communication. There are many interactions and overlaps between these three main elements of risk analysis. The first element of risk analysis is assessment: the process through which the chance or frequency of a loss and the magnitude of the loss (consequence), by or to a system is measured. Risk management is the process through which the potential (likelihood or frequency) magnitude and contributors to risk (loss) are evaluated, minimized, and controlled. Risk communication is the process through which information about the nature of risk (loss) and consequences, the risk-assessment approach, and the risk-management options are exchanged and discussed between the decision makers and other stakeholders.

Risk analysis measures the potential and magnitude of any loss from or to a system. If there are adequate historical data on such losses, the risk analysis can be directly measured from the statistics of the actual loss. This approach is often used for cases in which data on such losses are readily available, such as car accidents, cancer risk, or the frequency of certain storms. When there is no event on the actual losses, the loss is "modeled" in the risk analysis; that is, the potential loss (i.e., the risk) is predicted. There are not many cases, especially for complex engineering systems, for which data on losses are available. Therefore, often we must model and predict the risk. Risk analysts attempt to measure the magnitude of a loss and the consequences associated with complex systems, including evaluation and firming up policies. Generally, there are three types of risk analysis: quantitative, qualitative, and a mix of the two. Each of these widely used methods has different purposes, strengths, and weaknesses.

Risk analysis is widely used by private and government agencies to support regulatory and resource allocation decisions. Risk assessment consists of two distinct phases: a qualitative step of identifying, characterizing, and ranking hazards; and a quantitative step of risk evaluation, which includes estimating the likelihood (e.g., frequencies) and consequences of hazard occurrence. After risk assessment, appropriate risk-management options can be devised and considered; risk-benefit or CBA may be performed; and risk-management policies may be formulated and implemented. The main goals of risk management are to minimize the occurrence of accidents by reducing the likelihood of their occurrence (e.g., minimizing hazard occurrence); to reduce the impacts of uncontrollable accidents (e.g., prepare and adopt emergency responses); and to transfer risk (e.g., via insurance coverage). The estimation of likelihood or frequency of hazard occurrence depends greatly on the reliability of the system's components, the system as a whole, and human–system interactions. These topics have been extensively addressed in previous chapters of this book. In this chapter, we discuss how the reliability evaluation methods addressed in the preceding chapters are used, collectively, in a risk assessment. We will also discuss some relevant topics that are not discussed in the previous chapters (e.g., risk perception).

8.1 DETERMINATION OF RISK VALUES

There are two major parts of risk assessment:

- Determining the likelihood (e.g., probability P_i or frequency of occurrence, F_i) of an undesirable event, E_i. Sometimes the likelihood estimates are generated from a detailed analysis of past experience and available historical data; sometimes they are judgmental estimates based on an expert's belief of the situation; and sometimes they are simply a best guess. This assessment of event likelihood can be useful, but the confidence in such estimates depends on the quality and quantity of the data and the methods used to determine event likelihood.

- Evaluating the consequence, C_i, of this hazardous event. The choice of the type of consequence may affect the acceptability threshold and the tolerance level for the risk.

Risk assessment, generally, consists of the following three steps, sometimes called the "Risk Triplet," (which is represented by Equation 1.4):

1. Selecting a specific hazardous reference event or scenario i (sequence or chain of events) for quantitative analysis (hazard identification).
2. Estimating the likelihood or frequency of event or scenario, F_i.
3. Estimating the consequences of the event or scenario, C_i.

In most risk assessments the likelihood of event or scenario, i, is expressed in terms of the probability of that event. Alternatively, a frequency per year or per event (in units of time) may be used. Consequence, C_i, is a measure of the impacts of event or scenario i. This measurement may be in the form of mission loss, payload damage, damage to property, number of injuries, number of fatalities, or dollar loss.

The results of the risk estimation are then used to interpret the various contributors to risk, which are compared, ranked, and placed in perspective. This process consists of

1. Calculating and graphically displaying a risk profile based on individual failure event risks, similar to the process presented in Figure 8.1. This method will be discussed in more detail in this section.
2. Calculating a total expected risk value R from

$$R = \sum_i F_i \times C_i. \tag{8.1}$$

Naturally, all the calculations described involve some uncertainties, approximations, and assumptions. Therefore, uncertainties must be considered explicitly, as discussed in Section 7.3. Using expected losses and the risk profile, one can evaluate the amount of investment that is reasonable to control risks, alternative risk-management decisions to avoid risk (i.e., to decrease the risk probability), and alternative actions to mitigate consequences. Therefore, the following two additional planning steps are usually included in risk analysis:

1. Identifying cost-effective risk-management alternatives.
2. Adopting and implementing risk-management methods.

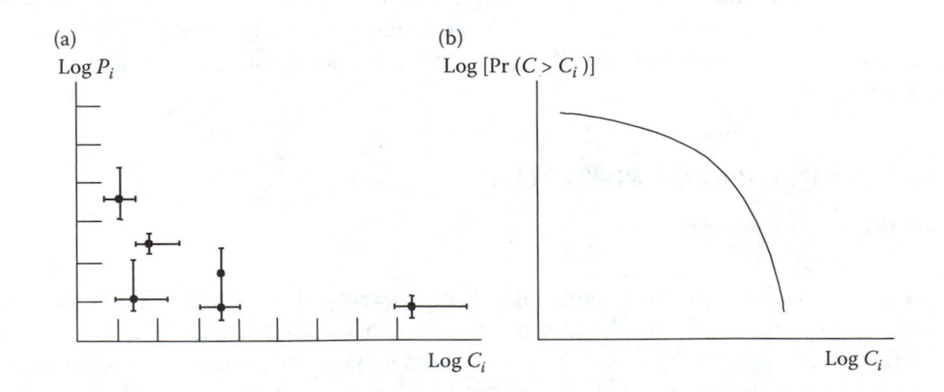

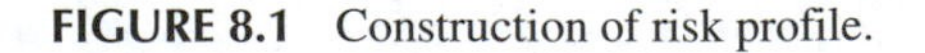

FIGURE 8.1 Construction of risk profile.

The risk estimation results are often shown in a general form similar to Equation 8.1. There are two useful ways to interpret such results: determining expected risk values, R_i, and constructing risk profiles. Both methods are used in quantitative risk analysis (Table 8.1).

Expected values are most useful when the consequences C_i are measured in financial terms or other directly measurable units. The expected risk value R_i (or expected loss) associated with event E_i is the product of its probability P_i and consequence values, as described by Equation 8.1. Thus, if the event occurs with a frequency of 0.01 per year, and if the associated loss is $1 million, then the expected loss (or risk value) is $R_i = 0.01 \times \$1{,}000{,}000 = \$10{,}000$. Conversely, if the frequency of event occurrence is 1 per year, but the loss is $10,000, the risk value is still $R_i = 1 \times \$10{,}000 = \$10{,}000$. Thus, the risk value for these two situations is the same; that is, both events are equally risky.

Since this is the expected annual loss, the total expected loss over 20 years (assuming a constant dollar value) would be $200,000. This assumes that the parameters do not vary significantly with time, and ignores the low probability of multiple losses over the period. Equation 8.1 can be used to obtain the total expected loss per year for a whole set of possible events. This expected loss value assumes that all events or scenarios contributing to risk exposure have equal weight. Occasionally, for risk decisions, value factors (weighting factors) are assigned to each event contributing to risk. The relative values of the terms associated with the different hazardous events give a useful measure of their relative importance, and the total risk value can be interpreted as the average or "expected" level of loss over a period of time.

As discussed earlier, another method for interpreting the results is to construct a risk profile. With this method, the probability values are plotted against the consequence values. Figure 8.1 illustrates these methods. Figure 8.1a shows the use of logarithmic scales, which are usually used because they can cover a wide range of values. The error brackets denote epistemic uncertainties in the probability estimate (vertical) and the consequences (horizontal). This approach provides a means of easily illustrating events with high probability, high consequence, or high uncertainty. It is useful when discrete probabilities and consequences are known. Figure 8.1b shows the construction of the complementary cumulative probability risk profile (sometimes known as a Farmer's curve [1]). In this case, the logarithm of the probability that the total consequence C exceeds C_i is plotted against the logarithm of C_i. The most notable application of this method was in the landmark Reactor Safety Study [2]. With this method, the low-probability/high-consequence risk values and high-probability/low-consequence risk values can be easily seen. That is, the extreme values of the estimated risk can be easily displayed.

TABLE 8.1
General Form of Output from the Analytic Phase of Risk Assessment

Undesirable Event/Scenario	Likelihood or Frequency	Consequences	Risk Level
E_1	F_1	C_1	$R_1 = F_1C_1$
E_2	F_2	C_2	$R_2 = F_2C_2$
E_3	F_3	C_3	$R_3 = F_3C_3$
.	.	.	.
.	.	.	.
.	.	.	.
E_n	F_n	C_n	$R_n = F_nC_n$

8.2 FORMALIZATION OF QUANTITATIVE RISK ASSESSMENT

The hazardous events E_i discussed in the previous section can occur as a result of a chain of some events. In combination, these events are called a "scenario." The risk-assessment process is therefore primarily one of scenario development, with the risk contribution from each possible scenario that leads to the outcome or event of interest. This concept is described in terms of the triplet represented by Equation 1.4. Because the risk-assessment process focuses on scenarios that lead to hazardous events, the general methodology allows the identification of all possible scenarios, calculation of their individual probabilities, and a consistent description of the consequences that result from each scenario. Scenario development requires a set of descriptions of how a barrier confining a hazard is threatened, how the barrier fails, and the effects on the subject when it is exposed to the uncontained hazard. This means that one needs to formally address the items described below.

8.2.1 Identification of Hazards

A survey of the process under analysis should be performed to identify the hazards of concern. These hazards can be categorized as follows:

- Chemical hazard (e.g., toxic chemicals released from a chemical process)
- Thermal hazard (e.g., a high-energy explosion from a chemical reactor)
- Mechanical hazard (e.g., kinetic or potential energy from a moving object)
- Electrical hazard (e.g., potential difference, electrical and magnetic fields, and electrical shock)
- Ionizing radiation (e.g., radiation released from a nuclear plant)
- Nonionizing radiation (e.g., radiation from a microwave or the sun)
- Biological hazard (e.g., genetically engineered food, resistance to antibiotics, and insects in imported products).

Presumably, each of these hazards will be part of the process, and normal process boundaries will be used as their containment. This means that, provided there is no disturbance in the process, the barrier that contains the hazard will be unchallenged. However, in a risk scenario one postulates the challenges to such barriers and tries to estimate the probability of these challenges.

8.2.2 Identification of Barriers

Each of these identified hazards must be examined to determine all the physical barriers that contain it or that can intervene to prevent or minimize exposure to the hazard. These barriers may physically surround the hazard (e.g., walls, pipes, valves, fuel clad, and structures); they can be based on a specified distance from a hazard source to minimize exposure to the hazard (e.g., to minimize exposure to radioactive materials); or they may provide direct shielding of the subject from the hazard (e.g., protective clothing and bunkers).

8.2.3 Identification of Challenges to Barriers

After identification of the individual barriers, a concise definition of the requirements for maintaining each one follows. Hierarchical analytical models provide a means to express these requirements. One can also simply identify what is needed to maintain the integrity of each barrier. Barrier failure is due to the degradation of strength of the barrier and high stress on the barrier.

Barrier strength degrades because of

- Reduced thickness (due to deformation, erosion, or corrosion)
- Changes in material properties (e.g., toughness and yield strength). Material properties may be affected by the local environment, e.g., temperature).

Stress on the barrier increases with

- Internal forces or pressure
- Penetration or distortion by external objects or forces.

The above causes of degradation are often the result of one or more of the following conditions:

- Malfunction of process equipment (e.g., the ECS in a nuclear plant)
- Problems with man–machine interface
- Poor design or maintenance
- Adverse natural phenomena
- Adverse human-made environment.

8.2.4 Estimation of Hazard Exposure

The next step in the risk-assessment procedure is to define those scenarios in which the barriers may be breached, and then make the best possible estimate of the probability or frequency for each sequence. Those scenarios that pose similar levels of hazard under similar conditions of hazard dispersal are grouped together, and the probabilities or frequencies of the respective event sequences associated with these groups are determined.

8.2.5 Consequences of Evaluation

The range of effects produced by exposure to the hazard may encompass harm to people, damage to equipment, and contamination of land or facilities. These effects are evaluated using the knowledge of the toxic behavior of the particular material(s) and the specific outcomes of the scenarios considered. In the case of the dispersal of toxic materials, the size of the release is combined with the potential dispersion mechanisms to calculate the outcome.

From the generic nature of risk analysis, there appears to be a common approach to understanding the ways in which hazard exposure occurs. This understanding is the key in the development of logical scenario models that can then be solved. Quantitative and qualitative solutions can provide estimates of barrier adequacy and methods of effective enhancement. This formalization provides a basis from which we can describe a commonly used practice in risk analysis called probability risk assessment (PRA). This technique, pioneered by the nuclear industry, is the basis of many formal risk assessments today. We describe this approach in Section 8.3 and provide two examples in Sections 8.4 and 8.5.

8.3 PROBABILITY RISK ASSESSMENT

PRA is a systematic procedure for investigating the ways in which complex systems are built and operated. The PRAs model how human, software, and hardware elements of the system interact with each other. Also, they assess the most significant contributors to the risks of the system and determine the value of the risk. PRA involves estimation of the degree or probability of loss. A formal definition proposed by Kaplan and Garrick [3] provides a simple and useful description of the elements of risk assessment, which involves addressing three basic questions:

1. What can go wrong that could lead to exposure of hazards?
2. How likely is this to happen?
3. If it happens, what consequences are expected?

The PRA procedure involves quantitative application of the above triplet in which probabilities (or frequencies) of scenarios of events leading to exposure of hazards are estimated and the corresponding magnitudes of health, safety, environmental, and economic consequences for each scenario are predicted. The risk value (i.e., expected loss) of each scenario is often measured as the product of the scenario frequency and its consequences. The most significant result of the PRA is not the actual value of the risk computed (the so-called bottom-line number); rather, it is the determination of the system elements that substantially contribute to the risks of that system, the uncertainties associated with such estimates, and the effectiveness of various risk-reduction strategies available. That is, the primary value of a PRA is to highlight the system design and operational deficiencies and to optimize resources that can be invested on improving the design and operation of the system.

In the remainder of this section, some strengths of PRA will be presented first, and subsequently major elements of PRA will be discussed.

8.3.1 Strengths of PRA

The most important strengths of the PRA as the formal engineering approach to risk assessment are as follows:

1. PRA provides an integrated and systematic examination of a broad set of design and operational features of an engineered system.
2. PRA incorporates the influence of system interactions and human–system interfaces.
3. PRA provides a model for incorporating operating experience with the engineered system and for updating risk estimates.
4. PRA provides a process for the explicit consideration of uncertainties.
5. PRA permits the analysis of competing risks (e.g., of one system vs. another or of possible modifications to an existing system).
6. PRA permits the analysis of (assumptions and data) issues via sensitivity studies.
7. PRA provides a measure of the absolute or relative importance of systems and components to the calculated risk value.
8. PRA provides a quantitative measure of the overall level of health and safety for the engineered system.

Major errors may result from weak or absent models or associated data of potentially important factors in the risk of the system, including cases in which

1. Initiating events occur very infrequently
2. Human performance models and interactions with the system are highly uncertain
3. Failures occurring from a CCF, such as an extreme operating environment, are difficult to identify and model.

8.3.2 Steps in Conducting a PRA

The following subsections provide a discussion of essential components of PRA as well as the steps that must be performed in a PRA analysis. *The NASA PRA Guide* by Stamatelatos et al. [4] describes the components of the PRA as shown in Figure 8.2. Each component of PRA is discussed in more detail below.

8.3.2.1 Objectives and Methodology

Preparing for a PRA begins with a review of the objectives of the analysis. Among the many objectives that are possible, the most common ones include design improvement, risk acceptability, decision support, regulatory and oversight support, and operations and life management. Once the objectives are clarified, an inventory of possible resources for the desired analyses should be developed. The

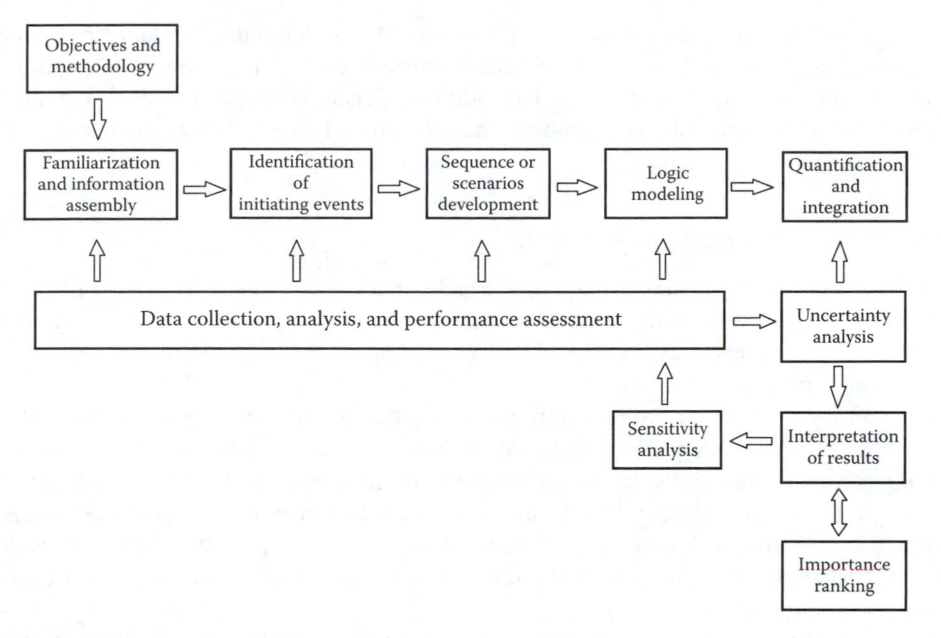

FIGURE 8.2 Components of the overall PRA process [4].

available resources range from required computer codes to system experts and analytical experts. This, in essence, provides a road map for the analysis. The resources required for each analytical method should be evaluated, and the most effective option selected. The basis for the selection should be documented and the selection process reviewed to ensure that the objectives of the analysis will be adequately met. See [5] for inventories of the methodological approaches to PRA. Also see ASME [6] for a standard on performing PRA for nuclear power plants.

8.3.2.2 Familiarization and Information Assembly

A general knowledge of the physical layout of the overall system (e.g., facility, design, process, aircraft, or spacecraft), administrative controls, maintenance and test procedures, as well as barriers and subsystems, whose job it is to protect, prevent, or mitigate hazard exposure conditions, is necessary to begin the PRA. All subsystems, structures, locations, and activities expected to play a role in the initiation, propagation, or arrest of a hazard exposure condition must be understood in sufficient detail to construct the models necessary to capture all possible scenarios. A detailed inspection of the overall system must be performed in the areas expected to be of interest to the analysis. The following items should be performed in this step:

1. Major critical barriers, structures, emergency safety systems, and human interventions should be identified.
2. Physical interactions among all major subsystems (or parts of the system) should be identified and explicitly described. The result should be summarized in a dependency matrix.
3. Past major failures and abnormal events that have been observed in the facility should be noted and studied. Such information would help ensure inclusion of important applicable scenarios.
4. Consistent documentation is critical to ensure the quality of the PRA. Therefore, a good filing system must be created at the outset and maintained throughout the study.

With the help of the designers, operators, and owners, the analysts should determine the ground rules for the analysis, the scope of the analysis, and the configuration and phases of the operation of the overall system to be analyzed. The analysts should also determine the faults and conditions to be included or excluded, the operating modes of concern, and the hardware configuration on the design freeze date (i.e., the date after which no additional changes in the overall system design and configuration will be modeled). The results of the PRA are only applicable to the overall system at the freeze date.

8.3.2.3 Identification of Initiating Events

This task involves identifying those events (abnormal events or conditions) that could, if not correctly and promptly responded to, result in hazard exposure. The first step involves identifying sources of hazard and barriers around these hazards. The next step involves identifying events that can lead to a direct threat to the integrity of the barriers.

A system may have one or more operational modes that produce its output. In each operational mode, specific functions are performed. Each function is directly realized by one or more systems by making certain actions and behaviors. These systems, in turn, are composed of more basic units (e.g., subsystems, components, and hardware) that accomplish the objective of the system. As long as a system is operating within its design parameter tolerances, there is little chance of challenging the system boundaries in such a way that hazards will escape those boundaries. These operational modes are called normal operation modes.

During normal operation mode, loss of certain functions or systems will cause the process to enter an off-normal (transient) state transition. Once in this transition, there are two possibilities. First, the state of the system could be such that no other function is required to maintain the process in a safe condition. (Safe refers to a mode where the chance of exposing hazards beyond the system boundaries is negligible.) The second possibility is a state wherein other functions (and thus systems) are required to prevent exposing hazards beyond the system boundaries. For this second possibility, the loss of the function or the system is considered an initiating event. Since such an event is related to the normally operating equipment, it is called an *operational initiating event*.

Operational initiating events can also apply to various modes of the system (if they exist). The terminology remains the same since, for each mode, certain equipment, people, or software must be functioning. For example, an operational initiating event found during the PRA of a test nuclear reactor was Low Primary Coolant System Flow. Flow is required to transfer heat produced in the reactor to heat exchanges and ultimately to the cooling towers and the outside environment. If this coolant flow function is reduced to the point where an insufficient amount of heat is transferred, core damage could result (and thus the possibility of exposing radioactive materials—the main source of hazard in this case). Therefore, another system must operate to remove the heat produced by the reactor (i.e., a protective barrier). By definition, then, the Low Primary Coolant System Flow is an operational initiating event.

One method for determining the operational initiating events begins with first drawing a functional block diagram of the system. From the functional block diagram, a hierarchical relationship is produced, with the process objective being successful completion of the desired system. Each function can then be decomposed into its subsystems and components, and can be combined in a logical manner to represent operations needed for the success of that function.

Potential initiating events are events that result in failures of particular functions, subsystems, or components, the occurrence of which causes the overall system to fail. These potential initiating events are "grouped" such that members of a group require similar subsystem responses to cope with the initiating event. These groupings are the operational initiator categories.

An alternative to the use of functional hierarchy for identifying initiating events is the use of failure mode and effect analysis (FMEA) (see [7]). The difference between these two methods is that the functional hierarchies are deductively and systematically constructed, whereas FMEA is an inductive and experiential technique. The use of FMEA for identifying initiating events consists of identifying failure events (modes of failures of equipment, software, and human) whose effect is a threat to the integrity and availability of the hazard barriers of the system. In both methods, one can always supplement the set of initiating events with generic initiating events (if known). For example, see [8] for these initiating events for nuclear reactors, and the *NASA Guide* [4] for space vehicles.

To simplify the process, after identifying all initiating events, it is necessary to combine those initiating events that pose the same threat to hazard barriers and require the same mitigating functions of the process to prevent hazard exposure.

The following inductive procedures should be followed when grouping initiating events:

1. Combine the initiating events that directly break all hazard barriers.
2. Combine the initiating events that break the same hazard barriers (not necessarily all the barriers).
3. Combine the initiating events that require the same group of mitigating human or automatic actions following their occurrence.
4. Combine the initiating events that simultaneously disable the normal operation as well as some of the available mitigating human, software, or automatic actions.

Events that cause off-normal operation of the overall system and require other systems to operate so as to maintain hazards within their desired boundaries, but that are not directly related to a hazard mitigation, protection, or prevention function, are called *nonoperational initiating events*. Nonoperational initiating events are identified with the same methods used to identify operational events. One class of such events of interest is those that are primarily external to the overall system or facility. These "external events" will be discussed later in more detail in this chapter. The following procedures should be followed in this step of the PRA:

1. Select a method for identifying specific operational and nonoperational initiating events. Two representative methods are functional hierarchy and FMEA. If a generic list of initiating events is available, it can be used as a supplement.
2. Using the method selected, identify a set of initiating events.
3. Group the initiating events having the same effect on the system. For example, those requiring the same mitigating functions to prevent hazard exposure are grouped together.

8.3.2.4 Sequence or Scenario Development

The goal of scenario development is to derive a complete set of scenarios that encompasses all of the potential exposure propagation paths that can lead to loss of containment or confinement of the hazards, following the occurrence of an initiating event. To describe the cause-and-effect relationship between initiating events and subsequent event progression, it is necessary to identify those functions (e.g., safety functions) that must be maintained to prevent loss of hazard barriers. The scenarios that describe the functional response of the process to the initiating events are frequently displayed by event trees. The event tree development techniques are discussed by Kumamoto and Henley [5].

Event trees order and depict (in an approximately chronological manner) the success or failure of key mitigating actions (e.g., human actions or mitigative hardware actions) that are required to act in response to an initiating event. In PRA, two types of event trees can be developed: functional and systemic. The functional event tree uses mitigating functions as its heading. The main purpose of the functional tree is to better understand the scenario of events at an abstract level, following the occurrence of an initiating event. The functional tree also guides the PRA analyst in the development of a more detailed systemic event tree. The systemic event tree reflects the scenarios of specific events (specific human actions, protective or mitigative subsystem operations, or failures) that lead to a hazard exposure. That is, the functional event tree can be further decomposed to show failure of specific hardware, software, or human actions that perform the functions described in the functional event tree. Therefore, a systemic event tree fully delineates the overall system response to an initiating event and serves as the main tool for further analyses in the PRA. For a detailed discussion on specific tools and techniques used for this purpose, see [9].

There are two kinds of external events. The first kind refers to events that originate from within the facility or the overall system (but outside of the physical boundary of the facility), which are called internal events external to the process of the system. Events that adversely affect the facility or overall system and occur external to its physical boundaries, but which can still be considered as part of the system, are defined as internal events external to the system. Typical internal events external to the system are internal conditions, such as fires from fuel stored within a facility or floods caused by the rupture of tank that is part of the overall system. The effects of these events should be modeled with event trees to show all possible scenarios.

The second kind of external events are those that originate outside of the overall system. These are called external events. Examples of external events are fires and floods that originate from outside of the system. Examples include seismic events, extreme heat, extreme drought, transportation events, volcanic events, high-wind events, terrorism, and sabotage. Again, this classification can be used in developing and grouping the event tree scenarios.

The following procedures should be followed in this step of the PRA:

1. Identify the mitigating functions for each initiating event (or group of events).
2. Identify the corresponding human actions, systems, or hardware operations associated with each function, along with their necessary conditions for success.
3. Develop a functional event tree for each initiating event (or group of events).
4. Develop a systemic event tree for each initiating event, delineating the success conditions, the initiating event progression phenomena, and the end effect of each scenario.

For specific examples of scenario development, see [4] and [5].

8.3.2.5 Logic Modeling

Event trees commonly involve branch points at which a given subsystem (or event) either works (or happens) or does not work (or does not happen). Sometimes, failure of these subsystems (or events) is rare, and there may not be an adequate record of observed failure events to provide a historical basis for estimating frequency of their failure. In such cases, other logic-based analysis methods, such as fault trees or MLDs, may be used, depending on the accuracy desired. The most common method used in PRA to calculate the probability of subsystem failure is FTA. This analysis involves developing a logic model in which the subsystem is broken down into its basic components or segments for which adequate data exist. For more details about how a fault tree can be developed to represent the event headings of an event tree, see Section 4.3.

Different event tree modeling approaches imply variations in the complexity of the logic models that may be required. If only main functions or systems are included as event tree headings, the fault trees become more complex and must accommodate all dependencies among the main and support functions (or subsystems) within the fault tree. If support functions (or systems) are explicitly included as event tree headings, more complex event trees and less complex fault trees will result. For more discussions on methods and techniques used for logic modeling, see Sections 4.2 through 4.4.

The following procedures should be followed as a part of developing the fault tree:

1. Develop a fault tree for each event in the event tree heading for which actual historical failure data does not exist.
2. Explicitly model dependencies of a subsystem on other subsystems and intercomponent dependencies (e.g., CCFs). For CCFs, see Mosleh et al. [10].
3. Include all potential reasonable and probabilistically quantifiable causes of failure, such as hardware, software, test and maintenance, and human errors, in the fault tree.

The following steps should be followed in the dependent failure analysis:

1. Identify the hardware, software, and human elements that are similar and could cause dependent or CCFs. For example, similar pumps, motor-operated valves, air-operated valves, human actions, software routines, diesel generators, and batteries are major components in process plants, and are considered important sources of CCFs.
2. Explicitly incorporate items that are potentially susceptible to CCF into the corresponding fault trees and event trees of the PRA where applicable.
3. Functional dependencies should be identified and explicitly modeled in the fault trees and event trees.

8.3.2.6 Failure Data Collection, Analysis, and Performance Assessment

A critical building block in assessing the reliability and availability of a complex system is the data on the performance of its barriers to contain hazards. In particular, the best resources for predicting future availability are past field experiences and tests. Hardware, software, and human reliability data are inputs to assess performance of hazard barriers, and the validity of the results depends highly on the quality of the input information. It must be recognized, however, that historical data have predictive value only to the extent that the conditions under which the data were generated remain applicable. Collection of the various failure data consists fundamentally of the following steps: collecting generic data, assessing generic data, statistically evaluating facility- or overall system-specific data, and developing failure probability distributions using test and/or facility- and system-specific data. Three types of events identified during the risk scenario definition and system modeling must be quantified for the event trees and fault trees to estimate the frequency of occurrence of sequences: initiating events, component failures, and human error.

Quantifying initiating events and hazard barriers and component failure probabilities involves two separate activities. First, the probabilistic failure model for each barrier or component failure event must be established; then the parameters of the model must be estimated. Typically the necessary data include time of failures, repair times, test frequencies, test downtimes, and CCF events. Further uncertainties associated with such data must also be characterized. Kapur and Lamberson [11] and Nelson [12] discuss available methods for analyzing data (see Sections 7.2 and 7.3) to obtain the probability of failure or the probability of occurrence of equipment failure. Also, Crow [13] and Ascher and Feingold [14] discuss analysis of data relevant to repairable systems. Finally, Mosleh et al. [10] discuss analysis of data for dependent failures, Poucet [15] reviews human reliability issues, and Smidts [16] examines SRMs. The databases generally include facility- or system-specific data combined with generic performance data when specific data are absent or sparse. For example, Section 3.7 describes generic data for electrical, electronic, and mechanical equipment.

To attain the very low levels of risk, the systems and hardware that comprise the barriers to hazard exposure must have very high levels of performance. This high performance is typically achieved through well-designed systems with adequate margins of safety in terms of uncertainties and redundancy and/or diversity in hardware, which provides multiple success paths. The problem then becomes one of ensuring the independence of the paths, since there is always some degree of coupling between agents of failures, such as those activated by failure mechanisms, either through the operating environment (events external to the system) or through functional and spatial dependencies. These dependencies should be carefully included in both the event tree and fault tree in the PRA. As the reliability of individual subsystems increases due to redundancy, the contribution from dependent failures becomes more important; in certain cases, dependent failures may dominate the value of overall reliability. Including the effects of dependent failures in the reliability models used in the PRA is a difficult process and requires sophisticated, fully integrated models to account for unique failure combinations that lead to failure of subsystems and ultimately exposure of hazards. The treatment of dependent failures is not a single step performed during the PRA; it must be considered throughout the analysis (e.g., in event trees, fault trees, and human reliability analyses).

The following procedures should be followed as part of the data analysis task:

1. Determine generic values of material strength or endurance, load or damage agents, failure times, failure occurrence rate, and failures on demand for each item (hardware, human action, or software) identified in the PRA models. This can be obtained either from facility- or system-specific experiences, from generic sources of data, or both.
2. Gather data on hazard barrier tests, repair, and maintenance data primarily from experience, if available. Otherwise use generic performance data.
3. Assess the frequency of initiating events and other probability of failure events from experience, expert judgment, or generic sources.
4. Determine the dependent or CCF probability for similar items, primarily from generic values. However, when significant specific data are available, they should be primarily used.

8.3.2.7 Quantification and Integration

Fault trees and event trees are integrated and their events are quantified to determine the frequencies of scenarios and associated uncertainties in the calculation of the final risk values. This integration depends somewhat on the manner in which system dependencies have been handled. We will describe the more complex situation, in which the fault trees are dependent—that is, in which there are physical dependencies (e.g., through support units of the main hazard barriers such as those providing motive, proper working environment and control functions).

Normally, the quantification will use a Boolean reduction process to arrive at a Boolean representation for each scenario. Starting with fault tree models for the various systems or event headings in the event trees, and using probabilistic estimates for each of the events modeled in the event trees and fault trees, the probability of each event tree heading (often representing failure of a hazard barrier) is calculated (if the heading is independent of other headings). The fault trees for the main subsystems and support units (e.g., lubricating and cooling units, power units) are merged where needed, and the equivalent Boolean expression representing each event in the event tree model is calculated. The Boolean expressions are reduced to arrive at the smallest combination of basic failure events (the so-called minimal cut sets) that lead to exposure of the hazards. These minimal cut sets for each of the main subsystems (barriers), which are often identified as headings on the event trees, are also obtained. The minimal cut sets for the event tree headings are then appropriately combined to determine the cut sets for the event tree scenarios. If possible, all minimal cut sets must be generated and retained during this process; unfortunately in complex systems and facilities this leads to an unmanageably large collection of terms and a combinatorial outburst. Therefore, the collection of cut sets is often truncated (i.e., probabilistically small and insignificant cut sets are discarded based on the number of terms in a cut set or on the probability of the cut set). This is usually a practical necessity because of the overwhelming number of cut sets that can result from the combination of a large number of failures, even though the probability of any of these combinations may be vanishingly small. The truncation process does not disturb the effort to determine the dominant scenarios, since we are discarding scenarios that are extremely unlikely.

Even though the discarded cut sets may individually be several orders of magnitude less probable than the average of those retained, the large number of them may add up to a significant part of the risk. The actual risk might thus be larger than what the PRA results indicate. This additional risk can be discussed as part of the modeling uncertainty characterization. A detailed examination of a few PRA studies of very complex systems, such as nuclear power plants, shows that cut set truncation will not introduce any significant error in the total risk-assessment results (see [17]).

Other methods for evaluating scenarios also exist, which directly estimate the frequency of the scenario without specifying cut sets. These methods are often used in highly dynamic systems whose configuration changes as a function of time, leading to dynamic event trees and fault trees. For more discussion on these systems, see [4,18,19]. Employing advanced computer programming concepts, one may directly simulate the operation of parts to mimic the real system for reliability and risk analysis (see [20]). The following procedures should be followed as part of the quantification and integration step in the PRA:

1. Merge corresponding fault trees associated with each failure or success event modeled in the event tree scenarios (i.e., combine them in a Boolean form). Develop a reduced Boolean function for each scenario (i.e., truncated minimal cut sets).
2. Calculate the total frequency of each sequence, using the frequency of initiating events, the probability of barrier failure including contributions from test and maintenance frequency (outage), CCF probability, and HEP.
3. Use the minimal cut sets of each sequence for the quantification process. If needed, simplify the process by truncating based on the cut sets or probability.
4. Calculate the total frequency of each scenario.
5. Calculate the total frequency of all scenarios of all event trees.

8.3.2.8 Uncertainty Analysis

Uncertainties are part of any assessment, model, or estimation. In engineering calculations, we routinely ignore the estimation of uncertainties associated with failure models and parameters, either because the uncertainties are very small or, more often, the analyses are done conservatively (e.g., by using high safety factors or design margins). Since PRAs are primarily used for decision making and management of risk, it is critical to incorporate uncertainties in all facets of the PRA.

Also, risk-management decisions that consider PRA results must consider estimated uncertainties. In PRAs, uncertainties are primarily shown in the form of probability distributions. For example, the probability of failure of a subsystem (e.g., a hazard barrier) may be represented by a probability distribution showing the range and likelihood of risk values.

The process involves characterization of the uncertainties associated with the frequency of initiating events, the probability of failure of subsystems (or barriers), the probability of all event tree headings, the strength or endurance of barriers, the applied load or incurred damage by the barriers, the amount of hazard exposures, the consequences of exposures to hazards, and the sustained total amount of losses. Other sources of uncertainties are in the models used, such as fault tree and event tree models, stress-strength and damage-endurance models used to estimate failure or capability of barriers, probabilistic failure models of hardware, software and human action, correlations between amount of hazard exposure and the consequence, exposure models and pathways, and models to treat inter- and intrabarrier failure dependencies. Another important source of uncertainty is incompleteness of the risk models and other failure models used in the PRAs, such as the level of detail used in decomposing subsystems using fault tree models, the scope of the PRA, and the omission of certain scenarios in the event tree simply because they are not known or have not been experienced before.

Once uncertainties associated with hazard barriers have been estimated and assigned to models and parameters, they must be "propagated" through the PRA model to find the uncertainties associated with the results of the PRA (see Sections 7.2 and 7.3), primarily with the bottom-line risk calculations, and with the list of risk of significant elements of the system. Propagation is done using one of several techniques, but the most popular method used is Monte Carlo simulation. The results are then shown and plotted in form of probability distributions. Steps in uncertainty analysis include the following:

1. Identify models and parameters that are uncertain and the method of uncertainty estimation to be used for each.

2. Describe the scope of the PRA and the significance and contribution of elements that are not modeled or considered.
3. Estimate and assign probability distributions depicting model and parameter uncertainties in the PRA.
4. Propagate uncertainties associated with the hazard barrier models and parameters to find the uncertainty associated with the risk value.

Present the uncertainties associated with risks and contributors to risk in a way that is easy to understand and visually straightforward to grasp. See Section 7.3 for more technical discussion on uncertainty analysis.

8.3.2.9 Sensitivity Analysis

Sensitivity analysis is the method for determining the significance of choice of a model or its parameters, the assumptions for including or not including a barrier, phenomenon or hazard, the performance of specific barriers, the intensity of hazards, and the significance of any highly uncertain input parameters or variables to the final risk value calculated. The process of sensitivity analysis is straightforward. The effects of the input variables and assumptions in the PRA are measured by modifying them one at a time by several folds, factors, or even one or more orders of magnitude, and measuring relative changes observed in the PRA's risk results. Those models, variables, and assumptions whose change leads to the highest change in the final risk values are determined as "sensitive." In such a case, revised assumptions, models, additional failure data, and more mechanisms of failure may be needed to reduce the uncertainties associated with sensitive elements of the PRA.

Sensitivity analysis helps focus resources and attention on those elements of the PRA that need better attention and characterization. A good sensitivity analysis strengthens the quality and validity of the PRA results. Usually elements of the PRA that could exhibit multiple impacts on the final results, such as certain phenomena (e.g., pitting corrosion, fatigue cracking, and CCF) and uncertain assumptions, are usually good candidates for sensitivity analysis. The steps involved in the sensitivity analysis are as follows:

1. Identify the elements of the PRA (including assumptions, failure probabilities, models, and parameters) that analysts believe might be sensitive to the final risk results.
2. Change the contribution or value of each sensitive item in either direction by several factors in the range of 2–100. Note that certain changes in the assumptions may require multiple changes of the input variables. For example, a change in the failure rate of similar equipment requires changing the failure rates of all the similar equipment in the PRA model.
3. Calculate the impact of the changes in step 2 one at a time and list the elements that are most sensitive.
4. Based on the results in step 3, propose additional data, any changes in the assumptions, the use of alternative models, or modification of the scope of the PRA analysis.

8.3.2.10 Risk Ranking and Importance Analysis

Ranking the elements of the system with respect to their risk or safety significance is one of the most important steps of a PRA. Ranking is simply arranging the elements of the system based on their increasing or decreasing contribution to the final risk value. Importance measures rank hazard barriers, subsystems, or more basic system elements usually based on their contribution to the total risk of the system. Ranking should be performed with much care. In particular, during the interpretation of the results, since formal importance measures are context dependent and their meaning varies depending on the intended application of the risk results, the choice of the ranking method is important.

There are several unique importance measures in PRAs. For example, the Fussell–Vesely [21], RRW, and RAW (see Section 6.4) measures are all appropriate measures for use in PRAs, and are all representatives of the level of contribution of various elements of the system as modeled in the PRA and entered in the calculation of the total risk of the system. For example, the Birnbaum importance measure represents changes in the total risk of the system as a function of changes in the basic event probability of one component at a time. If simultaneous changes in the basic event probabilities are being considered, a more complex representation would be needed.

Importance measures can be classified by their mathematical definitions. Some measures have fractional-type definitions and show changes (in the number of folds or factors) in system total risk under certain conditions with respect to the normal operating or use condition (e.g., as is the case in RRW and RAW measures). Some measures calculate the changes in the system total risk as the failure probability of hazard barriers and other elements of the system or conditions under which the system operates change. This difference can be normalized with respect to the total risk of the system or even expressed in a percentage change form. Other types of measures account for the rate of changes in the system risk with respect to changes in the failure probability of the elements of the system. These measures can be interpreted mathematically as partial derivatives of the risk as a function of failure probability of its elements (barriers, components, human actions, phenomena, etc.). For example, the Birnbaum measure falls under this category.

Another important set of importance measures focuses on ranking the elements of the system with the highest contribution to the total uncertainty of the risk results obtained from PRAs. This process is called "uncertainty ranking" and is different from component, subsystem, and barrier ranking. In this importance ranking, the analyst is interested only in knowing which of the system elements drive the final risk uncertainties, so that resources can be focused on reducing important uncertainties.

Another classification for importance measures can be divided into two major categories of absolute versus relative. Absolute measures express the fixed importance of one element of the system, independent of the importance of other elements, whereas relative importance measures express the significance of one element with respect to the importance of other elements. Absolute importance can be used to estimate the impact of component performance on the system regardless of how important other elements are, whereas relative importance estimates the significance of the risk impact of the component in comparison to the effect or contribution of others.

Absolute measures are useful when we speculate on improving actions, since they directly show the impact on the total risk of the system. Relative measures are preferred when resources or actions to improve or prevent failures are taken in a global and distributed manner. For additional discussions on the risk ranking methods and their implications in failure and success domains, see [22]. Applications of importance measures may be categorized into the following areas:

1. *(Re)design*: Supports decisions of the system design or redesign by adding or removing elements (barriers, subsystems, human interactions, etc.).
2. *Test and maintenance*: Addresses questions related to plant performance by changing the test and maintenance strategy for a given design.
3. *Configuration and control*: Measures the significance or the effect of failure of a component on risk or safety or temporarily taking a component out of service.
4. *Uncertainty*: Reduces uncertainties in the input variables of the PRAs.

The following are the major steps of importance ranking:

1. Determine the purpose of the ranking and select that appropriate ranking importance measure that provides consistent interpretation for the use of the ranked results.
2. Perform risk ranking and uncertainty ranking, as needed.
3. Identify the most critical and important elements of the system with respect to the total risk values and total uncertainty associated with the calculated risk values.

8.3.2.11 Interpretation of Results

Once the risk values are calculated, they must be interpreted to determine whether any revisions are necessary to refine the results and the conclusions. Two main steps are involved in the interpretation process. The first is to determine whether the final values and details of the scenarios are logically and quantitatively meaningful. This step verifies the adequacy of the PRA model and the scope of analysis. The second is to characterize the role of each element of the system in the final results. This step highlights additional analyses data and information gathering that would be considered necessary.

The interpretation process relies heavily on the examination of the details of the analysis to see whether the scenarios are logically meaningful (e.g., by examining the minimal cut sets of the scenarios), whether certain assumptions are significant and greatly control the risk results (using the sensitivity analysis results), and whether the absolute risk values are consistent with any historical data or expert opinion available. Based on the results of the interpretation, the details of the PRA logic or the assumptions and scope of the PRA may be modified to update the results into more realistic and dependable values.

The ranking and sensitivity analysis results may also be used to identify areas where gathering more information and performing better analysis (e.g., by using more accurate models) are warranted. The primary aim of the process is to reduce uncertainties in the risk results.

The interpretation step is a continuous process of receiving information from the quantification, sensitivity, uncertainty, and importance analysis activities of the PRA. The process continues until the final results can be best interpreted and used in the subsequent risk-management steps.

The basic steps of the PRA results interpretation are as follows:

1. Determine accuracy of the logic models and scenario structures, assumptions, and scope of the PRA.
2. Identify system elements for which better information is needed to reduce uncertainties in failure probabilities and models used to calculate performance.
3. Revise the PRA and reinterpret the results until stable and accurate results are obtained.

8.4 COMPRESSED NATURAL GAS POWERED BUSES: A PRA CASE STUDY

8.4.1 Primary CNG Fire Hazards

The fire safety hazards that should be considered in assessing the risks of using compressed natural gas (CNG) fuel are as follows:

- Fire potential from fuel leakage.
- Explosion potential from uncontrolled dispersion and mixing of CNG in the presence of an ignition source.
- Impact and missile-generated hazards due to fuel being stored at high pressure.
- Chemical hazards (the gas toxicity, asphyxiation potential, and higher hydrocarbons in CNG may cause it to be considered a neurotoxin, even though CNG is relatively nontoxic).
- ESD.

The issues of bulk transport and storage are completely different from most of the other fuel types that are typically transported to fleet storage via tanker trucks. For use as a fuel, natural gas is compressed and stored in high-pressure cylinders (tanks) on the vehicle at 2400–3000 psig. The containment of natural gas at such high pressures requires very strong storage tanks that are both heavy and costly. Furthermore, such tanks are subject to corrosion-fatigue, sustained load cracking, stress corrosion, and fatigue failure. This distinguishing feature of CNG has the most impact on safety issues. Some important fire–related incidents that include fatalities in CNG vehicles are summarized by Chamberlain and Modarres [23].

There are two distinct categories of CNG fuel tank designs. These are as follows:

Metal tanks: These tanks are made of aluminum or 4130X steel. There have been some cases of fragmentation rupture in this type of cylinder.

Composite-wrapped: These tanks are constructed with an aluminum/steel liner and E-glass wrap or a carbon fiber insert with an E-glass wrap. There have been several ruptures of this type of tank reported by the Gas Research Institute [24].

This PRA study only considers metal tank type.

8.4.2 PRA Approach

The standard PRA approach may be applied. Fault tree and event tree modeling techniques (see Chapter 4) describe scenarios of events leading to fire and explosion fatalities. The frequency of occurrence of such scenarios is then quantified using the generic failure data for similar basic components gathered by the American Institute of Chemical Engineers [25]. Engineering judgment and simple fire analysis methods are then used to determine the likelihood of occurrence of particular fire scenarios. The main source of uncertainty is the data used in the risk model (parameter uncertainty). The uncertainty from the failure data is factored into the PRA results. Model uncertainty is not considered since the methods discussed in the literature, such as the one described by Droquette [26], are still evolving; a universally acceptable methodology has yet to emerge. In the absence of such an approach, conservative modeling assumptions are used in the PRA to minimize error due to model uncertainty.

Uncertainties due to failure and consequence data are propagated using a standard Monte Carlo simulation technique. The sensitivity of individual elements of the risk model (e.g. failure data and assumptions) is measured by varying them individually by many factors and determining the resulting change in fire fatality risk.

8.4.3 System Description

The typical CNG bus system shown in Figure 8.3 was considered for this risk study. It is comprised of the following major subsystems.

8.4.4 Natural Gas Supply

Natural gas is supplied to the compressor station from a local distribution company through its pipeline system. Gas supply pressure in pipelines is typically around 25–60 psig.

8.4.5 Compression and Storage Station

The compression and storage station provides CNG at different pressures to the dispenser, depending on the fueling procedure. The compression stations are designed with the flexibility to "fast fill" or regularly fill CNG bus tanks. Additionally, refueling can take place directly from the compressor.

8.4.6 Storage Cascade

CNG is filtered after leaving the compressor and injected with methanol to reduce the moisture content. It is then sent through the priority and sequence valve panels to the low-pressure storage cascade bank, where it is stored at 1000 psig. Once the low-pressure storage bank is filled, medium- (1500 psig) and high- (3600 psig) pressure banks are sequentially filled.

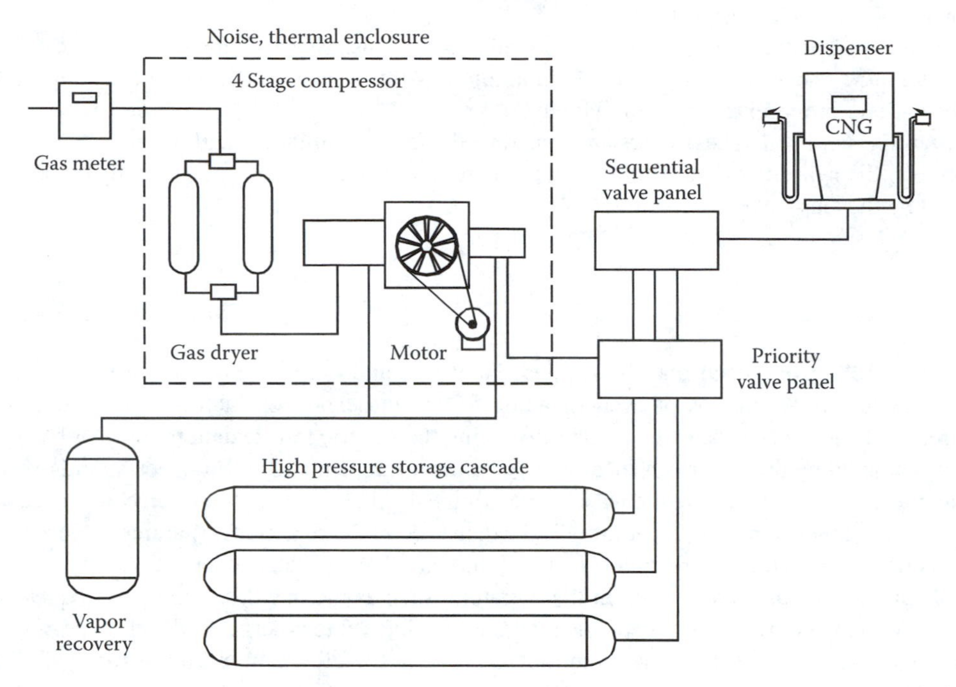

FIGURE 8.3 Components of a CNG station.

8.4.7 Dispensing Facility

The compressor and the storage tanks are connected to the dispenser through underground steel lines that are subject to corrosion and the possibility of damage from excavation or earth movements. The dispensing equipment draws gas either from the storage cascades or from the compressor. A master shut-off valve isolates the compressor and storage cascades from the dispensing equipment for maintenance or emergency situations.

8.4.8 CNG Bus

The typical fuel supply system of a CNG bus, shown in Figure 8.4, is used in this study. This design is a 40-ft-long bus with six undercarriage CNG storage tanks. Only the parts of the bus relevant to the usage of CNG as a fuel and contributing to fire hazards are identified and used in this study.

The PRA develops fire scenarios and consequences due to the occurrence of initiating events and subsequent hardware or human failures identified as important in the qualitative risk analysis. Generic and historical data obtained from various sources were used for quantifying the fault tree and event tree models, and were used to determine the frequency of occurrence of scenarios leading to fire-related fatalities. The risk was then computed from the frequency and consequence of each scenario. The steps are summarized in Figure 8.5.

8.4.9 Gas Release Scenarios

The failure modes were grouped into six initiating event groups, and subsequently, hardware, human, or software failures were identified to describe the scenarios of events leading to gas releases. Each scenario has an initiating event and subsequent failure events that can result in a fire or explosion with corresponding fatal consequences. The determination of the frequency of

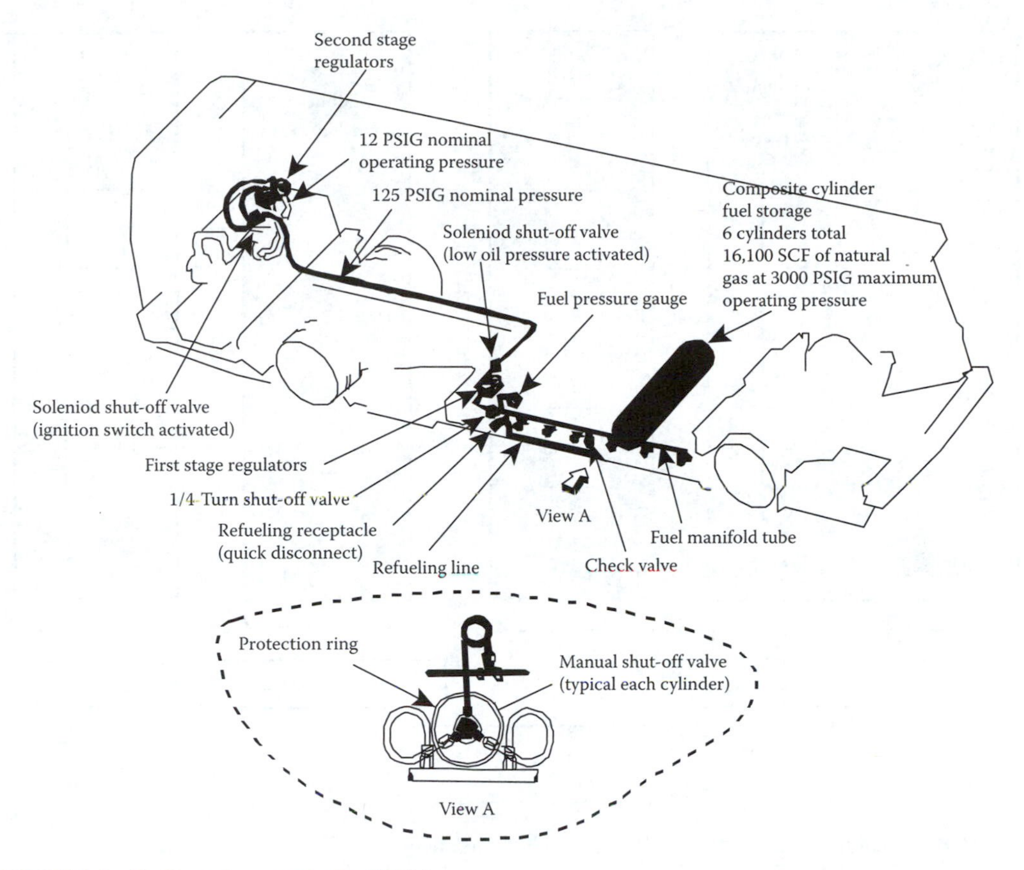

FIGURE 8.4 Fuel supply system of a CNG bus.

occurrence of each initiating event and the actual propagation of them into fires/explosions and consequences are part of the quantitative assessment. The classes of initiating events identified are as follows:

1. Hardware catastrophic failures due to intrinsic failure mechanisms, leading to instantaneous release of CNG in the presence of an ignition source.
2. Hardware-degraded failures resulting in gradual release of CNG in the presence of an ignition source.
3. Hardware or human failures resulting in a CNG release with potential for ESD ignition.
4. Accidental impact of hardware resulting in gas release in the presence of an ignition source.
5. Human error resulting in the release of CNG in the presence of an ignition source.
6. Non-CNG-related fire (e.g., fires due to oil or cargo burning) resulting in the release of CNG in the presence of an ignition source.

Fault trees were used to identify and compute the probability of failure leading to a sustained gas release in the presence of an ignition source, given the occurrence of the initial failure events. A constant failure rate model was used to represent the events modeled in the fault trees. From the fault tree, cut sets were generated. Rare event approximation, discussed in Sections 4.2 through 4.4, was used to combine scenarios. Component failures were assumed to be independent, and no CCF events were determined to be significant since very little redundancy in active components existed.

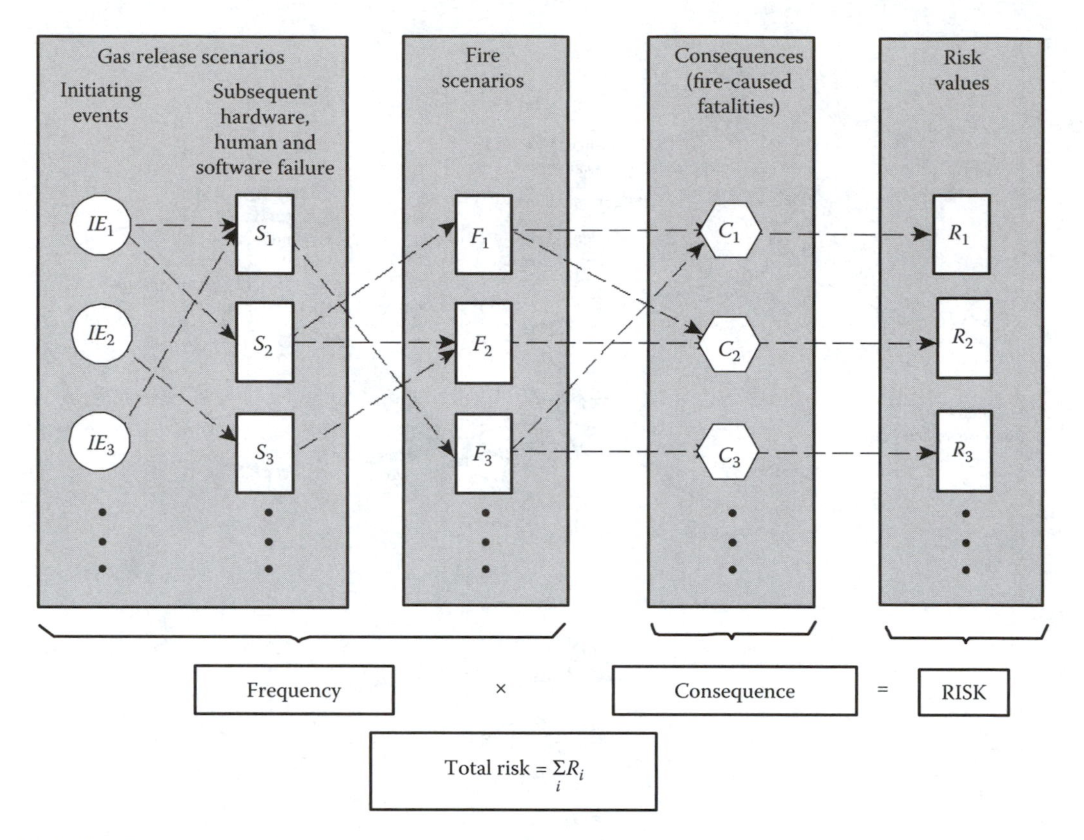

FIGURE 8.5 Summary of the overall PRA approach in this study.

8.4.10 Fire Scenario Description

Factors that determine the kind of fire that results from gas release include the following:

- Type of initial gas release (leak vs. sudden release)
- Gas dispersion
- Gas ignition likelihood.

8.4.11 Gas Release

The CNG release may be instantaneous or gradual. Adiabatic expansion of the gas occurs with an instantaneous release, for example, from a ruptured gas tank cylinder. A gradual release occurs from a leaking joint or cracked fuel line. Component defects with greater than 1/4″ opening in the containment system produce an instantaneous release of CNG. Conversely, a gradual release is produced by a crack or other defect with less than 1/4″ opening.

Adiabatic expansion of CNG results in the formation of a flammable air–gas mixture with explosive potential. If the mixture is ignited immediately it results in a fireball (see [27]). The pressure and the volume of the CNG containment determine the extent of the fireball and explosion. Delayed ignition of the released gas results in continued mixing with a vapor cloud forming. Ignition would result in a vapor cloud explosion or flash fire. Gradual release of CNG results in a limited mixing of gas with air, which, if ignited immediately, will result in a jet flame in the vicinity of the mixture. Delayed ignition leads to the accumulation of flammable air–gas mixture, which would produce a vapor cloud explosion or flash fire if ignited.

TABLE 8.2
Qualitative Ignition Potential

Qualitative Ranking	Quantitative Likelihood Range
Strong	0.25–1.00
Moderate	0.10–0.24
Weak	0.01–0.09

8.4.12 Gas Dispersion

The extent of mixing of the released CNG that is not immediately ignited is best determined with computer models. In the absence of an elaborate model, the dispersion scenarios considered are dense cloud, and neutral or buoyant dispersion.

If there is an instantaneous release of CNG and the ignition is delayed, then the gas will be dispersed as a dense cloud, which will form a flash fire or will explode when ignition occurs. Gradual release of CNG in a jet leads to buoyant dispersion, and delayed ignition results in a flash fire. Following the dispersion of CNG in a dense cloud, even if delayed ignition does not occur, there are thermodynamic effects on the human body that present additional risks apart from fire or explosion hazards.

8.4.13 Ignition Likelihood

If ignition occurs in less than 10 min of exposure, it is considered immediate. Delayed ignition is generally considered to be after 10–15 min of exposure. Table 8.2 shows the reference values discussed by Chamberlain and Modarres [23] and AIChE [28] used for qualitative judgment of the likelihood potential of ignition of a vapor cloud exposed to an ignition source.

In this study immediate ignition of a flammable gas mixture (fireball and jet flame) is assumed to occur with a conditional probability of 0.8. Delayed ignition and dispersion (flash fire) is assumed to occur with a conditional probability of 0.95. Sensitivity of the risk results to these assumptions were assessed later in the study.

Table 8.3 summarizes the different gas release, ignition or dispersion, and fire scenarios considered in this study.

8.4.14 Consequence Determination

An analytical assessment method was used to compute the number of fatalities due to the various fire scenarios identified above. The method estimates the heat release rate and flame height, exposure temperature from fire plume modeling, and the time required to reach critical damage thresholds causing fatalities. The consequences for each event in this study are computed by assuming the following:

TABLE 8.3
Summary of CNG Release, Ignition, and Fire Scenarios

CNG Release Mode	Ignition Mode	Expected Consequence
Instantaneous	Immediate	Fireball
	Delayed	Vapor cloud explosion or flash fire
Gradual	Immediate	Jet flame
	Delayed	Vapor cloud explosion or flash fire

- Worst-case fire and explosive intensities associated with each scenario.
- Fatalities occurring when exposed to a fire heat flux of 25.0 kW/m^2 for 1 min.
- Fatalities occurring when persons are present within the distance from a point source where a radiant flux of 25.0 kW/m^2 or more is being received.

8.4.15 Fire Location

To determine the lethality of each fire event, location of the vehicle is of paramount importance. The five vehicle fire locations chosen represent normal usage. A probability factor, representing the relative frequency of a vehicle's location during normal use, is applied to the consequences. The number of people present at each location was different and was represented by distributions consistent with current experience ranging from 1 to 30 people exposed (no evacuation or other mitigation measures were allowed). This allows determination of the expected fatalities in each location given a fire scenario.

The locations and probability factors are based on the following assumptions:

a. Garage or storage facility (0.5), bus is parked 12 h/day.
b. Fueling station (0.06), bus is refueled, maintained, and repaired 1.5 h/day.
c. Urban roadway (0.21), bus is in operation 5 h on urban roadway.
d. Rural roadway (0.21), bus is in operation 5 h near school, near homes, and on rural roadway.
e. Tunnel, under bridges, or other enclosed roadways (0.02), bus goes through these types of roadways for 0.5 h/day.

8.4.16 Risk Value Determination

Gas release and fire scenarios along with the consequences were combined for each initiating event to determine the risk associated with each event. The summation of each individual risk gave the total risk. Event trees of the gas release and fire scenarios were used to combine the elements of the risk model. For this calculation, any cut set-generating tool such as SAPHIRE [29] may be used.

8.4.17 Summary of PRA Results

The PRA fire safety risk results of this study are summarized in Table 8.4. The mean total fire fatality risk for CNG buses is estimated as 0.23 per 100 million miles of operation. The study also estimates

TABLE 8.4
PRA Risk Results

Scenarios Groups Leading to Fire/Fatality	Risk (Mean Fatalities/Bus/Year)	Risk (Mean Fatalities/100 M Miles)	% of Total Risk
Catastrophic failure of bus or station hardware components	2.4×10^{-6}	2.5×10^{-2}	10.84
Degraded failure of bus or station hardware components	8.7×10^{-6}	9.0×10^{-2}	38.77
ESD of CNG	2.7×10^{-6}	2.8×10^{-2}	12.21
Accidental impacts mainly due to collision	4.9×10^{-6}	5.1×10^{-2}	21.70
Non-CNG-related fires	3.4×10^{-6}	3.5×10^{-2}	14.94
Operator error	3.6×10^{-7}	3.7×10^{-3}	1.54
Total mean fire fatality risk	2.2×10^{-5}	2.3×10^{-1}	100

mean values of 0.16 fatalities per 100 million miles for passengers inside the CNG bus only. This would suggest that approximately 70% of the total fire-related fatalities are expected to be due to the passengers of the bus.

8.4.18 Overall Risk Results

The projected mean fatalities from a typical bus due to catastrophic failures resulting in an uncontained fire are 2.7×10^{-6} bus/year, and for all causes are 2.2×10^{-5} bus/year. For the 8500 CNG school buses in operation in the United States in year 2001, this would lead to a mean total risk value of approximately 0.19 deaths/year, or a mean time to occurrence of 5.4 years/fatality for all existing buses in operation. If all the existing buses were to be replaced with CNG buses, then the projected fatality would be 9.9 deaths/year or a mean time to occurrence of 0.1 year/fatality.

It should be noted that some actual cases of CNG-related explosions and fire from component failures have been recorded. One such incident was reported in Houston, Texas, in 1998. In this case, gradual gas release and delayed ignition resulted in the explosion and a subsequent flash fire. No fatalities were reported. A CNG cylinder rupture as a result of accidental impact caused a fireball in Nassau County, New York, in 2001. Seven individuals were killed in a CNG bus explosion/fire in Tajikistan and four in another accident in San Salvador in 2001. As such, the evidence of some of the scenarios considered in this study has already been observed.

Analysis of the projected fatalities from all the initiating events has revealed that while the number of fatalities from CNG bus fires is low, or has not been reported to date in the United States, this is only due to the small number of CNG buses in operation. Increasing the number of such buses will certainly increase the expected number of fatalities due to fires and explosions.

8.4.19 Uncertainty Analysis

The sources of uncertainty in the results of this study can be classified and characterized as follows:

- Generic bus system description and model used (model uncertainty).
- Major fire hazard scenarios considered (model/completeness uncertainty).
- Conservatism in fire scenario and consequence modeling techniques used (model/assumption uncertainty).
- Failure data for hardware/human failures and estimated frequency of fires and their location (lack of sufficient data/parameter uncertainty).
- Integration of all scenarios to estimate total risk (model uncertainty).

While model uncertainties are important, due to lack of a sound model uncertainty estimation methodology, at this point only the source of uncertainty due to failure data used to quantify the PRA models are quantified. Uncertainties in the models themselves were not considered, and as is the practice in PRA, model uncertainties were reduced by independent peer reviews and conservatism in constructing the models.

The uncertainty in failure data used in the PRA model was represented by probability distributions assigned to the component failure rate, probabilities for subsequent events, and consequences of individual scenarios. Where appropriate, normal, lognormal, and uniform distributions were used along with distribution parameters as suggested by the generic data sources and from engineering judgment to represent the uncertainties in failure data, parameters, and assumptions.

Propagation and combination of the uncertainties in the risk analysis were performed. The results of the uncertainty analysis are summarized in Table 8.5. The mean value of the fire fatality risk computed is 2.2×10^{-5} fatalities/bus/year (see Table 8.5). The probability bounds at 5% and 95% levels are calculated as 9.1×10^{-6} and 4.0×10^{-5}, respectively.

TABLE 8.5
Summary of Uncertainty Analysis Results

CNG Bus Fire Scenarios Resulting from the Following Causes	Mean Frequency of Occurrence/Bus/Year	Mean Risk (Fatalities/Bus/Year)	5%	95%
Catastrophic failure of bus or station hardware components	1.4×10^{-3}	2.4×10^{-6}	2.7×10^{-7}	6.1×10^{-6}
Degraded failure of bus or station hardware components	3.7×10^{-3}	8.7×10^{-6}	5.6×10^{-7}	2.5×10^{-5}
ESD of CNG	1.4×10^{-5}	2.7×10^{-6}	4.1×10^{-7}	6.6×10^{-6}
Accidental impacts mainly due to collision	3.6×10^{-2}	4.9×10^{-6}	4.3×10^{-7}	1.2×10^{-5}
Non-CNG-related fires	3.6×10^{-4}	3.4×10^{-6}	3.7×10^{-7}	8.5×10^{-6}
Operator error	4.0×10^{-2}	3.6×10^{-7}	2.4×10^{-8}	9.8×10^{-7}
Total fire fatality risk		2.2×10^{-5}	9.1×10^{-6}	4.0×10^{-5}

Note: Assuming 9598 miles of travel per bus per year.

8.4.20 Sensitivity and Importance Analysis

The sensitivity of the total risk to the input parameters was calculated by changing the failure rates of important components by a factor of ten (10) and changing the outcomes of each fire by a factor of one half. Each change was made while the other parameters are kept constant. This approach allows identification of the components that contribute most to the uncertainty of the overall risk results. The sensitivity of the risk to the presence of certain ignition sources was performed by similarly varying the probability of occurrence of each ignition source and comparing the effect on the total risk result.

Sensitivity analysis shows that the total fire fatality risk is relatively insensitive to ignition sources except to the introduction of an ignition source by an operator from such activities as smoking. This result is very important and should be included in any selection and training process for operators of CNG-powered equipment.

Fatality risk is sensitive to the location of the fire. This study shows that the risk is highly sensitive to flash fires in urban and rural areas. The third most important fire events are fireballs in urban areas. More detailed modeling of fire scenarios would be required so that critical adiabatic flame temperatures, flame speed, and burning rates can be determined analytically.

Birnbaum importance measures were also used to identify design weaknesses and component failures that are critical to the prevention of CNG release and fires. It is a good measure for identifying components and design features that contribute most to risk. Such importance measures provide insights into how one should focus risk reduction and management efforts. The components with the highest potential for risk reduction are the CNG cylinders, pressure relief devices, and bus fuel piping.

In conclusion, while the CNG buses are more susceptible to major fires, the total average expected fire risk from CNG buses over diesel power-buses is expected to be higher by a factor of only about two. However, since most of the fire fatalities are expected to occur among the passengers of CNG school buses, the fire risk for CNG bus passengers is expected to be larger than that of diesel school bus passengers by over two orders of magnitude. While on the average the total CNG fire safety risk is not much higher than the diesel buses, the worst possible fire scenarios could impose far higher fatality risk in CNG buses than the worst-case fire scenario from diesel-powered buses.

Reliance of this study on generic models and failure data is a good approximation to screen safety risks of CNG buses. However, a more accurate physics of failure-based model (such as fatigue- and corrosion-based life models) should supplement this study to provide more accurate results. This is part of ongoing research by the authors. Further, CNG-specific hardware failure data are also needed.

Three components have been identified from the sensitivity and importance analysis to contribute most to fire fatality risk and to the uncertainty of the overall fire fatality risk. These are pressure relief valves on cylinders, CNG storage cylinders, and bus fuel piping. The failure mechanisms of these components should be modeled using a physics of failure-based approach to analytically determine the failure rates.

To compare the results of CNG with diesel buses, the risk assessment should add the risk of nonfatal injuries to fatalities. For any policy decisions, the fire risk should be compensated for and properly characterized by integrating the safety risks of this study and the expected health and environmental benefits of using CNG-powered school buses.

8.5 A SIMPLE FIRE PROTECTION RISK ANALYSIS

Consider the fire protection system shown in Figure 8.6. This system is designed to extinguish all possible fires in a plant with toxic chemicals. Two physically independent water extinguishing nozzles are designed such that each is capable of controlling all types of fires in the plant. Extinguishing nozzle 1 is the primary method of injection. Upon receiving a signal from the detector/alarm/actuator device, pump 1 starts automatically, drawing water from the reservoir tank and injecting it into the fire area in the plant. If this pump injection path is not actuated, plant operators can start a second injection path manually. If the second path is not available, the operators will call for help from the local fire department, although the detector also sends a signal directly to the fire department. However, due to the delay in the arrival of the local fire department, the magnitude of damage would be higher than it would be if the local fire extinguishing nozzles were available to extinguish the fire.

Under all conditions, if the normal OSP is not available due to the fire or other reasons, a local generator would provide electric power to the pumps. The power to the detector/alarm/actuator system is provided through the batteries that are constantly charged by the OSP. Even if the AC power is not available, the DC power provided through the battery is expected to be available at all times. The manual valves on the two sides of pumps 1 and 2 are normally open, and only remain closed when they are being repaired. The entire fire system and generator are located outside of the reactor compartment, and are therefore not affected by an internal fire. The risk-analysis process for this situation consists of the steps explained below.

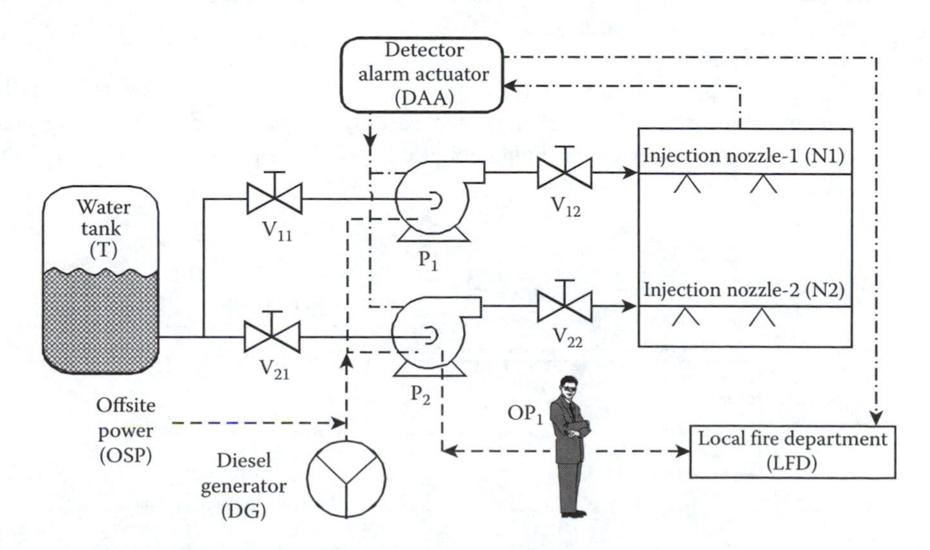

FIGURE 8.6 Fire protection system.

1. **Identification of Initiating Events**
   In this step, all events that lead to or promote a fire in the reactor compartment must be identified. These include equipment malfunctions, human errors, and facility conditions. The frequency of each event should be estimated. Assuming that all events would lead to the same magnitude of fire, the ultimate initiating event is a fire, the frequency of which is the sum of the frequencies of the individual fire-causing events. Assume for this example that the frequency of fire is estimated at 1×10^{-6} year^{-1}. Since fire is the only challenge to the plant in this example, we end up with only one initiating event. However, in more complex situations, a large set of initiating events can be identified, each posing a different challenge to the plant.
2. **Scenario Development**
   In this step, we explain the cause-and-effect relationship between the fire and the progression of events following the fire. We will use the event tree method to depict this relationship. Generally, this is done inductively, and the level of detail considered in the event tree is somewhat dependent on the analyst. Two protective measures have been considered in the event tree shown in Figure 8.7: on-site protective measures (on-site pumps, tanks, etc.) and off-site protective fire department measures. The selection of these measures is based on the fact that availability or unavailability of the on-site or off-site protective measures would lead to different plant damage states.
3. **System Analysis**
   In this step, we identify all failures (equipment or human) that lead to failure of the event tree headings (on-site or off-site protective measures).

   For example, Figure 8.8 shows the fault tree developed for the on-site fire protection system. In this fault tree, all basic events that lead to the failure of the two independent paths are described. Note that detector alarm actuator (DAA), electric power to the pumps, and the water tank are shared by the two paths. Clearly these are physical dependencies. This is taken into account in the quantification step of the risk analysis. In this tree, all external event failures and passive failures are neglected.

   Figure 8.9 shows the fault tree for the off-site fire protection system. This tree is simple since it includes only failures that do not lead to an on-time response from the local fire department.

   It is also possible to use the MLD for system analysis. An example of the MLD for this problem is shown in Figure 8.10. However, here only the fault trees are used for risk analysis, although MLD can also be used.
4. **Failure Data Analysis**
   It is important at this point to calculate the probabilities of the basic failure events described in the event trees and fault trees. As indicated earlier, this can be done by using plant-specific data, generic data, or expert judgment. Table 8.6 describes the data used and their

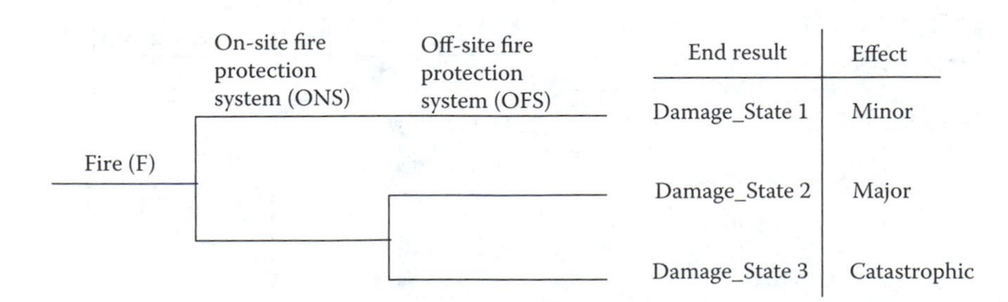

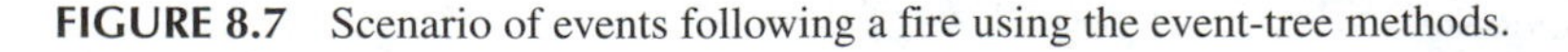
FIGURE 8.7 Scenario of events following a fire using the event-tree methods.

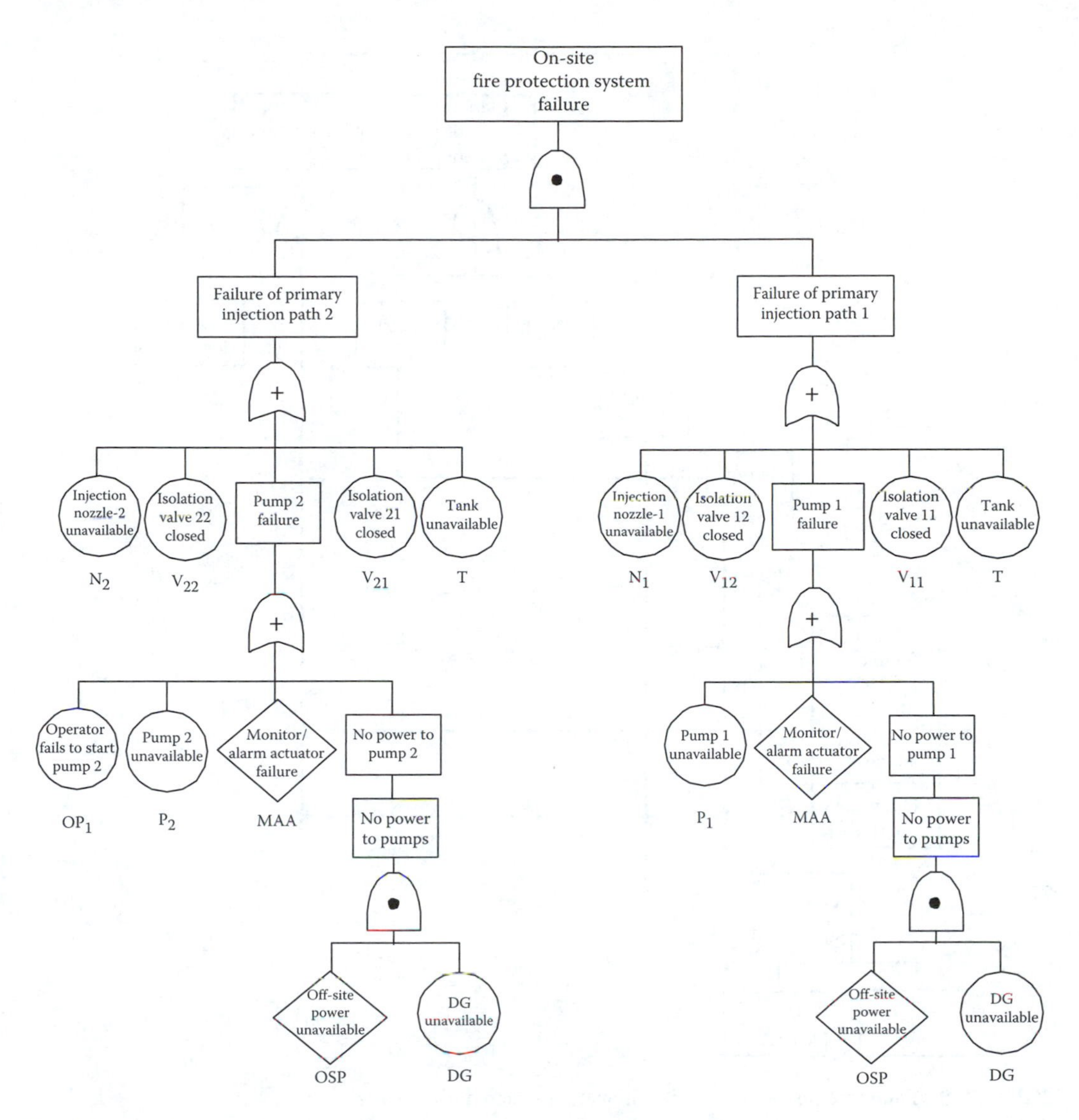

FIGURE 8.8 Fault tree for the on-site fire protection system failure.

sources. It is assumed that at least 10 h of operation is needed for the fire to be completely extinguished.

5. **Quantification**

To calculate the frequency of each scenario defined in Figure 8.7, we must first determine the cut sets of the two fault trees shown in Figures 8.8 and 8.9. From this, the cut sets of each scenario are determined, followed by calculation of the probabilities of each scenario based on the occurrence of one of its cut sets. These steps are described below.

1. The cut sets of the on-site fire protection system failure are obtained using the technique described in Section 4.2. These cut sets are listed in Table 8.7. Only cut set number 22, which is failure of both pumps, is subject to a CCF. This is shown by adding a new cut set (cut set number 24), which represents this CCF.
2. The cut sets of the off-site fire protection system failure are similarly obtained and listed in Table 8.8.

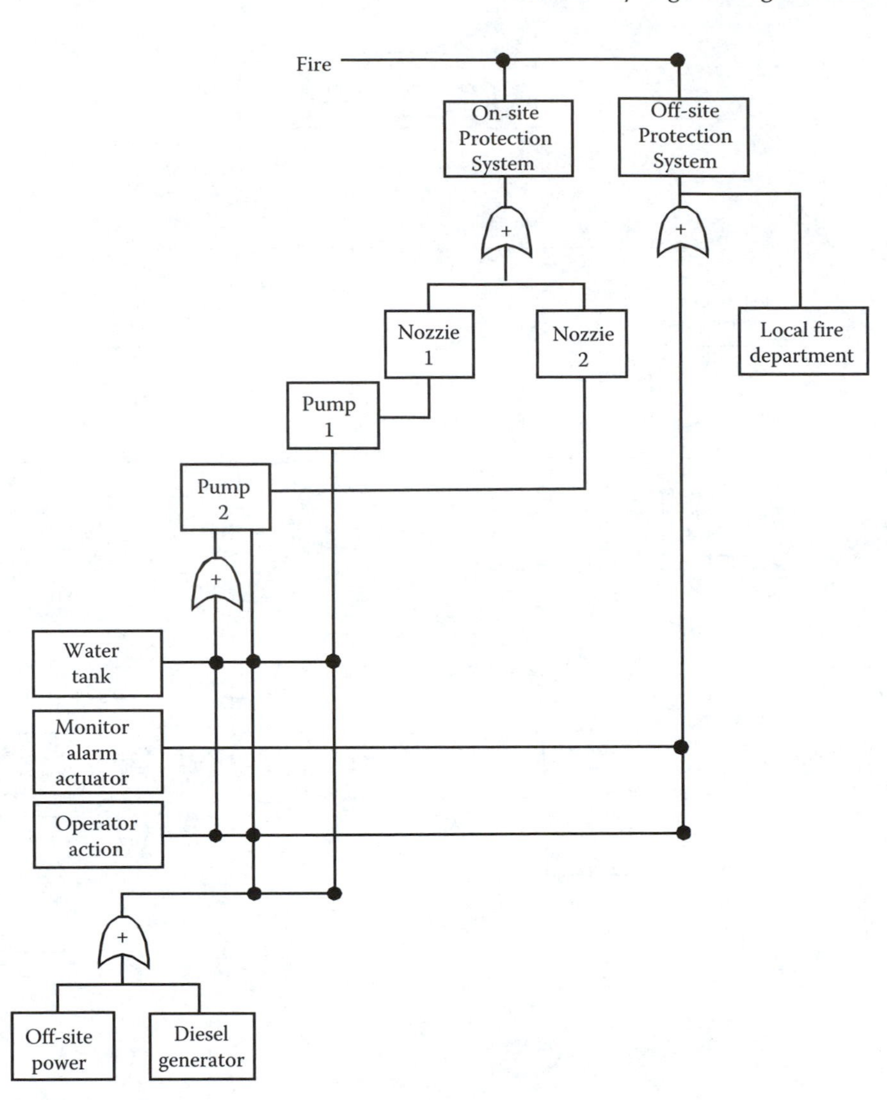

FIGURE 8.9 Fault tree for the off-site fire protection system failure.

3. The cut sets of the three scenarios are obtained using the following Boolean equations representing each scenario:

$$\text{Scenario} - 1 = F \cdot \overline{\text{ONS}}$$

$$\text{Scenario} - 2 = F \cdot \text{ONS} \times \overline{\text{OFS}}$$

$$\text{Scenario} - 3 = F \cdot \text{ONS} \times \text{OFS}$$

   The process is described in Section 4.3.2.
4. The frequency of each scenario is obtained using data listed in Table 8.6. These frequencies are shown in Table 8.9.
5. The total frequency of each scenario is calculated using the rare event approximation. These are also shown in Table 8.9.

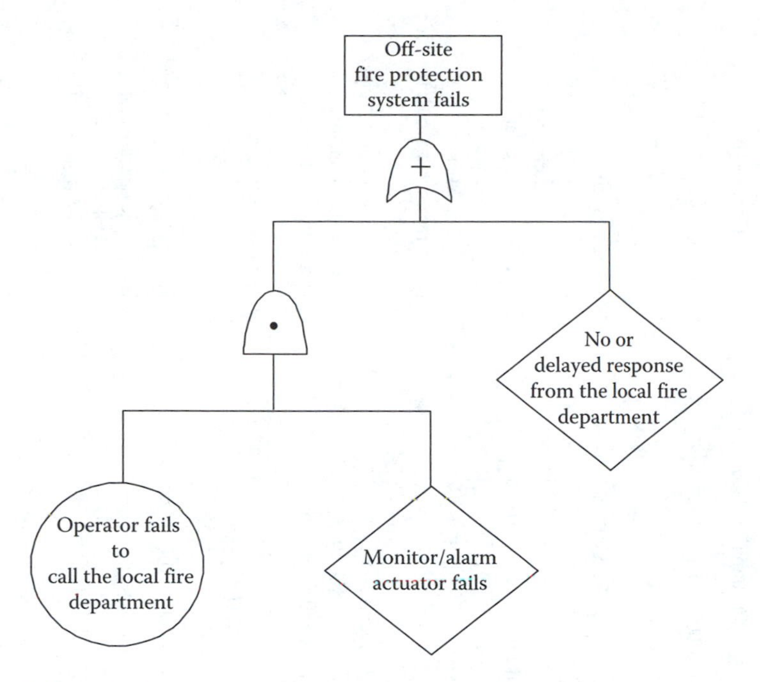

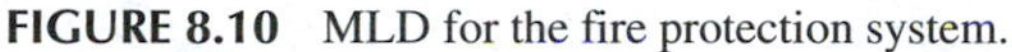
FIGURE 8.10 MLD for the fire protection system.

6. **Consequences**

In the scenario development and quantification tasks, we identified three distinct scenarios of interest, each with different outcomes and frequencies. The consequences associated with each scenario should be specified in terms of both economic and/or human losses. This part of the analysis is one of the most difficult for two reasons:

1. Each scenario poses different hazards and methods of hazard exposure, and requires careful monitoring. In this case, the model should include the ways the fire can spread through the plant, the ways people can be exposed, evacuation procedures, the availability of protective clothing, and so on.
2. The outcome of the scenario can be measured in terms of human losses. It can also be measured in terms of financial losses, that is, the total cost associated with the scenario. This involves assigning a dollar value to human life or casualties, which is a source of controversy.

Suppose a careful analysis of the spread of fire and fire exposure is performed, with consideration of the above issues, and ultimately results in damages measured only in terms of economic losses. These results are shown in Table 8.10.

The low value (in dollars) at risk indicates that fire risk is not important for this plant. However, Scenarios 1 and 2 are significantly more important than Scenario 3. Therefore, if the risk were high, one should improve those components that are major contributors to Scenarios 1 and 2. Scenario 1 is primarily due to CCF between pumps P_1 and P_2, so reducing this failure is a potential source of improvement.

7. **Risk Calculation and Evaluation**

Using values from Table 8.10, we can calculate the risk associated with each scenario. These risks are shown in Table 8.11.

Since this analysis shows that risk due to fire is rather low, uncertainty analysis is not very important. However, one of the methods described in Section 7.3 could be used to estimate the uncertainty associated with each component and the fire-initiating event if necessary.

TABLE 8.6
Sources of Data and Failure Probabilities

Failure Event	Plant-Specific Experience	Generic Data	Probability Used	Comments
Fire initiation frequency	No such experience in 10 years of operation	Five fires in similar plants. There are 70,000 plant-years of experience	$F = 5/70,000 = 7.1 \times 10^{-4}$ year^{-1}	Use generic data.
Pumps 1 and 2 failure	Four failures of two pumps to start. monthly tests are performed which takes negligible time. Repair time takes about 10 hs at a frequency of 1 per year. No experience of failure to run	Failure to run = 1×10^{-5} h^{-1}	$\frac{4}{2 \times 12 \times 10} = 1.7 \times 10^{-2}$/demand Unavailability $= 1.7 \times 10^{-2} + \frac{10}{8760}$ Unavailability $= 1.8 \times 10^{-2}$/demand Failure to run $= 1.0 \times 10^{-5}$/demand $P_1 = P_2 = 1.7 \times 10^{-2} + 1.0 \times 10^{-5} \times 10$ $P_1 = P_2 = 1.7 \times 10^{-2}$	For failure to start, use plant-specific data. For failure to run, use generic data. If possible, use Bayesian updating technique described in Section 3.6. Assume 10 years of experience and 8760 h in one year.
CCF between Pumps 1 and 2	No such experience	Using the β-factor method, β = 0.1 for failure of pumps to start	Unavailability due to CCF: CCF $= 0.1 \times 1.8 \times 10^{-2}$ CCF $= 1.8 \times 10^{-3}$/demand	Assume no significant CCF exists between valves and nozzles. See Section 7.2 for more detail.
Failure of isolation valves	One failure to leave the valve in open position following a pump test	Not used	$v_{11} = v_{12} = v_{21} = v_{22} = \frac{1}{10(12)(2)}$ $= 4.2 \times 10^{-3}$/demand	Plan-specific data used
Failure of nozzles	No such experience	1×10^{-5}/demand	$N_1 = N_2 = 1.0 \times 10^{-5}$/demand	Generic data used
Diesel generator failure	Three failures in monthly tests. 40 h of repair per year	3.0×10^{-2}/demand 3.0×10^{-3} h^{-1} 40 run	Failure on demand $= \frac{3}{12(10)}$ Failure on demand $= 2.5 \times 10^{-2}$/demand Failure on run $= 3.0 \times 10^{-3}$ h$^{-1} \times 10$ h $= 3.0 \times 10^{-2}$ Total failure of DG $= 2.5 \times 10^{-2} + 3.5 \times 10^{-2}$ DG $= 5.5 \times 10^{-2}$	Plant-specific data used for demand failure. Assume 10 years of experience

Loss of off-site power	No experience	0.1 year	$OSP = 0.1 \times \frac{10}{8760}$ $OSP = 1.1 \times 10^{-4}$/demand	Assume 104 h of operation for fire extinguisher and use generic data
Failure of MAA	No experience	No data available	$MAA = 1 \times 10^{-4}$/demand	This estimate is based on expert judgment. See Section 6.3 for the methods
Failure of operator to start Pump 2	No such experience	Using the THERP method for tasks of this kind, 1×10^{-2} is suggested	$OP_1 = 1 \times 10^{-2}$/demand	Use the THERP Handbook data discussed in Section 6.3
Failure of operator to call the fire department	No such experience	1×10^{-3}	$OP_2 = 1 \times 10^{-3}$/demand	This is based on experience from no response to similar situations. Generic probability is used
No or delayed response from fire department	No such experience	1×10^{-4}	$LFD = 1 \times 10^{-4}$/demand	This is based on response to similar cases from the fire department. Delayed/no arrival is due to accidents, traffic, communication problems, and so on
Tank failure	No such experience	1×10^{-5}	$T = 1 \times 10^{-5}$/demand	This is based on data obtained from rupture of the tank or insufficient water content

TABLE 8.7
Cut Sets of the On-Site Fire Protection System Failure

Cut Set No.	Cut Set	Probability/(% of Total)	Cut Set No.	Cut Set	Probability/(% of Total)
1	T	1.0×10^{-5} (0.35)	13	$V_{21} \cdot V_{12}$	1.8×10^{-5} (0.35)
2	MAA	1.0×10^{-4} (3.5)	14	$V_{22} \cdot P_1$	7.1×10^{-5} (2.5)
3	OSP · DG	6.0×10^{-6} (0.21)	15	$V_{21} \cdot V_{11}$	1.8×10^{-5} (0.64)
4	$N_2 \cdot N_1$	1.0×10^{-10} (~0)	16	$\text{OP}_1 \cdot N_1$	1.0×10^{-7} (~0)
5	$N_2 \cdot V_{12}$	4.2×10^{-8} (~0)	17	$\text{OP}_1 \cdot V_{12}$	4.2×10^{-5} (1.5)
6	$N_2 \cdot P_1$	1.7×10^{-7} (~0)	18	$\text{OP}_1 \cdot P_1$	1.7×10^{-4} (6.0)
7	$N_2 \cdot V_{11}$	4.2×10^{-8} (~0)	19	$\text{OP}_1 \cdot V_{11}$	4.2×10^{-5} (1.5)
8	$V_{22} \cdot N_1$	4.2×10^{-8} (~0)	20	$P_2 \cdot N_1$	1.7×10^{-7} (~0)
9	$V_{22} \cdot V_{12}$	1.8×10^{-5} (0.64)	21	$P_2 \cdot V_{12}$	7.1×10^{-5} (2.5)
10	$V_{22} \cdot P_1$	7.1×10^{-5} (2.5)	22	$P_2 \cdot P_1$	2.9×10^{-4} (0.3)
11	$V_{22} \cdot V_{11}$	1.8×10^{-5} (0.64)	23	$P_2 \cdot V_{11}$	7.1×10^{-5} (2.5)
12	$V_{21} \cdot N_1$	4.2×10^{-8} (~0)	24	CCF	1.8×10^{-3} (63.8)
				$\text{Pr(ON)} = \Sigma_i C_i = 2.8 \times 10^{-3}$	

TABLE 8.8
Cut Sets of the Off-Site Fire Protection System

Cut Set No.	Cut Set	Probability
1	LFD	1×10^{-4}
2	$\text{OP}_2 \cdot \text{DAA}$	1×10^{-7}
Total $\text{Pr}^{(\text{OFF})} \approx 1 \times 10^{-4}$		

The uncertainties should be propagated through the cut sets of each scenario to obtain the uncertainty associated with the frequency estimation of each scenario. The uncertainty associated with the consequence estimates can also be obtained. When uncertainty associated with the consequence values is combined with the scenario frequencies and their uncertainty, the uncertainty associated with the estimated risk can be calculated. Although this is not a necessary step in risk analysis, it is reasonable to make an estimate of the uncertainties when risk values are high.

Figure 8.11 shows the risk profile based on the values in Table 8.11.

8.6 PRECURSOR ANALYSIS

8.6.1 INTRODUCTION

Risk analysis may be carried out by completely hypothesizing scenarios of events that can lead to exposure of hazard, or it may be based on actuarial scenarios of events. Sometimes, however, certain actuarial scenarios of events may have occurred without leading to an exposure of hazard, but involve a substantial erosion of barriers that prevent or mitigate hazard exposure. These scenarios are considered as precursors to accidents (exposure of hazard).

Accident *precursor events* or simply *precursor events* (PEs), in the reliability context given, can be defined as those operational events that constitute important elements of accident sequences leading

TABLE 8.9
Cut Sets of the Scenarios

Scenario No.	Cut Sets	Frequency	Comment
1	$F \times \overline{ON}$	$7.1 \times 10^{-4}(1 - 2.0 \times 10^{-2}) = 7.1 \times 10^{-4}$	Since the probability can be directly evaluated for $\overline{ON}$ without evaluating the need to generate cut sets, only the probability is calculated
2	$F \times MAA \times LFD \times OP_2$	7.0×10^{-8}	1. Only cut sets from Table 7.6 that have a contribution greater than 1% are shown
	$F \times V_{22} \times P_1 \times \overline{LFD} \times \overline{OP_2}$	5.0×10^{-8}	2. Cut set $F \times MAA \times \overline{LFD} \times \overline{MAA}$ is eliminated since $MAA \times \overline{MAA} = \phi$
	$F \times V_{22} \times P_1 \times \overline{LFD} \times \overline{MAA}$	5.0×10^{-8}	
	$F \times V_{21} \times P_1 \times \overline{LFD} \times \overline{OP_2}$	5.0×10^{-8}	
	$F \times V_{22} \times P_1 \times \overline{LFD} \times \overline{MAA}$	5.0×10^{-8}	
	$F \times OP_1 \times V_{12} \times \overline{LFD} \times \overline{OP_2}$	2.9×10^{-9}	
	$F \times OP_1 \times V_{12} \times \overline{LFD} \times \overline{MAA}$	2.9×10^{-9}	
	$F \times OP_1 \times P_1 \times \overline{LFD} \times \overline{OP_2}$	1.1×10^{-7}	
	$F \times OP_1 \times P_1 \times \overline{LFD} \times \overline{MAA}$	1.1×10^{-7}	
	$F \times OP_1 \times V_{11} \times \overline{LFD} \times \overline{OP_2}$	2.9×10^{-9}	
2	$F \times OP_1 \times V_{11} \times \overline{LFD} \times \overline{MAA}$	2.9×10^{-9}	1. Only cut sets from Table 7.6 that have a contribution greater than 1% are shown
	$F \times P_2 \times V_{12} \times \overline{LFD} \times \overline{OP_2}$	5.0×10^{-8}	2. Cut set $F \times MAA \times \overline{LFD} \times \overline{MAA}$ is eliminated since $MAA \times \overline{MAA} = \phi$
	$F \times P_2 \times V_{12} \times \overline{LFD} \times \overline{MAA}$	5.0×10^{-8}	
	$F \times P_2 \times P_1 \times \overline{LFD} \times \overline{OP_2}$	5.0×10^{-8}	
	$F \times P_2 \times P_1 \times \overline{LFD} \times \overline{MAA}$	2.0×10^{-7}	
	$F \times P_2 \times V_{11} \times \overline{LFD} \times \overline{OP_2}$	2.0×10^{-7}	
	$F \times P_2 \times V_{11} \times \overline{LFD} \times \overline{MAA}$	5.0×10^{-8}	
	$F \times CCP \times \overline{LFD} \times \overline{OP_2}$	5.0×10^{-8}	
	$F \times CCP \times \overline{LFD} \times \overline{MAA}$	1.3×10^{-6}	
		$\Sigma_i = 1.3 \times 10^{-6}$	
3	$F \times MAA \times LFD$	7.1×10^{-12}	1. Only cut sets from Tables 7.6 and 7.7 that have the contribution to the scenario are shown
	$F \times V_{22} \times P_1 \times LFD$	5.0×10^{-12}	
	$F \times V_{21} \times P_1 \times LFD$	5.0×10^{-12}	
	$F \times OP_1 \times V_{12} \times LFD$	2.9×10^{-12}	
	$F \times OP_1 \times P_1 \times LFD$	2.8×10^{-12}	
	$F \times OP_1 \times V_{11} \times LFD$	2.9×10^{-12}	
	$F \times P_2 \times P_{12} \times LFD$	5.0×10^{-12}	
	$F \times P_2 \times P_1 \times LFD$	2.0×10^{-11}	
	$F \times P_2 \times V_{11} \times LFD$	5.0×10^{-12}	
	$F \times CCP \times LFD$	3.0×10^{-11}	
		$\Sigma_i = 8.4 \times 10^{-11}$	

TABLE 8.10
Economic Consequences of Fire Scenarios

Scenario No.	Economic Consequence
1	$1,000,000
2	$92,000,000
3	$210,000,000

TABLE 8.11
Risk Associated with Each Scenario

Scenario No.	Economic Consequence
1	(7.1×10^{-4}) ($1,000,000) = $710.000
2	(3.7) ($92,000,000) = $340.000
3	(8.4×10^{-11}) ($210,000,000) = $0.017

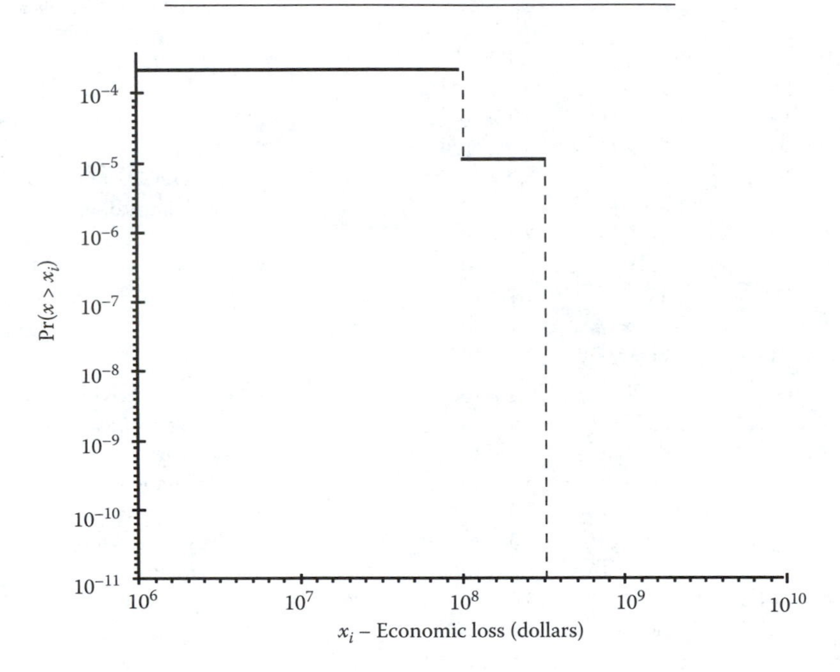

FIGURE 8.11 Risk profile.

to accidents (or hazard exposure) in complex systems, such as severe core damage in a nuclear power plant, severe aviation or marine accidents, or chemical plant accidents. The significance of a PE is measured through the conditional probability that the actual event or scenarios of events would result in exposure of hazard. In other words, PEs are those events that substantially reduce the margin of safety available for prevention of accidents.

Accident precursor analysis (APA) is a convenient tool for complex system safety and performance monitoring and analysis. The APA methodology considered in this section is mainly based on the methodology developed for nuclear power plants; nevertheless, its application to other complex systems is straightforward.

8.6.2 Basic Methodology

Considering a sequence of accidents in a system given as one following the HPP, the MLE for the rate of occurrence of accidents, λ, can be written as

$$\hat{\lambda} = \frac{n}{t}, \tag{8.2}$$

where n is the total number of accidents observed in nonrandom exposure (or cumulative exposure) time t. The total exposure time can be measured in such units as *reactor years* (for nuclear power plants), *aircraft hours flown*, *aircraft miles flown*, and so on.

Because a severe accident is a rare event (i.e., n is quite small), estimator (Equation 8.2) cannot be applied, so one must resort to postulated events whose occurrence would lead to severe accident. The marginal contribution from each PE in the numerator of Equation 8.2 can be counted as a positive number less than 1. For nuclear power plants, Apostolakis and Mosleh [30] have suggested using *conditional core damage probability* given a PE in the numerator of Equation 8.2. Obviously, this approach can be similarly used for other complex systems.

Considering all such PEs that have occurred in exposure time t, the estimator (Equation 8.2) is replaced by

$$\hat{\lambda} = \frac{\sum_i p_i}{t}, \tag{8.3}$$

where p_i is the conditional probability of severe accident given PE i.

The methodology of APA has two major components: screening (i.e., identification of events with anticipated high p_i values) and quantification (i.e., estimation of p_i and λ) and developing a corresponding trend analysis as an indicator of the overall system(s) safety. Bayesian interpretation of estimator λ is discussed in Chapter 3. For more discussion see Phimister et al. [31].

8.6.3 Differences between APA and PRAs

APA originated from the problems associated with nuclear power plant safety problems. Originally, its objective was to validate the PRA results, so that APA was traditionally viewed as a different approach from PRA. However, the two approaches are fundamentally the same but with different emphasis. For example, both approaches rely on event trees to postulate accident sequences, and both use plant-specific data to obtain failure probability of severe accidents (core damage in the case of nuclear power plants). The only difference between the two approaches is the process of identifying significant events. Readers are referred to Cooke et al. [32] who conclude that, while both approaches use the same models and data for analysis, PRA and APA differ only in the way the analysis is performed. Therefore, APA and PRA results cannot be viewed as totally independent, and one cannot validate the other.

Another small difference between the two approaches is the way in which dependent failures are treated. Dependent failures, such as CCFs, are considered in APA because a PE may include dependent failures. This is a favorable feature of APA calculations. One can also estimate the contribution that common-cause or other events make to the overall rate of occurrence of severe accidents. CCFs are explicitly modeled in PRA the same way as discussed in Section 7.2.

The last difference between the two methods is that PRAs limit themselves to a finite number of postulated events. However, some events that are not customarily included in PRA may occur as PEs, and these may be important contributions to risk. This is an important strength of APA methodology.

EXERCISES

8.1 The following table shows the data calculated in a risk study for assessing the risk of an liquified natural gas (LNG) terminal

   a. Based on the data in this table plot the risk profile in terms of "annual frequency of exceeding the given number of fatalities versus number of fatalities." That is, the so-called "Farmer's Curve."
   b. What is the frequency of exceeding 100 fatalities?

8.2 If an accident requires occurrence of event I followed by either of events B or C so that a major consequence happens:

   a. Develop an event tree to depict all scenarios that are possible.
   b. If frequency of event I is 0.1 year^{-1} and probabilities of occurrence of events B and C obtained from the following fault trees (with probabilities assigned to each basic event), determine the frequency of each scenario.

c. If consequence of each scenario is 100 injuries, what is the total risk of this accident?

Group	Expected Fatalities per Year	Number of People Sharing the Risk	Risk per Person per Year
Permanent population in Port O'Connor	2.0×10^{-8}	800	2.5×10^{-11}
Permanent population in Indianola	1.3×10^{-7}	80	1.7×10^{-9}
Transient daytime visitors	2.5×10^{-6}	2500	9.9×10^{-10}
Individuals in boats	1.35×10^{-5}	3000	4.5×10^{-9}
All individuals exposed to risk	1.7×10^{-5}	9000	1.9×10^{-9}

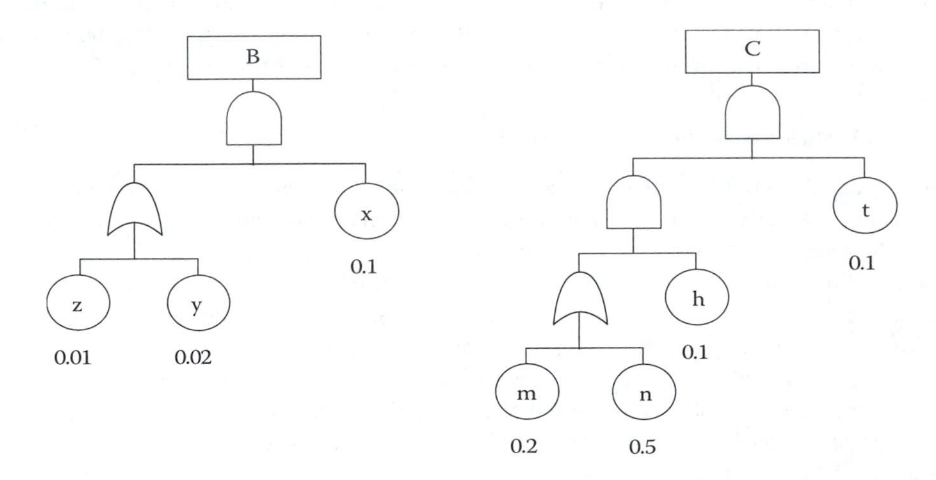

8.3 In a system, three subsystems A, B, and C must work when the system starts. The frequency, I, that the system is starred is 2 times per month: (i) When the system starts, subsystem A is turned on. If subsystem A properly works, subsystem B would not be needed, after which subsystem C must work. If subsystem C properly functions, the mission is successful, otherwise the mission fails. (ii) If subsystem A fails, then subsystem B can compensate for it. If subsystem B is successful then subsystem C would be needed. If subsystem C works, the mission is successful, otherwise the mission fails. If subsystem B fails following the failure of A, then the mission fails.

a. Draw an event tree representing the system above.
b. If Boolean expressions $A = x + yz$, $B = y + t$, $C = tz$, show the minimal cut sets of subsystems A, B, and C, then find the *minimal* cut sets of each scenario (sequence) of the event tree.
c. If the probability of the components or the subsystems are as follows, then calculate the frequency (per week) of each scenario: $\Pr(x) = \Pr(y) = 0.1$, $\Pr(z) = \Pr(t) = 0.2$

8.4 A turbine in a generator can cause a great deal of damage if it fails critically. The manager of a nuclear power plant can choose either to replace the old turbine with a new one, or to leave it in place.

The state of a turbine is categorized as either "good," "acceptable," or "poor." The probability of critical failure per quarter depends on the state of the turbine:

$$P(\text{failure}|\text{good}) = 0.0001,$$

$$P(\text{failure}|\text{acceptable}) = 0.001,$$

$$P(\text{failure}|\text{poor}) = 0.01.$$

The technical department has made a model of the degradation of the turbine. In this model it is assumed that, if the state is "good" or "acceptable," then at the end of the quarter it stays the same with probability 0.95 or degrades with probability 0.05 (i.e., good becomes "acceptable" and "acceptable" becomes "poor"). If the state is "poor," then it stays "poor."

a. Determine the probability that a turbine that is "good" becomes "good," "acceptable," and "poor" in the next three quarters, and does not fail.
   The cost of repairing the failure is \$1.5M (including replacement of parts). The cost of a new turbine is \$5K.
b. Consider a decision between two alternatives:

   a1: Install new turbine.
   a2: Continue to use old turbine.

   Consider a situation where the old turbine is categorized as "acceptable"; determine the optimal decision using expected monetary loss (risk). What would the decision be if the old turbine was "good" with 10% probability and "acceptable" with 90% probability?
   It is possible to carry out two sorts of inspection on the old turbine. A visual inspection costs \$200, but does not always give the right diagnosis. The probability of a particular outcome of the inspection given the actual state of the turbine is given in the table below:

	Actual State		
Inspection Outcome	**Good**	**Acceptable**	**Poor**
Good	0.9	0.1	0
Acceptable	0.1	0.8	0.1
Poor	0	0.1	0.9

   The second sort of inspection uses an x-ray and determines the state of the turbine *exactly*. The cost of the x-ray inspection is \$800.
c. Determine the optimal choice between no inspection, a visual inspection, and an x-ray inspection.

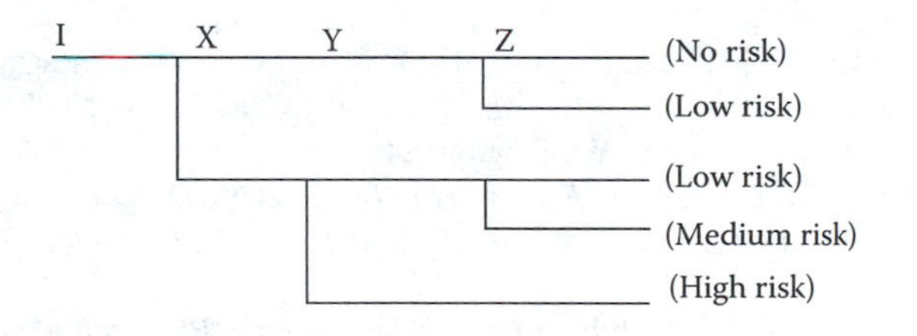

8.5 Consider the fault trees and event tree below.

a. Determine a Boolean equation representing each of the event tree scenarios in terms of the fault tree failure events (C_1, C_2, and C_3).

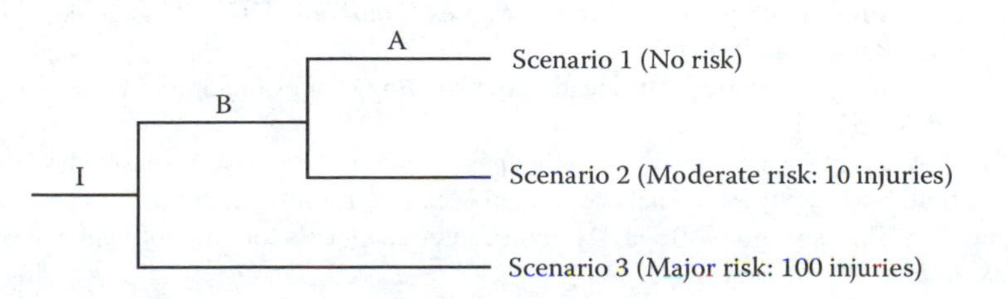

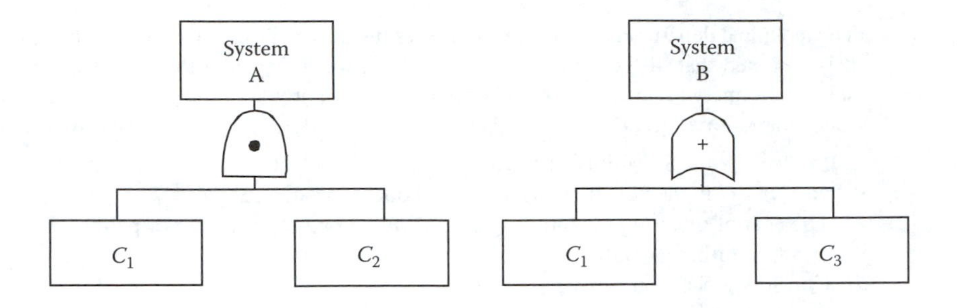

b. If the frequency of the initiating event I is 10^{-3} year^{-1}, and $\Pr(C_1) = 0.001$, $\Pr(C_2) = 0.008$, and $\Pr(C_3) = 0.005$, calculate the risk (expected injuries per year).
c. Plot the risk-profile curve (Farmer Curve) for this problem. That is to calculate complementary cumulative probability of exceeding certain number of injuries.

REFERENCES

1. Farmer, F. R., Reactor safety and siting: A proposed risk criterion, *Nuclear Safety*, 539, 23–32, 1961.
2. WASH-1400, Reactor Safety Study—An Assessment of Accident Risks in US Commercial Nuclear Power Plants, USNRC, October 1975.
3. Kaplan, S. and J. Garrick, On the quantitative definition of risk, *Risk Anal.*, 1 (1), 11–28, 1981.
4. Stamatelatos, M., G. Apostolakis, H. Dezfuli, C. Everline, S. Guarro, P. Moieni, A. Mosleh, T. Paulos, and R. Youngblood, *Probabilistic Risk Assessment Procedures Guide for NASA Managers and Practitioners*, Version 1.1, National Aeronautics and Space Administration, Washington, DC, 2002.
5. Kumamoto, H. and E. J. Henley, *Probabilistic Risk Assessment for Engineers and Scientists*, IEEE Press, New York, 1996.
6. ASME-American Society of Mechanical Engineers, Standard for Probabilistic Risk Assessment for Nuclear Power Plant Applications, ASME RA-S-2002.
7. Stamatis, D. H., *Failure Mode and Effect Analysis: FMEA from Theory to Execution*, 2nd edition, ASQ Quality Press, WI, 2003.
8. Sattison, M. B. and K. W. Hall, *Analysis of Core Damage Frequency: Zion*, Unit 1 Internal Events, US Nuclear Regulatory Commission NUREG/CR-4550, 1990.
9. Modarres, M., *Risk Analysis in Engineering, Techniques, Tools and Trends*. CRC Press, Boca Raton, FL, 2006.
10. Mosleh, A., G. W. Parry, H. M. Paula, D. H. Worledge, and D. M. Rasmuson, *Procedure for Treating Common Cause Failures in Safety and Reliability Studies*, US Nuclear Regulatory Commission, NUREG/CR-4780, Vols. I and II, Washington, DC, 1988.
11. Kapur, K. C. and L. R. Lamberson, *Reliability in Engineering Design*, Wiley, New York, 1977.
12. Nelson, W., *Accelerated Testing: Statistical Models, Test Plans and Data Analyses*, Wiley, New York, 1990.
13. Crow, L. H., Evaluating the reliability of repairable systems, *Proceedings of the Annual Reliability and Maintainability Symposium,* IEEE, 1990.
14. Ascher, H. and H. Feingold, *Repairable Systems Reliability: Modeling and Inference, Misconception and Their Causes*, Marcel Dekker, New York, 1984.
15. Poucet, A., Survey of methods used to assess human reliability in the human factors reliability benchmark exercise, *Reliab. Eng. Syst. Saf.*, 22, 257–268, 1988.
16. Smidts, C., Software reliability, in *The Electronics Handbook*, J. C. Whitaker, ed., CRC Press and IEEE Press, Boca Raton, FL, 1996.
17. Dezfuli, H. and M. Modarres, A truncation methodology for evaluation of large fault trees, *IEEE Trans. Reliab.*, R-33, 325–328, 1984.
18. Chang, Y. H., A. Mosleh, and V. Dang, Dynamic probabilistic risk assessment: Framework, tool, and application, Society for Risk Analysis Annual Meeting, Baltimore, 2003.
19. Dugan, J., S. Bavuso, and M. Boyd, Dynamic fault tree models for fault tolerant computer systems, *IEEE Trans. Reliab.*, 40 (3), 363, 1993.

20. Azarkhail, M. and M. Modarres, An intelligent-agent-oriented approach to risk analysis of complex dynamic systems with applications in planetary missions, *Proceedings of the 8th International Conference on Probabilistic Safety Assessment and Management*, ASME, New Orleans, May 2006.
21. Fussell, J., How to hand calculate system reliability and safety characteristics, *IEEE Trans. Reliab.*, R24 (3), 169–174, 1975.
22. Azarkhail, M. and M. Modarres, A study of implications of using importance measures in risk-informed decisions, PSAM-7, ESREL 04 Joint Conference, Berlin, Germany, June 2004.
23. Chamberlain, S., and M. Modarres, Compressed natural gas safety: A quantitative analysis, *Risk Anal., J.*, 25, 377–389, 2005.
24. Proper Care and Handling of Compressed Natural Gas Cylinders, *Gas Research Bulletin*, Gas Research Institute, Chicago, IL, 1996.
25. Process Equipment Reliability Data, Center for Chemical Process Safety, American Institute of Chemical Engineers, New York, 1989.
26. *Methodology for Treatment of Model Uncertainty*, Enrique Lopez Droquette, PhD Dissertation, University of Maryland, College Park, MD, 1999.
27. *Guidelines For Chemical Process Quantitative Risk Analysis*, Center For Chemical Process Safety, AIChE, New York, 1989.
28. *Bulk Storage of LPG—Factors Affecting Offsite Risk*, The Assessment of Major Hazards, AIChE, Paragon Press, New York, 1992.
29. SAPHIRE, Risk and Reliability Assessment Tools, http://saphire.inl.gov.
30. Apostolakis, G. and A. Mosleh, Expert opinion and statistical evidence. An application to reactor core melt frequency, *Nucl. Sci. Eng.*, 70 (2), 135, 1979.
31. Phimister, J., V. Bier, and C. Kunreuther, eds., *Accident Precursor Analysis and Management: Reducing Technological Risk through Diligence*, National Academics, 2004.
32. Cooke, R. M., L. Goossens, A. R. Hale, and J. Von Der Horst, Accident Sequence Precursor Methodology: A Feasibility Study for the Chemical process Industries, Technical University of Delft Report, 1987.

Appendix A

Statistical Tables

TABLE A.1
The Standard Normal Cumulative Distribution Function

z	Φ(z)	z	Φ(z)	z	Φ(z)	z	Φ(z)
−4.00	0.00003	−3.67	0.00012	−3.34	0.00042	−3.01	0.00131
−3.99	0.00003	−3.66	0.00013	−3.33	0.00043	−3.00	0.00135
−3.98	0.00003	−3.65	0.00013	−3.32	0.00045	−2.99	0.00139
−3.97	0.00004	−3.64	0.00014	−3.31	0.00047	−2.98	0.00144
−3.96	0.00004	−3.63	0.00014	−3.30	0.00048	−2.97	0.00149
−3.95	0.00004	−3.62	0.00015	−3.29	0.00050	−2.96	0.00154
−3.94	0.00004	−3.61	0.00015	−3.28	0.00052	−2.95	0.00159
−3.93	0.00004	−3.60	0.00016	−3.27	0.00054	−2.94	0.00164
−3.92	0.00004	−3.59	0.00017	−3.26	0.00056	−2.93	0.00169
−3.91	0.00005	−3.58	0.00017	−3.25	0.00058	−2.92	0.00175
−3.90	0.00005	−3.57	0.00018	−3.24	0.00060	−2.91	0.00181
−3.89	0.00005	−3.56	0.00019	−3.23	0.00062	−2.90	0.00187
−3.88	0.00005	−3.55	0.00019	−3.22	0.00064	−2.89	0.00193
−3.87	0.00005	−3.54	0.00020	−3.21	0.00066	−2.88	0.00199
−3.86	0.00006	−3.53	0.00021	−3.20	0.00069	−2.87	0.00205
−3.85	0.00006	−3.52	0.00022	−3.19	0.00071	−2.86	0.00212
−3.84	0.00006	−3.51	0.00022	−3.18	0.00074	−2.85	0.00219
−3.83	0.00006	−3.50	0.00023	−3.17	0.00076	−2.84	0.00226
−3.82	0.00007	−3.49	0.00024	−3.16	0.00079	−2.83	0.00233
−3.81	0.00007	−3.48	0.00025	−3.15	0.00082	−2.82	0.00240
−3.80	0.00007	−3.47	0.00026	−3.14	0.00084	−2.81	0.00248
−3.79	0.00008	−3.46	0.00027	−3.13	0.00087	−2.80	0.00256
−3.78	0.00008	−3.45	0.00028	−3.12	0.00090	−2.79	0.00264
−3.77	0.00008	−3.44	0.00029	−3.11	0.00094	−2.78	0.00272
−3.76	0.00008	−3.43	0.00030	−3.10	0.00097	−2.77	0.00280
−3.75	0.00009	−3.42	0.00031	−3.09	0.00100	−2.76	0.00289
−3.74	0.00009	−3.41	0.00032	−3.08	0.00104	−2.75	0.00298
−3.73	0.00010	−3.40	0.00034	−3.07	0.00107	−2.74	0.00307
−3.72	0.00010	−3.39	0.00035	−3.06	0.00111	−2.73	0.00317
−3.71	0.00010	−3.38	0.00036	−3.05	0.00114	−2.72	0.00326
−3.70	0.00011	−3.37	0.00038	−3.04	0.00118	−2.71	0.00336
−3.69	0.00011	−3.36	0.00039	−3.03	0.00122	−2.70	0.00347
−3.68	0.00012	−3.35	0.00040	−3.02	0.00126	−2.69	0.00357

continued

TABLE A.1 (continued)

z	Φ(z)	z	Φ(z)	z	Φ(z)	z	Φ(z)
−2.68	0.00368	−2.17	0.01500	−1.66	0.04846	−1.15	0.12507
−2.67	0.00379	−2.16	0.01539	−1.65	0.04947	−1.14	0.12714
−2.66	0.00391	−2.15	0.01578	−1.64	0.05050	−1.13	0.12924
−2.65	0.00402	−2.14	0.01618	−1.63	0.05155	−1.12	0.13136
−2.64	0.00415	−2.13	0.01659	−1.62	0.05262	−1.11	0.13350
−2.63	0.00427	−2.12	0.01700	−1.61	0.05370	−1.10	0.13567
−2.62	0.00440	−2.11	0.01743	−1.60	0.05480	−1.09	0.13786
−2.61	0.00453	−2.10	0.01786	−1.59	0.05592	−1.08	0.14007
−2.60	0.00466	−2.09	0.01831	−1.58	0.05705	−1.07	0.14231
−2.59	0.00480	−2.08	0.01876	−1.57	0.05821	−1.06	0.14457
−2.58	0.00494	−2.07	0.01923	−1.56	0.05938	−1.05	0.14686
−2.57	0.00508	−2.06	0.01970	−1.55	0.06057	−1.04	0.14917
−2.56	0.00523	−2.05	0.02018	−1.54	0.06178	−1.03	0.15150
−2.55	0.00539	−2.04	0.02068	−1.53	0.06301	−1.02	0.15386
−2.54	0.00554	−2.03	0.02118	−1.52	0.06426	−1.01	0.15625
−2.53	0.00570	−2.02	0.02169	−1.51	0.06552	−1.00	0.15866
−2.52	0.00587	−2.01	0.02222	−1.50	0.06681	−0.99	0.16109
−2.51	0.00604	−2.00	0.02275	−1.49	0.06811	−0.98	0.16354
−2.50	0.00621	−1.99	0.02330	−1.48	0.06944	−0.97	0.16602
−2.49	0.00639	−1.98	0.02385	−1.47	0.07078	−0.96	0.16853
−2.48	0.00657	−1.97	0.02442	−1.46	0.07215	−0.95	0.17106
−2.47	0.00676	−1.96	0.02500	−1.45	0.07353	−0.94	0.17361
−2.46	0.00695	−1.95	0.02559	−1.44	0.07493	−0.93	0.17619
−2.45	0.00714	−1.94	0.02619	−1.43	0.07636	−0.92	0.17879
−2.44	0.00734	−1.93	0.02680	−1.42	0.07780	−0.91	0.18141
−2.43	0.00755	−1.92	0.02743	−1.41	0.07927	−0.90	0.18406
−2.42	0.00776	−1.91	0.02807	−1.40	0.08076	−0.89	0.18673
−2.41	0.00798	−1.90	0.02872	−1.39	0.08226	−0.88	0.18943
−2.40	0.00820	−1.89	0.02938	−1.38	0.08379	−0.87	0.19215
−2.39	0.00842	−1.88	0.03005	−1.37	0.08534	−0.86	0.19489
−2.38	0.00866	−1.87	0.03074	−1.36	0.08691	−0.85	0.19766
−2.37	0.00889	−1.86	0.03144	−1.35	0.08851	−0.84	0.20045
−2.36	0.00914	−1.85	0.03216	−1.34	0.09012	−0.83	0.20327
−2.35	0.00939	−1.84	0.03288	−1.33	0.09176	−0.82	0.20611
−2.34	0.00964	−1.83	0.03362	−1.32	0.09342	−0.81	0.20897
−2.33	0.00990	−1.82	0.03438	−1.31	0.09510	−0.80	0.21186
−2.32	0.01017	−1.81	0.03515	−1.30	0.09680	−0.79	0.21476
−2.31	0.01044	−1.80	0.03593	−1.29	0.09853	−0.78	0.21770
−2.30	0.01072	−1.79	0.03673	−1.28	0.10027	−0.77	0.22065
−2.29	0.01101	−1.78	0.03754	−1.27	0.10204	−0.76	0.22363
−2.28	0.01130	−1.77	0.03836	−1.26	0.10383	−0.75	0.22663
−2.27	0.01160	−1.76	0.03920	−1.25	0.10565	−0.74	0.22965
−2.26	0.01191	−1.75	0.04006	−1.24	0.10749	−0.73	0.23270
−2.25	0.01222	−1.74	0.04093	−1.23	0.10935	−0.72	0.23576
−2.24	0.01255	−1.73	0.04182	−1.22	0.11123	−0.71	0.23885
−2.23	0.01287	−1.72	0.04272	−1.21	0.11314	−0.70	0.24196
−2.22	0.01321	−1.71	0.04363	−1.20	0.11507	−0.69	0.24510
−2.21	0.01355	−1.70	0.04457	−1.19	0.11702	−0.68	0.24825
−2.20	0.01390	−1.69	0.04551	−1.18	0.11900	−0.67	0.25143
−2.19	0.01426	−1.68	0.04648	−1.17	0.12100	−0.66	0.25463
−2.18	0.01463	−1.67	0.04746	−1.16	0.12302	−0.65	0.25785

continued

TABLE A.1 (continued)

z	Φ(z)	z	Φ(z)	z	Φ(z)	z	Φ(z)
−0.64	0.26109	−0.13	0.44828	+0.38	0.64803	+0.89	0.81327
−0.63	0.26435	−0.12	0.45224	+0.39	0.65173	+0.90	0.81594
−0.62	0.26763	−0.11	0.45620	+0.40	0.65542	+0.91	0.81859
−0.61	0.27093	−0.10	0.46017	+0.41	0.65910	+0.92	0.82121
−0.60	0.27425	−0.09	0.46414	+0.42	0.66276	+0.93	0.82381
−0.59	0.27760	−0.08	0.46812	+0.43	0.66640	+0.94	0.82639
−0.58	0.28096	−0.07	0.47210	+0.44	0.67003	+0.95	0.82894
−0.57	0.28834	−0.06	0.47608	+0.45	0.67364	+0.96	0.83147
−0.56	0.28774	−0.05	0.48006	+0.46	0.67724	+0.97	0.83398
−0.55	0.29116	−0.04	0.48405	+0.47	0.68082	+0.98	0.83646
−0.54	0.29460	−0.03	0.48803	+0.48	0.68439	+0.99	0.83891
−0.53	0.29806	−0.02	0.49202	+0.49	0.68793	+1.00	0.84134
−0.52	0.30153	−0.01	0.49601	+0.50	0.69146	+1.01	0.84375
−0.51	0.30503	+0.00	0.50000	+0.51	0.69497	+1.02	0.84614
−0.50	0.30854	+0.01	0.50399	+0.52	0.69847	+1.03	0.84850
−0.49	0.31207	+0.02	0.50798	+0.53	0.70194	+1.04	0.85083
−0.48	0.31561	+0.03	0.51197	+0.54	0.70540	+1.05	0.85314
−0.47	0.31918	+0.04	0.51595	+0.55	0.70884	+1.06	0.85543
−0.46	0.32276	+0.05	0.51994	+0.56	0.71226	+1.07	0.85769
−0.45	0.32636	+0.06	0.52392	+0.57	0.71566	+1.08	0.85993
−0.44	0.32997	+0.07	0.52790	+0.58	0.71904	+1.09	0.86214
−0.43	0.33360	+0.08	0.53188	+0.59	0.72240	+1.10	0.86433
−0.42	0.33724	+0.09	0.53586	+0.60	0.72575	+1.11	0.86650
−0.41	0.34090	+0.10	0.53983	+0.61	0.72907	+1.12	0.86864
−0.40	0.34458	+0.11	0.54380	+0.62	0.73237	+1.13	0.87076
−0.39	0.34827	+0.12	0.54776	+0.63	0.73565	+1.14	0.87286
−0.38	0.35197	+0.13	0.55172	+0.64	0.73891	+1.15	0.87493
−0.37	0.35569	+0.14	0.55567	+0.65	0.74215	+1.16	0.87698
−0.36	0.35942	+0.15	0.55962	+0.66	0.74537	+1.17	0.87900
−0.35	0.36317	+0.16	0.56356	+0.67	0.74857	+1.18	0.88100
−0.34	0.36693	+0.17	0.56749	+0.68	0.75175	+1.19	0.88298
−0.33	0.37070	+0.18	0.57142	+0.69	0.75490	+1.20	0.88493
−0.32	0.37448	+0.19	0.57535	+0.70	0.75804	+1.21	0.88686
−0.31	0.37828	+0.20	0.57926	+0.71	0.76115	+1.22	0.88877
−0.30	0.38209	+0.21	0.58317	+0.72	0.76424	+1.23	0.89065
−0.29	0.38591	+0.22	0.58706	+0.73	0.76730	+1.24	0.89251
−0.28	0.38974	+0.23	0.59095	+0.74	0.77035	+1.25	0.89435
−0.27	0.39358	+0.24	0.59483	+0.75	0.77337	+1.26	0.89617
−0.26	0.39743	+0.25	0.59871	+0.76	0.77637	+1.27	0.89796
−0.25	0.40129	+0.26	0.60257	+0.77	0.77935	+1.28	0.89973
−0.24	0.40517	+0.27	0.60642	+0.78	0.78230	+1.29	0.90147
−0.23	0.40905	+0.28	0.61206	+0.79	0.78524	+1.30	0.90320
−0.22	0.41294	+0.29	0.61409	+0.80	0.78814	+1.31	0.90490
−0.21	0.41683	+0.30	0.61791	+0.81	0.79103	+1.32	0.90658
−0.20	0.42074	+0.31	0.62172	+0.82	0.79389	+1.33	0.90824
−0.19	0.42465	+0.32	0.62552	+0.83	0.79673	+1.34	0.90988
−0.18	0.42858	+0.33	0.62930	+0.84	0.79955	+1.35	0.91149
−0.17	0.43251	+0.34	0.63307	+0.85	0.80234	+1.36	0.91309
−0.16	0.43644	+0.35	0.63683	+0.86	0.80511	+1.37	0.91466
−0.15	0.44038	+0.36	0.64958	+0.87	0.80785	+1.38	0.91621
−0.14	0.44433	+0.37	0.64431	+0.88	0.81057	+1.39	0.91774

continued

TABLE A.1 (continued)

z	Φ(z)	z	Φ(z)	z	Φ(z)	z	Φ(z)
+1.40	0.91924	+1.91	0.97193	+2.42	0.99224	+2.93	0.99831
+1.41	0.92073	+1.92	0.97257	+2.43	0.99245	+2.94	0.99836
+1.42	0.92220	+1.93	0.97320	+2.44	0.99266	+2.95	0.99841
+1.43	0.92364	+1.94	0.97381	+2.45	0.99286	+2.96	0.99846
+1.44	0.92507	+1.95	0.97441	+2.46	0.99305	+2.97	0.99851
+1.45	0.92647	+1.96	0.97500	+2.47	0.99324	+2.98	0.99856
+1.46	0.92785	+1.97	0.97558	+2.48	0.99343	+2.99	0.99861
+1.47	0.92922	+1.98	0.97615	+2.49	0.99361	+3.00	0.99865
+1.48	0.93056	+1.99	0.97670	+2.50	0.99379	+3.01	0.99869
+1.49	0.93189	+2.00	0.97725	+2.51	0.99396	+3.02	0.99874
+1.50	0.93319	+2.01	0.97778	+2.52	0.99413	+3.03	0.99878
+1.51	0.93448	+2.02	0.97831	+2.53	0.99430	+3.04	0.99882
+1.52	0.93574	+2.03	0.97882	+2.54	0.99446	+3.05	0.99886
+1.53	0.93766	+2.04	0.97932	+2.55	0.99461	+3.06	0.99889
+1.54	0.93822	+2.05	0.97982	+2.56	0.99477	+3.07	0.99893
+1.55	0.93943	+2.06	0.98030	+2.57	0.99492	+3.08	0.99897
+1.56	0.94062	+2.07	0.98077	+2.58	0.99506	+3.09	0.99900
+1.57	0.94179	+2.08	0.98124	+2.59	0.99520	+3.10	0.99903
+1.58	0.94295	+2.09	0.98169	+2.60	0.99534	+3.11	0.99906
+1.59	0.94408	+2.10	0.98214	+2.61	0.99547	+3.12	0.99910
+1.60	0.94520	+2.11	0.98257	+2.62	0.99560	+3.13	0.99913
+1.61	0.94630	+2.12	0.98300	+2.63	0.99573	+3.14	0.99916
+1.62	0.94738	+2.13	0.98341	+2.64	0.99585	+3.15	0.99918
+1.63	0.94845	+2.14	0.98382	+2.65	0.99698	+3.16	0.99921
+1.64	0.94950	+2.15	0.98422	+2.66	0.99609	+3.17	0.99924
+1.65	0.95053	+2.16	0.98461	+2.67	0.99621	+3.18	0.99926
+1.66	0.95154	+2.17	0.98500	+2.68	0.99632	+3.19	0.99929
+1.67	0.95254	+2.18	0.98537	+2.69	0.99643	+3.20	0.99931
+1.68	0.95352	+2.19	0.98574	+2.70	0.99653	+3.21	0.99934
+1.69	0.95449	+2.20	0.98610	+2.71	0.99664	+3.22	0.99936
+1.70	0.95543	+2.21	0.98645	+2.72	0.99674	+3.23	0.99938
+1.71	0.95637	+2.22	0.98679	+2.73	0.99683	+3.24	0.99940
+1.72	0.95728	+2.23	0.98713	+2.74	0.99693	+3.25	0.99942
+1.73	0.95818	+2.24	0.98745	+2.75	0.99702	+3.26	0.99944
+1.74	0.95907	+2.25	0.98778	+2.76	0.99711	+3.27	0.99946
+1.75	0.95994	+2.26	0.98809	+2.77	0.99720	+3.28	0.99948
+1.76	0.96080	+2.27	0.98840	+2.78	0.99728	+3.29	0.99950
+1.77	0.96164	+2.28	0.98870	+2.79	0.99736	+3.30	0.99952
+1.78	0.96246	+2.29	0.98999	+2.80	0.99744	+3.31	0.99953
+1.79	0.96327	+2.30	0.98928	+2.81	0.99752	+3.32	0.99955
+1.80	0.96407	+2.31	0.98956	+2.82	0.99760	+3.33	0.99957
+1.81	0.96485	+2.32	0.98983	+2.83	0.99767	+3.34	0.99958
+1.82	0.96562	+2.33	0.99010	+2.84	0.99774	+3.35	0.99960
+1.83	0.96638	+2.34	0.99036	+2.85	0.99781	+3.36	0.99961
+1.84	0.96712	+2.35	0.99061	+2.86	0.99788	+3.37	0.99962
+1.85	0.96784	+2.36	0.99086	+2.87	0.99795	+3.38	0.99964
+1.86	0.96856	+2.37	0.99111	+2.88	0.99801	+3.39	0.99965
+1.87	0.96926	+2.38	0.99134	+2.89	0.99807	+3.40	0.99966
+1.88	0.96995	+2.39	0.99158	+2.90	0.99813	+3.41	0.99968
+1.89	0.97062	+2.40	0.99180	+2.91	0.99819	+3.42	0.99969
+1.90	0.97128	+2.41	0.99202	+2.92	0.99825	+3.43	0.99970

continued

TABLE A.1 (continued)

z	Φ(z)	z	Φ(z)	z	Φ(z)	z	Φ(z)
+3.44	0.99971	+3.58	0.99983	+3.72	0.99990	+3.86	0.99994
+3.45	0.99972	+3.59	0.99983	+3.73	0.99990	+3.87	0.99995
+3.46	0.99973	+3.60	0.99984	+3.74	0.99991	+3.88	0.99995
+3.47	0.99974	+3.61	0.99985	+3.75	0.99991	+3.89	0.99995
+3.48	0.99975	+3.62	0.99985	+3.76	0.99992	+3.90	0.99995
+3.49	0.99976	+3.63	0.99986	+3.77	0.99992	+3.91	0.99995
+3.50	0.99977	+3.64	0.99986	+3.78	0.99992	+3.92	0.99996
+3.51	0.99978	+3.65	0.99987	+3.79	0.99992	+3.93	0.99996
+3.52	0.99978	+3.66	0.99987	+3.80	0.99993	+3.94	0.99996
+3.53	0.99979	+3.67	0.99988	+3.81	0.99993	+3.95	0.99996
+3.54	0.99980	+3.68	0.99988	+3.82	0.99993	+3.96	0.99996
+3.55	0.99981	+3.69	0.99989	+3.83	0.99994	+3.97	0.99996
+3.56	0.99981	+3.70	0.99989	+3.84	0.99994	+3.98	0.99997
+3.57	0.99982	+3.71	0.99990	+3.85	0.99994	+3.99	0.99997

TABLE A.2
Critical Values of Student's t Distribution

One-Sided Limit (Read Down)

ν	$t_{0.80}$	$t_{0.90}$	$t_{0.95}$	$t_{0.975}$	$t_{0.99}$	$t_{0.995}$	$t_{0.999}$	$t_{0.9995}$
1	1.3764	3.0777	6.3138	12.7062	31.8205	63.6567	318.3088	636.6192
2	1.0607	1.8856	2.9200	4.3027	6.9646	9.9248	22.3271	31.5991
3	0.9785	1.6377	2.3534	3.1824	4.5407	5.8409	10.2145	12.9240
4	0.9410	1.5332	2.1318	2.7764	3.7469	4.6041	7.1732	8.6103
5	0.9195	1.4759	2.0150	2.5706	3.3649	4.0321	5.8934	6.8688
6	0.9057	1.4398	1.9432	2.4469	3.1427	3.7074	5.2076	5.9588
7	0.8960	1.4149	1.8946	2.3646	2.9980	3.4995	4.7853	5.4079
8	0.8889	1.3968	1.8595	2.3060	2.8965	3.3554	4.5008	5.0413
9	0.8834	1.3830	1.8331	2.2622	2.8214	3.2498	4.2968	4.7809
10	0.8791	1.3722	1.8125	2.2281	2.7638	3.1693	4.1437	4.5869
11	0.8755	1.3634	1.7959	2.2010	2.7181	3.1058	4.0247	4.4370
12	0.8726	1.3562	1.7823	2.1788	2.6810	3.0545	3.9296	4.3178
13	0.8702	1.3502	1.7709	2.1604	2.6503	3.0123	3.8520	4.2208
14	0.8681	1.3450	1.7613	2.1448	2.6245	2.9768	3.7874	4.1405
15	0.8662	1.3406	1.7531	2.1314	2.6025	2.9467	3.7328	4.0728
16	0.8647	1.3368	1.7459	2.1199	2.5835	2.9208	3.6862	4.0150
17	0.8633	1.3334	1.7396	2.1098	2.5669	2.8982	3.6458	3.9651
18	0.8620	1.3304	1.7341	2.1009	2.5524	2.8784	3.6105	3.9216
19	0.8610	1.3277	1.7291	2.0930	2.5395	2.8609	3.5794	3.8834
20	0.8600	1.3253	1.7247	2.0860	2.5280	2.8453	3.5518	3.8495
21	0.8591	1.3232	1.7207	2.0796	2.5176	2.8314	3.5272	3.8193
22	0.8583	1.3212	1.7171	2.0739	2.5083	2.8188	3.5050	3.7921
23	0.8575	1.3195	1.7139	2.0687	2.4999	2.8073	3.4850	3.7676
24	0.8569	1.3178	1.7109	2.0639	2.4922	2.7969	3.4668	3.7454
25	0.8562	1.3163	1.7081	2.0595	2.4851	2.7874	3.4502	3.7251
26	0.8557	1.3150	1.7056	2.0555	2.4786	2.7787	3.4350	3.7066
27	0.8551	1.3137	1.7033	2.0518	2.4727	2.7707	3.4210	3.6896
28	0.8546	1.3125	1.7011	2.0484	2.4671	2.7633	3.4082	3.6739
29	0.8542	1.3114	1.6991	2.0452	2.4620	2.7564	3.3962	3.6594
30	0.8538	1.3104	1.6973	2.0423	2.4573	2.7500	3.3852	3.6460
31	0.8534	1.3095	1.6955	2.0395	2.4528	2.7440	3.3749	3.6335
32	0.8530	1.3086	1.6939	2.0369	2.4487	2.7385	3.3653	3.6218
33	0.8526	1.3077	1.6924	2.0345	2.4448	2.7333	3.3563	3.6109
34	0.8523	1.3070	1.6909	2.0322	2.4411	2.7284	3.3479	3.6007
35	0.8520	1.3062	1.6896	2.0301	2.4377	2.7238	3.3400	3.5911
36	0.8517	1.3055	1.6883	2.0281	2.4345	2.7195	3.3326	3.5821
37	0.8514	1.3049	1.6871	2.0262	2.4314	2.7154	3.3256	3.5737
38	0.8512	1.3042	1.6860	2.0244	2.4286	2.7116	3.3190	3.5657
38	0.8512	1.3042	1.6860	2.0244	2.4286	2.7116	3.3190	3.5657
40	0.8507	1.3031	1.6839	2.0211	2.4233	2.7045	3.3069	3.5510
	$t_{0.60}$	$t_{0.80}$	$t_{0.90}$	$t_{0.95}$	$t_{0.98}$	$t_{0.99}$	$t_{0.998}$	$t_{0.999}$

Two-Sided Limit (Read UP)

continued

TABLE A.2 (continued)

ν	$t_{0.80}$	$t_{0.90}$	$t_{0.95}$	$t_{0.975}$	$t_{0.99}$	$t_{0.995}$	$t_{0.999}$	$t_{0.9995}$
						One-Sided Limit (Read Down)		
41	0.8505	1.3025	1.6829	2.0195	2.4208	2.7012	3.3013	3.5442
42	0.8503	1.3020	1.6820	2.0181	2.4185	2.6981	3.2960	3.5377
43	0.8501	1.3016	1.6811	2.0167	2.4163	2.6951	3.2909	3.5316
44	0.8499	1.3011	1.6802	2.0154	2.4141	2.6923	3.2861	3.5258
45	0.8497	1.3006	1.6794	2.0141	2.4121	2.6896	3.2815	3.5203
46	0.8495	1.3002	1.6787	2.0129	2.4102	2.6870	3.2771	3.5150
47	0.8493	1.2998	1.6779	2.0117	2.4083	2.6846	3.2729	3.5099
48	0.8492	1.2994	1.6772	2.0106	2.4066	2.6822	3.2689	3.5051
49	0.8490	1.2991	1.6766	2.0096	2.4049	2.6800	3.2651	3.5004
50	0.8489	1.2987	1.6759	2.0086	2.4033	2.6778	3.2614	3.4960
55	0.8482	1.2971	1.6730	2.0040	2.3961	2.6682	3.2451	3.4764
60	0.8477	1.2958	1.6706	2.0003	2.3901	2.6603	3.2317	3.4602
65	0.8472	1.2947	1.6686	1.9971	2.3851	2.6536	3.2204	3.4466
70	0.8468	1.2938	1.6669	1.9944	2.3808	2.6479	3.2108	3.4350
75	0.8464	1.2929	1.6654	1.9921	2.3771	2.6430	3.2025	3.4250
80	0.8461	1.2922	1.6641	1.9901	2.3739	2.6387	3.1953	3.4163
85	0.8459	1.2916	1.6630	1.9883	2.3710	2.6349	3.1889	3.4087
90	0.8456	1.2910	1.6620	1.9867	2.3685	2.6316	3.1833	3.4019
95	0.8454	1.2905	1.6611	1.9853	2.3662	2.6286	3.1782	3.3959
100	0.8452	1.2901	1.6602	1.9840	2.3642	2.6259	3.1737	3.3905
110	0.8449	1.2893	1.6588	1.9818	2.3607	2.6213	3.1660	3.3812
120	0.8446	1.2886	1.6577	1.9799	2.3578	2.6174	3.1595	3.3735
130	0.8444	1.2881	1.6567	1.9784	2.3554	2.6142	3.1541	3.3669
140	0.8442	1.2876	1.6558	1.9771	2.3533	2.6114	3.1495	3.3614
150	0.8440	1.2872	1.6551	1.9759	2.3515	2.6090	3.1455	3.3566
∞	0.8416	1.2816	1.6449	1.9600	2.3263	2.5758	3.0902	3.2905
	$t_{0.60}$	$t_{0.80}$	$t_{0.90}$	$t_{0.95}$	$t_{0.98}$	$t_{0.99}$	$t_{0.998}$	$t_{0.999}$
	Two-Sided Limit (Read Up)							

For $t < 0$ values, $F(t)$ can be obtained from $F(-t) = 1 - F(t)$.

TABLE A.3
Critical Values of Chi-Square χ^2_α Distribution Function

	α							
ν	$\chi^2_{0.0005}$	$\chi^2_{0.001}$	$\chi^2_{0.005}$	$\chi^2_{0.01}$	$\chi^2_{0.025}$	$\chi^2_{0.05}$	$\chi^2_{0.1}$	$\chi^2_{0.25}$
1	$0.0^6 39$	$0.0^5 16$	$0.0^4 393$	$0.0^3 157$	$0.0^3 982$	0.00393	0.0158	0.1015
2	0.0010	0.0020	0.0100	0.0201	0.0506	0.1026	0.2107	0.5754
3	0.0153	0.0243	0.0717	0.1148	0.2158	0.3518	0.5844	1.2125
4	0.0639	0.0908	0.2070	0.2971	0.4844	0.7107	1.0636	1.9226
5	0.1581	0.2102	0.4117	0.5543	0.8312	1.1455	1.6103	2.6746
6	0.2994	0.3811	0.6757	0.8721	1.2373	1.6354	2.2041	3.4546
7	0.4849	0.5985	0.9893	1.2390	1.6899	2.1673	2.8331	4.2549
8	0.7104	0.8571	1.3444	1.6465	2.1797	2.7326	3.4895	5.0706
9	0.9717	1.1519	1.7349	2.0879	2.7004	3.3251	4.1682	5.8988
10	1.2650	1.4787	2.1559	2.5582	3.2470	3.9403	4.8652	6.7372
11	1.5868	1.8339	2.6032	3.0535	3.8157	4.5748	5.5778	7.5841
12	1.9344	2.2142	3.0738	3.5706	4.4038	5.2260	6.3038	8.4384
13	2.3051	2.6172	3.5650	4.1069	5.0088	5.8919	7.0415	9.2991
14	2.6967	3.0407	4.0747	4.6604	5.6287	6.5706	7.7895	10.1653
15	3.1075	3.4827	4.6009	5.2293	6.2621	7.2609	8.5468	11.0365
16	3.5358	3.9416	5.1422	5.8122	6.9077	7.9616	9.3122	11.9122
17	3.9802	4.4161	5.6972	6.4078	7.5642	8.6718	10.0852	12.7919
18	4.4394	4.9048	6.2648	7.0149	8.2307	9.3905	10.8649	13.6753
19	4.9123	5.4068	6.8440	7.6327	8.9065	10.1170	11.6509	14.5620
20	5.3981	5.9210	7.4338	8.2604	9.5908	10.8508	12.4426	15.4518
21	5.8957	6.4467	8.0337	8.8972	10.2829	11.5913	13.2396	16.3444
22	6.4045	6.9830	8.6427	9.5425	10.9823	12.3380	14.0415	17.2396
23	6.9237	7.5292	9.2604	10.1957	11.6886	13.0905	14.8480	18.1373
24	7.4527	8.0849	9.8862	10.8564	12.4012	13.8484	15.6587	19.0373
25	7.9910	8.6493	10.5197	11.5240	13.1197	14.6114	16.4734	19.9393
26	8.5379	9.2221	11.1602	12.1981	13.8439	15.3792	17.2919	20.8434
27	9.0932	9.8028	11.8076	12.8785	14.5734	16.1514	18.1139	21.7494
28	9.6563	10.3909	12.4613	13.5647	15.3079	16.9279	18.9392	22.6572
29	10.2268	10.9861	13.1211	14.2565	16.0471	17.7084	19.7677	23.5666
30	10.8044	11.5880	13.7867	14.9535	16.7908	18.4927	20.5992	24.4776
35	13.7875	14.6878	17.1918	18.5089	20.5694	22.4650	24.7967	29.0540
40	16.9062	17.9164	20.7065	22.1643	24.4330	26.5093	29.0505	33.6603
44	19.4825	20.5763	23.5837	25.1480	27.5746	29.7875	32.4871	37.3631
50	23.4610	24.6739	27.9907	29.7067	32.3574	34.7643	37.6886	42.9421
60	30.3405	31.7383	35.5345	37.4849	40.4817	43.1880	46.4589	52.2938
70	37.4674	39.0364	43.2752	45.4417	48.7576	51.7393	55.3289	61.6983
80	44.7910	46.5199	51.1719	53.5401	57.1532	60.3915	64.2778	71.1445
90	52.2758	54.1552	59.1963	61.7541	65.6466	69.1260	73.2911	80.6247
100	59.8957	61.9179	67.3276	70.0649	74.2219	77.9295	82.3581	90.1332
120	75.4665	77.7551	83.8516	86.9233	91.5726	95.7046	100.6236	109.2197
> ν	$\frac{1}{2}[A - 3.29]^2$ $A = (2\nu - 1)^{1/2}$	$\frac{1}{2}[A - 3.09]^2$ $A = (2\nu - 1)^{1/2}$	$\frac{1}{2}[A - 2.58]^2$ $A = (2\nu - 1)^{1/2}$	$\frac{1}{2}[A - 2.33]^2$ $A = (2\nu - 1)^{1/2}$	$\frac{1}{2}[A - 1.96]^2$ $A = (2\nu - 1)^{1/2}$	$\frac{1}{2}[A - 1.64]^2$ $A = (2\nu - 1)^{1/2}$	$\frac{1}{2}[A - 1.28]^2$ $A = (2\nu - 1)^{1/2}$	$\frac{1}{2}[A - 0.67]^2$ $A = (2\nu - 1)^{1/2}$

continued

TABLE A.3 (continued)

	α							
ν	$\chi^2_{0.75}$	$\chi^2_{0.90}$	$\chi^2_{0.95}$	$\chi^2_{0.025}$	$\chi^2_{0.99}$	$\chi^2_{0.995}$	$\chi^2_{0.999}$	$\chi^2_{0.9995}$
1	1.3233	2.7055	3.8415	5.0239	6.6349	7.8794	10.8276	12.1157
2	2.7726	4.6052	5.9915	7.3778	9.2103	10.5966	13.8155	15.2018
3	4.1083	6.2514	7.8147	9.3484	11.3449	12.8382	16.2662	17.7300
4	5.3853	7.7794	9.4877	11.1433	13.2767	14.8603	18.4668	19.9974
5	6.6257	9.2364	11.0705	12.8325	15.0863	16.7496	20.5150	22.1053
6	7.8408	10.6446	12.5916	14.4494	16.8119	18.5476	22.4577	24.1028
7	9.0371	12.0170	14.0671	16.0128	18.4753	20.2777	24.3219	26.0178
8	10.2189	13.3616	15.5073	17.5345	20.0902	21.9550	26.1245	27.8680
9	11.3888	14.6837	16.9190	19.0228	21.6660	23.5894	27.8772	29.6658
10	12.5489	15.9872	18.3070	20.4832	23.2093	25.1882	29.5883	31.4198
11	13.7007	17.2750	19.6751	21.9200	24.7250	26.7568	31.2641	33.1366
12	14.8454	18.5493	21.0261	23.3367	26.2170	28.2995	32.9095	34.8213
13	15.9839	19.8119	22.3620	24.7356	27.6882	29.8195	34.5282	36.4778
14	17.1169	21.0641	23.6848	26.1189	29.1412	31.3193	36.1233	38.1094
15	18.2451	22.3071	24.9958	27.4884	30.5779	32.8013	37.6973	39.7188
16	19.3689	23.5418	26.2962	28.8454	31.9999	34.2672	39.2524	41.3081
17	20.4887	24.7690	27.5871	30.1910	33.4087	35.7185	40.7902	42.8792
18	21.6049	25.9894	28.8693	31.5264	34.8053	37.1565	42.3124	44.4338
19	22.7178	27.2036	30.1435	32.8523	36.1909	38.5823	43.8202	45.9731
20	23.8277	28.4120	31.4104	34.1696	37.5662	39.9968	45.3147	47.4985
21	24.9348	29.6151	32.6706	35.4789	38.9322	41.4011	46.7970	49.0108
22	26.0393	30.8133	33.9244	36.7807	40.2894	42.7957	48.2679	50.5111
23	27.1413	32.0069	35.1725	38.0756	41.6384	44.1813	49.7282	52.0002
24	28.2412	33.1962	36.4150	39.3641	42.9798	45.5585	51.1786	53.4788
25	29.3389	34.3816	37.6525	40.6465	44.3141	46.9279	52.6197	54.9475
26	30.4346	35.5632	38.8851	41.9232	45.6417	48.2899	54.0520	56.4069
27	31.5284	36.7412	40.1133	43.1945	46.9629	49.6449	55.4760	57.8576
28	32.6205	37.9159	41.3371	44.4608	48.2782	50.9934	56.8923	59.3000
29	33.7109	39.0875	42.5570	45.7223	49.5879	52.3356	58.3012	60.7346
30	34.7997	40.2560	43.7730	46.9792	50.8922	53.6720	59.7031	62.1619
35	40.2228	46.0588	49.8018	53.2033	57.3421	60.2748	66.6188	69.1986
40	45.6160	51.8051	55.7585	59.3417	63.6907	66.7660	73.4020	76.0946
45	50.9849	57.5053	61.6562	65.4102	69.9568	73.1661	80.0767	82.8757
50	56.3336	63.1671	67.5048	71.4202	76.1539	79.4900	86.6608	89.5605
60	66.9815	74.3970	79.0819	83.2977	88.3794	91.9517	99.6072	102.6948
70	77.5767	85.5270	90.5312	95.0232	100.4252	104.2149	112.3169	115.5776
80	88.1303	96.5782	101.8795	106.6286	112.3288	116.3211	124.8392	128.2613
90	98.6499	107.5650	113.1453	118.1359	124.1163	128.2989	137.2084	140.7823
100	109.1412	118.4980	124.3421	129.5612	135.8067	140.1695	149.4493	153.1670
120	130.0546	140.2326	146.5674	152.2114	158.9502	163.6482	173.6174	177.6029
$> \nu$	$\frac{1}{2}[A+0.67]^2$ $A=(2\nu-1)^{1/2}$	$\frac{1}{2}[A+1.28]^2$ $A=(2\nu-1)^{1/2}$	$\frac{1}{2}[A+1.64]^2$ $A=(2\nu-1)^{1/2}$	$\frac{1}{2}[A+1.96]^2$ $A=(2\nu-1)^{1/2}$	$\frac{1}{2}[A+2.33]^2$ $A=(2\nu-1)^{1/2}$	$\frac{1}{2}[A+2.58]^2$ $A=(2\nu-1)^{1/2}$	$\frac{1}{2}[A+3.09]^2$ $A=(2\nu-1)^{1/2}$	$\frac{1}{2}[A+3.29]^2$ $A=(2\nu-1)^{1/2}$

Notes: 0.0639 = 0.00000039, ν is the number of degrees of freedom.

TABLE A.4
Critical Values of the Kolmogorov-Smirnov Statistic $D_n(\alpha)$

	α				
n	**0.20**	**0.15**	**0.10**	**0.05**	**0.01**
1	0.900	0.925	0.950	0.975	0.995
2	0.684	0.725	0.776	0.842	0.929
3	0.565	0.597	0.636	0.708	0.829
4	0.493	0.525	0.565	0.624	0.734
5	0.447	0.474	0.510	0.563	0.669
6	0.410	0.436	0.468	0.519	0.617
7	0.381	0.405	0.436	0.483	0.576
8	0.358	0.381	0.410	0.454	0.542
9	0.339	0.360	0.387	0.430	0.513
10	0.323	0.342	0.369	0.409	0.489
11	0.308	0.326	0.352	0.391	0.468
12	0.296	0.313	0.338	0.375	0.449
13	0.285	0.302	0.325	0.361	0.432
14	0.275	0.292	0.314	0.349	0.418
15	0.266	0.283	0.304	0.338	0.404
16	0.258	0.274	0.295	0.327	0.392
17	0.250	0.266	0.286	0.318	0.381
18	0.244	0.259	0.279	0.309	0.371
19	0.237	0.252	0.271	0.301	0.361
20	0.232	0.246	0.265	0.294	0.352
21	0.226	0.249	0.259	0.287	0.344
22	0.221	0.243	0.253	0.281	0.337
23	0.216	0.238	0.247	0.275	0.330
24	0.212	0.233	0.242	0.269	0.323
25	0.208	0.228	0.238	0.264	0.317
26	0.204	0.224	0.233	0.259	0.311
27	0.200	0.219	0.229	0.254	0.305
28	0.197	0.215	0.225	0.250	0.300
29	0.193	0.212	0.221	0.246	0.295
30	0.190	0.208	0.218	0.242	0.290
31	0.187	0.205	0.214	0.238	0.285
32	0.184	0.201	0.211	0.234	0.281
33	0.182	0.198	0.208	0.231	0.277
34	0.179	0.195	0.205	0.227	0.273
35	0.177	0.193	0.202	0.224	0.269
36	0.174	0.190	0.199	0.221	0.265
37	0.172	0.187	0.196	0.218	0.262
38	0.170	0.185	0.194	0.215	0.258
39	0.168	0.182	0.191	0.213	0.255
40	0.165	0.180	0.189	0.210	0.252
> 40	$1.07/(n)^{1/2}$	$1.14/(n)^{1/2}$	$1.22/(n)^{1/2}$	$1.36/(n)^{1/2}$	$1.63/(n)^{1/2}$

n is the number of trials.

TABLE A.5
F-Cumulative Distribution Function, Upper 10 Percentage Points

$\nu_2 \downarrow$ / $\nu_1 \rightarrow$	1	2	3	4	5	6	7	8	9	10	12	15	20	24	30	40	60	120	∞
1	39.9	49.5	53.6	55.8	57.2	58.2	58.9	59.4	59.9	60.2	60.7	61.2	61.7	62.0	62.3	62.5	62.8	63.1	63.3
2	8.53	9.00	9.16	9.24	9.29	9.33	9.35	9.37	9.38	9.39	9.41	9.42	9.44	9.45	9.46	9.47	9.47	9.48	9.49
3	5.54	5.46	5.39	5.34	5.31	5.28	5.27	5.25	5.24	5.23	5.22	5.20	5.18	5.18	5.17	5.16	5.15	5.14	5.13
4	4.54	4.32	4.19	4.11	4.05	4.01	3.98	3.95	3.94	3.92	3.90	3.87	3.84	3.83	3.82	3.80	3.79	3.78	3.76
5	4.06	3.78	3.62	3.52	3.45	3.40	3.37	3.34	3.32	3.30	3.27	3.24	3.21	3.19	3.17	3.16	3.14	3.12	3.11
6	3.78	3.46	3.29	3.18	3.11	3.05	3.01	2.98	2.96	2.94	2.90	2.87	2.84	2.82	2.80	2.78	2.76	2.74	2.72
7	3.59	3.26	3.07	2.96	2.88	2.83	2.78	2.75	2.72	2.70	2.67	2.63	2.59	2.58	2.56	2.54	2.51	2.49	2.47
8	3.46	3.11	2.92	2.81	2.73	2.67	2.62	2.59	2.56	2.54	2.50	2.46	2.42	2.40	2.38	2.36	2.34	2.32	2.29
9	3.36	3.01	2.81	2.69	2.61	2.55	2.51	2.47	2.44	2.42	2.38	2.34	2.30	2.28	2.25	2.23	2.21	2.18	2.16
10	3.29	2.92	2.73	2.61	2.52	2.46	2.41	2.38	2.35	2.32	2.28	2.24	2.20	2.18	2.16	2.13	2.11	2.08	2.06
11	3.23	2.86	2.66	2.54	2.45	2.39	2.34	2.30	2.27	2.25	2.21	2.17	2.12	2.10	2.08	2.05	2.03	2.00	1.97
12	3.18	2.81	2.61	2.48	2.39	2.33	2.28	2.24	2.21	2.19	2.15	2.10	2.06	2.04	2.01	1.99	1.96	1.93	1.90
13	3.14	2.76	2.56	2.43	2.35	2.28	2.23	2.20	2.16	2.14	2.10	2.05	2.01	1.98	1.96	1.93	1.90	1.88	1.85
14	3.10	2.73	2.52	2.39	2.31	2.24	2.19	2.15	2.12	2.10	2.05	2.01	1.96	1.94	1.91	1.89	1.86	1.83	1.80
15	3.07	2.70	2.49	2.36	2.27	2.21	2.16	2.12	2.09	2.06	2.02	1.97	1.92	1.90	1.87	1.85	1.82	1.79	1.76
16	3.05	2.67	2.46	2.33	2.24	2.18	2.13	2.09	2.06	2.03	1.99	1.94	1.89	1.87	1.84	1.81	1.78	1.75	1.72
17	3.03	2.64	2.44	2.31	2.22	2.15	2.10	2.06	2.03	2.00	1.96	1.91	1.86	1.84	1.81	1.78	1.75	1.72	1.69
18	3.01	2.62	2.42	2.29	2.20	2.13	2.08	2.04	2.00	1.98	1.93	1.89	1.84	1.81	1.78	1.75	1.72	1.69	1.66
19	2.99	2.61	2.40	2.27	2.18	2.11	2.06	2.02	1.98	1.96	1.91	1.86	1.81	1.79	1.76	1.73	1.70	1.67	1.63
20	2.97	2.59	2.38	2.25	2.16	2.09	2.04	2.00	1.96	1.94	1.89	1.84	1.79	1.77	1.74	1.71	1.68	1.64	1.61
21	2.96	2.57	2.36	2.23	2.14	2.08	2.02	1.98	1.95	1.92	1.87	1.83	1.78	1.75	1.72	1.69	1.66	1.62	1.59
22	2.95	2.56	2.35	2.22	2.13	2.06	2.01	1.97	1.93	1.90	1.86	1.81	1.76	1.73	1.70	1.67	1.64	1.60	1.57
23	2.94	2.55	2.34	2.21	2.11	2.05	1.99	1.95	1.92	1.89	1.84	1.80	1.74	1.72	1.69	1.66	1.62	1.59	1.55
24	2.93	2.54	2.33	2.19	2.10	2.04	1.98	1.94	1.91	1.88	1.83	1.78	1.73	1.70	1.67	1.64	1.61	1.57	1.53
25	2.92	2.53	2.32	2.18	2.09	2.02	1.97	1.93	1.89	1.87	1.82	1.77	1.72	1.69	1.66	1.63	1.59	1.56	1.52
26	2.91	2.52	2.31	2.17	2.08	2.01	1.96	1.92	1.88	1.86	1.81	1.76	1.71	1.68	1.65	1.61	1.58	1.54	1.50
27	2.90	2.51	2.30	2.17	2.07	2.00	1.95	1.91	1.87	1.85	1.80	1.75	1.70	1.67	1.64	1.60	1.57	1.53	1.49
28	2.89	2.50	2.29	2.16	2.06	2.00	1.94	1.90	1.87	1.84	1.79	1.74	1.69	1.66	1.63	1.59	1.56	1.52	1.48
29	2.89	2.50	2.28	2.15	2.06	1.99	1.93	1.89	1.86	1.83	1.78	1.73	1.68	1.65	1.62	1.58	1.55	1.51	1.47
30	2.88	2.49	2.28	2.14	2.05	1.98	1.93	1.88	1.85	1.82	1.77	1.72	1.67	1.64	1.61	1.57	1.54	1.50	1.46
40	2.84	2.44	2.23	2.09	2.00	1.93	1.87	1.83	1.79	1.76	1.71	1.66	1.61	1.57	1.54	1.51	1.47	1.42	1.38
60	2.79	2.39	2.18	2.04	1.95	1.87	1.82	1.77	1.74	1.71	1.66	1.60	1.54	1.51	1.48	1.44	1.40	1.35	1.29
120	2.75	2.35	2.13	1.99	1.90	1.82	1.77	1.72	1.68	1.65	1.60	1.55	1.48	1.45	1.41	1.37	1.32	1.26	1.19
∞	2.71	2.30	2.08	1.94	1.85	1.77	1.72	1.67	1.63	1.60	1.55	1.49	1.42	1.38	1.34	1.30	1.24	1.17	1.00

continued

TABLE A.5 (continued)
F-Cumulative Distribution Function, Upper 5 Percentage Points

$\nu_1 \rightarrow$ / $\nu_2 \downarrow$	1	2	3	4	5	6	7	8	9	10	12	15	20	24	30	40	60	120	∞
1	161	200	216	225	230	234	237	239	241	242	244	246	248	249	250	251	252	253	254
2	18.50	19.00	19.16	19.25	19.30	19.33	19.35	19.37	19.38	19.40	19.41	19.43	19.45	19.45	19.46	19.47	19.48	19.49	19.50
3	10.13	9.55	9.28	9.12	9.01	8.94	8.89	8.85	8.81	8.79	8.74	8.70	8.66	8.64	8.62	8.59	8.57	8.55	8.53
4	7.71	6.94	6.59	6.39	6.26	6.16	6.09	6.04	6.00	5.96	5.91	5.86	5.80	5.77	5.75	5.72	5.69	5.66	5.63
5	6.61	5.79	5.41	5.19	5.05	4.95	4.88	4.82	4.77	4.74	4.68	4.62	4.56	4.53	4.50	4.46	4.43	4.40	4.37
6	5.99	5.14	4.76	4.53	4.39	4.28	4.21	4.15	4.10	4.06	4.00	3.94	3.87	3.84	3.81	3.77	3.74	3.70	3.67
7	5.59	4.74	4.35	4.12	3.97	3.87	3.79	3.73	3.68	3.64	3.57	3.51	3.44	3.41	3.38	3.34	3.30	3.27	3.23
8	5.32	4.46	4.07	3.84	3.69	3.58	3.50	3.44	3.39	3.35	3.28	3.22	3.15	3.12	3.08	3.04	3.01	2.97	2.93
9	5.12	4.26	3.86	3.63	3.48	3.37	3.29	3.23	3.18	3.14	3.07	3.01	2.94	2.90	2.86	2.83	2.79	2.75	2.71
10	4.96	4.10	3.71	3.48	3.33	3.22	3.14	3.07	3.02	2.98	2.91	2.85	2.77	2.74	2.70	2.66	2.62	2.58	2.54
11	4.84	3.98	3.59	3.36	3.20	3.09	3.01	2.95	2.90	2.85	2.79	2.72	2.65	2.61	2.57	2.53	2.49	2.45	2.40
12	4.75	3.89	3.49	3.26	3.11	3.00	2.91	2.85	2.80	2.75	2.69	2.62	2.54	2.51	2.47	2.43	2.38	2.34	2.30
13	4.67	3.81	3.41	3.18	3.03	2.92	2.83	2.77	2.71	2.67	2.60	2.53	2.46	2.42	2.38	2.34	2.30	2.25	2.21
14	4.60	3.74	3.34	3.11	2.96	2.85	2.76	2.70	2.65	2.60	2.53	2.46	2.39	2.35	2.31	2.27	2.22	2.18	2.13
15	4.54	3.68	3.29	3.06	2.90	2.79	2.71	2.64	2.59	2.54	2.48	2.40	2.33	2.29	2.25	2.20	2.16	2.11	2.07
16	4.49	3.63	3.24	3.01	2.85	2.74	2.66	2.59	2.54	2.49	2.42	2.35	2.28	2.24	2.19	2.15	2.11	2.06	2.01
17	4.45	3.59	3.20	2.96	2.81	2.70	2.61	2.55	2.49	2.45	2.38	2.31	2.23	2.19	2.15	2.10	2.06	2.01	1.96
18	4.41	3.55	3.16	2.93	2.77	2.66	2.58	2.51	2.46	2.41	2.34	2.27	2.19	2.15	2.11	2.06	2.02	1.97	1.92
19	4.38	3.52	3.13	2.90	2.74	2.63	2.54	2.48	2.42	2.38	2.31	2.23	2.16	2.11	2.07	2.03	1.98	1.93	1.88
20	4.35	3.49	3.10	2.87	2.71	2.60	2.51	2.45	2.39	2.35	2.28	2.20	2.12	2.08	2.04	1.99	1.95	1.90	1.84
21	4.32	3.47	3.07	2.84	2.68	2.57	2.49	2.42	2.37	2.32	2.25	2.18	2.10	2.05	2.01	1.96	1.92	1.87	1.81
22	4.30	3.44	3.05	2.82	2.66	2.55	2.46	2.40	2.34	2.30	2.23	2.15	2.07	2.03	1.98	1.94	1.89	1.84	1.78
23	4.28	3.42	3.03	2.80	2.64	2.53	2.44	2.37	2.32	2.27	2.20	2.13	2.05	2.01	1.96	1.91	1.86	1.81	1.76
24	4.26	3.40	3.01	2.78	2.62	2.51	2.42	2.36	2.30	2.25	2.18	2.11	2.03	1.98	1.94	1.89	1.84	1.79	1.73
25	4.24	3.39	2.99	2.76	2.60	2.49	2.40	2.34	2.28	2.24	2.16	2.09	2.01	1.96	1.92	1.87	1.82	1.77	1.71
26	4.23	3.37	2.98	2.74	2.59	2.47	2.39	2.32	2.27	2.22	2.15	2.07	1.99	1.95	1.90	1.85	1.80	1.75	1.69
27	4.21	3.35	2.96	2.73	2.57	2.46	2.37	2.31	2.25	2.20	2.13	2.06	1.97	1.93	1.88	1.84	1.79	1.73	1.67
28	4.20	3.34	2.95	2.71	2.56	2.45	2.36	2.29	2.24	2.19	2.12	2.04	1.96	1.91	1.87	1.82	1.77	1.71	1.65
29	4.18	3.33	2.93	2.70	2.55	2.43	2.35	2.28	2.22	2.18	2.10	2.03	1.94	1.90	1.85	1.81	1.75	1.70	1.64
30	4.17	3.32	2.92	2.69	2.53	2.42	2.33	2.27	2.21	2.16	2.09	2.01	1.93	1.89	1.84	1.79	1.74	1.68	1.62
40	4.08	3.23	2.84	2.61	2.45	2.34	2.25	2.18	2.12	2.08	2.00	1.92	1.84	1.79	1.74	1.69	1.64	1.58	1.51
60	4.00	3.15	2.76	2.53	2.37	2.25	2.17	2.10	2.04	1.99	1.92	1.84	1.75	1.70	1.65	1.59	1.53	1.47	1.39
120	3.92	3.07	2.68	2.45	2.29	2.18	2.09	2.02	1.96	1.91	1.83	1.75	1.66	1.61	1.55	1.50	1.43	1.35	1.25
∞	3.84	3.00	2.60	2.37	2.21	2.10	2.01	1.94	1.88	1.83	1.75	1.67	1.57	1.52	1.46	1.39	1.32	1.22	1.00

continued

TABLE A.5 (continued)
F-Cumulative Distribution Function, Upper 2.5 Percentage Points

$\nu_1 \rightarrow$ $\nu_2 \downarrow$	1	2	3	4	5	6	7	8	9	10	12	15	20	24	30	40	60	120	∞
1	648	800	864	900	922	937	948	957	963	969	977	985	993	997	1001	1006	1010	1014	1018
2	38.51	39.00	39.17	39.25	39.30	39.33	39.36	39.37	39.39	39.40	39.41	39.43	39.45	39.46	39.47	39.47	39.48	39.49	39.50
3	17.44	16.04	15.44	15.10	14.88	14.73	14.62	14.54	14.47	14.42	14.34	14.25	14.17	14.12	14.08	14.04	13.99	13.95	13.90
4	12.22	10.65	9.98	9.60	9.36	9.20	9.07	8.98	8.90	8.84	8.75	8.66	8.56	8.51	8.46	8.41	8.36	8.31	8.26
5	10.01	8.43	7.76	7.39	7.15	6.98	6.85	6.76	6.68	6.62	6.52	6.43	6.33	6.28	6.23	6.18	6.12	6.07	6.02
6	8.81	7.26	6.60	6.23	5.99	5.82	5.70	5.60	5.52	5.46	5.37	5.27	5.17	5.12	5.07	5.01	4.96	4.90	4.85
7	8.07	6.54	5.89	5.52	5.29	5.12	4.99	4.90	4.82	4.76	4.67	4.57	4.47	4.42	4.36	4.31	4.25	4.20	4.14
8	7.57	6.06	5.42	5.05	4.82	4.65	4.53	4.43	4.36	4.30	4.20	4.10	4.00	3.95	3.89	3.84	3.78	3.73	3.67
9	7.21	5.71	5.08	4.72	4.48	4.32	4.20	4.10	4.03	3.96	3.87	3.77	3.67	3.61	3.56	3.51	3.45	3.39	3.33
10	6.94	5.46	4.83	4.47	4.24	4.07	3.95	3.85	3.78	3.72	3.62	3.52	3.42	3.37	3.31	3.26	3.20	3.14	3.08
11	6.72	5.26	4.63	4.28	4.04	3.88	3.76	3.66	3.59	3.53	3.43	3.33	3.23	3.17	3.12	3.06	3.00	2.94	2.88
12	6.55	5.10	4.47	4.12	3.89	3.73	3.61	3.51	3.44	3.37	3.28	3.18	3.07	3.02	2.96	2.91	2.85	2.79	2.73
13	6.41	4.97	4.35	4.00	3.77	3.60	3.48	3.39	3.31	3.25	3.15	3.05	2.95	2.89	2.84	2.78	2.72	2.66	2.60
14	6.30	4.86	4.24	3.89	3.66	3.50	3.38	3.29	3.21	3.15	3.05	2.95	2.84	2.79	2.73	2.67	2.61	2.55	2.49
15	6.20	4.77	4.15	3.80	3.58	3.41	3.29	3.20	3.12	3.06	2.96	2.86	2.76	2.70	2.64	2.59	2.52	2.46	2.40
16	6.12	4.69	4.08	3.73	3.50	3.34	3.22	3.12	3.05	2.99	2.89	2.79	2.68	2.63	2.57	2.51	2.45	2.38	2.32
17	6.04	4.62	4.01	3.66	3.44	3.28	3.16	3.06	2.98	2.92	2.82	2.72	2.62	2.56	2.50	2.44	2.38	2.32	2.25
18	5.98	4.56	3.95	3.61	3.38	3.22	3.10	3.01	2.93	2.87	2.77	2.67	2.56	2.50	2.45	2.38	2.32	2.26	2.19
19	5.92	4.51	3.90	3.56	3.33	3.17	3.05	2.96	2.88	2.82	2.72	2.62	2.51	2.45	2.39	2.33	2.27	2.20	2.13
20	5.87	4.46	3.86	3.51	3.29	3.13	3.01	2.91	2.84	2.77	2.68	2.57	2.46	2.41	2.35	2.29	2.22	2.16	2.09
21	5.83	4.42	3.82	3.48	3.25	3.09	2.97	2.87	2.80	2.73	2.64	2.53	2.42	2.37	2.31	2.25	2.18	2.11	2.04
22	5.79	4.38	3.78	3.44	3.22	3.05	2.93	2.84	2.76	2.70	2.60	2.50	2.39	2.33	2.27	2.21	2.15	2.08	2.00
23	5.75	4.35	3.75	3.41	3.18	3.02	2.90	2.81	2.73	2.67	2.57	2.47	2.36	2.30	2.24	2.18	2.11	2.04	1.97
24	5.72	4.32	3.72	3.38	3.15	2.99	2.87	2.78	2.70	2.64	2.54	2.44	2.33	2.27	2.21	2.15	2.08	2.01	1.94
25	5.69	4.29	3.69	3.35	3.13	2.97	2.85	2.75	2.68	2.61	2.51	2.41	2.30	2.24	2.18	2.12	2.05	1.98	1.91
26	5.66	4.27	3.67	3.33	3.10	2.94	2.82	2.73	2.65	2.59	2.49	2.39	2.28	2.22	2.16	2.09	2.03	1.95	1.88
27	5.63	4.24	3.65	3.31	3.08	2.92	2.80	2.71	2.63	2.57	2.47	2.36	2.25	2.19	2.13	2.07	2.00	1.93	1.85
28	5.61	4.22	3.63	3.29	3.06	2.90	2.78	2.69	2.61	2.55	2.45	2.34	2.23	2.17	2.11	2.05	1.98	1.91	1.83
29	5.59	4.20	3.61	3.27	3.04	2.88	2.76	2.67	2.59	2.53	2.43	2.32	2.21	2.15	2.09	2.03	1.96	1.89	1.81
30	5.57	4.18	3.59	3.25	3.03	2.87	2.75	2.65	2.57	2.51	2.41	2.31	2.20	2.14	2.07	2.01	1.94	1.87	1.79
40	5.42	4.05	3.46	3.13	2.90	2.74	2.62	2.53	2.45	2.39	2.29	2.18	2.07	2.01	1.94	1.88	1.80	1.72	1.64
60	5.29	3.93	3.34	3.01	2.79	2.63	2.51	2.41	2.33	2.27	2.17	2.06	1.94	1.88	1.82	1.74	1.67	1.58	1.48
120	5.15	3.80	3.23	2.89	2.67	2.52	2.39	2.30	2.22	2.16	2.05	1.95	1.82	1.76	1.69	1.61	1.53	1.43	1.31
∞	5.02	3.69	3.12	2.79	2.57	2.41	2.29	2.19	2.11	2.05	1.94	1.83	1.71	1.64	1.57	1.48	1.39	1.27	1.00

continued

TABLE A.5 (continued)
F-Cumulative Distribution Function, Upper 1 Percentage Points

$\nu_1 \rightarrow$ / $\nu_2 \downarrow$	1	2	3	4	5	6	7	8	9	10	12	15	20	24	30	40	60	120	∞
1	4052	5000	5403	5625	5764	5859	5928	5981	6022	6056	6106	6157	6209	6235	6261	6287	6313	6339	6366
2	98.50	99.00	99.17	99.25	99.30	99.33	99.36	99.37	99.39	99.40	99.42	99.43	99.45	99.46	99.47	99.47	99.48	99.49	99.50
3	34.12	30.82	29.46	28.71	28.24	27.91	27.67	27.49	27.35	27.23	27.05	26.87	26.69	26.60	26.51	26.41	26.32	26.22	26.13
4	21.20	18.00	16.69	15.98	15.52	15.21	14.98	14.80	14.66	14.55	14.37	14.20	14.02	13.93	13.84	13.75	13.65	13.56	13.46
5	16.26	13.27	12.06	11.39	10.97	10.67	10.46	10.29	10.16	10.05	9.89	9.72	9.55	9.47	9.38	9.29	9.20	9.11	9.02
6	13.75	10.93	9.78	9.15	8.75	8.47	8.26	8.10	7.98	7.87	7.72	7.56	7.40	7.31	7.23	7.14	7.06	6.97	6.88
7	12.25	9.55	8.45	7.85	7.46	7.19	6.99	6.84	6.72	6.62	6.47	6.31	6.16	6.07	5.99	5.91	5.82	5.74	5.65
8	11.26	8.65	7.59	7.01	6.63	6.37	6.18	6.03	5.91	5.81	5.67	5.52	5.36	5.28	5.20	5.12	5.03	4.95	4.86
9	10.56	8.02	6.99	6.42	6.06	5.80	5.61	5.47	5.35	5.26	5.11	4.96	4.81	4.73	4.65	4.57	4.48	4.40	4.31
10	10.04	7.56	6.55	5.99	5.64	5.39	5.20	5.06	4.94	4.85	4.71	4.56	4.41	4.33	4.25	4.17	4.08	4.00	3.91
11	9.65	7.21	6.22	5.67	5.32	5.07	4.89	4.74	4.63	4.54	4.40	4.25	4.10	4.02	3.94	3.86	3.78	3.69	3.60
12	9.33	6.93	5.95	5.41	5.06	4.82	4.64	4.50	4.39	4.30	4.16	4.01	3.86	3.78	3.70	3.62	3.54	3.45	3.36
13	9.07	6.70	5.74	5.21	4.86	4.62	4.44	4.30	4.19	4.10	3.96	3.82	3.67	3.59	3.51	3.43	3.34	3.26	3.17
14	8.86	6.52	5.56	5.04	4.70	4.46	4.28	4.14	4.03	3.94	3.80	3.66	3.51	3.43	3.35	3.27	3.18	3.09	3.00
15	8.68	6.36	5.42	4.89	4.56	4.32	4.14	4.00	3.90	3.81	3.67	3.52	3.37	3.29	3.21	3.13	3.05	2.96	2.87
16	8.53	6.23	5.29	4.77	4.44	4.20	4.03	3.89	3.78	3.69	3.55	3.41	3.26	3.18	3.10	3.02	2.93	2.85	2.75
17	8.40	6.11	5.19	4.67	4.34	4.10	3.93	3.79	3.68	3.59	3.46	3.31	3.16	3.08	3.00	2.92	2.84	2.75	2.65
18	8.29	6.01	5.09	4.58	4.25	4.02	3.84	3.71	3.60	3.51	3.37	3.23	3.08	3.00	2.92	2.84	2.75	2.66	2.57
19	8.19	5.93	5.01	4.50	4.17	3.94	3.77	3.63	3.52	3.43	3.30	3.15	3.00	2.93	2.84	2.76	2.67	2.58	2.49
20	8.10	5.85	4.94	4.43	4.10	3.87	3.70	3.56	3.46	3.37	3.23	3.09	2.94	2.86	2.78	2.70	2.61	2.52	2.42
21	8.02	5.78	4.87	4.37	4.04	3.81	3.64	3.51	3.40	3.31	3.17	3.03	2.88	2.80	2.72	2.64	2.55	2.46	2.36
22	7.95	5.72	4.82	4.31	3.99	3.76	3.59	3.45	3.35	3.26	3.12	2.98	2.83	2.75	2.67	2.58	2.50	2.40	2.31
23	7.88	5.66	4.77	4.26	3.94	3.71	3.54	3.41	3.30	3.21	3.07	2.93	2.78	2.70	2.62	2.54	2.45	2.35	2.26
24	7.82	5.61	4.72	4.22	3.90	3.67	3.50	3.36	3.26	3.17	3.03	2.89	2.74	2.66	2.58	2.49	2.40	2.31	2.21
25	7.77	5.57	4.68	4.18	3.86	3.63	3.46	3.32	3.22	3.13	2.99	2.85	2.70	2.62	2.54	2.45	2.36	2.27	2.17
26	7.72	5.53	4.64	4.14	3.82	3.59	3.42	3.29	3.18	3.09	2.96	2.82	2.66	2.59	2.50	2.42	2.33	2.23	2.13
27	7.68	5.49	4.60	4.11	3.79	3.56	3.39	3.26	3.15	3.06	2.93	2.78	2.63	2.55	2.47	2.38	2.29	2.20	2.10
28	7.64	5.45	4.57	4.07	3.75	3.53	3.36	3.23	3.12	3.03	2.90	2.75	2.60	2.52	2.44	2.35	2.26	2.17	2.06
29	7.60	5.42	4.54	4.05	3.73	3.50	3.33	3.20	3.09	3.01	2.87	2.73	2.57	2.50	2.41	2.33	2.23	2.14	2.03
30	7.56	5.39	4.51	4.02	3.70	3.47	3.30	3.17	3.07	2.98	2.84	2.70	2.55	2.47	2.39	2.30	2.21	2.11	2.01
40	7.31	5.18	4.31	3.83	3.51	3.29	3.12	2.99	2.89	2.80	2.67	2.52	2.37	2.29	2.20	2.11	2.02	1.92	1.81
60	7.08	4.98	4.13	3.65	3.34	3.12	2.95	2.82	2.72	2.63	2.50	2.35	2.20	2.12	2.03	1.94	1.84	1.73	1.60
120	6.85	4.79	3.95	3.48	3.17	2.96	2.79	2.66	2.56	2.47	2.34	2.19	2.04	1.95	1.86	1.76	1.66	1.53	1.38
∞	6.64	4.61	3.78	3.32	3.02	2.80	2.64	2.51	2.41	2.32	2.19	2.04	1.88	1.79	1.70	1.59	1.47	1.33	1.00

Appendix B

Generic Failure Probability Data

TABLE B.1
Generic Failure Data for Mechanical Components[a]

Component	Operation	Failure Mode	5%	Median	Mean	95%	Pdf Type[b]	α	β	Component Description and Comments
Air-operated valve	Standby	FTO/C	5.75E−05	7.69E−04	1.11E−03	3.31E−03	Beta	1.005	9.075E+02	The air-operated valve component boundary includes the valve, the valve operator (including the associated solenoid operated valves), local circuit breaker, and local instrumentation and control circuitry.
	Standby	SO	1.95E−11	4.43E−08	1.82E−07	8.31E−07	Gamma	0.300	1.651E+06	
	Control	FC	3.21E−10	7.31E−07	3.00E−06	1.37E−05	Gamma	0.300	1.000E+05	
Battery	Running	FTOP	2.94E−09	7.26E−07	1.86E−06	7.57E−06	Gamma	0.427	2.290E+05	The battery boundary includes the battery cells.
Battery charger	Running/alternating	FTOP	6.51E−07	4.06E−06	5.08E−06	1.30E−05	Gamma	1.585	3.121E+05	The battery charger boundary includes the battery charger and its breakers.
Electric bus	Running	FTOP	1.74E−09	1.98E−07	4.34E−07	1.67E−06	Gamma	0.502	1.155E+06	The bus boundary includes the bus component itself. Associated circuit breakers and stepdown transformers are not included.
Circuit breaker	All	FTO/C	4.40E−05	1.49E−03	2.55E−03	8.68E−03	Beta	0.698	2.729E+02	The circuit breaker is defined as the breaker itself and local instrumentation and control circuitry. External equipment used to monitor under voltage, ground faults, differential faults, and other protection schemes for individual breakers are considered part of the breaker.
	All	SO	3.00E−08	1.43E−07	1.71E−07	4.06E−07	Gamma	1.983	1.163E+07	

Check valve	Standby	FTO	5.10E−08	5.90E−06	1.30E−05	4.98E−05	Beta	0.500	3.855E+04	The check valve component boundary includes the valve and no other supporting components.
	Standby	FTC	4.08E−07	4.72E−05	1.04E−04	3.99E−04	Beta	0.500	4.816E+03	
Diesel-driven pump	Standby	FTS	4.17E−07	9.50E−04	3.88E−03	1.77E−02	Beta	0.300	7.728E+01	The diesel-driven pump boundary includes the pump, diesel engine, local lubrication or cooling systems, and local instrumentation and control circuitry.
	Standby	FTR<1h	1.70E−07	3.86E−04	1.58E−03	7.25E−03	Gamma	0.300	1.893E+02	
	Standby	FTR>1h	1.01E−08	2.31E−05	9.48E−05	4.34E−04	Gamma	0.300	3.165E+03	
Emergency diesel generator	Standby	FTS	2.77E−04	3.24E−03	4.53E−03	1.32E−02	Beta	1.075	2.363E+02	The emergency diesel generators (EDGs) covered are those used at U.S. commercial nuclear power plants.
	Standby	FTLR	3.07E−04	2.25E−03	2.90E−03	7.69E−03	Beta	1.411	4.856E+02	
	Standby	FTR>1h	1.52E−04	7.12E−04	8.48E−04	2.01E−03	Gamma	2.010	2.371E+03	
Filter	All	PG	3.88E−10	4.49E−08	9.86E−08	3.79E−07	Gamma	0.500	5.069E+06	The filter boundary includes the filter.
Heat exchanger	All	PG	6.86E−08	5.01E−07	6.45E−07	1.71E−06	Gamma	1.416	2.195E+06	The heat exchanger boundary includes the heat exchanger shell and tubes.
Inverter	Running	FTOP	4.12E−07	3.91E−06	5.28E−06	1.48E−05	Gamma	1.203	2.278E+05	The inverter boundary includes the inverter unit.
Motor-driven pump	Standby	FTS	5.87E−05	9.77E−04	1.47E−03	4.54E−03	Beta	0.909	6.198E+02	The motor-driven pump boundary includes the pump, motor, local circuit breaker, local lubrication or cooling systems, and local instrumentation and control circuitry.
	Standby	FTR<1h	5.40E−05	3.07E−04	3.78E−04	9.43E−04	Gamma	1.703	4.509E+03	
	Standby	FTR>1h	2.28E−08	2.63E−06	5.79E−06	2.22E−05	Gamma	0.500	8.640E+04	

continued

TABLE B.1 (continued)

Component	Operation	Failure Mode	5%	Median	Mean	95%	Pdf Type[b]	α	β	Component Description and Comments
	Running/ alternating	FTS	8.18E−05	1.47E−03	2.23E−03	6.98E−03	Beta	0.881	3.942E+02	
	Running/ alternating	FTR	6.21E−07	3.66E−06	4.54E−06	1.14E−05	Gamma	1.655	3.649E+05	
Motor-operated valve	Standby	FTO/C	9.42E−05	8.08E−04	1.07E−03	2.94E−03	Beta	1.277	1.192E+03	The motor-operated valve (MOV) component boundary includes the valve, the valve operator, local circuit breaker, and local instrumentation and control circuitry.
	Standby	SO	1.75E−10	2.02E−08	4.45E−08	1.71E−07	Gamma	0.500	1.124E+07	
	Control	FC	3.21E−10	7.31E−07	3.00E−06	1.37E−05	Gamma	0.300	1.000E+05	
Manual switch	Running	FTO/C	4.97E−07	5.75E−05	1.26E−04	4.85E−04	Beta	0.500	3.958E+03	The manual switch boundary includes the switch itself.
Large pipe	All	ELL (1/h-ft)	1.48E−14	3.36E−11	1.38E−10	6.31E−10	Gamma	0.300	2.176E+09	The pipe boundary includes piping and pipe welds in each system. Estimates are for piping with 2-inch or larger diameter. The flanges connecting piping segments are not included in the pipe component.
Small pipe	All	ELL (1/h-ft)	2.71E−15	6.16E−12	2.53E−11	1.16E−10	Gamma	0.300	1.187E+10	
Power-operated relief valve	All	FTO	1.30E−05	2.91E−03	7.25E−03	2.92E−02	Beta	0.435	5.955E+01	The boundary includes the valve, the valve operator, local circuit breaker, and local instrumentation and control circuitry.
	All	FTC	4.29E−06	4.96E−04	1.09E−03	4.18E−03	Beta	0.500	4.591E+02	
	All	SO	4.95E−11	1.13E−07	4.63E−07	2.12E−06	Gamma	0.300	6.481E+05	

Safety relief valve	All	FTO	8.33E−07	1.89E−03	7.71E−03	3.50E−02	Beta	0.300	3.891E+01	The safety relief valve component boundary includes the valve, the valve operator, and local instrumentation and control circuitry.
	All	FTC	3.13E−06	3.62E−04	7.95E−04	3.05E−03	Beta	0.500	6.282E+02	
	All	SO	5.44E−11	1.24E−07	5.08E−07	2.33E−06	Gamma	0.300	5.900E+05	
	All	FTCL	4.62E−04	5.20E−02	1.00E−01	3.62E−01	Beta	0.500	4.500E−00	
Sensor/transmitter components	Running	STF FTOP	3.21E−06	3.71E−04	8.15E−04	3.13E−03	Beta	0.500	6.132E+02	The sensor/transmitter flow (STF), sensor/transmitter level (STL), sensor/transmitter pressure (STP), and sensor/transmitter temperature (STT) boundaries includes the sensor and transmitter.
	Running	STF FTOP	4.00E−10	4.63E−08	1.02E−07	3.91E−07	Gamma	0.500	4.916E+06	
	Running	STL FTOP	3.21E−06	3.71E−04	8.15E−04	3.13E−03	Beta	0.500	6.132E+02	
	Running	STL FTOP	4.00E−10	4.63E−08	1.02E−07	3.91E−07	Gamma	0.500	4.916E+06	
	Running	STP FTOP	4.60E−07	5.32E−05	1.17E−04	4.49E−04	Beta	0.500	4.278E+03	
	Running	STP FTOP	3.23E−09	3.74E−07	8.22E−07	3.16E−06	Gamma	0.500	6.083E+05	
	Running	STT FTOP	1.70E−06	1.97E−04	4.32E−04	1.66E−03	Beta	0.500	1.157E+03	
	Running	STT FTOP	3.30E−09	3.82E−07	8.40E−07	3.23E−06	Gamma	0.500	5.950E+05	
Strainer	All	PG	7.89E−10	1.80E−06	7.38E−06	3.37E−05	Gamma	0.300	4.067E+04	The strainer component boundary includes the strainer. This data is based on only emergency service water system.
Traveling screen assembly	All	PG	1.87E−08	2.14E−06	4.68E−06	1.80E−05	Gamma	0.502	1.072E+05	The traveling screen component boundary includes the traveling screen, motor, and drive mechanism. This data is based on circulating water system and emergency cooling water system.

continued

TABLE B.1 (continued)

Component	Operation	Failure Mode	5%	Median	Mean	95%	Pdf Type[b]	α	β	Component Description and Comments
Manual Valve	Standby	FTO/C	2.93E−06	3.39E−04	7.43E−04	2.86E−03	Beta	0.500	6.720E+02	The manual valve component boundary includes the valve and valve operator.
	Standby	PLG	2.50E−11	2.90E−09	6.36E−09	2.45E−08	Gamma	0.500	7.855E+07	

ELL = External leakage large
FC = Fail to control
FTC = Failure to close
FTCL = Failure to close after passing liquid
FTLR = Failure to load and run
FTO = Failure to open
FTO/C = Failure to open or close
FTOP = Failure to operate
FTR = Failure To run
FTR > 1h = Failure to run after 1 hour of operation
FTR < 1h = Failure to run for 1 hour of operation
FTS = Failure to start
PG = Plugged
SO = Spurious operation

[a] Reference: Industry-Average Performance for Components and Initiating Events at U.S. Nuclear Power Plants, NUREG/CR-6928, U.S. Nuclear Regulatory Commission, Feb. 2007, Washington, DC.

[b] For standby components that start upon demand with a binomial variable p, the uncertainty associated with p is represented by the beta pdf: $f(p) = \frac{\Gamma(\alpha+\beta)}{\Gamma(\alpha)\Gamma(\beta)} p^{\alpha-1}(1-p)^{\beta-1}$ for $0 \le p \le 1$ and α and $\beta > 0$. Also,

$$p_{\text{mean}} = \frac{\alpha}{\alpha+\beta} \quad \text{and} \quad p_{\text{variance}} = \frac{\alpha\beta}{(\alpha+\beta)^2(\alpha+\beta+1)}.$$

For operating components with a failure rate or rate of occurrence of failure λ (units of events/time) the uncertainty associated with λ is represented by the gamma pdf: $f(\lambda) = \frac{(\beta)^\alpha}{\Gamma(\alpha)} \lambda^{\alpha-1} e^{-\lambda\beta}$ where λ, α and $\beta > 0$. Also, $\lambda_{\text{mean}} = \frac{\alpha}{\beta}$ and $\lambda_{\text{variance}} = \frac{\alpha}{\beta^2}$.

Index

A

A priori failure modes, 191
Absolute importance measures, 399
Accelerated degradation, 284
Accelerated life (AL) testing, 327
  basic notions, 327–329
  data analysis
    coefficient of variation, checking, 333
    exploratory data analysis, 330
    quantile–quantile plot, 333
    reliability model fitting, 334–335
    time-to-failure variance, logarithm of, 333
    two-sample criterion, 330–333
  proportional hazard model data analysis, 339, 340–342
    automotive tire reliability, 339–340
    reliability test procedure, 340
  popular models, 329–330
  for time-dependent stress, 335–336
    exploratory data analysis, 338
    and Palmgren–Miner's rule, 336–337
Acceptance region, 54
Accident precursor analysis (APA), 418
  and PRAs, differences between, 419
Aging system, 222
AL testing. *See* Accelerated life testing
α-factor model, 349
Alternative hypothesis, 53
AMSAA method. *See* Army material systems analysis activity method
Analytical methods, 301
  HCR correlation, 304
  SPAR-H HRA method, 301–302
  technique for human error rate prediction (THERP), 302–304
  time-reliability correlation, 304–305
APA. *See* Accident precursor analysis
Army material systems analysis activity (AMSAA) method, 320–321
Arrhenius Reaction Model, 329, 330
As-good-as-new restoration, 72, 86, 219, 223
Asymptotic availability of repairable system, 259
Automotive tire reliability, 339–340
Availability, definition of, 12, 219, 256
Average availability, 256, 260

B

Barlow and Proschan model, 329
Barriers identification, in risk assessment, 388
  challenges, 388–389
Basic execution time model (BETM), 293
Bathtub curve, 74
  burn-in early failure region, 74
  chance failure region, 74
  effect of stress in, 75
  for electrical devices, 74
  for mechanical devices, 75
  wearout region, 74
Bayesian estimation, 113, 114
  Bayesian probability papers, 134
    Bayesian simple linear regression and Bayesian probability papers, 137–138
    classical probability papers, 136–137
    classical simple linear regression, 135–136
    prior information about model parameters, 138–141
    prior information about reliability function or cdf, 141–142
  binomial distribution, parameter of, 120
    beta prior distribution, 123–125
    lognormal prior distribution, 125–126
    standard uniform prior distribution, 121–122
    truncated standard uniform prior distribution, 122–123
  exponential distribution, parameter of, 115
    parameters of prior distribution, selecting, 116–118
    uniform prior distribution, 118–120
  of Weibull distribution, 126
    joint posterior distribution, 130–132
    particular case, 132–134
    posterior distributions, 130
    prior distribution, 129
Bayes's Monte Carlo simulation, 361
Bayes's theorem, 23–24
BDD. *See* Binary decision diagrams
Beta distribution, 41
Beta prior distribution, 123–125, 129
BETM. *See* Basic execution time model
BFR model. *See* Binomial failure rate model
Binary decision diagrams (BDD), 175–178
Binomial distribution, 28, 106–107
  parameters, 120
    beta prior distribution, 123–125
    lognormal prior distribution, 125–126
    standard uniform prior distribution, 121–122
    truncated standard uniform prior distribution, 122–123
Binomial failure rate (BFR) model, 349–350
Birnbaum measure of importance, 306–307
Boolean algebra, 18
  in logic trees, 169–172
Bootstrap method, 361–365
Bounds and inequalities for IFR (DFR)
  coefficient of variation, inequality for, 76
  mean, bounds based on, 76
  quantile, bounds based on, 76
Burn-in early failure region, 74

C

CCFs. *See* Common-cause failures
cdf. *See* Cumulative distribution function
CES. *See* Cognitive environment simulation
Censored data
  left and right censoring, 97
  random censoring, 98
  reliability tests, types of, 98
  Type I censoring, 97
  Type II censoring, 97
Central limit theorem, 81
Central moment, 46
Centroid test. *See* Laplace's test
Challenge–response model, 2–3
Chance failure region, 74
Chi-square test, 56–59
CIF. *See* Cumulative intensity function
Classical Monte Carlo simulation, 360–361
Classical nonparametric distribution estimation, 108–113
Classical probability papers, 136–137
Classical simple linear regression, 135–136
Clopper–Pearson procedure, 359
CNG. *See* Compressed natural gas
Cognitive environment simulation (CES), 300
Common-cause failures (CCFs), 343, 344
  data analysis for, 350
Competing risks (or series) system, 98
Complement set, 15
Complete sample, 96–97
Completeness uncertainty, 354
Complex parallel-series systems, 161–164
Component reliability, elements of, 71
  Bayesian estimation procedures, 113
    Bayesian probability papers, 134–142
    parameter of binomial distribution, 120–126
    parameter of exponential distribution, 115–120
    of Weibull distribution, 126–134
  classical nonparametric distribution estimation, 108–113
  distributions in, 78
    exponential distribution, 78
    extreme value distributions, 81
    gamma distribution, 79–80
    lognormal distribution, 81
    normal distribution, 81
    Weibull distribution, 78–79
  generic failure rate determination, methods of, 142–143
  model selection, 85
    graphical nonparametric procedures, 86–88
    probability plotting, 89–94
    total-time-on-test plot, 94–96
  reliability
    bounds and inequalities for IFR (DFR), 76
    failure rate, 73–76, 77
    function, 71–72
    meaning of, 71
  reliability distribution parameters, estimation of, 96
    binomial distribution, 106–107
    censored data, 97–98
    exponential distribution interval estimation, 100–103
    exponential distribution point estimation, 98–100
    lognormal distribution, 103–104
    Weibull distribution, 104–106
Component reliability model selection
  graphical nonparametric procedures, 86
    large samples, 87–88
    small samples, 86
  probability plotting, 89
    exponential distribution, 89–90
    normal and lognormal distribution, 92–94
    Weibull distribution, 90–92
  total-time-on-test plot, 94–96
Composite-wrapped tanks, 401
Compressed natural gas (CNG) powered buses
  compression and storage station, 401
  consequence determination, 405–406
  dispensing facility, 402
  fire hazards, 400–401
  fire location, 406
  fire scenario description, 404
  fuel supply system of, 403
  gas dispersion, 405
  gas release scenarios, 402–404
  ignition likelihood, 405
  natural gas supply, 401
  overall risk results, 407
  PRA approach, 401
  PRA results, summary of, 406–407
  risk value determination, 406
  sensitivity and importance analysis, 408–409
  storage cascade, 401
  system description, 401, 402
  uncertainty analysis, 407–408
Conditional pdfs, 43
Confidence coefficient, 52
Confidence intervals for cdf and reliability function
    for complete and singly censored data, 108–110
    for multiply censored data, 110–113
Confidence level, 52
Constant failure rate model, 403
Continuous distribution, 33
  beta distribution, 41
  exponential distribution, 38
  gamma distribution, 39–40
  lognormal distribution, 36–38
  normal distribution, 35–36
  Weibull distribution, 39
Continuous random variable, 27
Correlation coefficient, 47
Counting processes, 221
Covariance, of random variables, 47
Cox–Lewis model, 227
Critical values, 54
Criticality importance, 307–308
Criticality matrix, 203
Crow-AMSAA model, 227
Cumulative damage models, 329
Cumulative distribution function (cdf), 33–34, 59
Cumulative hazard function, 73
Cumulative intensity function (CIF), 221, 235
  confidence limits for, 240–242
  estimation, 236
  for IRP, 236

natural estimate, 236
of point process, 238
for RP, HPP, NHPP and GRP with rejuvenation parameter, 234
Cumulative rate of occurrence of failures (ROCOF), 221
Cut set, 162
of off-site fire protection system, 416
of on-site fire protection system failure, 416

D

Damage distribution function, 329
Damage–endurance model, 2
Daniels model, 159
Decreasing failure rate (DFR), 74
Decreasing failure rate average (DFRA) distribution, 74
Degrading system. *See* Aging system
Degrees of freedom, 40
Dependent events, types of, 343
Dependent failures analysis, 342–345
CCFs, data analysis for, 350
multiple-parameter models, 347
α-factor model, 349
BFR model, 349–350
multiple Greek letter (MGL) model, 347–348
single-parameter models, 345
Detection, 194
Detection ratings, based on design control criteria, 198
DFR. *See* Decreasing failure rate
DFRA. *See* Decreasing failure rate average distribution
Direct numerical estimation, 301
Discrete distribution
binomial distribution, 28
discrete uniform distribution, 28
geometric distribution, 23
hypergeometric distribution, 30
Poisson distribution, 31
Discrete random variable, 27
cumulative distribution function of, 33
Discrete uniform distribution, 28
Disjoint sets, 17
Distributions of minima
Freshet distribution, 84
Gumbel distribution, 84
Weibull distribution, 84
Duane method, 318–320

E

edf. *See* Empirical cdf
EF. *See* Error factor
Efficient, 50
Electric Power Research Institute study, 366
Electrical failure mechanisms, 4, 6–7
electrical stress failure, 4
extrinsic failure mechanisms, 6
intrinsic failure mechanisms, 4, 6
Electrical stress failure, 4
Empirical cdf (edf), 59, 108
Empirical cumulative intensity function (ECIF), 236, 237, 238
Empirical reliability function, 108
Empty set, 16
Erlangian distribution, 80
Error factor (EF), 369
Estimate, 50
Estimation, 49
interval estimation, 38–39
point estimation, 50
maximum likelihood method, 51–52
method of moments, 50–51
Estimator, 50
Evaluation consequences, in risk assessment, 389
Event, 18–19
Event tree, 180, 393, 396
construction, 181–182
evaluation, 182–183
Exclusive cut sets, 173–174
Expectation, 44
algebra of, 46
Expected value, of random variable, 44
Expert judgment methods, 300
direct numerical estimation, 301
paired comparison, 301
success likelihood index method (SLIM), 301
Expert opinion, for reliability parameters estimation, 366–367
accuracy, statistical evidence on, 369–370
Bayesian model, 369
geometric averaging technique, 368–369
Exponential distribution, 38
in reliability analysis, 78
interval estimation, 100–103
parameters, 115
prior distribution, 116–118
uniform prior distribution, 118–120
Exponential distribution point estimation
Type I life test with replacement, 98–99
Type I life test without replacement, 99
Type II life test with replacement, 99
Type II life test without replacement, 99–100
Exponential distribution probability plotting, 89–90
Exponential underlying distribution, 223
External events, kinds of, 394
Extreme value distributions, 81
concepts and definitions, 82
limit distributions, types of, 84–85
maxima and minima, asymptotic distributions of, 83
random sample size, statistics from, 82–83
Extrinsic failure mechanisms, 6, 8

F

Failure analysis. *See* Probabilistic failure analysis
Failure count model, 293
Failure data analysis, of fire protection system, 410, 414–415
Failure effect probability, 197–198
Failure mechanisms, 3–4
categorization, 4–6
electrical failure mechanisms, 4, 6
mechanical failure mechanism, 4
Failure mode and effect analysis (FMEA), 189, 194
for aerospace application
compensating provision, 193–194
failure detection method, 193

Failure mode and effect analysis (FMEA) (*Continued*)
failure effects, 193
failure modes and causes, 191–193
item/functional identification, 191
severity, 194
system description and block diagrams, 191
FMEA/FMECA procedure, 190–191
FMECA procedure, 197–205
vs. functional hierarchy, 392–393
for transportation applications, 194
current design controls, 194–197
types, 190
Failure mode and effect criticality analysis (FMECA), 197–205
Failure mode criticality number, 198
Failure mode ratio, 198
Failure models, 1–2
challenge–response model, 2–3
damage–endurance model, 2
performance–requirement model, 3
stress–strength model, 2
Failure rate, 73–76, 77, 197, 198
Failure set (FS), 174
Failure-terminated life test, 98
Fault seeding model, 291–292
Fault tree analysis (FTA), 394
Fault tree method, 164–168, 396, 403
for off-site fire protection system failure, 412
for on-site fire protection system failure, 411
Fire protection risk analysis
consequences, 413
failure data analysis, 410, 414–415
initiating events, identification of, 410
quantification, 411–412, 416, 417
risk calculation and evaluation, 413, 416, 417, 418
scenario development, 410
steps in, 409–416
system analysis, 410, 411, 412, 413
Fire scenarios, economic consequences of, 417
First moment about origin, 44
Fisher Information Matrix, 105
FMEA. *See* Failure mode and effect analysis
FMECA. *See* Failure mode and effect criticality analysis
Frequency tables and histograms, 54–55
Frequentist method of treatment, 113
Freshet distribution
distributions of minima, 84
nondegenerated distributions for maxima, 84
FS. *See* Failure set
FTA. *See* Fault tree analysis
Functional event tree, 393–394
Functional hierarchy vs. failure mode and effect analysis, 392–393
Fussell–Vesely importance, 308–309

G

Gamma distribution, 39–40
in reliability analysis, 79–80
Gamma underlying distribution, 223
Gate symbols, in logic trees, 165
Gaussian distribution. *See* Normal distribution
General Renewal Process (GRP), 228–231
data analysis for, 254–256
Generic failure rate determination, methods of, 142–143
design factors, 142
environmental factors, 142
operating factors, 142
Geometric averaging technique, 368–369
Geometric distribution, 23
Goodness-of-fit tests, 56
Chi-square test, 56–59
Kolmogorov–Smirnov test, 59–62
Graphical method, 318
Graphical nonparametric procedures, of component reliability, 86
large samples, 87–88
small samples, 86
Grouped or interval data, 97
GRP. *See* General Renewal Process
Gumbel distribution
distributions of minima, 84
nondegenerated distributions for maxima, 84

H

Hazard exposure estimation, in risk assessment, 389
Hazard rate, 74
Hazards identification, in risk assessment, 388
HCR. *See* Human cognitive reliability
Histograms and frequency tables, 54–55
Homogeneous Poisson process (HPP), 222
data analysis for
exponential distribution of time intervals, procedures based on, 245–246
Poisson distribution, procedures based on, 243–245
HPP. *See* Homogeneous Poisson process
HRA. *See* Human reliability analysis
Human cognitive reliability (HCR), 304
Human reliability analysis (HRA), 295, 299
analytical methods, 301
HCR correlation, 304
SPAR-H HRA method, 301–302
technique for human error rate prediction (THERP), 302–304
time–reliability correlation (TCR), 304–305
expert judgment methods, 300
direct numerical estimation, 301
paired comparison, 301
success likelihood index method (SLIM), 301
human reliability data, 305–306
process, 296–299
simulation methods, 300
cognitive environment simulation (CES), 300
Human reliability data, 305–306
Hypergeometric distribution, 30
Hyperparameters, 115
Hypothesis testing, 53–54

I

Identical and independently distributed (IID) random variable, 220
IFR. *See* Increasing failure rate
IFRA. *See* Increasing failure rate average

IID r.v. *See* Identical and independently distributed random variable
Importance measures, 399
absolute measures, 399
applications of, 399
relative measures, 399
steps in, 399
Increasing failure rate (IFR), 74
Increasing failure rate average (IFRA), 74
Initiating events
identification of, 392–393, 410
inductive procedures, 393
Input-domain-based model, 292
Instantaneous (point) availability, 256
Interarrival time. *See* Time between successive failures
Intermediate event symbols, in logic trees, 165
Interpretation process
of PRA, 400
steps in, 400
Intersection of two sets, 16
Interval estimation, 38–39
Intrinsic failure mechanisms, 4, 6, 7
IRP. *See* increasing ROCOF process, 236

J

Joint distribution, 42–43
Joint pdf, 42
Joint posterior distribution, of Weibull parameters, 130–132

K

Kolmogorov–Smirnov test, 59–62
K-out-of-*N* system, 156, 169

L

Laplace's test, 250–254
Largest extreme value distribution, 85
Left and right censoring, 97
Life test terminations, 98
Limit distributions, 84–85
Limiting availability, 256, 257–259
Limiting average availability, 256
Limiting point availability, 259
Lindstrom–Madden method, 358, 360
Lloyd–Lipow method, 358–359
Load-sharing systems, 159–161
Location parameter, 79
Logic modeling, 394–395
Logic trees, evaluation of, 168
binary decision diagrams, 175–178
Boolean algebra analysis, 169–172
combinatorial technique for, 172–175
Lognormal distribution, 36–38, 81, 103–104
Lognormal prior distribution, 125–126
Long-term behavior, 224, 225
of GRP, 231, 232
Low Primary Coolant System Flow, 392

M

Maintenance personnel performance simulation (MAPPS), 300
Maintenance process, 219
MAPPS. *See* Maintenance personnel performance simulation
Marginal distribution, 42–43
Marginal pdfs, 42
Markov processes, for system availability determination, 260–265
Master logic diagram, 183–189
MAUD. *See* Multi-Attribute Utility Decomposition
Maxima and minima, asymptotic distributions of, 83
Maximum likelihood method, 51–52
Maximus method, 360
MCF. *See* Mean cumulative function
Mean, 44
Mean cumulative function (MCF), 221
Mean time between failures (MTBF) vs. mean time to failure (MTTF), 72
Mean time to failure (MTTF), 154, 155, 156
vs. mean time between failures (MTBF), 72
Mechanical failure mechanism, 4
Mechanistic models of failure, 279
Metal tanks, 401
Method of moments, 50–51
Method of quantiles, 117
MFR model. *See* Multinomial failure rate model
MGL. *See* Multiple Greek letter model
MIL-STD-1629A security levels, 193
Model uncertainties, 353–354
Monitored testing, 96
MTBF. *See* Mean time between failures
MTTF. *See* Mean time to failure
Multi-Attribute Utility Decomposition (MAUD), 301
Multinomial failure rate (MFR) model, 349–350
Multiple Greek letter (MGL) model, 347–348
Multiple-parameter models, 347
Musa–Okumoto's logarithmic Poisson time model, 293–295
Musa's basic execution time model (BETM), 293
Mutually exclusive sets, 17

N

Natural gas supply, 401
Nelson–Aalen procedure, 237
Nelson's procedure. *See* Nelson–Aalen procedure
NHPP. *See* Nonhomogeneous Poisson process
Nil combination, 174
Nondegenerated distributions for maxima
Freshet distribution, 84
Gumbel distribution, 84
Weibull distribution, 84
Nonhomogeneous Poisson process (NHPP), 225–228
data analysis for, 246
Laplace's test, 250–254
maximum likelihood procedures, 248–250
time intervals, regression analysis of, 246–248
Nonoperational initiating events, 393
Nonstaggered testing, 269
Normal and lognormal distribution probability plotting, 92–94

Normal distribution, 35–36, 81
Normal operation mode, 392
Null hypothesis, 53
Null set, 16

O

Occurrence, definition of, 194
One-sided lower confidence limit, 52
One-sided upper confidence limit, 52
Operating time, 198
Operational initiating event, 392
OR logic, 169
Order statistics, 82

P

Paired comparison, 301
Palmgren–Miner's rule, 336
Parallel systems, 155–157
Parallel-series systems, 161
Parameter uncertainties, 353
Parent distribution, 82
Perfect repair assumption, 223
Performance measures, 7, 9–10
Performance shaping factors (PSFs), 298
Performance–requirement model, 3
PH model. *See* Proportional hazard
Point estimation, 50
  maximum likelihood method, 51–52
  method of moments, 50–51
Point process, definition of, 221
Poisson distribution, 31
Poisson limits, 240
Posterior probability, 24, 130
Potential initiating events, 392
Power law model, 227
Power of the test, 54
Power Rule Model, 329–330
PRA. *See* Probability risk assessment
Practical aspects of importance measures, 312–314
Precursor analysis, 416
  APA and PRAs, differences between, 419
  methodology, 418–419
Precursor events (PEs)
  definition of, 416
  significance of, 418
Primary event symbols, in logic trees, 165
Prior distribution, parameters of, 116–118
Prior information, 138–142
Prior probability, 24
Probabilistic failure analysis, 370–371
  failure probability estimation, 372–373
  failure rate estimation, 372–373
  root-cause analysis, 374–375
  trends detection, in observed failure events, 372
Probability, 15, 85
  basic laws, 18
  Bayes's theorem, 23–24
  calculus of, 20–23
  classical interpretation of, 19
  frequency interpretation of, 19
  sets and Boolean algebra, 15–18
  subjective interpretation of, 19
Probability density function (pdf), 33
  conditional pdfs, 43
  joint pdf, 42
  marginal pdfs, 42
Probability distribution, 27
  continuous distribution, 33
    beta distribution, 41
    exponential distribution, 38
    gamma distribution, 39–40
    lognormal distribution, 36–38
    normal distribution, 35–36
    Weibull distribution, 39
  discrete distribution
    binomial distribution, 28
    discrete uniform distribution, 28
    geometric distribution, 23
    hypergeometric distribution, 30
    Poisson distribution, 31
  joint and marginal distribution, 42–43
  random variable, 27
Probability element, 33
Probability plotting, 89
  exponential distribution, 89–90
  normal and lognormal distribution, 92–94
  Weibull distribution, 90–92
Probability risk assessment (PRA), 389, 401, 406–407
  and APAs, differences between, 419
  components of, 391
  steps in, 390
    failure data collection, analysis, and performance assessment, 395–396
    familiarization and information assembly, 391–392
    initiating events, identification of, 392–393
    interpretation of results, 400
    logic modeling, 394–395
    objectives and methodology, 390–391
    quantification and integration, 396–397
    risk ranking and importance analysis, 398–399
    sensitivity analysis, 398
    sequence or scenario development, 393–394
    uncertainty analysis, 397–398
  strengths of, 390
Proportional hazard (PH) model, 327, 329
  data analysis, 339, 340
    automotive tire reliability, 339–340
    reliability test procedure, 340–342
PSFs. *See* Performance shaping factors
Pumping system, 167
  fault tree for, 167
  functional block diagram for, 193

Q

Quantification, of fire protection system, 411–412, 416, 417
Quantile–quantile plot, 333
Quantitative risk assessment, formalization of, 388
  barriers identification, 388
  challenges to barriers identification, 388–389
  evaluation, consequences of, 389
  hazard exposure estimation, 389
  hazards identification, 388

R

Random censoring, 98
Random variable, 27
  characteristics of, 44–49
Rare event approximation, 22, 171
Rate of occurrence of failures (ROCOF), 221, 222
RAW importance. *See* Risk-achievement worth importance measures
Real accelerated fatigue test data, 141–142
Real age, 228
Realizations, definition of, 220, 236
Reduced order binary decision diagrams (BDD), 175
Regression analysis, 62–66
Regression coefficients/parameters, 63
Rejuvenating (or improving) system, 222
Relative importance measures, 399
Reliability
  bounds and inequalities for IFR (DRF), 76
  failure rate, 73–76, 77
  function, 71–72
  meaning of, 71
Reliability block diagram method, 153
  complex systems, 161–164
  load-sharing systems, 159–161
  parallel systems, 155–157
  series system, 153–154
  standby redundant systems, 157–159
Reliability-centered maintenance, 314
  history and current procedures, 314–315
  optimal preventive maintenance scheduling, 315–316
  optimization, economic benefits of, 317
Reliability distribution parameters, estimation of, 96
  binomial distribution, 106–107
  censored data
    left and right censoring, 97
    random censoring, 98
    reliability tests, types of, 98
    Type I censoring, 97
    Type II censoring, 97
  exponential distribution interval estimation, 100–103
  exponential distribution point estimation
    Type I life test with replacement, 98–99
    Type I life test without replacement, 99
    Type II life test with replacement, 99
    Type II life test without replacement, 99–100
  lognormal distribution, 103–104
  Weibull distribution, 104–106
Reliability engineering, 1
  availability, definition of, 12
  failure models, 1–2
    challenge-response model, 2–3
    damage-endurance model, 2
    performance-requirement model, 3
    stress-strength model, 2
  failure mechanisms, 3–4
    electrical failure mechanisms, 4, 6
    mechanical failure mechanism, 4
  performance measures, 7, 9–10
  formal definition of, 11–12
  risk, definition of, 12–13
Reliability function
  and confidence intervals for cdf
    for complete and singly censored data, 108–110
    for multiply censored data, 110–113
Reliability growth, 227, 317
  army material systems analysis activity (AMSAA) method, 320–321
  Duane method, 318–320
  graphical method, 318
Reliability modeling, 279
  Birnbaum measure of importance, 306–307
  criticality importance, 307–308
  Fussell–Vesely importance, 308–309
  human reliability, 295
    HRA models, 299–306
    HRA process, 296–299
  physics-of-failure modeling, 279
    damage–endurance model, 283–288
    performance–requirement model, 288–289
    stress–strength model, 280–283
  practical aspects of importance measures, 312–314
  RAW importance, 309–312
  reliability-centered maintenance, 314
    history and current procedures, 314–315
    optimal preventive maintenance scheduling, 315–316
    optimization, economic benefits of, 317
  reliability growth, 317
    army material systems analysis activity (AMSAA) method, 320–321
    Duane method, 318–320
    graphical method, 318
  RRW importance, 309
  software analysis
    life-cycle models, 295
    models, 290–294
Reliability parameters estimation, expert opinion for, 366–367
  accuracy, statistical evidence on, 369–370
  Bayesian model, 369
  geometric averaging technique, 368–369
Reliability tests, types of, 98
Renewal equation, 223
Renewal process (RP), 222–225
Repairable components and systems, 219, 256–257
  General Renewal Process (GRP), 228–231
    data analysis for, 254–256
  homogeneous Poisson process, 222
  HPP, data analysis for
    exponential distribution of time intervals, procedures based on, 245–246
    Poisson distribution, procedures based on, 243–245
  instantaneous (point) availability, 257–259
    average availability, 260
    point availability, limiting, 259
  nonhomogeneous Poisson process (NHPP), 225–228, 246
    Laplace's test, 250–254
    maximum likelihood procedures, 248–250
    time intervals, regression analysis of, 246–248
  nonparametric data analysis
    cumulative intensity function, confidence limits for, 240–242
    cumulative intensity function, estimation of, 236
    estimation based on one realization, 236–237

Repairable components and systems (*Continued*)
estimation based on several realizations, 237–240
point process, basics of, 219–222
probabilistic bounds, 231–236
renewal process, 222–225
system analysis techniques, use of, 265–271
system availability determination, Markov processes for, 260–265
Repairable item, 219
Residual mean time to failure (MTTF), 73
Residual sum of squares, 64
Residual variance, 64
Risk, definition of, 12–13
Risk-achievement worth (RAW) importance measures, 309–312
Risk analysis
compressed natural gas (CNG) powered buses, 402, 403, 404
compression and storage station, 401
consequence determination, 405–406
dispensing facility, 402
fire hazards, 400–401
fire location, 406
fire scenario description, 404
gas dispersion, 405
gas release scenarios, 402–404
ignition likelihood, 405
natural gas supply, 401
overall risk results, 407
PRA approach, 401
PRA results, summary of, 406–407
risk value determination, 406
sensitivity and importance analysis, 408–409
storage cascade, 401
system description, 401, 402
uncertainty analysis, 407–408
elements, 385
fire protection risk analysis, 409–416
calculation and evaluation, 413, 416, 417, 418
planning steps, 386
precursor analysis, 416
APA and PRAs, differences between, 419
methodology, 418–419
probability risk assessment (PRA), 389
steps in conducting, 390–400
strengths of, 390
quantitative risk assessment, formalization of, 388
barriers identification, 388
challenges to barriers, identification of, 388–389
evaluation, consequences of, 389
hazard exposure estimation, 389
hazards identification, 388
risk values determination, 385–387
risk assessment parts, 385–386
risk assessment phases, 385
types of, 385
Risk communication, 385
Risk management goals, 385
Risk priority number (RPN), 194
Risk profile, construction of, 386
Risk ranking and importance analysis, of PRA, 398–399
Risk-reduction worth (RRW) importance measure, 309
Risk Triplet, 386
Risk values, determination of, 385–387
ROCOF. *See* Rate of occurrence of failures
Root-cause analysis, 374–375
RP. *See* Renewal process, 234
RPN. *See* Risk priority number
RRW importance. *See* Risk-reduction worth importance measure

S

SAE J1739 FMEA, 194
Safe, meaning of, 392
Same-as-new restoration, 223
Same-as-old restoration, 226
Sample, 49
Sample cdf. *See* Empirical cdf
Sample mean, 50
Sample size, 49
Sample space, of experiment, 18
Sample variance, 50–51
SARAH. *See* Systematic Approach to the Reliability Assessment of Humans
Scale parameter, 39
Scenario
definition, 388
development, for fire protection system, 410
Second moment about mean, 46
Sensitivity analysis
of PRA, 398
steps in, 397–398
Sensitivity and importance analysis, of CNG buses, 408–409
Sequence or scenario development, 393–394
goal of, 393
Series system, 153–154
Series–parallel system, average system unavailability of, 269–271
Set, 15
Severity, 197
ten-grade scale of, 197
Shape parameter, 39
SHARP. *See* Systematic Human Action Reliability Procedure
Short-term behavior, 224
of GRP, 230, 231
Simulation methods, 300
cognitive environment simulation (CES), 300
maintenance personnel performance simulation (MAPPS), 300
Single-parameter models, 345
SLIM. *See* Success likelihood index method
Smallest extreme value distribution, 85
Software life-cycle models, 295
Waterfall model, 295
Software reliability models (SRM), 290
failure count model, 293
fault seeding model, 291–292
input-domain-based model, 292
Musa's basic execution time model (BETM), 293
Musa–Okumoto's logarithmic Poisson time model, 293–295
time-between-failure models, 291
SPAR-H HRA method. *See* Standardized Plant Analysis Risk Model-Human (SPAR-H) HRA method
SRM. *See* Software reliability models

Staggered testing, 269
Standard deviation, 46
Standard error of estimate, 64
Standard uniform prior distribution, 121–122
Standardized Plant Analysis Risk Model-Human (SPAR-H) HRA method, 301–302
Standby redundant systems, 157–159
Static, 50
Stationary point, 222
Stress–strength model, 2
Subset, 15
Success likelihood index method (SLIM), 301
Success path, 180
Success tree method, 179–180
Sufficient statistic, 50
System analysis
  in availability calculations, 265–271
  of fire protection system, 410, 411, 412, 413
System availability determination, Markov processes for, 260–265
System block diagram, 171
System reliability analysis, 153
  event tree, 180
    construction, 181–182
    evaluation, 182–183
  fault tree method, 164–168
  FMEA, 189
    implementation, 191–197
    types, 190
  FMEA/FMECA procedure, 190–191
  FMECA procedure, 197–205
  logic trees, evaluation of, 168
    binary decision diagrams, 175–178
    Boolean algebra analysis, 169–172
    combinatorial technique for, 172–175
  master logic diagram, 183–189
  reliability block diagram method, 153
    complex systems, 161–164
    load-sharing systems, 159–161
    parallel systems, 155–157
    series system, 153–154
    standby redundant systems, 157–159
  success tree method, 179–180
Systematic Approach to the Reliability Assessment of Humans (SARAH), 301
Systematic Human Action Reliability Procedure (SHARP), 296
  definition, 297–298
  documentation, 299
  impact assessment, 298
  qualitative analysis, 298
  quantification, 299
  representation, 298
  screening, 298
Systemic event tree, 393–394

T

TAAF. *See* Test Analyze And Fix
TBF. *See* Time between successive failures
TBS. *See* Tread and belt separation
TCR. *See* Time-reliability correlation
Technique for human error rate prediction (THERP), 302–304
Test Analyze And Fix (TAAF), 317
THERP. *See* Technique for human error rate prediction
Time between successive failures (TBF), 222
Time-between-failure models, 291
Time-dependent load-sharing system model, 160
Time-dependent stress
  AL model for, 335–336
  exploratory data analysis for, 338
  reliability model for, 336–337
Time-independent reliability models, 159
Time–reliability correlation (TCR), 304–305
Time to the first failure (TTFF), 220, 222, 227
Time-terminated life test, 98
Time-to-failure distributions, and characteristics, 77
Time transformation function, 327
Tolerance–requirement model. *See* Performance–requirement model
Top event, 164
Total-time-on-test plot, 94–96
Transfer symbols, in logic trees, 165
Tread and belt separation (TBS), 339–340
Truncated Gumbel distribution, 227
Truncated standard uniform prior distribution, 122–123
TTFF. *See* Time to the first failure
Type I censoring, 97
Type I error, probability of, 54
Type I life test
  with replacement, 98–99
  without replacement, 99
Type II censoring, 97
Type II error, probability of, 54
Type II life test
  with replacement, 99
  without replacement, 99–100

U

Unavailability
  definition of, 219
  of parallel system, 269
Unbiased estimator, 50
Uncertainty analysis, 352
  of CNG buses, 407–408
  component failure model, system reliability confidence limits based on, 358–360
  graphic representation, 365–366, 367, 368
  Maximus method, 360
    Bayes's Monte Carlo simulation, 361
    bootstrap method, 361–365
    classical Monte Carlo simulation, 360–361
  of PRA, 397–398
  steps in, 397–398
  types, 352
    completeness uncertainty, 354
    model uncertainties, 353–354
    parameter uncertainties, 353
  uncertainty propagation methods, 354
    method of moments, 354–358
Uncertainty ranking, 399
Underlying distribution, 222
Uniform prior distribution, 118–120

Union of two sets, 15–16
Universal set, 15

V

Variance, 46
Virtual age, 228

W

Waterfall model, 295
Wear (damage)-stress life modeling, 284
Wearout region, 74
Weibull distribution, 39, 104–106
  Bayesian estimation of, 126
    joint posterior distribution, 130–132
    particular case, 132–134
    posterior distributions, 130
    prior distribution, 129
  distributions of minima, 84
  nondegenerated distributions for maxima, 84
  probability plotting, 90–92
  in reliability analysis, 78–79